W9-ABI-346

GPS SATELLITE SURVEYING

GPS SATELLITE SURVEYING
Third Edition

ALFRED LEICK

WILEY

JOHN WILEY & SONS, INC.

This book is printed on acid-free paper. ∞

Published by John Wiley & Sons, Inc., Hoboken, New Jersey
Published simultaneously in Canada

For general information on our other products and services or for technical support, please contact our Customer Care Department within the United States at (800) 762-2974, outside the United States at (317) 572-3993 or fax (317) 572-4002.

Wiley also publishes its books in a variety of electronic formats. Some content that appears in print may not be available in electronic books. For more information about Wiley products, visit our web site at www.wiley.com.

Library of Congress Cataloging-in-Publication Data:

Leick, Alfred.
GPS satellite surveying / Alfred Leick.—3rd ed.
p. cm.
Includes bibliographical references and index.
ISBN 0-471-05930-7 (cloth)
1. Artificial satellites in surveying. 2. Global Positioning System. I. Title.
TA595.5.L45 2004
526.9'82—dc21 2003049651

Printed in the United States of America

10 9 8 7 6 5 4 3 2 1

125.00

CONTENTS

APPENDIXES

PREFACE

Many people are experiencing the "waves of GPS" for the first time as new applications of GPS positioning and timing continue to surface. The domain of applications is so diverse and open to entrepreneurs' imagination that it is almost risky to give examples because of fear they might not convey the "awe" they once did. Examples are GPS receivers in cellular phones, location commerce, and the use of GPS in geo-encryption for data security. Experts on the subject of positioning are well aware that GPS has been in existence for quite a while and that it is in the midst of a modernization phase while it continues to perform flawlessly. New and better technology on the ground and in space, as well as requirements for the system to function as a dual-use technology for civilians and the military, drives the modernization. The expected European Galileo satellite system that is currently on the drawing board and that, naturally, will incorporate the latest technology will also increase the need for GPS to be modernized to stay competitive. This edition of *GPS Satellite Surveying* presents a major reorganization of the material, keeping in mind that many developments in GPS and its uses have matured while new capability is added. This made it possible to delete some material contained in previous editions. However, it is also recognized that other topics such as adjustments and geodetic models remain essentially unchanged and relevant and are, therefore, included with considerable detail as a permanent reference. This new edition, like previous editions, attempts to provide a comprehensive treatment on GPS as used in applications that require high positional accuracy.

As the tools for positioning become more refined with time, geodesy remains relevant as the basic science that provides a unified foundation for spatial referencing. Whereas geodesy and the geodetic reference frame are always involved, users will need a deeper understanding of geodetic theory when higher position accuracy is required and when the area of coverage increases. The "global approach" is embedded in GPS and geodesy. For these reasons, Chapter 2 is now dedicated to geodetic reference systems such as the conventional terrestrial reference frame, the conventional celestial reference frame, and the datum. In order to cover the wide spectrum of accuracy, we include the topics of polar motion, tectonic plate motion, solid ocean tides, and ocean loading. As in previous editions, we assume that the geoid is available and, therefore, do not address theories for accurate geoid determination. The 3D geodetic model is presented in this chapter to emphasize its importance as a unifying model applicable to geodetic or surveying problems of any size and to all levels of accuracy. It naturally applies to terrestrial and GPS observations because it is three dimensional. In addition, it is mathematically simple because it does not require the geodesic line.

GPS, the Russian GLONASS, and the European Galileo satellite systems are discussed in Chapter 3. The chapter begins with an elementary discussion of satellite motions, the Kepler elements that describe such motions, and the particularly simple theory of normal orbits, i.e., motion in a central gravity field. The disturbing forces that cause satellites to deviate from normal orbits are discussed as well. However, the material is not presented at the level of detail needed for accurate satellite orbit determination. We assume that orbit determination will continue to be handled by existing expert groups and that respective products will be available either through the broadcast navigation message or the International GPS Service (IGS) and other agencies in the form of precise and/or ultra rapid ephemeris and satellite clock data. Consequently, the section on orbital relaxation, contained in previous editions, has been eliminated. Chapter 3 includes new sections on GPS modernization, GLONASS, and the forthcoming Galileo system. Receiver technology is not addressed in detail. The reader is advised to consult other textbooks that have recently become available and that expertly address the subject.

Chapter 4 offers a comprehensive treatment of least-squares adjustments. Except for a new and small section on Kalman filtering, the material is essentially the same as published in previous editions. Least-squares estimation unquestionably plays a central role in computing positions with GPS and dealing with surveying data. The use of networks with redundant observations is a particularly characteristic feature of the geodesy and surveying profession, and it is central to the chapter. Networks are particularly suitable to implement data quality control strategies that use minimal constraint, reliability measures, and procedures for blunder detection. Such uses of networks and the respective analysis strategies apply equally well to low- or high-accuracy surveys. Chapter 4 and the supplementary material of Appendix A are a suitable reference for a full-fledged course on least-squares estimation, even if such a course does not deal with the particularities of GPS data processing. As in previous editions, the treatment begins with the mixed model and develops the observation equation model by means of simple specifications.

Chapter 5 provides equations for the pseudoranges and carrier phases. These equations can readily be used to set up the observation equations needed in least-squares estimation. The single-, double-, and triple-difference equations are given and discussed as to their ability to cancel common-mode errors. The Initial Evaluation section addresses several aspects related to using pseudoranges and carrier phases. For example, in positioning with satellites it is necessary to compute the geometric distance between the receiver's antenna and the satellite's antenna at the instant of signal transmission, taking into account the finite travel time of light. Initial considerations are given to the phenomenon of cycle slip, which is a sudden jump of the carrier phase observation by an unknown number of cycles. The expression that relates achievable baseline accuracy to a priori knowledge of the absolute station location and ephemeris accuracy is discussed.

The chapter on the troposphere and ionosphere, Chapter 6, has been significantly expanded compared to previous editions in recognition of the major contribution of GPS in sensing the troposphere and ionosphere. In addition to dealing with tropospheric refraction and various models for zenith delays and vertical angle dependencies, new material has been added for tropospheric absorption and water vapor radiometers. The impact of the ionosphere on GPS observations is discussed in terms of the ionospheric-free and ionospheric carrier phase and pseudorange functions. The chapter ends with a brief discussion on global ionospheric models.

Chapter 7 addresses the processing of pseudorange and carrier phases. The chapter has been reorganized to reflect the importance of newly developed dual-frequency precise point positioning techniques. These techniques are now available in the static and kinematic mode and in real time, in addition to legacy baseline determination. Because these new techniques require the precise ephemeris and accurate satellite clock data, the chapter begins with a summary of products available from the IGS. These new techniques are also sensitive to errors that tend to cancel in traditional relative positioning. Examples discussed are the phase windup correction and the impact of the satellite antenna phase center offsets.

The solutions are grouped into geometry-free solutions, point positioning, precise point positioning, real-time precise point positioning, relative positioning, and real-time relative positioning. The geometry-free solutions are a useful tool for analyzing observations and possibly even determining integer ambiguities.

Point positioning refers to single receiver positioning using the broadcast ephemeris. This technique is often called, simply, the navigation solutions. The linearized and nonlinear solutions are provided and the dilutions of precision factors are given.

Precise point positioning refers to single receiver positioning using dual-frequency ionospheric-free pseudorange and carrier phase functions, the precise ephemeris and accurate satellite clock corrections. This technique offers centimeter-accurate results for long station occupation times. In the kinematic mode, the positioning accuracy decreases to about 10 cm. In real-time applications, the necessary corrections are received via geostationary communication satellite or even the Internet. Common to all point positioning techniques is that the achievable position accuracy does not depend on the station separation.

The section on relative positioning refers to using single, double, or triple differences that eliminate or reduce the impact of common-mode errors and determine baselines very accurately. These techniques are most commonly used in geodesy and surveying. If one is able to reliably estimate the integer values of the double-difference ambiguities one speaks of a "fixed solution." If the integers cannot be determined and the estimated ambiguities remain noninteger, one speaks of a "float solution". Various approaches are available to fix the ambiguities reliably and do that for an observation time that is as short as possible. The LAMBDA technique is one of the most popular techniques for fixing ambiguities and is, therefore, presented in considerable detail in this edition.

Network adjustment of independent and redundant GPS baselines as a means of quality control traces its roots back to the first applications of GPS in surveying. When many stations are involved, minimal constraint network adjustments are still important as an objective measure of quality control of the survey. Network solutions are useful for detecting blunders and other error sources by analyzing the residuals and provide objective statements on the relative accuracy achieved. Chapter 8 gives the formulation for a baseline or vector adjustment, as well as three network examples from previous editions.

Surveyors frequently use conformal mapping coordinates. An example of a conformal mapping system is the State Plane Coordinate system in the United States. Since these mapping planes are conformal images of the ellipsoidal surface, it becomes necessary to also address computations on the ellipsoidal surface. The ellipsoidal and conformal models are discussed in Chapter 9. These models require use of the geodesic line. As is well known, dealing with the geodesic line on the ellipsoid and its conformal image is a mathematical challenge. As in previous editions we have, therefore, only summarized the relevant expressions and procedures needed for using the ellipsoidal or conformal mapping models. However, recognizing that these models are the "classics" in the geodetic tool box, that they essentially remain invariant with time, are still very relevant, and that mathematical literature about these models is increasingly difficult to find, this edition offers more detail in Appendix B (The Ellipsoid) and Appendix C (Conformal Mapping). These appendixes contain enough mathematical detail to ensure that students understand the limitations of these models and will, hopefully, never confuse conformal mapping and similarity or equate the conformal mapping plane with the geodetic or astronomical horizon at some station.

Large portions of this revision were completed while the author was on sabbatical leave at the Jet Propulsion Laboratory in Pasadena, from January to July 2002. I very much appreciate the support by Yoaz Bar-Sever, Anthony Mannucci, Frank Webb, Larry Young, and Jim Zumberge, who made a memorable sabbatical possible at a location with such singular expertise in GPS. The generous support by the companies Bowne AE&T Group, Coler & Colantonio, DeLorme, and Navcom Technology is gratefully acknowledged. It is especially reassuring to me to see a commitment of private industry in our field to sponsor academic endeavors such as writing a book. I appreciate Tamrah Brown's assistance in editing the draft. I am especially indebted to Tomás Soler of the National Geodetic Survey for the technical reading and the many suggestions he volunteered. Such support is invaluable.

ACKNOWLEDGMENTS

Financial support from the following companies is gratefully acknowledged:

ABBREVIATIONS

COMMONLY USED GPS ABBREVIATIONS

ACSM	American Congress on Surveying and Mapping
ARNS	Aeronautical radio navigation service
ARP	Antenna reference point
AS	Antispoofing
ASK	Amplitude shift keying
BOC	Binary offset carrier
BPSK	Binary phase shift keying
C/A-code	Coarse/acquisition code (1.023 MHz)
CCRF	Conventional celestial reference frame
CEP	Celestial ephemeris pole
CORS	Continuously operating reference stations
CTP	Conventional terrestrial pole
CTRS	Conventional terrestrial reference system
DGPS	Differential GPS
DMA	Defense Mapping Agency
DOP	Dilution of precision
DORIS	Doppler orbitography and radiopositioning integrated on satellite
DOY	Day of year
ECEF	Earth-centered earth-fixed coordinate system
EOP	Earth orientation parameter
FAA	Federal Aviation Administration
FBSR	Feedback shift register
FSK	Frequency shift keying
FOC	Full operational capability
GAST	Greenwich apparent sidereal time
GDOP	Geometric dilution of precision

GIM	Global ionospheric model
GIS	Geographic information system
GLONASS	Global'naya Navigatsionnaya Sputnikkovaya Sistema
GNSS	Global navigation satellite system
GML	Gauss midlatitude functions
GMST	Greenwich mean sidereal time
GPS	Global positioning system
GPSIC	GPS Information Center
GPST	GPS time
GRS80	Geodetic reference system of 1980
HDOP	Horizontal dilution of precision
HOW	Handover word
IAG	International Association of Geodesy
IAU	International Astronomical Union
ICRF	International celestial reference frame
IERS	International Earth Rotation Service
IGDG	Internet-based dual-frequency global differential GPS
IGS	International GPS Service
ITRF	International terrestrial reference frame
IUGG	International Union of Geodesy and Geophysics
IOC	Initial operational capability
ION	Institute of Navigation
IWV	Integrated water vapor
JD	Julian date
JPL	Jet Propulsion Laboratory
L1	L1 carrier (1575.42 MHz)
L2	L2 carrier (1227.6 MHz)
L5	L5 carrier (1176.45 MHz)
LAMBDA	Least-squares ambiguity decorrelation adjustment
LC	Lambert conformal mapping
LEO	Low-earth orbiting satellite
NAD83	North American datum of 1983
NAVSTAR	Navigation Satellite Timing and Ranging
NEP	North ecliptic pole
NGS	National Geodetic Survey
NIMA	National Imagery and Mapping Agency
NIST	National Institute of Standards and Technology
NOAA	National Oceanic and Atmospheric Administration
OPUS	Online processing user service
OTF	On-the-fly ambiguity resolution
P-code	Precision code (10.23 MHz)
PCV	Phase center variation
PDOP	Positional dilution of precision
ppb	parts per billion
ppm	parts per million

PPP	Precise point positioning
PPS	Precise positioning service
PRN	Pseudorandom noise
PSK	Phase-shift keying
PWV	Precipitable water vapor
QPSK	Quadri phase-shift keying
RINEX	Receiver independent exchange format
RNSS	Radio navigation satellite services
RTCM	Radio Technical Commission for Maritime Services
RTK	Real-time kinematic positioning
SA	Selective availability
SINEX	Solution independent exchange format
SLR	Satellite laser ranging
SP3	Standard product #3 for ECEF orbital files
SPC	State plane coordinate system
SPS	Standard positioning service
SRP	Solar radiation pressure
SVN	Space vehicle launch number
SWD	Slant wet delay
TAI	International atomic time
TDOP	Time dilution of precision
TEC	Total electron content
TECU	TEC unit
TLM	Telemetry word
TM	Transverse Mercator mapping
TOW	Time of week
TRANSIT	Navy navigation satellite system
URE	User range error
USNO	U.S. Naval Observatory
UT1	Universal time corrected for polar motion
UTC	Coordinate universal time
VDOP	Vertical dilution of precision
VLBI	Very long baseline interferometry
VRS	Virtual reference station
WAAS	Wide area augmentation service
WADGPS	Wide area differential GPS
WGS84	World Geodetic System of 1984
WVR	Water vapor radiometer
WRC	World Radio Conference
Y-code	Encrypted P-code
ZHD	Zenith hydrostatic delay
ZWD	Zenith wet delay

NOTATION

COORDINATES SYSTEMS NOTATION

The notation distinguishes between the name of a coordinate system, i.e., (X)—upright, not bold, and in parentheses—and the 3D position, i.e., **X**—upright, bold, and no parentheses. For example, the position of the satellite in (X) is **X**. The coordinates labels are X, Y, and Z for this particular coordinate system. The coordinate systems frequently used are

$(\bar{\mathrm{X}}) = (\bar{X}, \bar{Y}, \bar{Z})$ Mean celestial Cartesian coordinate system. The $\bar{X}$ axis points toward the vernal equinox; $\bar{X}$ and $\bar{Y}$ span the mean celestial equator.

$(\mathrm{X}) = (X, Y, Z)$ True celestial Cartesian coordinate system. The X axis points toward the vernal equinox; X and Y span the true celestial equator.

$(\mathrm{x}) = (x, y, z)$ ECEF coordinate Cartesian system. The x axis points toward the zero meridian; x and y span the terrestrial equator.

$(\breve{\mathrm{x}}) = (\breve{x}, \breve{y}, \breve{z})$ Intermediary terrestrial Cartesian coordinate system. The $\breve{x}$ axis points toward the zero meridian; $\breve{x}$ and $\breve{y}$ span the instantaneous terrestrial equator.

$(w) = (n, e, u)$ Local geodetic Cartesian coordinate system with coordinates n (northing), e (easting), and u (up).

$(q) = (q_1, q_2, q_3)$ An auxiliary orbital-plane Cartesian coordinate system centered at the focal point of the orbital ellipse. The axis q_1 is in direction of perigee; q_1 and q_2 span the orbital plane.

$(\xi) = (\xi, \eta, \zeta)$ An auxiliary orbital-plane Cartesian coordinate system centered at the origin of orbital ellipse. The axis ξ is in direction of perigee; ξ and η span the orbital plane.

$(p) = (p_1, p_2, p_3)$ An auxiliary orbital-plane Cartesian coordinate system centered at the focal point of the orbital ellipse. The axis p_1 is in direction of the ascending node, p_1 and p_2 span the orbital plane.

GEODETIC MODELS NOTATION

3D Model	2D Ellipsoidal	Conformal Map
α geodetic azimuth	$\widehat{\alpha}$ geodesic asimuth	
β geodetic vertical angle ϑ geodetic zenith angle		
δ geodetic horizontal angle between normal sections	$\widehat{\delta}$ geodesic angle between geodesics	$\bar{\delta}$ map angle between chords
		$\bar{t}$ grid north $\bar{T}$ geodesic north $\bar{\gamma}$ meridian convergence
s slant distance	$\widehat{s}$ geodesic length	$\bar{s}$ length of mapped geodesic $\bar{d}$ length of chord

The symbol N is used for geoid undulation and for radius of curvature of prime vertical section on the ellipsoid. The symbol e is used for easting (coordinate in geodetic horizon) and for the eccentricity of ellipsoid.

SURFACE AND LINE NOTATION

Appendixes B and C contain material from differential geometry. We use the notation [ ] to stress that a surface or line on a surface is concerned. For example $[E]$ indicates the ellipsoidal surface.

CHAPTER 1

INTRODUCTION

A new and exciting era of positioning on land, on the sea, and in space began with the launch of the first global positioning system (GPS) satellite on February 22, 1978. The primary purpose of the satellite system was to meet the needs of the military and national security, in regards to positioning and timing, on a 24-hour per day basis all around the world and under all weather conditions. Very soon, however, the potential benefits of GPS for civilian applications became apparent, with that number rapidly increasing and no end in sight twenty plus years later.

The satellites transmit at frequencies L1 (1575.42 MHz) and L2 (1227.6 MHz) modulated with two types of codes and the navigation message. The codes are the civilian C/A-code and the encrypted military P(Y)-codes. At present the L1 carrier is modulated with both types of codes, whereas L2 is modulated with a P-code only. Modernized GPS will transmit a second civil code on L2 and a third civil code on a new carrier L5 (1176.45 MHz).

There are two types of observables of interest to users. One of them is the pseudorange, which equals the distance between the satellite and the receiver plus small corrective terms due to receiver and satellite clock errors, the impact of the ionosphere and troposphere on signal propagation, and multipath. Given the geometric positions of the satellites as a function of time, i.e., satellite ephemeris, four pseudoranges are in principle sufficient to compute the location of the receiver and its clock correction. Pseudoranges are a measure of the travel time of the codes, C/A or P(Y). The second observable, the carrier phase, is the difference between the received phase and the phase of the receiver oscillator at the epoch of measurement. Receivers are programmed to make phase observations at the same equally spaced epochs. In addition, receivers keep track of the number of complete cycles received since the beginning of a measurement. Thus, the actual output is the accumulated phase observable at pre-set epochs.

Government policies (SPS, 2001) currently define a standard positioning service (SPS) based on the C/A-code observations and a precise positioning service (PPS) based on P(Y)-code observations. SPS and PPS address "classical satellite" navigation methods where one receiver observes several satellites in order to determine its geocentric position, using the broadcast ephemeris. Typically, a position is computed for every epoch of observation. The advantages of relative positioning have long been recognized as a way to satisfy the high accuracy requirements of geodesy, surveying, and other geosciences. In relative positioning, also called differential positioning, the relative location between co-observing receivers is determined. In this case many common errors cancel, or their impact is significantly reduced. During the pioneering years of GPS, there appeared to be a clear distinction between applications in navigation and surveying. This distinction, if ever real, has rapidly disappeared. Whereas navigation solutions used to incorporate primarily pseudorange observations, surveying solutions have always been based on the millimeter-accurate carrier phase observations. Modern approaches combine both types of observables in an optimal manner. This leads to a unified GPS positioning theory for both surveying and navigation. The availability of precise, postprocessed ephemerides—even predicted precise ephemerides—allows for single-point positioning that is better than specified for SPS or even PPS. Powerful processing algorithms reduce the time required for data collection, so as to render even the distinction between static (both receivers are static) and kinematic (at least one receiver moves) techniques unnecessary.

The achievable accuracy very much depends on many factors that will be detailed throughout this book. In order to emphasize the characteristic difference between geocentric and relative position accuracy, let us simply state that geocentric position accuracy ranges from meters to decimeters, whereas the relative position accuracy is at the centimeters to millimeters level. The secrets that make GPS such a powerful positioning device can be readily explained. At the center is the ability to measure carrier phases to about 1/100 of a cycle, which equals about 2 mm in linear distance. The high frequencies (L1 and L2) penetrate the ionosphere relatively well. Because the time delay caused by the ionosphere is inversely proportional to the square of the frequency, carrier phase observations at both frequencies can be used to model and, thus, eliminate most ionospheric effects. Dual-frequency observations are particularly useful when the station separation is large and when shortening the observation time is important. There has been significant progress in the design of stable clocks and their miniaturization, providing precise timing at the satellite. The GPS satellite orbits are stable because at such high satellite altitudes only the major gravitational forces affect their motion. There are no atmospheric drag effects acting on satellites. The impact of the sun and the moon on the orbits is significant but can be computed accurately. The remaining worrisome physical aspects are solar radiation pressure on the satellites, as well as the tropospheric delay and multipath effects on signal propagation. On the algorithmic side, much is gained by using linear combinations of the basic phase observables. For example, unwanted parameters are eliminated and certain effects need not be modeled. Let the receivers k and m observe satellite p at the same time. The difference between these two phase observations is called a single-difference observable. It can be readily shown that single differences are largely independent of satellite frequency offset and linear drift. Next, assume that two single differences are available, one referring to satellite p and one to satellite q. The difference between these two single differences, called the double-difference observable, is largely independent of receiver clock errors. Finally, taking the difference of two double differences that refer to different epochs yields the triple-difference observable. This last type of observation is useful for initial processing and screening of the data.

Single-, double-, or triple-difference processing yields the relative location between the co-observing receivers and is usually referred to as the vector between the stations. Because the satellites are at a finite distance from the earth, there is also a "geocentric positioning component" to these observables which is, as a matter of fact, a function of the baseline length. In practice, the absolute location of the baseline must be sufficiently known in order not to degrade the relative positioning capability. This topic will be discussed later. By itself, one accurate vector between stations is generally not of much use, at least in surveying. Of course, one can add the vector to the geocentric position of the "known" station and formally compute the geocentric position of the new station. The problem with this procedure is that the uncertainty of the "known" station is transferred in full to the new station. Also, despite all of modern technology, the vectors themselves can still be in error. Possibilities of misidentifying ground marks, centering errors, misreading antenna heights, etc., can never be completely avoided. Like other observations, the GPS vector observations are most effectively controlled by a least-squares network adjustment consisting of a set of redundant vectors. Such network solutions make it possible to assess the quality of the observations, validate the correctness of statistical data, and detect (and possibly remove) existing blunders. Therefore, the primary result of a GPS survey is a polyhedron of stations whose accurate relative locations have been controlled by a least-squares adjustment.

1.1 HISTORICAL PERSPECTIVE

A summary of GPS development and performance to date is detailed in Table 1.1. Because the scope of GPS research and application development is so broad and conducted by researchers all over the globe, it is impossible to give a comprehensive listing. Table 1.1, therefore, merely demonstrates the extraordinarily rapid development of the GPS positioning system.

GPS made its debut in surveying and geodesy with a big bang. During the summer of 1982, the testing of the Macrometer receiver, developed by C. C. Counselman at M.I.T., verified a GPS surveying accuracy of 1–2 parts per million (ppm) of the station separation. Baselines were measured repeatedly using several hours of observations to study this new surveying technique and to gain initial experience with GPS. During 1983 a thirty (plus)-station first-order network densification in the Eifel region of Germany was observed (Bock et al., 1985). This project was a joint effort by the State Surveying Office of North Rhein-Westfalia, a private U.S. firm, and scientists from M.I.T. In early 1984, the geodetic network densification of Montgomery County (Pennsylvania) was completed. The sole guidance of this project rested with a private

TABLE 1.1 GPS Development and Performance at a Glance

1978	Launch of first GPS satellite
1982	Prototype Macrometer testing at M.I.T.
1983	Geodetic network densification (Eifel, Germany) President Reagan offers GPS to the world "free of charge"
1984	Geodetic network densification (Montgomery County, Pennsylvania) Engineering survey at Stanford Remondi's dissertation
1985	Precise geoid undulation differences for Eifel network Codeless dual band observations Kinematic GPS surveying Antenna swap for ambiguity initialization First international symposium on precise positioning with GPS
1986	*Challenger* accident (January 28) 10 cm aircraft positioning
1987	JPL baseline repeatability tests to 0.2–0.04 ppm
1989	Launch of first Block II satellite OTF solution Wide area differential GPS (WADGPS) concepts U.S. Coast Guard GPS Information Center (GPSIC)
1990	GEOID90 for NAD83 datum
1991	NGS ephemeris service GIG 91 experiment (January 22–February 13)
1992	IGS campaign (June 21–September 23) Initial solutions to deal with antispoofing (AS) Narrow correlator spacing C/A-code receiver Attitude determination system
1993	Real-time kinematic GPS ACSM ad hoc committee on accuracy standards Orange County GIS/cadastral densification Initial operational capability (IOC) on December 8 1–2 ppb baseline repeatability LAMBDA
1994	IGS service beginning January 1 Antispoofing implementation (January 31) RTCM recommendations on differential GPS (Version 2.1) National Spatial Reference System Committee (NGS) Multiple (single-frequency) receiver experiments for OTF Proposal to monitor the earth's atmosphere with GPS (occultations)
1995	Full operational capability (FOC) on July 17 Precise point positioning (PPP) at JPL

TABLE 1.1 (*Continued*)

1996	Presidential Decision Directive, first U.S. GPS policy
1998	Vice president announces second GPS civil signal at 1227.60 MHz JPL's automated GPS data analysis service via Internet
1999	Vice president announces GPS modernization initiative and third civil GPS signal at 1176.45 MHz IGDG (Internet-based global differential GPS) at JPL
2000	Selective availability set to zero GPS JPO begins modifications to IIR-M and IIF satellites

GPS surveying firm (Collins and Leick, 1985). Also in 1984, GPS was used at Stanford University for a high-precision GPS engineering survey to support construction for extending the Stanford Linear Accelerator (SLAC). Terrestrial observations (angles and distances) were combined with GPS vectors. The Stanford project yielded a truly millimeter-accurate GPS network, thus demonstrating, among other things, the high quality of the Macrometer antenna. This accuracy could be verified through comparison with the alignment laser at the accelerator, which reproduces a straight line within one-tenth of a millimeter (Ruland and Leick, 1985). Therefore, by the middle of 1984, 1–2 ppm GPS surveying had been demonstrated beyond any doubt. No visibility was required between the stations. Data processing could be done on a microcomputer. Hands-on experience was sufficient to acquire most of the skills needed to process the data—i.e., first-order geodetic network densification suddenly became within the capability of individual surveyors.

President Reagan offered GPS free of charge for civilian aircraft navigation in 1983 once the system became fully operational. This announcement was made after the Soviet downing of the Korean Air flight 007 over the Korea Eastern Sea. This announcement can be viewed as the beginning of sharing arrangements of GPS for military and civilian users.

Engelis et al. (1985) computed accurate geoid undulation differences for the Eifel network, demonstrating how GPS results can be combined with orthometric heights, as well as what it takes to carry out such combinations accurately. New receivers became available—e.g., the dual-frequency P-code receiver TI-4100 from Texas Instruments—which was developed with the support of several federal agencies. Ladd et al. (1985) reported on a survey using codeless dual-frequency receivers and claimed 1 ppm in all three components of a vector in as little as 15 minutes of observation time. Thus, the move toward rapid static surveying had begun. Around 1985, kinematic GPS became available (Remondi, 1985). Kinematic GPS refers to ambiguity-fixed solutions that yield centimeter (and better) relative accuracy for a moving antenna. The only constraint on the path of the moving antenna is visibility of the same four (at least) satellites at both receivers. Remondi introduced the antenna swapping technique for the rapid initialization of ambiguities. Antenna swapping made kinematic positioning in surveying more efficient.

The deployment of GPS satellites came to a sudden halt due to the tragic January 28, 1986 *Challenger* accident. Several years passed until the Delta II launch vehicle was modified to carry GPS satellites. However, the theoretical developments continued at full speed. They were certainly facilitated by the publication of Remondi's (1984) dissertation, the very successful First International Symposium on Precise Positioning with the Global Positioning System (Goad, 1985), and a specialty conference on GPS held by the American Society of Civil Engineers in Nashville in 1988.

Kinematic GPS was used for decimeter positioning of airplanes relative to receivers on the ground (Mader, 1986; Krabill and Martin, 1987). The goal of these tests was to reduce the need for traditional and expensive ground control in photogrammetry. These early successes not only made it clear that precise airplane positioning would play a major role in photogrammetry, but they also highlighted the interest in positioning other remote sensing devices in airplanes and spacecraft.

Lichten and Border (1987) report repeatability of 2–5 parts in 10^8 in all three components for static baselines. Note that 1 part in 10^8 corresponds to 1 mm in 100 km. Such highly accurate solutions require satellite positions of about 1 m and better. Because such accurate orbits were not yet available at the time, researchers were forced to estimate improved GPS orbits simultaneously with baseline estimation. The need for a precise orbital service became apparent. Other limitations, such as the uncertainty in the tropospheric delay over long baselines, also became apparent and created an interest in exploring water vapor radiometers to measure the wet part of the troposphere along the path of the satellite transmissions. The geophysical community requires high baseline accuracy for obvious reasons; e.g., slow-moving crustal motions can be detected earlier with more accurate baseline observations. However, the GPS positioning capability of a few parts in 10^8 was also noticed by surveyors for its potential to change well-established methods of spatial referencing and geodetic network design.

Perhaps the year 1989 could be labeled the year when "modern GPS" positioning began in earnest. This was the year when the first production satellite, Block II, was launched. Seeber and Wübbena (1989) discussed a kinematic technique that used carrier phases and resolved the ambiguity "on-the-way." This technique is today usually called "on-the-fly" (OTF) ambiguity resolution (fixing), meaning there is no static initialization required to resolve the ambiguities. The technique works for postprocessing and real-time applications. OTF is one of the modern techniques that applies to both navigation and surveying. The navigation community began in 1989 to take advantage of relative positioning, in order to eliminate errors common to co-observing receivers, and to make attempts to extend the distance in relative positioning. Brown (1989) referred to it as extended differential GPS, but it is more frequently referred to as wide area differential GPS (WADGPS). Many efforts were made to standardize real-time differential GPS procedures, resulting in several publications by the Radio Technical Commission for Maritime Services. The U.S. Coast Guard established the GPS Information Center (GPSIC) to serve nonmilitary user needs for GPS information.

The introduction of the geoid model GEOID90 in reference to the NAD83 datum represented a major advancement for combining GPS (ellipsoidal) and orthometric height differences. The most recent version is GEOID99.

During 1991 and 1992, the geodetic community embarked on major efforts to explore the limits of GPS on a global scale. The efforts began with the GIG91 campaign and continued the following year with the International GPS Service (IGS) campaign. GIG91 (GPS experiment for International Earth Rotation Service [IERS] and Geodynamics) resulted in very accurate polar motion coordinates and earth rotation parameters. Geocentric coordinates were obtained that agreed with those derived from satellite laser ranging within 10 to 15 cm, and ambiguities could be fixed on a global scale providing daily repeatability of about 1 part in 10^9. Such results are possible because of the truly global distribution of the tracking stations. The primary purpose of the IGS campaign was to prove that the scientific community is able to produce high-accuracy orbits on an operational basis. The campaign was successful beyond all expectations, confirming that the concept of IGS is possible. The IGS service formally began January 1, 1994.

For many years, users worried about what impact antispoofing (AS) would have on the practical uses of GPS. AS implies switching from the known P-code to the encrypted Y-code, expressed by the notation P(Y)-code. The purpose of AS is to make the P-codes available only to authorized (military) users. The anxiety about AS was considerably relieved when Hatch et al. (1992) reported on the code-aided squaring technique to be used when AS is active. Most manufacturers developed proprietary solutions for dealing with AS. When AS was actually implemented on January 31, 1994, it presented no insurmountable hindrance to the continued use of GPS and, particularly, the use of modern techniques such as OTF. GPS users became even less dependent on AS with the introduction of accurate narrow correlator spacing C/A-code receivers (van Dierendonck et al., 1992), since the C/A-code is not subject to AS measures. By providing a second civil code on L2, and eventually a third one on L5, and adding new military codes, GPS modernization will make the P(Y)-code encryption a nonissue for civilian applications, and at the same time, provide enhanced performance to civilian and military users.

A major milestone in the development of GPS was achieved on December 8, 1993, when the initial operational capability (IOC) was declared when twenty-four satellites (Blocks I, II, IIA) became successfully operational. The implication of IOC was that commercial, national, and international civil users could henceforth rely on the availability of the SPS. Full operational capability (FOC) would be declared July 17, 1995, when twenty-four satellites of the type Blocks II and IIA became operational. Teunissen (1993) introduced the least-squares ambiguity decorrelation adjustment (LAMBDA), which is now widely used.

The determination of attitude/orientation using GPS has drawn attention for quite some time. Qin et al. (1992) report on a commercial product for attitude determination. Talbot (1993) reports on a real-time kinematic centimeter-accuracy surveying system. Lachapelle et al. (1994) experiment with multiple (single-frequency) receiver configurations, in order to accelerate the on-the-fly ambiguity resolution by means of imposing length constraints and conditions between the ambiguities. While much attention was given to monitoring the ionosphere with dual-frequency and single-frequency code or carrier phase observations, Kursinski (1994) discusses the applicability of radio occultation techniques to use GPS in a general earth's atmospheric monitoring system (which could provide high vertical-resolution profiles of atmospheric temperature across the globe).

The surveying community promptly responded to the opportunities and challenges that came with GPS. The American Congress on Surveying and Mapping (ACSM) tasked an ad hoc committee in 1993 to study the accuracy standards to be used in the era of GPS. The committee addressed questions concerning relative and absolute accuracy standards. The National Geodetic Survey (NGS) enlisted the advice of experts regarding the shape and content of the geodetic reference frame; these efforts eventually resulted in the continuously operating reference stations (CORS). Orange County (California) established 2000 plus stations to support geographic information systems (GIS) and cadastral activities. There are many other examples.

Zumberge et al. (1998a,b) report single-point positioning at the couple of centimeters level for static receivers and at the subdecimeter level for moving receivers. This technique became available at the Jet Propulsion Laboratory (JPL) around 1995. The technique that requires dual-frequency observations, a precise ephemeris, and precise clock corrections is referred to as precise point positioning (PPP). These remarkable results were achieved with postprocessed ephemerides at a time when selective availability (SA) was still active. Since 1998 JPL has offered automated data processing and analysis for PPP on the Internet (Zumberge, 1998). Users submit the observation file over the Internet and retrieve the results via FTP soon thereafter. Since 1999 JPL has operated an Internet-based dual-frequency global differential GPS system (IGDG). This system determines satellite orbits, satellite clock corrections, and earth orientation parameters in real-time and makes corrections available via the Internet for real time positioning. A website at JPL demonstrates real-time kinematic positioning at the subdecimeter of a receiver located at JPL's facilities in Pasadena.

Finally, during 1998 and 1999, major decisions were announced regarding the modernization of GPS. In 2000, SA was set to zero as per Presidential Directive. When active, SA entails an intentional falsification of the satellite clock (SA-dither) and of the broadcast satellite ephemeris (SA-epsilon); when active it is effectively an intentional denial to civilian users of the full capability of GPS.

1.2 GEODETIC ASPECTS

The three-dimensional (3D) geodetic model is definitely the preferred model for adjusting three-dimensional GPS vector observations and combining them with classical terrestrial observations such as slant distance, horizontal angle, azimuth, vertical angle, and, with some restrictions, leveled height differences. The three-dimensional model is applicable with equal ease to the following: small surveys the size of a parcel or smaller, large surveys covering whole regions and nations, three-dimensional surveys for measuring and monitoring engineering structures, and the "pseudo three-dimensional" surveys typical of classical geodetic networks or in "plane surveying." Application of simple concepts from the theory of adjustments, such as "weighted parameters" and "significance of parameters," make it possible to use the three-dimensional model in all of these applications in a uniform manner. Perhaps the most important point in favor of the three-dimensional model is that the geodesic line on the ellipsoid is not needed at all. Anyone who has studied the mathematics related to geodesics will certainly appreciate this simplification of surveying theory.

The 3D geodetic model requires that the observations are reduced for polar motion and deflection of the vertical. It is well known that the theodolite senses the local plumb line and, thus, measures with respect to the local vertical and the local astronomic horizon. It is further known that astronomic observations depend on the position of the instantaneous pole of rotation. The goal is to reduce angular observations measured with the theodolite to the ellipsoidal normal (deflection of vertical reduction) and to reduce the astronomic quantities to the conventional terrestrial pole (CTP). Having said this, I would like to comfort worried surveyors by reminding them that the most popular observations do not depend critically on polar motion and deflection of the vertical; e.g., horizontal angles depend very little on the deflection of the vertical (because horizontal angles are the difference between two azimuths, the largest deflection term cancels). The GPS vector observations (which refer to a crust-fixed coordinate system, whose third axis coincides with the CTP) and distances measured with the electronic distance meter (EDM) do not depend on either polar motion or deflection of the vertical. Furthermore, modern surveyors are unlikely to make astronomic observations in view of GPS surveying capability.

In surveying applications, there will typically be no need to improve on the deflection of the vertical already available from, e.g., the NGS. Besides, surveyors can

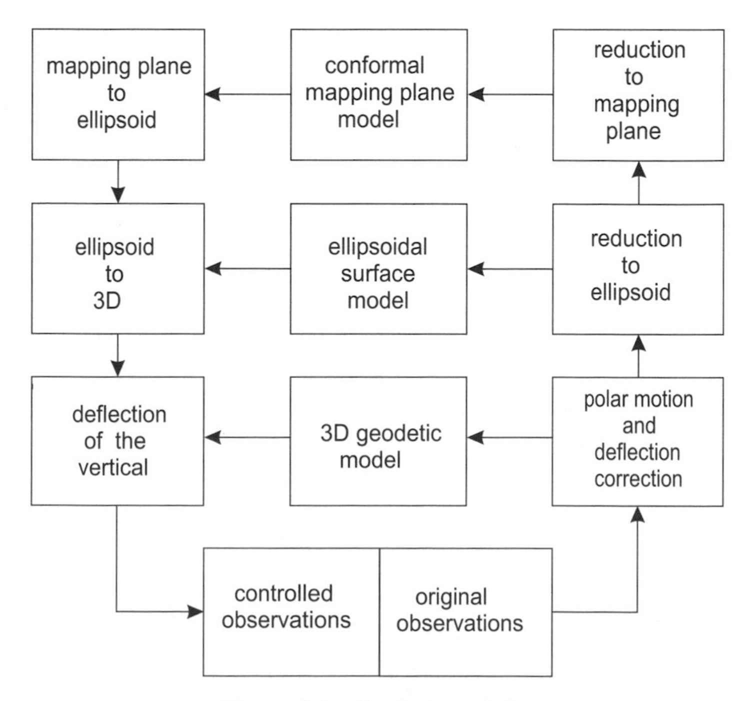

Figure 1.1 Geodetic models.

conveniently introduce their own local ellipsoid that is tangent to the equipotential surface at the center of the survey area. The deflections of the vertical with respect to the local ellipsoid are then zero for all practical purposes within the small geographical region of interest. The adjustment and the quality control of the observations can be carried out in this system. The controlled observations can be deposited in a database.

The approach followed in this book is shown in Figure 1.1. The scheme starts with observations, which are reduced for polar motion and deflection of the vertical (if applicable), adjusted in the three-dimensional model, and then corrected (with the opposite sign) for deflection of the vertical. The results are quality-controlled observations that refer to the local plumb line and the conventional terrestrial coordinate system. The two remaining loops in Figure 1.1 are actually redundant when viewed from a "narrow geodetic" perspective, but they are still of much interest to surveyors because of conformal mappings such as the state plane coordinate (SPC) system. In this book, the expressions for the ellipsoidal surface model and the conformal mapping model are only summarized.

Aspects of GPS satellite surveying can be found in several excellent publications, i.e., Hoffmann-Wellenhof et al. (2001), Kaplan (1996), Misra and Enge (2001), Parkinson et al. (1996), Seeber (2003), and Strang and Borre (1997). *Navigation,* published by the Institute of Navigation, and *GPS Solutions,* published by Springer Verlag, are journals that focus on GPS.

CHAPTER 2

GEODETIC REFERENCE SYSTEMS

It becomes increasingly important to focus on the definition of reference frames as accuracy of geodetic space techniques increases. There are three types of frames we are concerned with—the earth-fixed (international terrestrial reference frame, ITRF) frame, the space-fixed (international conventional reference frame, ICRF) frame, and the geodetic datum. Of course, we also need to be able to transform between the frames. To satisfy the needs of scientists for a clear definition of coordinates, as well as to explore fully the phenomenal increase in accuracy of geodetic space techniques, the definition and maintenance of these reference frames in connection with the deformable earth has become a science in itself. Current solutions have evolved over many years, with contributions from the best scientific minds. The literature is rich in contributions that document the interdisciplinary spectrum and depth needed to arrive at solutions.

The International Earth Rotation Service (IERS) is responsible for establishing and maintaining the ITRF and ICRF frames, whereas typically a national geodetic agency is responsible for establishing and maintaining the datum. The IERS relies on the cooperation of many research groups and national agencies to accomplish its tasks. The International Astronomical Union (IAU) and the International Union of Geodesy and Geophysics (IUGG) established the service in 1988. The IERS maintains a central bureau that is responsible for the general management of the IERS and is governed by a directing board. The conventions underlying the ITRF and the ICRF are published in McCarthy (1996). They are currently completing revision and will become available as IERS Conventions 2000. The old and new conventions are posted at the IERS (2002). McCarthy (1996) is the principal reference for this chapter. For additional details, please consult the many references listed in that publication.

Accurate positioning within the ITRF and ICRF frames requires application of a number of complex mathematical expressions to account for phenomena, such as polar motion, plate tectonic movements, solid earth tides, and ocean loading displacements, as well as precession and nutations. The respective software for these corrections is available, generally on the web. Because the names of computer directories often change, we do not list the full URLs at which the specific software resides. Instead, it is recommended that the reader simply navigate to key agencies and research groups and follow the link to the appropriate levels and directories. A recommended starting point is IERS (2002). Other important sites are of the International GPS Service (IGS), IGS (2002), the U.S. Naval Observatory, USNO (2002), and the National Geodetic Survey, NGS (2002). Because the software is readily available at these sites, we only list mathematical expressions to the extent needed for a conceptual presentation of the topics. However, users striving to achieve complete clarity in definition and the ultimate in positional accuracy must make sure that the software components are mutually consistent and be aware of reductions that might already have been applied to observations.

Most scientists prefer to work with geocentric Cartesian coordinates. In many cases, however, it is easier to interpret results in terms of ellipsoidal coordinates such as geodetic latitude, longitude, and height. It then becomes important to specify the location of the origin of the ellipsoid and its orientation. Ideally, one would like to see the origin coincide with the center of mass and the axes coincide with the directions of the ITRF. The location and orientation of the ellipsoid, as well as its size and shape, are part of the definition of a datum. Below we discuss the details for converting between Cartesian coordinates and geodetic latitude, longitude, and height.

GPS observation such as pseudoranges and carrier phases depend only indirectly on gravity. For example, once the orbit of the satellites has been computed and the ephemeris is available, there is no need to further consider gravity. To make the use of GPS even easier, the GPS ephemeris is typically provided in a well-defined earth-centered earth-fixed (ECEF) coordinate system to which the user can directly relate. In contrast, astronomic latitude, longitude, and azimuth determinations with a theodolite using star observations refer to the instantaneous rotation axis, the instantaneous terrestrial equator of the earth, and the local astronomic horizon (the plane perpendicular to the local plumb line). For applications where accuracy matters, it is typically the responsibility of the user to apply the necessary reductions or corrections to obtain positions in an ECEF coordinate system. Even vertical and horizontal angles as measured by surveyors with a theodolite or total station refer to the plumb line and the local astronomic horizon. Another type of observation that depends on the plumb line is leveling. To deal with types of observations that depend on the direction of gravity (plumb line, horizon), we introduce the geoid.

The goal is to reduce observations that depend on the direction of gravity and to model observations that refer to the ellipsoid. This is accomplished by applying geoid undulations and deflection of the vertical correction. These "connecting elements" are part of the definition of the datum. For a modern datum these elements are readily available, typically on the web (for an example, see NGS, 2002). The reduced observations are the model observation of the 3D geodetic model.

2.1 CONVENTIONAL TERRESTRIAL REFERENCE SYSTEM

A conventional terrestrial reference system (CTRS) must allow the products of various geodetic space techniques, such as coordinates and orientation parameters of the deformable earth, to be combined into a unified data set. Such a reference system should (a) be geocentric (whole earth, including oceans and atmosphere), (b) incorporate corrections or procedures stemming from the relativistic theory of gravitation, (c) maintain consistency in orientation with earlier definitions, and (d) have no residual global rotation with respect to the crust as viewed over time. This section deals with the major phenomena such as polar motion, plate tectonic motions, solid earth tides, and ocean loading that cause variations of coordinates in a terrestrial reference frame. To appreciate the demand placed on a modern reference system, consider the following statement: "GPS data are used to compute daily estimates of the earth's center of mass and scale. Recent center of mass estimates have daily repeatability at the level of 1 cm in x, 1 cm in y, and 1.5 cm in z. Seasonal variations in the center of mass occur at the 3–4 mm level, due primarily to global water mass redistribution. Recent scale estimates repeat daily at the level of 0.3 parts per billion" (Heflin, JPL, private communication).

2.1.1 Polar Motion

The intersection of the earth's instantaneous rotation axis and the crust moves with time. This motion is called polar motion. Figure 2.1 shows polar motion for the time 2001–2003. This motion is somewhat periodic. There is a major constituent of about 434 days, called the Chandler period. The amplitude varies but does not seem to exceed 10 m. Several of the polar motion features can be explained satisfactorily from a geophysical model of the earth; however, the fine structures in polar motion are still subject to research.

To avoid variations in latitude and longitude of about 10 m due to polar motion, we need to define a conventional terrestrial pole (CTP) that is fixed to the crust. Originally, this pole was defined as the center of figure of polar motion for the years 1900–1905. This definition required several refinements as the observation techniques improved. The instantaneous rotation axis can be referenced to the CTP by the polar motion coordinates (x_p, y_p). The origin of the polar motion coordinate system is at the CTP, the x axis is along the conventional zero meridian, and the y axis is positive along the 270° meridian. The center of figure of today's polar motion does not contain the CTP. There appears to be "polar wander" (gradual shifting of the center of figure away from the CTP).

The CTP represents the direction of the third axis of the conventional terrestrial reference system. The definition of the CTRS becomes increasingly complicated because of plate tectonic motions that cause observable station drifts and other temporal variations in the coordinates of a "crust-fixed" coordinate system. As the plates move, the fixed station coordinates become inconsistent with each other. The solution is to define the reference frame by a consistent set of coordinates and their velocities of a global network of stations at a specific epoch. The center of mass of the earth is

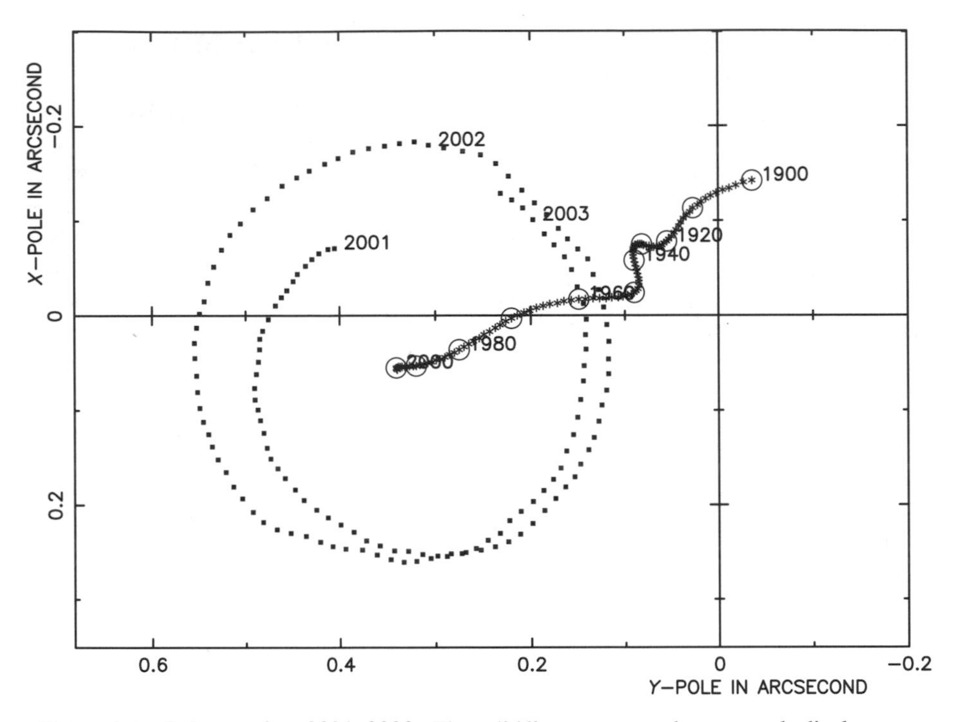

Figure 2.1 Polar motion, 2001–2003. The solid line represents the mean pole displacement, 1900–2000. (Courtesy of the International Earth Orientation Service [IERS], Paris Observatory.)

the natural choice for the origin of the CTRS because satellite dynamics are sensitive to the center of mass (whole earth plus oceans and atmosphere). A particular realization of a CTRS is the ITRF. The IERS maintains the ITRF using extraterrestrial data from various sources, such as GPS, very long baseline interferometry (VLBI), satellite laser ranging (SLR), and Doppler orbitography and radiopositioning integrated on satellite (DORIS). GPS is a viable tool for defining a global reference frame either alone or in combination with the other systems (Heflin et al., 2002). Because the motions of the deformable earth are so complex, there is a need to identify the sites that are part of the solution and, because of evolving data reduction techniques, the IERS publishes updated ITRF solutions. These are designated by adding the year; e.g., ITRF96, ITRF97, and ITRF00. Transformation parameters for the family of ITRFs have been estimated and are available from the IERS. Details on ITRF transformations are found in Soler and Marshall (2002) and the literature listed therein.

An ITRF-type of reference frame is also called an ECEF frame. We denote an ECEF frame by (x) and the coordinate triplet by (x, y, z). The z axis as defined by the IERS is the origin of the polar motion coordinate system. The x and y axes define the terrestrial equatorial plane. In order to maintain continuity with older realizations, the x axis lies in what may be loosely called the Greenwich meridian.

Historically speaking, the International Latitude Service (ILS) was created in 1895, shortly after polar motion had been verified observationally. It was the first international group using globally distributed stations to monitor the reference frame. This group evolved into the International Polar Motion Service (IPMS) in 1962. The IERS was established in 1988 as a single international authority that henceforth uses modern geodetic space techniques to establish and maintain the reference frames. GPS increasingly contributes to the definition and maintenance of the terrestrial reference frame, largely due to the excellent cooperation of international research groups and agencies with the IGS. The IGS began routine operation in 1994, providing GPS orbits, tracking data, and offering other data products in support of geodetic and geophysical research.

2.1.2 Tectonic Plate Motion

The tectonic plate rotations can be approximated by spherical geophysical models such as NNR-NUVELL1A (DeMets et al., 1994). This model is an improved version of the original NUVEL-1 (Argus and Gordon, 1991). Table 2.1 lists the Cartesian angular velocity components for each of the thirteen major plates. At the edges of some of these plates, the motions can be as much as 5 cm per year. Denoting the vector of rotation velocities by $\boldsymbol{\Omega} = [\Omega_x \quad \Omega_y \quad \Omega_z]^T$ and specifying the matrix $\mathbf{R}$ as

$$\mathbf{R}(\boldsymbol{\Omega}) \equiv \begin{bmatrix} 0 & -\Omega_z & \Omega_y \\ \Omega_z & 0 & -\Omega_x \\ -\Omega_y & \Omega_x & 0 \end{bmatrix} \tag{2.1}$$

TABLE 2.1 The NNR-NUVEL1A Kinematic Plate Model

Plate Name	Ω_x (mas/y)	Ω_y (mas/y)	Ω_z (mas/y)	$\lVert\boldsymbol{\Omega}\rVert$ (mas/y)
Africa	0.1837	−0.6392	0.8090	1.047283
Antarctica	−0.1693	−0.3508	0.7644	0.857922
Arabia	1.3789	−0.1075	1.3943	1.963923
Australia	1.6169	1.0569	1.2957	2.325992
Caribbean	−0.0367	−0.6982	0.3261	0.771473
Cocos	−2.1503	−4.4563	2.2534	5.436930
Eurasia	−0.2023	−0.4940	0.6503	0.841339
India	1.3758	0.0082	1.4005	1.407265
Nazca	−0.3160	−1.7691	1.9820	2.675424
North America	0.0532	−0.7423	−0.0316	0.744874
Pacific	−0.3115	0.9983	−2.0564	2.307036
South America	−0.2141	−0.3125	−0.1794	0.419141
Philippines	2.0812	−1.4768	−1.9946	3.238944

Sources: McCarthy, 1996, p. 14, and Soler and Marshall (2002).

Note: The units were changed to milliarc seconds per year for easier visualization.

the transformation between two epochs is accomplished by (McCarthy, 1996, p. 16)

$$\mathbf{x}(t) = \left[\mathbf{I} + 4.84813681 * 10^{-9}\, \mathbf{R}(\boldsymbol{\Omega})\,(t - t_0)\right]\mathbf{x}(t_0) \tag{2.2}$$

Expression (2.2) propagates the position vector $\mathbf{x}$ from epoch t_0 to epoch t within the same reference frame. The NNR-NUVELL1A model can be applied to reference station coordinates to update them as closely as possible to the actual epoch of observations. For consistency, the reference frame for all fiducial points should be the one implicit in the precise ephemeris used. The resulting coordinates would then refer to the reference system of the precise ephemeris and the epoch of the observations. Long-term station motions can readily be appreciated from Figures 2.2 and 2.3.

Because the definition of the frame ultimately involves stations that move with the crust, one must take the time dependency of transformation parameters into consideration when transforming between frames. For example, the parameters listed in Table 2.2 refer to the IGS realization of the ITRF, which is expressed by the designation IGS(ITRFxx). The epoch for these transformation parameters happens to

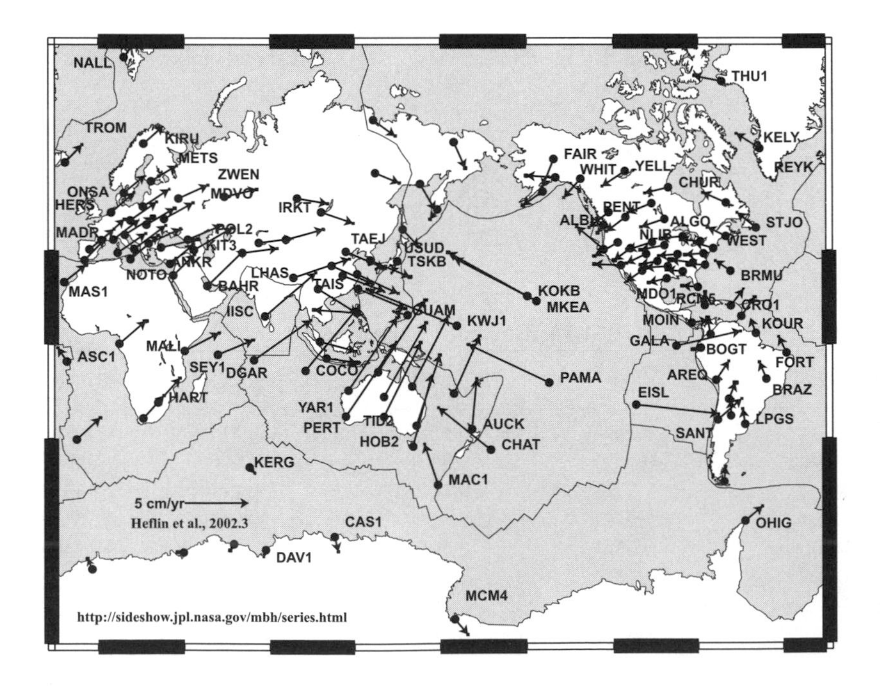

Figure 2.2 Observed motions of globally distributed stations. Velocities for each site were determined from more than eleven years of GPS data. Results are shown in the ITRF00 reference frame with no-net rotation of the crust. Rigid plate motion is clearly visible and describes roughly 80% of the observed motion. The remaining 20% is nonrigid motion in plate boundary zones associated with seismic and volcanic activity. The most visible plate boundary zone on the map is southern California. (Courtesy of Mike Heflin, JPL.)

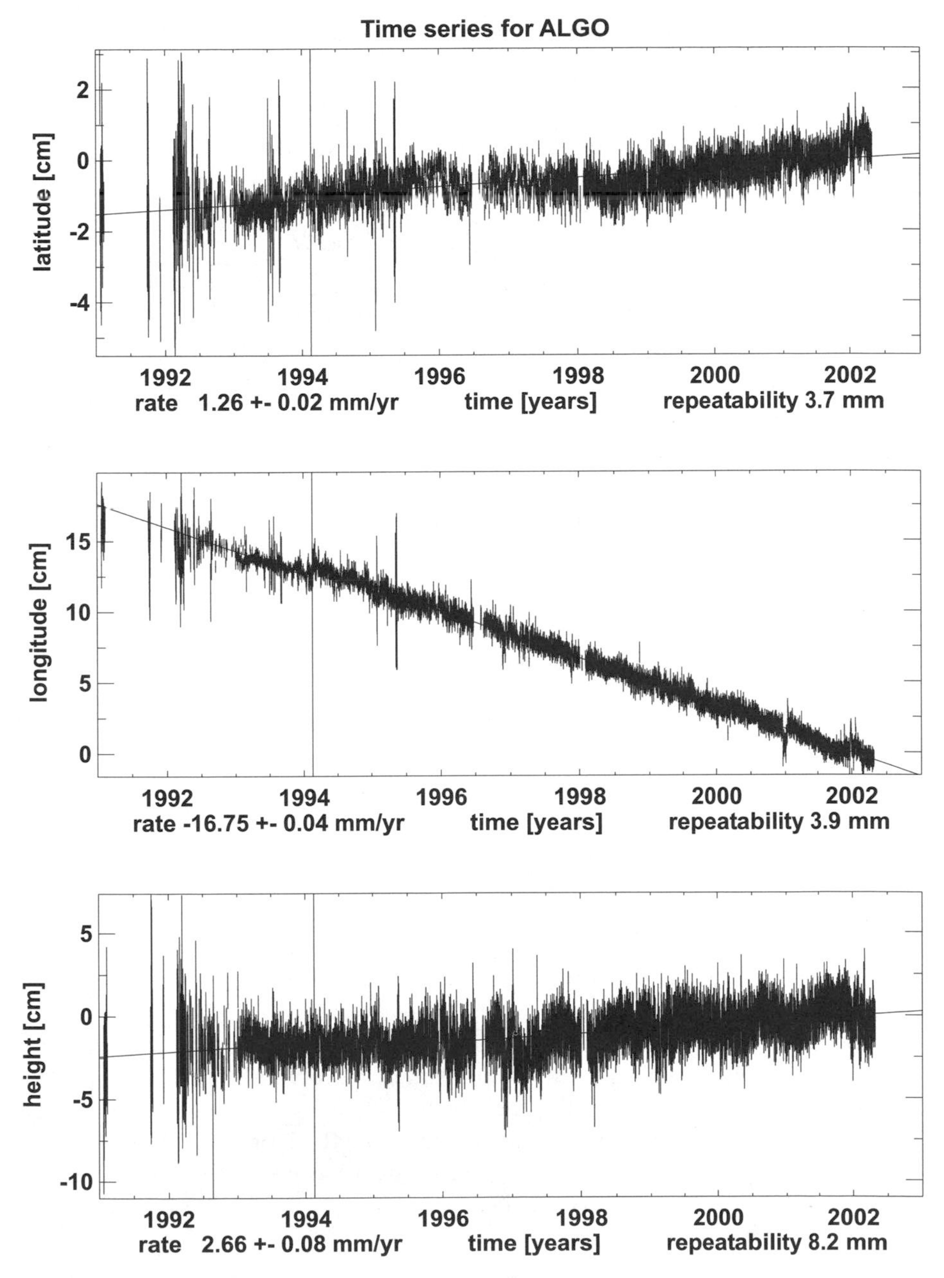

Figure 2.3 Observed motion of station ALGO. The GPS data are used to compute daily estimates of latitude, longitude, and height at each site. Velocity estimates are derived from the time series and typically improve with the time span T in years according to 3.6 mm/T, 4.5 mm/T, and 9.1 mm/T for the north, east, and vertical components, respectively. Recent comparisons of the GPS velocities with ITRF00 show agreement at the level of 0.7 mm/yr for north and east, and 1.5 mm/yr for the vertical. (Courtesy of Mike Heflin, JPL.)

TABLE 2.2 Example of Fourteen-Parameter Transformation between Geocentric Frames

T_x (m)	T_y (m)	T_z (m)	ε_x (mas)	ε_y (mas)	ε_z (mas)	s (ppb)
0.0047	0.0028	−0.0256	−0.030	−0.003	−0.140	1.48
±0.0005	±0.0006	±0.0008	±0.025	±0.021	±0.021	±0.09

$\dot{T}_x$ (m/y)	$\dot{T}_y$ (m/y)	$\dot{T}_z$ (m/y)	$\dot{\varepsilon}_x$ (mas/y)	$\dot{\varepsilon}_y$ (mas/y)	$\dot{\varepsilon}_z$ (mas/y)	$\dot{s}$ (ppb/y)
−0.0004	−0.0008	−0.0016	0.003	−0.001	−0.030	0.03
±0.0003	±0.0003	±0.0004	±0.012	±0.011	±0.011	±0.05

Note: Transformation from IGS(ITRF00) to IGS(ITRF97) at epoch $t_k = 2001.5$ (Ferland, 2002, p. 26). Anticlockwise rotations are positive (mas = milliarc seconds, ppb = part per billion).

be 2001.5 (the fraction of the year is given to one decimal). Soler and Marshall (2002) derive the following fourteen-parameter transformation for transforming ITRFyy to ITRFzz

$$\begin{aligned}\mathbf{x}_{t,\text{ITRFzz}} &= \mathbf{t}_{t_k} + \left(1 + s_{t_k}\right)\left(\mathbf{I} - \mathbf{R}\left(\boldsymbol{\varepsilon}_{t_k}\right)\right)\mathbf{x}_{t,\text{ITRFyy}} \\ &\quad + (t - t_k)\left\{\dot{\mathbf{t}} + \left[-\left(1 + s_{t_k}\right)\mathbf{R}(\dot{\boldsymbol{\varepsilon}}) + \dot{s}\left(\mathbf{I} - \mathbf{R}\left(\boldsymbol{\varepsilon}_{t_k}\right)\right)\right]\mathbf{x}_{t,\text{ITRFyy}}\right\}\end{aligned} \tag{2.3}$$

The $\mathbf{x}_{t,\text{ITRFyy}}$ positions on the right side of (2.3) can be computed from

$$\mathbf{x}_{t,\text{IRTFyy}} = \mathbf{x}_{t_0,\text{IRTFyy}} + (t - t_0)\,\mathbf{v}_{t_0,\text{IRTFyy}} \tag{2.4}$$

In terms of notation, t_k is the epoch at which the transformation parameters are given, t_0 is the epoch of the initial frame IRTFyy, and t is the epoch of the final transformed frame ITRFzz (t could be the actual time of the GPS observations). The vector $\mathbf{t}_k$ contains the Cartesian coordinates of the origin of ITRFyy in the frame ITRFzz, i.e., it is the shift between the two frames. $\boldsymbol{\varepsilon} = [\varepsilon_x \ \varepsilon_y \ \varepsilon_z]^{\mathrm{T}}$ denotes three differential counterclockwise rotations around the x, y, and z axes of the ITRFyy frame, to establish parallelism with the ITRFzz frame. The symbol s denotes the differential scale change. When applying (2.3), the units must be conformable. The simplified form of Equation (2.3) assumes that the velocities $\mathbf{v}_{t_k} = \mathbf{v}_{t_0}$ are in the same frame. The transformation parameters are available from the IERS or research institutions that maintain their own realization of the ITRF. Respective software is also readily available on the web, e.g., Kouba (2001).

2.1.3 Solid Earth Tides

Tides are caused by the temporal variation of the gravitational attraction of the sun and the moon on the earth due to orbital motion. While the ocean tides are very much influenced by the coastal outlines and the shape of the near-coastal ocean floor, the solid earth tides are accurately computable from relatively simple earth models. Their periodicities can be directly derived from the motion of the celestial bodies, similarly to nutation (see below). The solid earth tides generate periodic site displacements of stations that depend on latitude. The tidal variation can be as much as 30 cm in the vertical and 5 cm in the horizontal. McCarthy (1996, p. 61) lists the following expression:

$$\Delta \mathbf{x} = \sum_{j=2}^{3} \frac{GM_j}{GM_E} \frac{\|\mathbf{r}_E\|^4}{\|\mathbf{r}_j\|^3} \left\{ h_2 \mathbf{e} \left(\frac{3}{2} (\mathbf{r}_j \cdot \mathbf{e})^2 - \frac{1}{2} \right) + 3l_2 (\mathbf{r}_j \cdot \mathbf{e}) \left[\mathbf{r}_j - (\mathbf{r}_j \cdot \mathbf{e}) \mathbf{e} \right] \right\} \tag{2.5}$$

In this expression, GM_E is the gravitational constant of the earth, GM_j is the one for the moon (subscript $j = 2$) and the sun ($j = 3$), $\mathbf{e}$ is the unit vector of the station in the geocentric coordinate system (x), and $\mathbf{r}$ denotes the unit vector of the celestial body. h_2 and l_2 are the nominal degree 2 Love and Shida numbers that describe elastic properties of the earth model. Equation (2.5) gives the solid earth tides accurate to at least 5 mm. For additional expressions concerning higher-order terms or expressions for the permanent tide, see McCarthy (1996).

2.1.4 Ocean Loading

Ocean loading refers to the deformation of the sea floor and coastal land that results from the redistribution of ocean water that takes place during the ocean tide. The earth's crust yields under the weight of the tidal water. McCarthy (1996, p. 53) lists the following expression for the site displacement components Δc (where the c refers to the radial, west, and south component) at a particular site at time t,

$$\Delta c = \sum_j f_j A_{cj} \cos\left(\omega_j t + \chi_j + u_j - \Phi_{cj}\right) \tag{2.6}$$

The summation over j represents eleven tidal waves traditionally designated as semidiurnal M_2, S_2, N_2, K_2, diurnal K_1, O_1, P_1, and long-periodic M_f, M_m, S_{sa}. The symbols ω_j and χ_j denote the angular velocities and the astronomic arguments at time $t = 0^h$. The fundamental arguments χ_j reflect the position of the sun and the moon (see nutations below). f_j and u_j depend on the longitude of the lunar node. The station-specific amplitudes A_{cj} and phases Φ_{cj} can be computed using ocean tide models and coastal outline data. The IERS makes these values available for most ITRF reference stations. Typically the M_2 loading deformations are largest, but they do not exceed 5 cm in the vertical and 2 cm in the horizontal.

2.2 CONVENTIONAL CELESTIAL REFERENCE SYSTEM

Dynamical equations of motion are solved in this inertial frame. The equator, ecliptic, and pole of the rotation of the earth historically defined the celestial reference frame. Two-dimensional coordinates of a large number of stars realized it. Present-day ICRF is defined by coordinates of a smaller set of essentially stationary quasars whose positions are accurately known.

We denote the directions of the instantaneous rotation axis by the celestial ephemeris pole (CEP) and the normal of the ecliptic by the north ecliptic pole (NEP). The angle between both directions, or the obliquity, is about 23.5°, which, by virtue of geometry, is also the angle between the instantaneous equator and the ecliptic. As shown in Figure 2.4, the rotation axis can be viewed as moving on a mantle of a cone whose axis coincides with the ecliptic normal.

Mathematically, the motion is split into a smooth long-periodic motion called lunisolar precession and short-periodic motions called nutations. Precession and nutation therefore refer to the motion of the earth's instantaneous rotation axis in space. It takes about 26,000 years for the rotation axis to complete one motion around the cone. The nutations can be viewed as ripples on the circular cone. The longest nutation is 18.6 years and has the largest amplitude of about 20″. The cause of precession and nutation is the ever-changing gravitational attraction of the sun, the moon, and the planets on the earth. Newton's law of gravitation states that the gravitational force between two bodies is proportional to their masses and is inversely proportional to the square of their separation. Because of the earth's and the moon's orbital motions, the separation between the sun, the moon, and the earth changes continuously. Since these changes are periodic, the resulting precession and nutations are periodic in time as well, reflecting the periodic orbital motions. The only exception is a small planetary precession stemming from a motion of the ecliptic. Because of Newton's law of gravitation, the distribution of the earth's mass also critically impacts precession and nutation. Important features are the flattening of the earth, the noncoincidence of the equatorial plane with the ecliptic, and the noncoincidence of the orbital plane of the moon with the ecliptic. Nonrigidity effects of the earth on the nutations can be observed with today's high-precision measurement systems. A spherical earth with homogeneous density distribution would neither precess nor nutate.

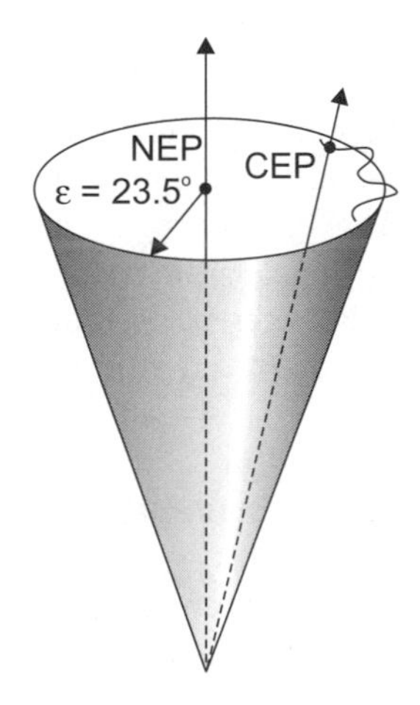

Figure 2.4 Lunisolar precession and nutation. The spatial motion of the CEP is parameterized in terms of precession and nutation.

Because the rotation axis moves in space, the coordinates of stars or extragalactic radio sources change with time due to the motion of the coordinate system. A conventional celestial reference frame (CCRF) has been defined for the fundamental epoch

$$\text{J2000.0} \equiv \text{January 1, 2000, 12h TT} \tag{2.7}$$

The letter "J" in J2000.0 indicates "Julian." In a separate section below, we treat the subject of time in greater detail. Let it suffice here to simply state that TT represents terrestrial time (McCarthy, 1996, p. 83), which is realized by the international atomic time (TAI) as

$$\text{TT} \simeq \text{TAI} + 32^{\text{s}}.184 \tag{2.8}$$

We denote the respective coordinate system, called the mean celestial coordinate system at epoch J2000.0, by $(\bar{\text{X}})$. The $\bar{Z}$ axis coincides with the mean pole. This is the direction of a fictitious rotation axis that has been corrected for nutation, i.e., the fictitious rotation axis that is "driven" by precession only. The mean celestial equatorial plane is the plane perpendicular to the direction of $\bar{Z}$. The $\bar{X}$ axis lies in the equatorial plane and points toward the vernal equinox (intersection of mean celestial equatorial plane and ecliptic). In reality, the precise definition of the first axis takes earlier definitions into consideration that were based on fundamental star catalogues in order to maintain consistency.

Because the CCRF is defined for the epoch J2000.0, the directions of the axis are stable in space per definition. The practical realization of the celestial frame, and therefore the directions of the coordinate axes, is based on a set of celestial radio source coordinates. The IERS selects the celestial radio sources and specifies the observation techniques and analysis procedures. The outcome of this coordinated effort is the ICRF. Extragalactic radio sources, such as quasars, whose signals can be observed with VLBI, play a key role in the establishment and maintenance of the ICRF. Consider two widely separated VLBI antennas on the surface of the earth observing the signals from a quasar. Because of the large distance to quasars, their direction is the same to observers regardless of where the observer is on the earth's surface, as well as where the earth is on its orbit around the sun. The VLBI observations allow one to relate the orientation of the baseline, and therefore the orientation of the earth, to the inertial directions to the quasars.

Any variation in the earth's daily rotation around the instantaneous rotation axis, in polar motion, or any deficiencies in the adopted mathematical model of nutations, can be detected. Today, many quasars and a global network of VLBI antennas are used to measure and monitor these variations. The current ICRF solution includes more than 600 extragalactic radio sources. The details of VLBI are not discussed here, but they are available in the specialized literature. VLBI techniques are very similar to those used in GPS. In fact, the early developments in accurate GPS baseline determination very much benefited from existing knowledge of and experience with VLBI.

2.2.1 Transforming between ITRF and ICRF

The transformation from the ITRF coordinate system (x) to the ICRF coordinate system ($\bar{\mathrm{X}}$) at epoch t is (McCarthy, 1996, p. 21; Mueller, 1969, p. 65):

$$\bar{\mathbf{X}} = \bar{\mathbf{R}}(t)\ \hat{\mathbf{R}}(t)\ \breve{\mathbf{R}}(t)\mathbf{x} \tag{2.9}$$

where

$$\bar{\mathbf{R}}(t) = \mathbf{P}(t)\mathbf{N}(t) \tag{2.10}$$

$$\hat{\mathbf{R}}(t) = \mathbf{R}_3(-\text{GAST}) \tag{2.11}$$

$$\breve{\mathbf{R}}(t) = \mathbf{R}_1(y_p)\mathbf{R}_2(x_p) \tag{2.12}$$

$$\mathbf{P}(t) = \mathbf{R}_3(\zeta)\mathbf{R}_2(-\theta)\mathbf{R}_3(z) \tag{2.13}$$

$$\mathbf{N}(t) = \mathbf{R}_1(-\varepsilon)\mathbf{R}_3(\Delta\psi)\mathbf{R}_1(\varepsilon + \Delta\varepsilon) \tag{2.14}$$

with

$$\zeta = 2306''.2181t + 0''.30188t^2 + 0''.017998t^3 \tag{2.15}$$

$$z = 2306''.2181t + 1''.09468t^2 + 0''.018203t^3 \tag{2.16}$$

$$\theta = 2004''.3109t - 0''.42665t^2 - 0''.041833t^3 \tag{2.17}$$

$$\begin{aligned} \Delta\psi = &-17''.1996\ \sin(\Omega) + 0''.2062\ \sin(2\Omega) \\ &- 1''.3187\ \sin(2F - 2D + 2\Omega) + \cdots + d\psi \end{aligned} \tag{2.18}$$

$$\begin{aligned} \Delta\varepsilon = &\ 9''.2025\ \cos(\Omega) - 0''.0895\ \cos(2\Omega) \\ &+ 0''.5736\ \cos(2F - 2D + 2\Omega) + \cdots + d\varepsilon \end{aligned} \tag{2.19}$$

$$\varepsilon = 84381''.448 - 46''.8150t - 0''.00059t^2 + 0''.001813t^3 \tag{2.20}$$

where t is the time since J2000.0, expressed in Julian centuries of 36,525 days. The arguments of the trigonometric terms in (2.18) and (2.19) are integer multiples of the fundamental periodic elements l, l', F, D, and Ω, resulting in nutation periods that vary from 18.6 years to about 5 days. Of particular interest is Ω, which appears as an argument in the first term of these equations. The largest nutation, which also has the longest period (18.6 years), is a function of Ω, which represents the rotation of the lunar orbital plane around the ecliptic pole. The complete set of nutations contains more than 100 entries. The amplitudes of the nutations are based on geophysical models of the earth. Currently, the IAU 2000 precession and nutation

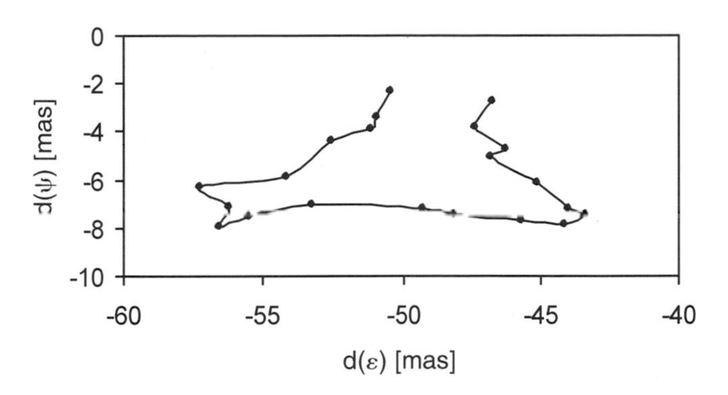

Figure 2.5 Celestial pole offset for 1999 with respect to the IAU 1980 Nutation Model. (Data from 1999 IERS Annual Report.)

model is replacing the IAU 1980 theory of nutations. Because any model is imperfect and imperfections become noticeable as the observation accuracy increases, the so-called celestial pole offsets $d\psi$ and $d\varepsilon$ have been added to (2.18) and (2.19). These offsets are determined and reported by the IERS. An example is seen in Figure 2.5.

The element Ω also describes the 18.6-year tidal period. Because tides and nutation are caused by the same gravitational attraction, it is actually possible to transform the mathematical series of nutations into the corresponding series of tides. Therefore, Expression (2.5) could be developed into a series of sine and cosine terms with the fundamental periodic elements as arguments. These elements are

$$\begin{aligned} l &= \text{Mean Anomaly of the Moon} \\ &= 134^\circ.96340251 + 1717915923''.2178t + 31''.8792t^2 + 0''.051635t^3 + \cdots \end{aligned} \tag{2.21}$$

$$\begin{aligned} l' &= \text{Mean Anomaly of the Sun} \\ &= 357^\circ.52910918 + 12596581''.0481t - 0''.5532t^2 - 0''.000136t^3 + \cdots \end{aligned} \tag{2.22}$$

$$\begin{aligned} F &= L - \Omega \\ &= 93^\circ.27209062 + 1739527262''.8478t - 12''.7512t^2 - 0''.001037t^3 + \cdots \end{aligned} \tag{2.23}$$

$$\begin{aligned} D &= \text{Mean Elongation of the Moon from the Sun} \\ &= 297^\circ.85019547 + 1602961601''.2090t - 6''.3706t^2 + 0''.006593t^3 + \cdots \end{aligned} \tag{2.24}$$

$$\begin{aligned} \Omega &= \text{Mean Longitude of the Ascending Node of the Moon} \\ &= 125^\circ.04455501 - 6962890''.2665t + 7''.4722t^2 + 0''.007702t^3 + \cdots \end{aligned} \tag{2.25}$$

L is mean longitude of the moon. In these equations, the time t is measured in Julian centuries of 36,525 days since J2000.0,

$$t = (\mathrm{TT} - \mathrm{J2000.0})_{[\mathrm{days}]}/36{,}525 \tag{2.26}$$

The Julian date (JD) of the fundamental epoch is

$$\mathrm{JD(J2000.0)} = 2{,}451{,}545.0\mathrm{TT} \tag{2.27}$$

It follows that t can be computed as

$$t = \frac{\mathrm{JD} + \mathrm{TT}_{[\mathrm{h}]}/24 - 2{,}451{,}545.0}{36{,}525} \tag{2.28}$$

The Julian date is a convenient counter for mean solar days. Conversion of any Gregorian calendar date (Y = year, M = month, D = day) to JD is accomplished by (van Flandern and Pulkkinen, 1979)

$$\mathrm{JD} = 367 \times \mathrm{Y} - 7 \times [\mathrm{Y} + (\mathrm{M} + 9)/12]/4 + 275 \times \mathrm{M}/9 + \mathrm{D} + 1{,}721{,}014 \tag{2.29}$$

for Greenwich noon. This expression is valid for dates since March 1900. The expression is read as a Fortran-type statement; division by integers implies truncation of the quotients of integers (no decimals are carried). Note that D is an integer.

In order to compute the Greenwich apparent sidereal time (GAST) needed in (2.11), we must have the universal time (UT1) for the epoch of observation. The latter time is obtained from UTC (coordinate universal time) of the epoch of observation and the UT1-UTC correction. UTC and UT1 will be discussed below. Suffice to say that the correction UT1-UTC is a byproduct of the observations; in other words, it is available from IERS publications. GAST is best computed in three steps. First, we compute Greenwich mean sidereal time (GMST) at the epoch 0^{h}UT1,

$$\begin{aligned}\mathrm{GMST}_{0^{\mathrm{h}}\mathrm{UT1}} = {} & 6^{\mathrm{h}}41^{\mathrm{m}}50^{\mathrm{s}}.54841 + 8640184^{\mathrm{s}}.812866\mathrm{T}_u + 0^{\mathrm{s}}.093104\mathrm{T}_u^2 \\ & - 6^{\mathrm{s}}.2 \times 10^{-6}\mathrm{T}_u^3\end{aligned} \tag{2.30}$$

where $\mathrm{T}_u = \mathrm{d}_u/36525$ and d_u is the number of days elapsed since January 1, 2000, 12^{h}UT1 (taking on values ± 0.5, ± 1.5, etc.). In the second step, we add the difference in sidereal time that corresponds to UT1 hours of mean time,

$$\mathrm{GMST} = \mathrm{GMST}_{0^h\mathrm{UT1}} + r[(\mathrm{UT1} - \mathrm{UTC}) + \mathrm{UTC}] \tag{2.31}$$

$$r = 1.002737909350795 + 5.9006 \times 10^{-11}\mathrm{T}_u - 5.9 \times 10^{-15}\mathrm{T}_u^2 \tag{2.32}$$

In step 3, we apply the nutation to convert the mean sidereal time to apparent sidereal time,

$$\mathrm{GAST} = \mathrm{GMST} + \Delta\psi \cos\varepsilon + 0''.00264 \sin\Omega + 0''.000063 \sin 2\Omega \tag{2.33}$$

The true celestial coordinate system (X), whose third axis coincides with instantaneous rotation axis and X and Y axes span true celestial equator, follows from

$$\mathbf{X} = \mathbf{R}_3(-\text{GAST})\mathbf{R}_1(y_p)\mathbf{R}_2(x_p)\mathbf{x} \tag{2.34}$$

The intermediary coordinate system $(\breve{\mathrm{x}})$,

$$\breve{\mathbf{x}} = \mathbf{R}_1(y_p)\mathbf{R}_2(x_p)\mathbf{x} \tag{2.35}$$

is not completely crust-fixed, because the third axis moves with polar motion. $(\breve{\mathrm{x}})$ is sometimes referred to as the instantaneous terrestrial coordinate system.

Using (X), the apparent right ascension and declination are computed from the expression

$$\alpha = \tan^{-1}\frac{Y}{X} \tag{2.36}$$

$$\delta = \tan^{-1}\frac{Z}{\sqrt{X^2 + Y^2}} \tag{2.37}$$

with $0° \leq \alpha < 360°$. Applying (2.36) and (2.37) to (x) gives the spherical longitude λ and latitude ϕ, respectively. Whereas the zero right ascension is at the vernal equinox and zero longitude is at the reference meridian, both increase counterclockwise when viewed from the third axis.

2.2.2 Time Systems

The GAST relates the terrestrial and celestial reference frames, as far as the earth's daily rotation is concerned, as is seen from (2.34). Twenty-four hours of GAST represents the time for two consecutive transits of the same meridian over the vernal equinox (the direction of the X axis). Unfortunately, these "twenty-four" hours are not suitable to define a constant time interval. As seen from (2.33), GAST depends on the nutation in longitude, $\Delta\psi$, which in turn is a function of time according to (2.18). The vernal equinox reference direction moves along the celestial equator by the time-varying amount $\Delta\psi \cos\varepsilon$. In addition, the earth's daily rotation rate slows down or speeds up. This rate variation can affect the length of day by about 1 ms, corresponding to a length of 4.5 m on the equator.

Let us assume that a geodetic space technique is available for which the mathematical function between observations ℓ and parameters is known,

$$\ell = f\left(\mathbf{X}, \mathbf{x}, \text{GAST}, x_p, y_p\right) \tag{2.38}$$

While we do not go into the details of such solutions, one can readily imagine different types of solutions depending on which parameters are unknown and the type of observations available. For simplicity, let $\mathbf{X}$ (space object) and $\mathbf{x}$ (observing station) be known. Then, given sufficient observational strength, it is conceptually possible to solve (2.38) for GAST, and polar motion x_p, and y_p, given ℓ. We could then compute

GMST from (2.33) and substitute it in (2.31). Finally, assuming that the observations ℓ were taken at known UTC epochs, Expression (2.31) can be solved for the correction

$$\Delta \text{UT1} = \text{UT1} - \text{UTC} \tag{2.39}$$

UTC is related to TAI as established by atomic clocks. Briefly, at the 13th General Conference of Weights and Measures (CGPM) in Paris in 1967, the definition of the atomic second, also called the international system (SI) second, was defined as the duration of 9,192,631,770 periods of the radiation corresponding to the state-energy transition between two hyperfine levels of the ground state of the cesium-133 atom. This definition made the atomic second agree with the length of the ephemeris time (ET) second, to the extent that measurement allowed. ET was the most stable time available around 1960 but is no longer in use. ET was derived from orbital positions of the earth around the sun. Its second was defined as a fraction of the year 1900. Because of the complicated gravitational interactions between the earth and the moon, potential loss of energy due to tidal frictions, etc., the realization of ET was difficult. Its stability eventually did not meet the demands of emerging measurement capabilities. It served as an interim time system. Prior to ET, time was defined in terms of the earth rotation, the so-called earth rotational time scales such as GMST. The rotational time scales were even less constant because of the earth's rotational variations. It takes a good cesium clock 20 to 30 million years to gain or lose one second. Under the same environmental conditions, atomic transitions are identical from atom to atom and do not change their properties. Clocks based on such transitions should generate the same time. Bergquist et al. (2001) offer up-to-date insight on modern atomic clocks.

TAI is based on the SI second; its epoch is such that ET $-$ TAI $= 32^{s}.184$ on January 1, 1977. TAI is related to state transitions of atoms and not to the rotation of the earth. Even though atoms are suitable to define an extremely constant time scale, it could in principle happen that in the distant future we would have noon, i.e., lunchtime at midnight TAI. The hybrid time scale UTC avoids a possible divergence by using the SI second but changing the epoch labeling such that

$$|\Delta \text{UT1}| < 0^{s}.9 \tag{2.40}$$

UTC is the time that is broadcast on TV, on radio, and by other time services.

To visualize the mean universal time (UT1), consider a mean (mathematical) earth traveling in the ecliptic in a circular orbit at constant angular rate. Let this mean earth begin its motion at the time when the true earth is in the direction of the vernal equinox. At each consecutive annual rotation, the mean earth and the true earth should arrive at the vernal equinox at the same time. One often adopts the view as seen from the center of the earth. In that case, one speaks about a mean sun moving around the earth at a constant rate. Twenty-four hours of UT1, i.e., a mean solar day, equals the time it takes for two consecutive transits of the sun over a meridian of the mean earth, or equivalently, two consecutive transits of the mean sun when viewed from the earth-fixed reference frame. If we consider the actual earth or sun, as opposed to their mean motions, we speak of true solar time. Astronomers call the difference between mean time and true time the equation of time. Geometrically, it represents the angle between the true earth and the mean earth as viewed from the sun. Simple graphics shows that the mean solar day is longer than the sidereal day by about $24^h/365 \approx 4^m$. The accurate ratio of universal day over sidereal day is given in (2.32). The condition (2.40) underscores the compromising role of UTC. The precise time and frequency users get the most uniform and accurate time available, and yet the epoch closely adjusts to the rotational behavior of the earth.

Let's consider Equation (2.31) once again. If the earth were to rotate with constant speed, and if the SI second would be absolutely equal to the theoretical value of the ET (or UT1) second, then the difference UT1 − UTC in (2.31) would be constant. Any variation in this difference is therefore attributable to variations in the earth's rotation and the definition of the SI second. UTC is adjusted in steps of a full second (leap second) if the difference (2.40) exceeds the specified limit. Adjustments are made on either June 30 or December 31, if a change is warranted. The IERS determines the need for a leap second and announces any forthcoming step adjustment. Figure 2.6 shows the history of leap second adjustments. The trend seen in Figure 2.6 could be removed by changing the definition of the SI second, i.e., adopting a different number of energy state transitions. However, changing the definition of a fundamental constant has many implications (Mohr and Taylor, 2001). Figure 2.7 shows the total variation of UT1 − UTC. This includes the seasonal variations (annual and semiannual), as well as variations due to zonal tides. Similarly to the effect on the nutations and solid earth tides, the solar and lunar gravitational attractions cause periodic variations in

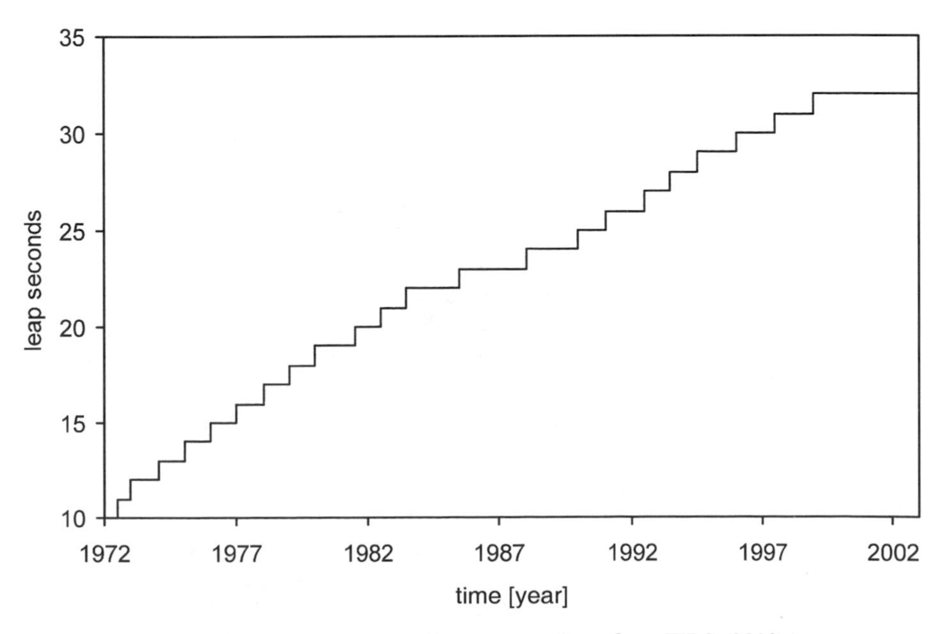

Figure 2.6 Leap second adjustments. (Data from IERS (2002).)

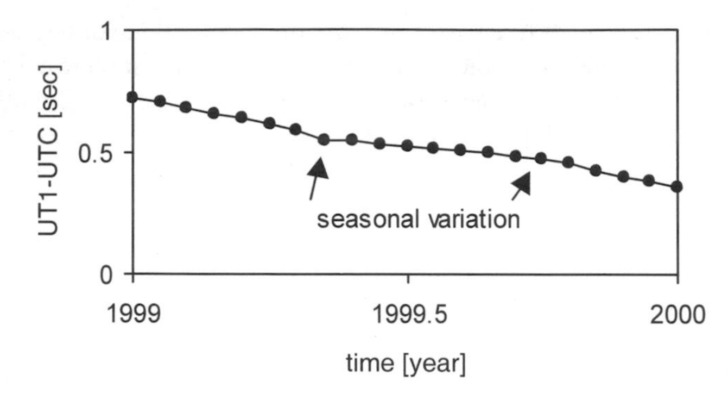

Figure 2.7 UT1-UTC variation during 1999. (Data from 1999 IERS Annual Report.)

the earth's rotation. These forced variations are computable. The currently adopted model includes terms with periods up to thirty-five days.

The five corrections, UT1 − UTC, polar motion x_p, and y_p, and the celestial pole offsets $d\psi$ and $d\varepsilon$, are required to transform the terrestrial reference frame to the celestial one and vice versa. The IERS monitors and publishes these values. They are the earth orientation parameters (EOP). Modern space techniques allow these parameters to be determined with centimeter accuracy.

Various laboratories and agencies operate several atomic clocks and produce their own independent atomic time. For example, the time scale of the U.S. Naval Observatory is called UTC(USNO), and the National Institute of Standards and Technology (NIST) produces UTC(NIST). The IERS, which uses input from 200 plus clocks from sixty plus different laboratories scattered around the world, computes TAI. UTC and TAI differ by the integer leap seconds. TAI is not adjusted, but UTC is adjusted for leap seconds.

The GPS satellites follow GPS time (GPST). This time scale is steered to be within one microsecond of UTC(USNO). The initial epoch of GPST is 0^hUTC January 6, 1980. Since that epoch, GPST has not been adjusted to account for leap seconds. It follows that GPST − TAI $= -19^s$, i.e., equal to the offset of TAI and UTC at the initial GPST epoch. Each satellite carries several atomic clocks, including the spare clock. These clocks establish the space vehicle time. The control center synchronizes the clocks of the various space vehicles to GPST.

The Julian day date (JD) used in (2.29) is but a convenient continuous counter of mean solar days from the beginning of the year 4713 B.C. By tradition, the Julian day date begins at Greenwich noon, i.e., 12^hUT1. As such, the JD has nothing to do with the Julian calendar, which was created by Julius Caesar. It provided for the leap year rule that declared a leap year of 366 days if the year's numerical designation is divisible by 4. This rule was later supplemented in the Gregorian calendar by specifying that the centuries that are not divisible by 400 are not leap years. Accordingly, the year 2000 was a leap year but the year 2100 will not be. The Gregorian calendar reform also included that the day following October 4 (Julian calendar), 1582, was labeled October 15 (Gregorian calendar). The proceedings of the conference to commemorate the 400th anniversary of the Gregorian calendar (Coyne et al., 1983) give background information on the Gregorian calendar. The astronomic justification for the leap year rules stems from the fact that the tropical year consists of $365^{d}.24219879$ mean solar days. The tropical year equals the time it take the mean (fictitious) sun to make two consecutive passages over the mean vernal equinox.

2.3 DATUM

The complete definition of a geodetic datum includes the size and shape of the ellipsoid, its location and orientation, and its relation to the geoid by means of geoid undulations and deflection of the vertical. The datum currently used in the United States is NAD83, which was developed by the NGS (NGS, 2002). In the discussion below we briefly introduce the geoid and the ellipsoid. A discussion of geoid undulations and deflection of the vertical follows, with emphasis on how to use these elements to reduce observations to the ellipsoidal normal and the geodetic horizon. Finally, the 3D geodetic model is introduced as a general and unified model that not only deals with observations in all three dimensions, but is also mathematically the simplest of all.

2.3.1 Geoid

The geoid is a fundamental physical reference surface to which all observations refer if they depend on gravity. Because its shape is a result of the mass distribution inside the earth, the geoid is not only of interest to the measurement specialist but also to scientists who study the interior of the earth. Consider two point masses m_1 and m_2 separated by a distance s. According to Newton's law of gravitation, they attract each other with the force

$$F = \frac{k^2 m_1 m_2}{s^2} \tag{2.41}$$

where k^2 is the universal gravitational constant. The attraction between the point masses is symmetric and opposite in direction. As a matter of convenience, we consider one mass to be the "attracting" mass and the other to be the "attracted" mass. Furthermore, we assign to the attracted mass the unit mass ($m_2 = 1$) and denote the attracting mass with m. The force equation then becomes

$$F = \frac{k^2 m}{s^2} \tag{2.42}$$

and we speak about the force between an attracting mass and a unit mass as being attracted. Introducing an arbitrary coordinate system, as seen in Figure 2.8, we decompose the force vector into Cartesian components. Thus,

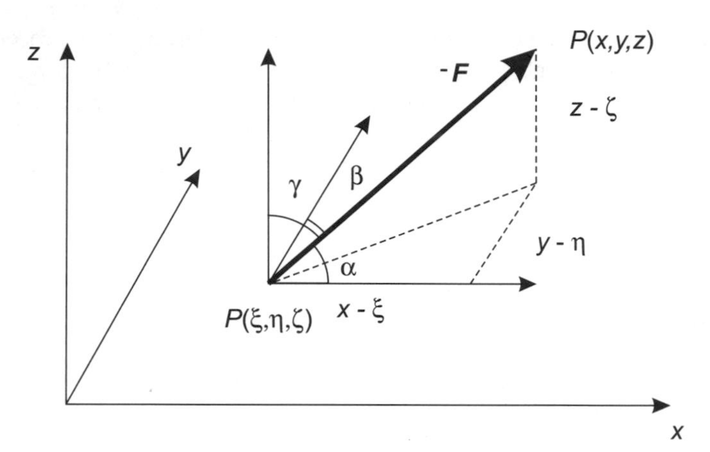

Figure 2.8 Components of the gravity vector.

$$\mathbf{F} = \begin{bmatrix} F_x \\ F_y \\ F_z \end{bmatrix} = -F \begin{bmatrix} \cos\alpha \\ \cos\beta \\ \cos\gamma \end{bmatrix} = -\frac{k^2 m}{s^2} \begin{bmatrix} \dfrac{x-\xi}{s} \\ \dfrac{y-\eta}{s} \\ \dfrac{z-\zeta}{s} \end{bmatrix} \tag{2.43}$$

where

$$s = \sqrt{(x-\xi)^2 + (y-\eta)^2 + (z-\zeta)^2} \tag{2.44}$$

The negative sign in the decomposition indicates the convention that the force vector points from the attracted mass toward the attracting mass. The coordinates (x, y, z) identify the location of the attracted mass in the specified coordinate system, and (ξ, η, ζ) denote the location of the attracting mass. The expression

$$V = \frac{k^2 m}{s} \tag{2.45}$$

is called the potential of gravitation. It is a measure of the amount of work required to transport the unit mass from its initial position, a distance s from the attracting mass, to infinity. Integrating the force equation (2.42) gives

$$V = \int_s^\infty F\, ds = \int_s^\infty \frac{k^2 m}{s^2} ds = -\frac{k^2 m}{s}\bigg|_s^\infty = \frac{k^2 m}{s} \tag{2.46}$$

In vector notation, the potential of gravitation V and the gravitational force vector $\mathbf{F}$ are related by

$$F_x = \frac{\partial V}{\partial x} = k^2 m \frac{\partial}{\partial x}\left(\frac{1}{s}\right) = -\frac{k^2 m}{s^2}\frac{\partial s}{\partial x} = -\frac{k^2 m}{s^2}\frac{x - \xi}{s} \tag{2.47}$$

Similar expressions can be written for F_y and F_z. Thus, the gradient V is

$$\text{grad } V \equiv \begin{bmatrix} \frac{\partial V}{\partial x} & \frac{\partial V}{\partial y} & \frac{\partial V}{\partial z} \end{bmatrix}^T = \begin{bmatrix} F_x & F_y & F_z \end{bmatrix}^T \tag{2.48}$$

From (2.45) it is apparent that the gravitational potential is a function only of the separation of the masses and is independent of any coordinate system used to describe the position of the attracting mass and the direction of the force vector **F**. The gravitational potential, however, completely characterizes the gravitational force at any point by use of (2.48).

Because the potential is a scalar, the potential at a point is the sum of the individual potentials,

$$V = \sum V_i = \sum \frac{k^2 m_i}{s_i} \tag{2.49}$$

Considering a solid body M rather than individual masses, a volume integral replaces the discrete summation over the body,

$$V(x, y, z) = k^2 \iiint_M \frac{dm}{s} = k^2 \iiint_v \frac{\rho \, dv}{s} \tag{2.50}$$

where ρ denotes a density that varies throughout the body and v denotes the mass volume.

When deriving (2.50), we assumed that the body is at rest. In the case of the earth, we must consider the earth's rotation. Let the vector **f** denote the centrifugal force acting on the unit mass. If the angular velocity of the earth's rotation is ω, then the centrifugal force vector can be written

$$\mathbf{f} = \omega^2 \mathbf{p} = \begin{bmatrix} \omega^2 x & \omega^2 y & 0 \end{bmatrix}^T \tag{2.51}$$

The centrifugal force acts parallel to the equatorial plane and is directed away from the axis of rotation. The vector **p** is the distance from the rotation axis. Using the definition of the potential and having the z axis coincide with the rotation axis, we obtain the centrifugal potential:

$$\Phi = \frac{1}{2}\omega^2 \left(x^2 + y^2\right) \tag{2.52}$$

Equation (2.52) can be verified by taking the gradient to get (2.51). Note again that the centrifugal potential is a function only of the distance from the rotation axis and is not affected by a particular coordinate system definition. The potential of gravity W is the sum of the gravitational and centrifugal potentials:

$$W(x, y, z) = V + \Phi = k^2 \iiint_v \frac{\rho \, dv}{s} + \frac{1}{2}\omega^2 \left(x^2 + y^2\right) \tag{2.53}$$

The gravity force vector **g** is the gradient of the gravity potential,

$$\mathbf{g}(x, y, z) = \text{grad } W = \begin{bmatrix} \dfrac{\partial W}{\partial x} & \dfrac{\partial W}{\partial y} & \dfrac{\partial W}{\partial z} \end{bmatrix}^{\mathrm{T}} \tag{2.54}$$

and represents the total force acting at a point as a result of the gravitational and centrifugal forces. The magnitude $\|\mathbf{g}\| = g$ is called gravity. It is traditionally measured in units of gals where 1 gal = 1 cm/sec^2. The gravity increases as one moves from the equator to the poles because of the decrease in centrifugal force. Approximate values for gravity are $g_{\text{equator}} \cong 978$ gal and $g_{\text{poles}} \cong 983$ gal. The units of gravity are those of acceleration, implying the equivalence of force per unit mass and acceleration. Because of this, the gravity vector **g** is often termed gravity acceleration. The direction of **g** at a point and the direction of the plumb line or the vertical are the same.

Surfaces on which $W(x, y, z)$ is a constant are called equipotential surfaces, or level surfaces. These surfaces can principally be determined by evaluating (2.53) if the density distribution and angular velocity are known. Of course, the density distribution of the earth is not precisely known. Physical geodesy deals with theories that allow estimation of the equipotential surface without explicit knowledge of the density distribution. The geoid is defined to be a specific equipotential surface having gravity potential

$$W(x, y, z) = W_0 \tag{2.55}$$

In practice, this reference gravity potential is chosen such that on the average it coincides with the global ocean surface. This is a purely arbitrary specification chosen for ease of the physical interpretation. The geoid is per definition an equipotential surface, not some ideal ocean surface.

There is an important relationship between the direction of the gravity force and equipotential surfaces, demonstrated by Figure 2.9. The total differential of the gravity potential at a point is

$$\begin{aligned} dW &= \frac{\partial W}{\partial x} dx + \frac{\partial W}{\partial y} dy + \frac{\partial W}{\partial z} dz \\ &= \left[\text{grad } W\right]^{\mathrm{T}} \cdot d\mathbf{x} = \mathbf{g} \cdot d\mathbf{x} \end{aligned} \tag{2.56}$$

The quantity dW is the change in potential between two differentially separated points $P(x, y, z)$ and $P'(x + dx, y + dy, z + dz)$. If the vector, $d\mathbf{x}$ is chosen such that P and P' occupy the same equipotential surface, then $dW = 0$ and

$$\mathbf{g} \cdot d\mathbf{x} = 0 \tag{2.57}$$

Expression (2.57) implies that the direction of the gravity force vector at a point is normal or perpendicular to the equipotential surface passing through the point.

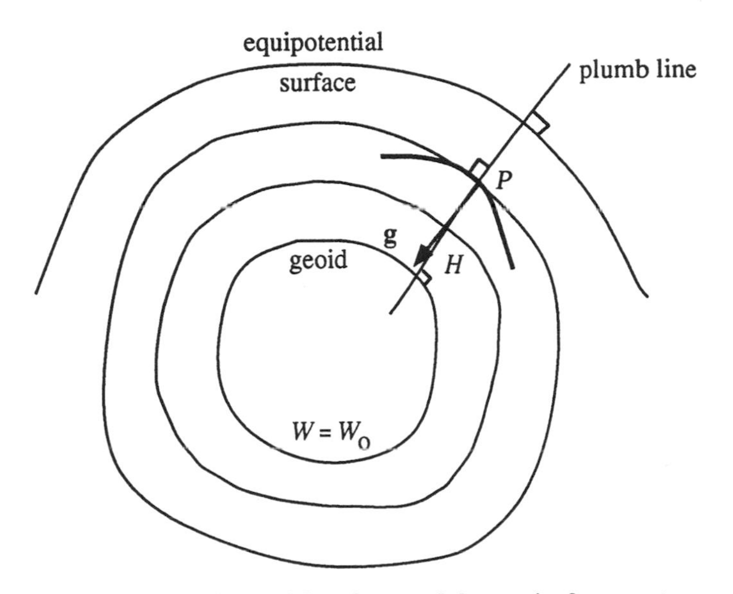

Figure 2.9 Equipotential surfaces and the gravity force vector.

The shapes of equipotential surfaces, which are related to the mass distribution within the earth through (2.53), have no simple analytic expressions. The plumb lines are normal to the equipotential surfaces and are space curves with finite radii of curvature and torsion. The distance along a plumb line from the geoid to a point is called the orthometric height, H. The orthometric height is often misidentified as the "height above sea level." Possibly, confusion stems from the specification that the geoid closely approximates the global ocean surface.

Consider a differential line element $d\mathbf{x}$ along the plumb line, $\|d\mathbf{x}\| = dH$. By noting that H is reckoned positive upward and $\mathbf{g}$ points downward, we can rewrite (2.56) as

$$\begin{aligned} dW &= \mathbf{g} \cdot d\mathbf{x} \\ &= g\, dH \cos(\mathbf{g}, d\mathbf{x}) = g\, dH \cos(180°) = -g\, dH \end{aligned} \tag{2.58}$$

This expression relates the change in potential to a change in the orthometric height. This equation is central in the development of the theory of geometric leveling. Writing (2.58) as

$$g = -\frac{dW}{dH} \tag{2.59}$$

it is obvious that the gravity g cannot be constant on the same equipotential surface because the equipotential surfaces are neither regular nor concentric with respect to the center of mass of the earth. This is illustrated in Figure 2.10, which shows two differentially separate equipotential surfaces. It is observed that

$$g_1 = -\frac{dW}{dH_1} \neq g_2 = -\frac{dW}{dH_2} \tag{2.60}$$

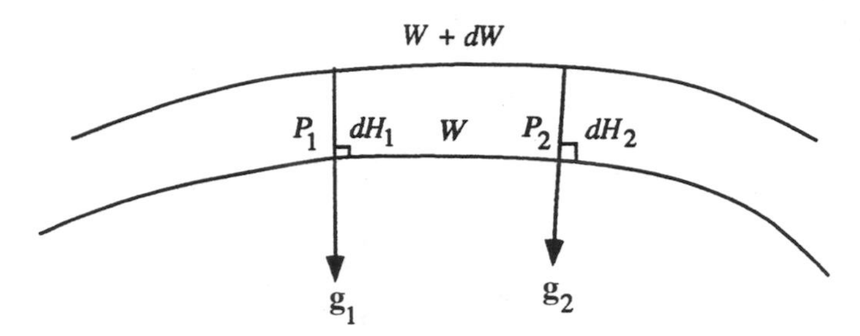

Figure 2.10 Gravity on the equipotential surface.

The astronomic latitude, longitude, and azimuth refer to the plumb line at the observing station. Figure 2.11 shows an equipotential surface through a surface point P and the instantaneous rotation axis and equator. The astronomic normal at point P, also called the local vertical, is identical to the direction of the gravity force at that point, which in turn is tangent to the plumb line. The astronomic latitude Φ at P is the angle subtended on the instantaneous equator by the astronomic normal. The astronomic normal and the parallel line to the instantaneous rotation axis span the astronomic meridian plane at point P. Note that the instantaneous rotation axis and the astronomic normal may or may not intersect. The astronomic longitude Λ is the angle subtended in the instantaneous equatorial plane between this astronomic meridian and a reference meridian, nominally the Greenwich meridian.

The geopotential number C is simply the algebraic difference between the potentials at the geoid and point P,

$$C = W_0 - W \tag{2.61}$$

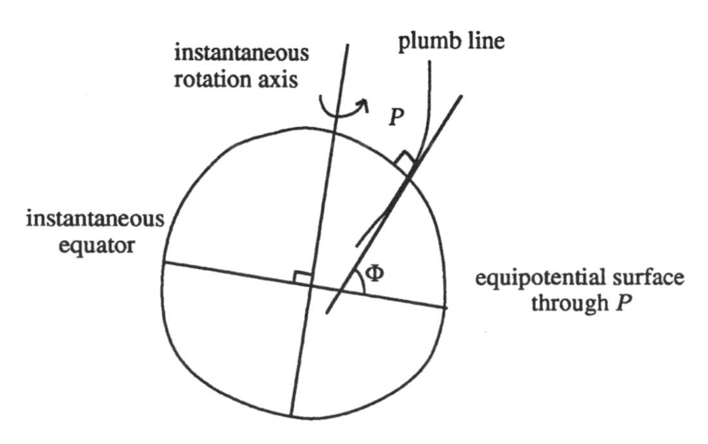

Figure 2.11 Astronomic latitude.

From (2.58) it follows that

$$W = W_0 - \int_0^H g\, dH \tag{2.62}$$

or

$$C = W_0 - W = \int_0^H g\, dH \tag{2.63}$$

or

$$H = -\int_{W_0}^{W} \frac{dW}{g} = \int_0^C \frac{dC}{g} \tag{2.64}$$

Equation (2.63) shows how combining gravity observations and leveling yields potential differences. The increment dH is obtained from spirit leveling, and the gravity g is measured along the leveling path. Consider a leveling loop as an example. Because one returns to the same point when leveling a loop, i.e., one returns to the same equipotential surface, (2.63) implies that the integral (or the sum) of the products $g\, dH$ adds up to zero. Because g varies along the loop, the sum over the leveled differences dH does not necessarily add up to zero.

The difference between the orthometric heights and the leveled heights is called the orthometric correction. Expressions for computing the orthometric correction from gravity are available in the specialized geodetic literature. An excellent introduction to height systems is Heiskanen and Moritz (1967, Chapter 4). Guidelines for accurate leveling are available from the NGS (Schomaker and Berry, 1981).

2.3.2 Ellipsoid of Revolution

The ellipsoid of revolution, called here simply the ellipsoid, is a relatively simple mathematical figure that closely approximates the actual geoid. When using an ellipsoid for geodetic purposes, we need to specify its shape, location, and orientation with respect to the earth. The size and shape of the ellipsoid is defined by two parameters: the semimajor axis a and the flattening f. The flattening is related to the semiminor axis b by

$$f = \frac{a-b}{a} \tag{2.65}$$

Appendix B contains the details of the mathematics of the ellipsoid and common values for a and b. The orientation and location of the ellipsoid often depends on when and how it was established. In the presatellite era, the goal often was to establish a local ellipsoid that best fitted the geoid in a well-defined region, i.e., the border of a nation-state. The third axis, of course, always pointed toward the North Pole and the first axis in the direction of the Greenwich meridian. Using local ellipsoids as a reference does have the advantage that some of the reductions (geoid undulation, deflection of the vertical) can possibly be neglected, which is an important consideration when the geoid is not known accurately. With today's advanced geodetic satellite techniques, in particular GPS, and accurate knowledge of the geoid, one prefers so-called global ellipsoids that fit the geoid globally (whose center of figure is at the center of mass, and whose axes coincide with the directions of the ITRF). The relationship between the Cartesian coordinates $(x) = (x, y, z)$ and the geodetic coordinates $(\varphi) = (\varphi, \lambda, h)$ is according to B.9 to B.11,

$$x = (N + h)\cos\varphi\cos\lambda \tag{2.66}$$

$$y = (N + h)\cos\varphi\sin\lambda \tag{2.67}$$

$$z = [N(1 - e^2) + h]\sin\varphi \tag{2.68}$$

where the auxiliary quantities N and e are

$$N = \frac{a}{\sqrt{1 - e^2\sin^2\varphi}} \tag{2.69}$$

$$e^2 = 2f - f^2 \tag{2.70}$$

The transformation from (x) to (φ) is given in Appendix B. It is typically performed iteratively.

The expression (2.3) can be applied to transform between a local datum and a geocentric datum provided the transformation parameters are known. It is best to contact the responsible agency for the latest set of parameters because the transformation parameters are continuously updated, particularly for older datums. For example, the large collection that includes probably all known datums is available through the National Imagery and Mapping Agency, NIMA (2002). The NGS makes the transformation software regarding the NAD83 available at NGS (2002). Both agencies provide software that in some cases considers the geodetic network distortions and crustal motions to achieve a more accurate transformation. A difficulty in using (2.3) is that in the past, one dealt with a horizontal and vertical datum separately and that the respective connecting elements, the geoid undulations, might not be available.

2.3.3 Geoid Undulations and Deflections of the Vertical

One approach to estimate the geoid undulation is by measuring gravity or gravity gradients at the surface of the earth. At least in principle, any observable that is a function of the gravity field is suitable for determining the geoid. Low-earth orbiting satellites (LEOs) have successfully been used to determine the large structure of the geoid. Satellite-to-satellite tracking is being used to determine the temporal variations of the gravity field, and thus the geoid. The reader may want to check the results of the Gravity Recovery and Climate Experiment (GRACE) mission launched in early 2002.

The gravity field or functions of the gravity field are typically expressed in terms of spherical harmonic expansions. For example, the expression for the geoid undulation N is (Lemoine et al., 1998, pp. 5–11),

$$N = \frac{GM}{\gamma r} \sum_{n=2}^{\infty} \left(\frac{a}{r}\right)^n \sum_{m=0}^{n} (\bar{C}_{nm} \cos m\lambda + \bar{S}_{nm} \sin m\lambda)\, \bar{P}_{nm}(\cos\,\theta) \tag{2.71}$$

In this equation the following notation is used:

N	Geoid undulation. There should not be cause for confusion using the same symbol for the geoid undulation (2.71) and the radius of curvature of the prime vertical (2.69); both notations are traditional in the geodetic literature.
φ, λ	Latitude and longitude of station where the undulation is computed.
$\bar{C}_{nm}, \bar{S}_{nm}$	Normalized spherical harmonic coefficients (geopotential coefficients), of degree n and order m. A set degree and order 360 is currently published by the Goddard Space Flight Center (GSFC, 2002). In this notation, $\bar{C}_{nm}$ denotes the difference between the spherical harmonics of the geopotential and the normal gravity field harmonics.
$\bar{P}_{nm}(\cos\theta)$	Associated Legendre functions. $\theta = 90 - \varphi$ is the colatitude.
r	Geocentric distance of the station.
GM	Product of the gravitational constant and the mass of the earth. GM is identical to k^2M used elsewhere in this book. Unfortunately, the symbolism is not unique in the literature. We retain the symbols typically used within the respective context.
γ	Normal gravity. Details are given below.
a	Semimajor axis of the ellipsoid.

Geoid undulation computed from an expression like (2.71) refers to a geocentric ellipsoid with semimajor axis a. The coefficients $\bar{C}_{nm}$ are computationally adjusted to the specific flattening of the reference ellipsoid. The summation starts with $n = 2$. Figure 2.12 shows a map of a global geoid.

There is a simple mathematical relationship between the geoid undulation and the deflection of the vertical. The deflections of the vertical are related to the undulations as follows (Heiskanen and Moritz, 1967, p. 112):

$$\xi = -\frac{1}{r}\frac{\partial N}{\partial \theta} \tag{2.72}$$

$$\eta = -\frac{1}{r \sin\theta}\frac{\partial N}{\partial \lambda} \tag{2.73}$$

Differentiating (2.71) gives

$$\xi = -\frac{GM}{\gamma r^2} \sum_{n=2}^{\infty} \left(\frac{a}{r}\right)^n \sum_{m=0}^{n} (\bar{C}_{nm} \cos m\lambda + \bar{S}_{nm} \sin m\lambda)\, \frac{d\bar{P}_{nm}(\cos\,\theta)}{d\theta} \tag{2.74}$$

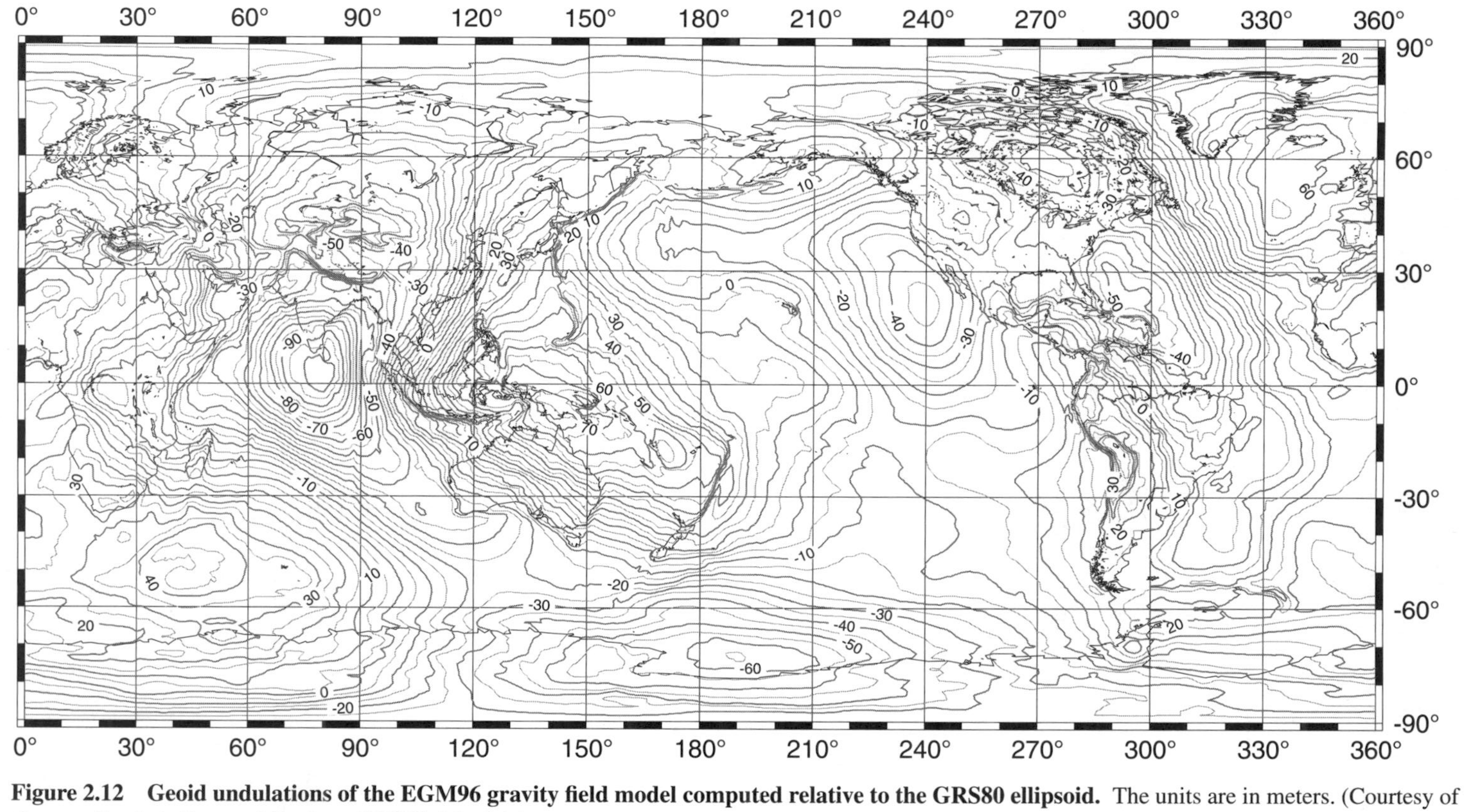

Figure 2.12 Geoid undulations of the EGM96 gravity field model computed relative to the GRS80 ellipsoid. The units are in meters. (Courtesy of German Geodetic Research Institute [DGFI], Munich.)

$$\eta = -\frac{GM}{\gamma r^2 \sin\theta} \sum_{n=2}^{\infty} \left(\frac{a}{r}\right)^n \sum_{m=0}^{n} m \left(-\bar{C}_{nm} \sin m\lambda + \bar{S}_{nm} \cos m\lambda\right) \bar{P}_{nm}(\cos\,\theta) \tag{2.75}$$

Geoid and deflection of the vertical maps specifically adjusted to the NAD83 datum can be viewed at NGS (2002). NGS also provides software for convenient computation of these gravity functions.

The ellipsoid of revolution provides a simple model for the geometric shape of the earth. It is the reference for geometric computations in two and three dimensions. Assigning a gravitational field that approximates the actual gravitational field of the earth extends the functionality of the ellipsoid. Merely a few specifications are needed to fix the gravity and potential of the ellipsoid of revolution. We need an appropriate mass for the ellipsoid and assume that the ellipsoid rotates with the earth. Furthermore, by means of mathematical conditions, the surface of the ellipsoid is defined to be an equipotential surface of its own gravity field. Therefore, the plumb lines of this gravity field intersect the ellipsoid perpendicularly. Because of this property, this gravity field is called the normal gravity field, and the ellipsoid itself is sometimes referred to as the level ellipsoid.

It can be shown that the normal gravity potential U is completely specified by four defining constants, which are symbolically expressed by

$$U = f(a, J_2, GM, \omega) \tag{2.76}$$

In addition to a and GM, which have already been introduced above, we need the dynamical form factor J_2 and the angular velocity of the earth ω. The dynamic form factor is a function of the principal moments of inertia of the earth (polar and equatorial moment of inertia) and is functionally related to the flattening of the ellipsoid. One important definition of the four constants in (2.76) comprises the Geodetic Reference System of 1980 (GRS80). The defining constants are listed in Table 2.3. A full documentation on this reference system is available in Moritz (1984).

The normal gravitational potential does not depend on the longitude and is given by a series of zonal spherical harmonics

TABLE 2.3 Constants for GRS80

Defining Constants	Derived Constants
$a = 6378138$ m	$b = 6356752.3141$ m
$\mathrm{GM} = 3986005 \times 10^8\ \mathrm{m^3/s^2}$	$1/f = 298.257222101$
$J_2 = 108263 \times 10^{-8}$	$m = 0.00344978600308$
$\omega = 7292115 \times 10^{-11}$ rad/s	$\gamma_e = 9.7803267715\ \mathrm{m/s^2}$
	$\gamma_p = 9.8321863685\ \mathrm{m/s^2}$

$$V = \frac{GM}{r}\left[1 - \sum_{n=1}^{\infty} J_{2n}\left(\frac{a}{r}\right)^{2n} P_{2n}(\cos\theta)\right] \tag{2.77}$$

Note that the subscript $2n$ is to be read "2 times n." P_{2n} denotes Legendre polynomials. The coefficients J_{2n} are a function of J_2 that can be readily computed. Several useful expressions can be derived from (2.77). For example, the normal gravity, defined as the magnitude of the gradient of the normal gravity field (normal gravitational potential plus centrifugal potential), is given by Somigliana's closed formula (Heiskanen and Moritz, 1967, p. 70):

$$\gamma = \frac{a\gamma_e \cos^2\varphi + b\gamma_p \sin^2\varphi}{\sqrt{a^2\cos^2\varphi + b^2\sin^2\varphi}} \tag{2.78}$$

The normal gravity at height h above the ellipsoid is given by (Heiskanen and Moritz, 1967, p. 70)

$$\gamma_h - \gamma = -\frac{2\gamma_e}{a}\left[1 + f + m + \left(-3f + \frac{5}{2}m\right)\sin^2\varphi\right]h + \frac{3\gamma_e}{a^2}h^2 \tag{2.79}$$

Equations (2.78) and (2.79) are often useful approximations of the actual gravity. The value for the auxiliary quantity m in (2.79) is given in Table 2.3. The normal gravity values for the poles and the equator, γ_p and γ_e are also listed in Table 2.3.

2.3.4 Reductions to the Ellipsoid

The relationship between the ellipsoidal height h, the orthometric height H, and the geoid undulation is

$$h = H + N \tag{2.80}$$

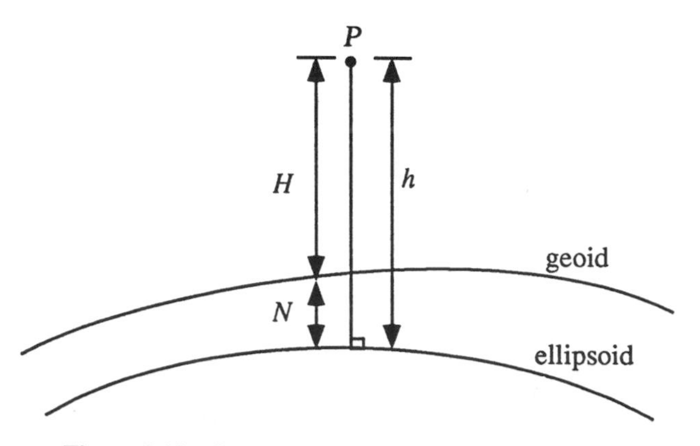

Figure 2.13 Orthometric versus ellipsoidal heights.

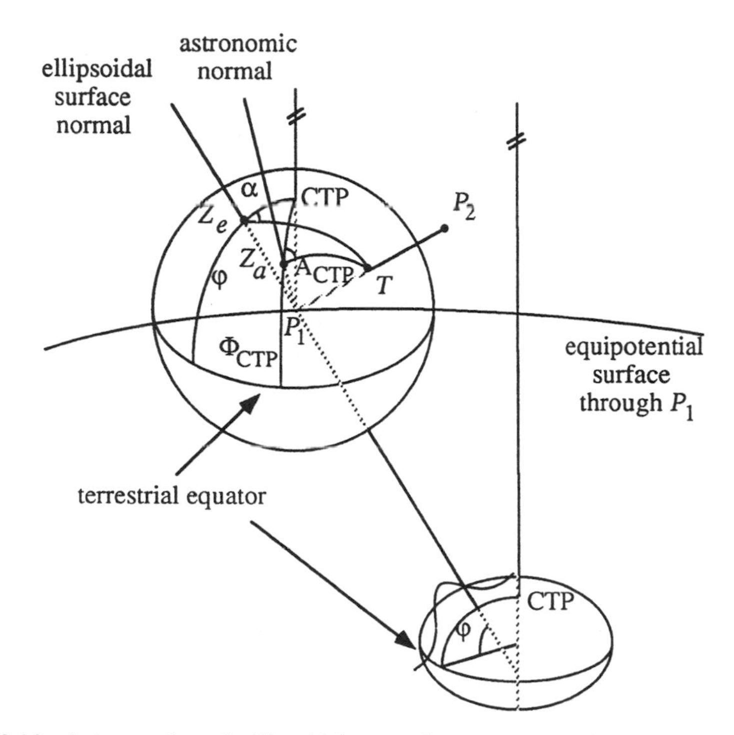

Figure 2.14 Astronomic and ellipsoidal normal on a topocentric sphere of direction. The astronomic normal is perpendicular to the equipotential surface at P_1. The ellipsoidal normal passes through P_1.

where N is the geoid undulation with respect to the specific ellipsoid. See Figure 2.13.

Parallelism of the semiminor axis of the ellipsoid and the direction of the CTP leads to important relationships between the reduced astronomic quantities (Φ_{CTP}, Λ_{CTP}, A_{CTP}) and the corresponding ellipsoidal or geodetic quantities (φ, λ, α). The geometric relationships are shown in Figures 2.14 and 2.15. The following symbols are used:

Z_a	Astronomic zenith (= intersection of local vertical with the sphere direction)
CTP	Position of the conventional terrestrial pole
Z_e	Ellipsoidal zenith (= intersection of the ellipsoidal normal through P_1 with the sphere of direction)
T	Target point to which the azimuth is measured
A_{CTP}	Reduced astronomic azimuth
Φ_{CTP}, Λ_{CTP}	Reduced astronomic latitude and longitude
ϑ'	Observed zenith angle
φ, λ	Ellipsoidal (geodetic) latitude and longitude
α	Ellipsoidal (geodetic) azimuth between two normal planes

ϑ	Ellipsoidal (geodetic) zenith angle
θ	Total deflection of the vertical (not colatitude)
ε	Deflection of the vertical in the direction of azimuth
ξ, η	Deflection of the vertical components along the meridian and the prime vertical

By applying spherical trigonometry to the various triangles in Figure 2.15, we can eventually derive the following relations:

$$A_{\mathrm{CTP}} - \alpha = (\Lambda_{\mathrm{CTP}} - \lambda)\sin\varphi + (\xi\sin\alpha - \eta\cos\alpha)\cot\vartheta \tag{2.81}$$

$$\xi = \Phi_{\mathrm{CTP}} - \varphi \tag{2.82}$$

$$\eta = (\Lambda_{\mathrm{CTP}} - \lambda)\cos\varphi \tag{2.83}$$

$$\vartheta = \vartheta' + \xi\cos\alpha + \eta\sin\alpha \tag{2.84}$$

The derivations of these classical equations can be found in most of the geodetic literature, e.g., Heiskanen and Moritz (1967, p. 186). They are also given in Leick (2002). Equation (2.81) is the Laplace equation. It relates the reduced astronomic azimuth and the geodetic azimuth of the normal section containing the target point. Equations (2.82) and (2.83) define the deflection of the vertical components. The deflection of the vertical is simply the angle between the directions of the plumb line and the ellipsoidal normal at the same point. By convention, the deflection of the vertical is decomposed into two components, one lying in the meridian and one lying in the prime vertical, or orthogonal to the meridian. The deflection components depend directly on the shape of the geoid in the region. Because the deflections of the vertical are merely another manifestation of the irregularity of

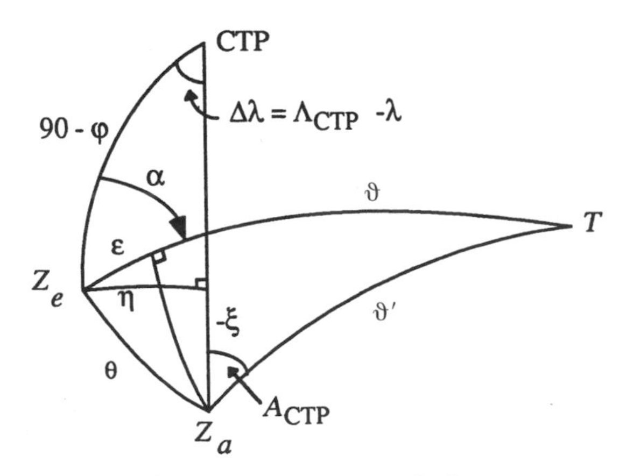

Figure 2.15 Deflection of the vertical components.

the gravity field or the geoid, they are mathematically related to the geoid undulation. Equation (2.84) relates the ellipsoidal and observed zenith angle (refraction not considered).

Equations (2.81) to (2.83) can be used to correct the reduced astronomic latitude, longitude, and azimuth and thus to obtain the ellipsoidal latitude, longitude, and azimuth. It is important to note that the reduction of a horizontal angle due to deflection of the vertical is obtained from the difference of (2.81) as applied to both legs of the angle. If the zenith angle to the endpoints of both legs is close to 90°, then the corrections are small and can possibly be neglected. Historically, Equation (2.81) was used as a condition between the reduced astronomic azimuth and the computed geodetic azimuth to control systematic errors. This can best be accomplished now with GPS. However, if surveyors were to check the orientation of a GPS vector with the astronomic azimuth from the sun or polaris, they must expect a discrepancy indicated by (2.81).

Equations (2.81) to (2.83) also show how to specify a local ellipsoid that is tangent to the geoid at some centrally located station called the initial point, and whose semiminor axis is still parallel to the CTP. If we specify that at the initial point the reduced astronomic latitude, longitude, and azimuth equal the ellipsoidal latitude, longitude, and azimuth, respectively, then we ensure parallelism of the semimajor axis and the direction of the CTP; the geoid normal and the ellipsoidal normal coincide at that initial point. If, in addition, we set the undulation to zero, then the ellipsoid touches the geoid tangentially at the initial point. Thus the local ellipsoid will have at the initial point:

$$\varphi = \Phi_{CTP} \tag{2.85}$$

$$\lambda = \Lambda_{CTP} \tag{2.86}$$

$$\alpha = A_{CTP} \tag{2.87}$$

$$N = 0 \tag{2.88}$$

Other possibilities for specifying a local ellipsoid exist.

The local ellipsoid can serve as a convenient computation reference for least-squares adjustments of networks typically encountered in local and regional surveys. In these cases, it is not at all necessary to determine the size and shape of a best-fitting local ellipsoid. It is sufficient to adopt the size and shape of any of the currently valid geocentric ellipsoids. Because the deflections of the vertical will be small in the region around the initial point, they can often be neglected completely. This is especially true for the reduction of angles. The local ellipsoid is even more useful than it appears at first sight. So long as typical observations, such as horizontal directions, angles, and slant distances, are adjusted, the accurate position of the initial point in (2.85) and (2.86) is not needed. In fact, if the (local) undulation variation is negligible, the coordinate values for the position of the initial point are arbitrary. The same is true for the azimuth condition (2.87). These simplifications make it attractive to use an ellipsoid as a reference for the adjustment of even the smallest survey, thus providing a unified adjustment approach for surveys of large and small areas.

2.3.5 The 3D Geodetic Model

Once the angular observations have been corrected for the deflection of the vertical, it is a simple matter to develop the mathematics for the 3D geodetic model. The reduced observations, i.e., the observables of the 3D geodetic model, are the geodetic azimuth α, the geodetic horizontal angle δ, the geodetic vertical angle β (or the geodetic zenith angle ϑ), and the slant distance s. Geometrically speaking, these observables refer to the geodetic horizon and the ellipsoidal normal. The reduced horizontal angle is an angle between two normal planes, defined by the target points and the ellipsoidal normal at the observing stations. The geodetic vertical angle is the angle between the geodetic horizon and the line of sight to the targets.

We assume that the vertical angle has been corrected for atmospheric refraction. The model can be readily extended to include refraction parameters if needed. Thanks to the availability of GPS, we no longer depend on vertical angle observations to support the vertical dimension, except possibly for applications that call for first-order leveling accuracy. The primary purpose of vertical angles in most cases is to support the vertical dimension when adjusting slant distances (because slant distances contribute primarily horizontal information, at least in flat terrain).

Figure 2.16 shows the local geodetic coordinate system $(\mathrm{w}) = (n, e, u)$, which plays a central role in the development of the mathematical model. The axes n and e span the local geodetic horizon (plane perpendicular to the ellipsoidal normal through the point P_1 on the surface of the earth). The n axis points north, the e axis points east, and the u axis coincides with the ellipsoidal normal (with the positive end outward of the ellipsoid). The spatial orientation of the local geodetic coordinate system is completely specified by the geodetic latitude φ and the geodetic longitude λ. Recall that the z axis coincides with the direction of the CTP.

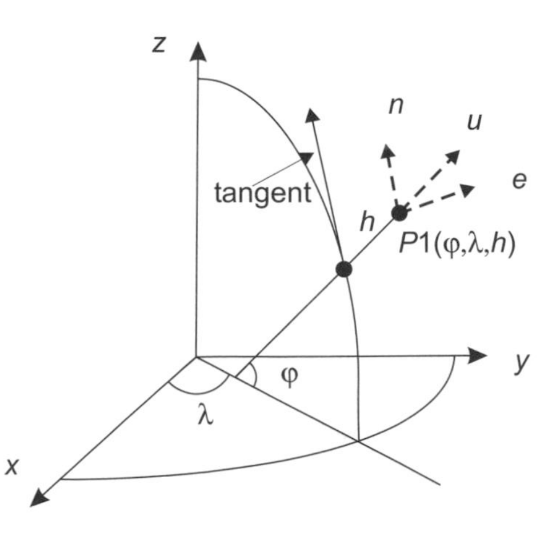

Figure 2.16 The local geodetic coordinate system.

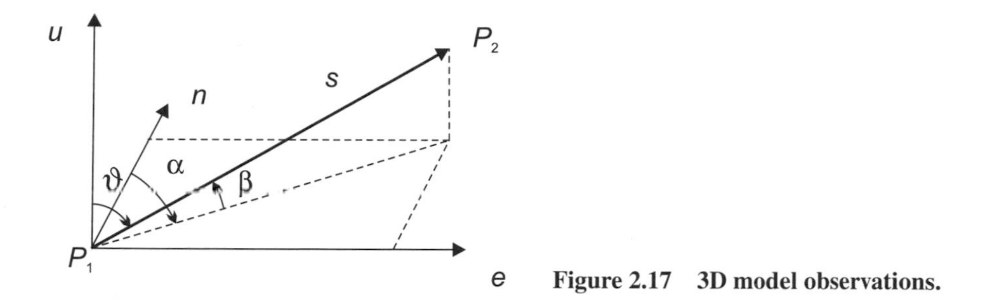

Figure 2.17 3D model observations.

Figure 2.17 shows the geodetic azimuth and vertical angle (or zenith angle) between points P_1 and P_2 in relation to the local geodetic coordinate system. One should keep in mind that the symbol h still denotes the geodetic height of a point above the ellipsoid, whereas the u coordinate refers to the height of the second station P_2 above the local geodetic horizon of P_1. It follows that

$$n = s \cos \beta \cos \alpha \tag{2.89}$$

$$e = s \cos \beta \sin \alpha \tag{2.90}$$

$$h = s \sin \beta \tag{2.91}$$

The inverses of (2.89) to (2.91) are

$$\alpha = \tan^{-1}\left(\frac{e}{n}\right) \tag{2.92}$$

$$\beta = 90° - \vartheta = \sin^{-1}\left(\frac{u}{s}\right) \tag{2.93}$$

$$s = \sqrt{n^2 + e^2 + u^2} \tag{2.94}$$

The relationship between the local geodetic coordinate system and the geocentric Cartesian system (x) is illustrated in Figure 2.16:

$$\begin{bmatrix} n \\ -e \\ u \end{bmatrix} = \mathbf{R}_2\,(\varphi - 90°)\,\mathbf{R}_3\,(\lambda - 180°)\begin{bmatrix} \Delta x \\ \Delta y \\ \Delta z \end{bmatrix} \tag{2.95}$$

where $\mathbf{R}_2$ and $\mathbf{R}_3$ denote the rotation matrices given in Appendix A, and

$$\Delta \mathbf{X} \equiv \begin{bmatrix} \Delta x \\ \Delta y \\ \Delta z \end{bmatrix} = \begin{bmatrix} x_2 - x_1 \\ y_2 - y_1 \\ z_2 - z_1 \end{bmatrix} \tag{2.96}$$

Subscripts will be used when needed to clarify the use of symbols. For example, the differencing operation Δ in (2.95) implies $\Delta x \equiv \Delta x_{12} = x_2 - x_1$. The same convention is followed for other differences. A more complete notation for the local geodetic coordinates is (n_1, e_1, u_1) instead of (n, e, u), to emphasize that these components refer to the geodetic horizon at P_1. Similarly, a more unambiguous notation is $(\alpha_{12}, \beta_{12}, \vartheta_{12})$ instead of just $(\alpha, \beta, \vartheta)$ or even $(\alpha_1, \beta_1, \vartheta_1)$, to emphasize that these observables are taken at station P_1 with foresight P_2. For slant distance, the subscripts do not matter because $s = s_1 = s_{12} = s_{21}$.

Changing the sign of e in (2.95) and combining the rotation matrices $\mathbf{R}_2$ and $\mathbf{R}_3$ one obtains

$$\mathbf{w} = \mathbf{R}(\varphi, \lambda)\, \Delta\mathbf{x} \tag{2.97}$$

with

$$\mathbf{R} = \begin{bmatrix} -\sin\varphi\cos\lambda & -\sin\varphi\sin\lambda & \cos\varphi \\ -\sin\lambda & \cos\lambda & 0 \\ \cos\varphi\cos\lambda & \cos\varphi\sin\lambda & \sin\varphi \end{bmatrix} \tag{2.98}$$

Substituting (2.97) and (2.98) into (2.92) to (2.94) gives expressions for the geodetic observables as functions of the geocentric Cartesian coordinate differences and the geodetic position of P_1:

$$\alpha_1 = \tan^{-1}\left(\frac{-\sin\lambda_1\,\Delta x + \cos\lambda_1\,\Delta y}{-\sin\varphi_1\cos\lambda_1\,\Delta x - \sin\varphi_1\sin\lambda_1\,\Delta y + \cos\varphi_1\,\Delta z}\right) \tag{2.99}$$

$$\beta_1 = \sin^{-1}\left(\frac{\cos\varphi_1\cos\lambda_1\,\Delta x + \cos\varphi_1\sin\lambda_1\,\Delta y + \sin\varphi_1\,\Delta z}{\sqrt{\Delta x^2 + \Delta y^2 + \Delta z^2}}\right) \tag{2.100}$$

$$s = \sqrt{\Delta x^2 + \Delta y^2 + \Delta z^2} \tag{2.101}$$

Equations (2.99) to (2.101) are the backbone of the 3D geodetic model. Other observations such as horizontal angles, heights, and height differences—even GPS vectors—can be readily implemented. Equation (2.100) assumes that the vertical angle has been corrected for refraction. One should take note of the fact how little mathematics is required to derive these equations. Differential geometry is not required, and neither is the geodesic line.

2.3.5.1 *Partial Derivatives* Because (2.99) to (2.101) expressed the geodetic observables explicitly as a function of the coordinates, the observation equation adjustment model $\boldsymbol{\ell}_a = \mathbf{f}(\mathbf{x}_a)$ can be readily used. The 3D nonlinear model has the general form

$$\alpha_1 = \alpha\,(x_1, y_1, z_1, x_2, y_2, z_2) \tag{2.102}$$

$$\beta_1 = \beta(x_1, y_1, z_1, x_2, y_2, z_2) \tag{2.103}$$

$$s = s(x_1, y_1, z_1, x_2, y_2, z_2) \tag{2.104}$$

The observables and parameters are $\{\alpha_1, \beta_1, s\}$ and $\{x_1, y_1, z_1, x_2, y_2, z_2\}$, respectively. To find the elements of the design matrix, we require the total partial derivatives with respect to the parameters. The general form is

$$\begin{bmatrix} d\alpha_1 \\ d\beta_1 \\ ds \end{bmatrix} = \begin{bmatrix} g_{11} & g_{12} & g_{13} & \vdots & g_{14} & g_{15} & g_{16} \\ g_{21} & g_{22} & g_{23} & \vdots & g_{24} & g_{25} & g_{26} \\ g_{31} & g_{32} & g_{33} & \vdots & g_{34} & g_{35} & g_{36} \end{bmatrix} \begin{bmatrix} dx_1 \\ dy_1 \\ dz_1 \\ \cdots \\ dx_2 \\ dy_2 \\ dz_2 \end{bmatrix} = [\mathbf{G}_1 \vdots \mathbf{G}_2] \begin{bmatrix} d\mathbf{x}_1 \\ \cdots \\ d\mathbf{x}_2 \end{bmatrix} \tag{2.105}$$

with $d\mathbf{x}_i = [dx_i \quad dy_i \quad dz_i]^T$. The partial derivatives are listed in Table 2.4. This particular form of the partial derivatives follows from those of Wolf (1963), after some additional algebraic manipulations.

TABLE 2.4 Partial Derivatives with Respect to Cartesian Coordinates

$$g_{11} = \frac{\partial \alpha_1}{\partial x_1} = -g_{14} = \frac{-\sin\varphi_1 \cos\lambda_1 \sin\alpha_1 + \sin\lambda_1 \cos\alpha_1}{s\cos\beta_1} \tag{a}$$

$$g_{12} = \frac{\partial \alpha_1}{\partial y_1} = -g_{15} = \frac{-\sin\varphi_1 \sin\lambda_1 \sin\alpha_1 - \cos\lambda_1 \cos\alpha_1}{s\cos\beta_1} \tag{b}$$

$$g_{13} = \frac{\partial \alpha_1}{\partial z_1} = -g_{16} = \frac{\cos\varphi_1 \sin\alpha_1}{s\cos\beta_1} \tag{c}$$

$$g_{21} = \frac{\partial \beta_1}{\partial x_1} = -g_{24} = \frac{-s\cos\varphi_1 \cos\lambda_1 + \sin\beta_1 \,\Delta x}{s^2\cos\beta_1} \tag{d}$$

$$g_{22} = \frac{\partial \beta_1}{\partial y_1} = -g_{25} = \frac{-s\cos\varphi_1 \sin\lambda_1 + \sin\beta_1 \,\Delta x}{s^2\cos\beta_1} \tag{e}$$

$$g_{23} = \frac{\partial \beta_1}{\partial z_1} = -g_{26} = \frac{-s\sin\varphi_1 + \sin\beta_1 \,\Delta z}{s^2\cos\beta_1} \tag{f}$$

$$g_{31} = \frac{\partial s}{\partial x_1} = -g_{34} = \frac{-\Delta x}{s} \tag{g}$$

$$g_{32} = \frac{\partial s}{\partial y_1} = -g_{35} = \frac{-\Delta y}{s} \tag{h}$$

$$g_{33} = \frac{\partial s}{\partial z_1} = -g_{36} = \frac{-\Delta z}{s} \tag{i}$$

2.3.5.2 Reparameterization Often the geodetic latitude, longitude, and height are preferred as parameters instead of the Cartesian components of (x). One reason for such a reparameterization is that humans can visualize changes more readily in latitude, longitude, and height than changes in geocentric coordinates. The required transformation is given by (B.16).

$$
\begin{aligned}
d\mathbf{x} &= \begin{bmatrix} -(M+h)\cos\lambda\sin\varphi & -(N+h)\cos\varphi\sin\lambda & \cos\varphi\cos\lambda \\ -(M+h)\sin\lambda\sin\varphi & (N+h)\cos\varphi\cos\lambda & \cos\varphi\sin\lambda \\ (M+h)\cos\varphi & 0 & \sin\varphi \end{bmatrix} \begin{bmatrix} d\varphi \\ d\lambda \\ dh \end{bmatrix} \\
&= \mathbf{J} \begin{bmatrix} d\varphi \\ d\lambda \\ dh \end{bmatrix}
\end{aligned} \tag{2.106}
$$

The expressions for the radius of curvatures $M$and N are given in (B.7) and (B.6). The matrix $\mathbf{J}$ must be evaluated for the geodetic latitude and longitude of the point under consideration; thus, $\mathbf{J}_1(\varphi_1, \lambda_1, h_1)$ and $\mathbf{J}_2(\varphi_2, \lambda_2, h_2)$ denote the transformation matrices for points P_1 and P_2, respectively. Substituting (2.106) into (2.105), we obtain the parameterization in terms of geodetic latitude, longitude, and height:

$$
\begin{bmatrix} d\alpha_1 \\ d\beta_1 \\ ds \end{bmatrix} = [\mathbf{G}_1\mathbf{J}_1 \quad : \quad \mathbf{G}_2\mathbf{J}_2] \begin{bmatrix} d\varphi_1 \\ d\lambda_1 \\ dh_1 \\ \cdots \\ d\varphi_2 \\ d\lambda_2 \\ dh_2 \end{bmatrix} \tag{2.107}
$$

To achieve a parameterization that is even easier to interpret, we transform the differential changes in geodetic latitude and longitude parameters $(d\varphi, d\lambda)$ into corresponding changes (dn, de) in the local geodetic horizon. Keeping the geometric interpretation of the radii of curvatures M and N as detailed in Appendix B one can further deduce that

$$
d\mathbf{w} = \begin{bmatrix} M+h & 0 & 0 \\ 0 & (N+h)\cos\varphi & 0 \\ 0 & 0 & 1 \end{bmatrix} \begin{bmatrix} d\varphi \\ d\lambda \\ dh \end{bmatrix} = \mathbf{H}(\varphi, h) \begin{bmatrix} d\varphi \\ d\lambda \\ dh \end{bmatrix} \tag{2.108}
$$

The components $d\mathbf{w} = [dn \quad de \quad du]^{\mathrm{T}}$ intuitively related to the "horizontal" and "vertical" and because the units are in length, the standard deviations of the parameters can be readily visualized. The matrix $\mathbf{H}$ is evaluated for the station under consideration. The final parameterization becomes

$$\begin{bmatrix} d\alpha_1 \\ d\beta_1 \\ ds \end{bmatrix} = \mathbf{A} \begin{bmatrix} d\mathbf{w}_1 \\ \cdots \\ d\mathbf{w}_2 \end{bmatrix} \tag{2.109}$$

with

$$\mathbf{A} = \left[\mathbf{G}_1\mathbf{J}_1\mathbf{H}_1^{-1} \quad \vdots \quad \mathbf{G}_2\mathbf{J}_2\mathbf{H}_2^{-1}\right] = \begin{bmatrix} a_{11} & a_{12} & a_{13} & & a_{14} & a_{15} & a_{16} \\ a_{21} & a_{22} & a_{23} & \vdots & a_{24} & a_{25} & a_{26} \\ a_{31} & a_{32} & a_{33} & & a_{34} & a_{35} & a_{36} \end{bmatrix} \tag{2.110}$$

The partial derivatives are given in Table 2.5 (Wolf, 1963; Heiskanen and Moritz, 1967; Vincenty, 1979). Some of the partial derivatives have been expressed in terms of the back azimuth $\alpha_2 \equiv \alpha_{21}$ and the back vertical angle $\beta_2 \equiv \beta_{21}$, meaning azimuth and vertical angle from station 2 to station 1.

2.3.5.3 Implementation Considerations It is not only easy to derive the 3D geodetic model; it is also easy to implement it in software. Normally, the observations will be uncorrelated and their contribution to the normal equations can be added one by one. The following are some useful things to keep in mind when using this model:

- **Point of Expansion:** As in any nonlinear adjustment, the partial derivatives must be evaluated at the current point of expansion (adjusted positions of the previous iteration). This applies to coordinates and azimuths and angles used to express the mathematical functions for the partial derivatives.
- **Reduction to the Mark:** An advantage of the 3D geodetic model is that the observations do not have to be reduced to the marks on the ground. When computing $\boldsymbol{\ell}_0$ from (2.99) to (2.101), use $h + \Delta h$ instead of h for the station heights. The symbol Δh denotes the height of the instrument or that of the target above the mark on the ground. $\boldsymbol{\ell}_b$ always denotes the measured value, i.e., the geodetic observable that is not further reduced. After completion of the adjustment, the adjusted observations $\boldsymbol{\ell}_a$, with respect to the marks on the ground, can be computed from the adjusted positions using h in (2.99) to (2.101).
- **Minimal Constraints:** The (φ) or (w) parameterizations are particularly useful for introducing height observations, height difference observations, or minimal constraints by fixing or weighting individual coordinates. The set of minimal constraints depends on the type of observations available and where the observations are located within the network. One choice for the minimal constraints might be to fix the coordinates (φ, λ, h) of one station (translation), the azimuth or the longitude of another station (rotation in azimuth), and the heights of two additional stations.

TABLE 2.5 Partial Derivatives with Respect to Local Geodetic Coordinates

$$a_{11} = \frac{\partial \alpha_1}{\partial n_1} = \frac{\sin \alpha_1}{s \cos \beta_1} \quad \text{(a)} \qquad a_{12} = \frac{\partial \alpha_1}{\partial e_1} = -\frac{\cos \alpha_1}{s \cos \beta_1} \quad \text{(b)}$$

$$a_{13} = \frac{\partial \alpha_1}{\partial u_1} = 0 \quad \text{(c)}$$

$$a_{14} = \frac{\partial \alpha_1}{\partial n_2} = -\frac{\sin \alpha_1}{s \cos \beta_1} [\cos(\varphi_2 - \varphi_1) + \sin \varphi_2 \sin(\lambda_2 - \lambda_1) \cot \alpha_1] \quad \text{(d)}$$

$$a_{15} = \frac{\partial \alpha_1}{\partial e_2} = \frac{\cos \alpha_1}{s \cos \beta_1} [\cos(\lambda_2 - \lambda_1) - \sin \varphi_1 \sin(\lambda_2 - \lambda_1) \tan \alpha_1] \quad \text{(e)}$$

$$a_{16} = \frac{\partial \alpha_1}{\partial u_2} = \frac{\cos \alpha_1 \cos \varphi_2}{s \cos \beta_1} [\sin(\lambda_2 - \lambda_1) + (\sin \varphi_1 \cos(\lambda_2 - \lambda_1) - \cos \varphi_1 \tan \varphi_2) \tan \alpha_1] \quad \text{(f)}$$

$$a_{21} = \frac{\partial \beta_1}{\partial n_1} = \frac{\sin \beta_1 \cos \alpha_1}{s} \quad \text{(g)} \qquad a_{22} = \frac{\partial \beta_1}{\partial e_1} = \frac{\sin \beta_1 \sin \alpha_1}{s} \quad \text{(h)}$$

$$a_{23} = \frac{\partial \beta_1}{\partial u_1} = -\frac{\cos \beta_1}{s} \quad \text{(i)}$$

$$a_{24} = \frac{\partial \beta_1}{\partial n_2} = \frac{-\cos \varphi_1 \sin \varphi_2 \cos(\lambda_2 - \lambda_1) + \sin \varphi_1 \cos \varphi_2 + \sin \beta_1 \cos \beta_2 \cos \alpha_2}{s \cos \beta_1} \quad \text{(j)}$$

$$a_{25} = \frac{\partial \beta_1}{\partial e_2} = \frac{-\cos \varphi_1 \sin(\lambda_2 - \lambda_1) + \sin \beta_1 \cos \beta_2 \sin \alpha_2}{s \cos \beta_1} \quad \text{(k)}$$

$$a_{26} = \frac{\partial \beta_1}{\partial u_2} = \frac{\sin \beta_1 \sin \beta_2 + \sin \varphi_1 \sin \varphi_2 + \cos \varphi_1 \cos \varphi_2 \cos(\lambda_2 - \lambda_1)}{s \cos \beta_1} \quad \text{(l)}$$

$$a_{31} = \frac{\partial s}{\partial n_1} = -\cos \beta_1 \cos \alpha_1 \quad \text{(m)} \qquad a_{32} = \frac{\partial s}{\partial e_1} = -\cos \beta_1 \sin \alpha_1 \quad \text{(n)}$$

$$a_{33} = \frac{\partial s}{\partial u_1} = -\sin \beta_1 \quad \text{(o)} \qquad a_{34} = \frac{\partial s}{\partial n_2} = -\cos \beta_2 \cos \alpha_2 \quad \text{(p)}$$

$$a_{35} = \frac{\partial s}{\partial e_2} = -\cos \beta_2 \sin \alpha_2 \quad \text{(q)} \qquad a_{36} = \frac{\partial s}{\partial u_2} = -\sin \beta_2 \quad \text{(r)}$$

- **Transforming Postadjustment Results:** If the adjustment happens to have been carried out with the (x) parameterization, and it is, subsequently, determined necessary to transform the result into the (φ) or (w) coordinates, then the transformations (2.106) and (2.108) can be used; i.e.,

$$d\mathbf{w} = \mathbf{R}\, d\mathbf{x} \tag{2.111}$$

where

$$\mathbf{R} = \mathbf{H}\,\mathbf{J}^{-1} \tag{2.112}$$

according to (2.98). The law of variance-covariance propagation provides the 3×3 covariance submatrices:

$$\boldsymbol{\Sigma}_{(w)} = \mathbf{R}\,\boldsymbol{\Sigma}_{(x)}\mathbf{R}^{\mathrm{T}} \tag{2.113}$$

$$\boldsymbol{\Sigma}_{(\varphi,\lambda,h)} = \mathbf{J}^{-1}\boldsymbol{\Sigma}_{(x)}\left(\mathbf{J}^{-1}\right)^{\mathrm{T}} \tag{2.114}$$

- **Leveled Height Differences:** If geoid undulation differences are available, then leveled height differences can be corrected for the undulation differences to yield ellipsoidal height differences. The respective elements of the design matrix are 1 and -1. The accuracy of incorporating leveling data in this manner is limited by our ability to compute accurate undulation differences.
- **Refraction:** If vertical angles are observed for providing an accurate vertical dimension, it may be necessary to introduce and estimate vertical refraction parameters. If this is done, we must be careful to avoid overparameterization by introducing too many refraction parameters that could potentially absorb other systematic effects not caused by refraction and/or result in an ill-conditioned solution. However, it may be sufficient to correct the observations for refraction using a standard model for the atmosphere.

  In view of GPS capability, the importance of high-precision vertical angle measurement is diminishing. The primary purpose of vertical angles is to give sufficient height information to process the slant distances. Therefore, the types of observations most likely to be used by the modern surveyors are horizontal angles, slant distances, and GPS vectors.
- **Horizontal Angles:** Horizontal angles, of course, are simply the difference of azimuths. Using the 2-1-3 subscript notation to identify an angle measured at station 1 from station 2 to station 3 in a clockwise sense the mathematical model for the geodetic angle δ_{213} is

$$\begin{aligned}\delta_{213} = {} & \tan^{-1}\left(\frac{-\sin\lambda_1\,\Delta x_{12} + \cos\lambda_1\,\Delta y_{12}}{-\sin\varphi_1\cos\lambda_1\,\Delta x_{12} - \sin\varphi_1\sin\lambda_1\,\Delta y_{12} + \cos\varphi_1\,\Delta z_{12}}\right)\\ & - \tan^{-1}\left(\frac{-\sin\lambda_1\,\Delta x_{13} + \cos\lambda_1\,\Delta y_{13}}{-\sin\varphi_1\cos\lambda_1\,\Delta x_{13} - \sin\varphi_1\sin\lambda_1\,\Delta y_{13} + \cos\varphi_1\,\Delta z_{13}}\right)\end{aligned} \tag{2.115}$$

  The partial derivatives can be readily obtained from the coefficients a_{2i} listed in Table 2.5 by applying them to both legs of the angles and then subtracting.
- **Height-Controlled 3D Adjustment:** If the observations contain little or no vertical information, i.e., if zenith angles and leveling data are not available, it is still possible to adjust the network in three dimensions. The height parameters h can be weighted using reasonable estimates for their a priori variances. This

is the so-called height-controlled three-dimensional adjustment. In the extreme case, the height parameters can even be eliminated altogether from the list of parameters.

A priori weights can also be assigned to the geodetic latitude and longitude or to the local geodetic coordinates n and e. Weighting of parameters is a convenient method for incorporating existing information about control stations into the adjustment.

CHAPTER 3

SATELLITE SYSTEMS

Satellite motions are introduced by means of normal orbits and the derivation and discussion of the three Kepler laws. A summary of the major effects that cause orbital perturbed motions follows. The description of the global positioning system (GPS) takes up most of this chapter. A brief summary of GPS modernization is offered. The chapter concludes with a description of GLONASS and the forthcoming Galileo system. Readers interested in signal processing inside receivers are encouraged to consult the excellent references Kaplan (1986), Misra and Enge (2001), and Parkinson et al. (1996).

3.1 MOTION OF SATELLITES

The orbital motion of a satellite is a result of the earth's gravitational attraction, as well as a number of other forces acting on the satellite. The attraction of the sun and the moon and the pressure on the satellite caused by impacting solar radiation particles are examples of these forces. For high-orbiting satellites, the atmospheric drag is negligible. Mathematically, the equations of motion for satellites are differential equations that are solved by numerical integration over time. The integration begins with initial conditions, such as the position and velocity of the satellite at some initial epoch. The computed (predicted) satellite positions can be compared with actual observations. Possible discrepancies are useful to improve the force function or the station position of the observer.

3.1.1 Kepler Elements

Six Kepler elements are often used to describe the position of satellites in space. To simplify attempts to study satellite motions, we study so-called normal orbits. For normal orbits the satellites move in an orbital plane that is fixed in space; the actual path of the satellite in the orbital plane is an ellipse in the mathematically strict sense. One focal point of the orbital ellipse is at the center of the earth. The conditions leading to such a simple orbital motion are as follows:

1. The earth is treated as a point mass, or, equivalently, as a sphere with spherically symmetric density distribution. The gravitational field of such a body is radially symmetric; i.e., the plumb lines are all straight lines and point toward the center of the sphere.
2. The mass of the satellite is negligible compared to the mass of the earth.
3. The motion of the satellite takes place in a vacuum; i.e., there is no atmospheric drag acting on the satellite and no solar radiation pressure.
4. No sun, moon, or other celestial body exerts a gravitational attraction on the satellite.

The orbital plane of a satellite moving under such conditions is shown in Figure 3.1. The ellipse denotes the path of the satellite. The shape of the ellipse is determined by the semimajor axis a and the semiminor axis b. The symbol e denotes the eccentricity of the ellipse. The ellipse is enclosed by an auxiliary circle with radius a. The principal axes of the ellipse form the coordinate system (ξ, η). S denotes the current position of the satellite; the line SS' is in the orbital plane and is parallel to the η axis. The coordinate system (q_1, q_2) is in the orbital plane with its origin at the focal point F of the ellipse that coincides with the center of the earth. The third axis q_3, not shown in Figure 3.1, completes the right-handed coordinate system. The geocentric distance from the center of the earth to the satellite is denoted by r. The orbital locations closest to and farthest from the focal point are called the perigee and apogee, respectively. The true anomaly f and the eccentric anomaly E are measured counterclockwise, as shown in Figure 3.2.

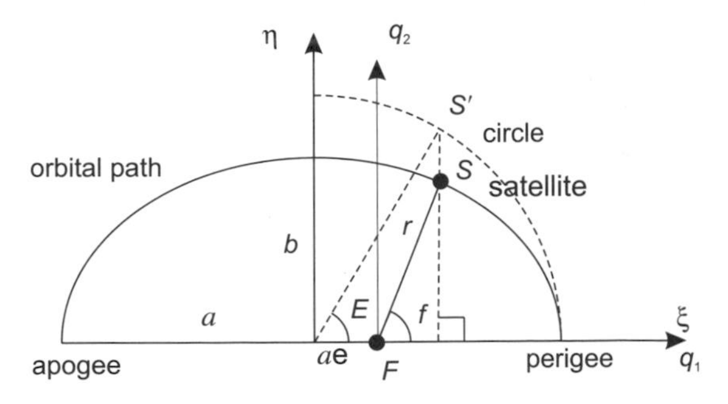

Figure 3.1 Coordinate systems in the orbital plane.

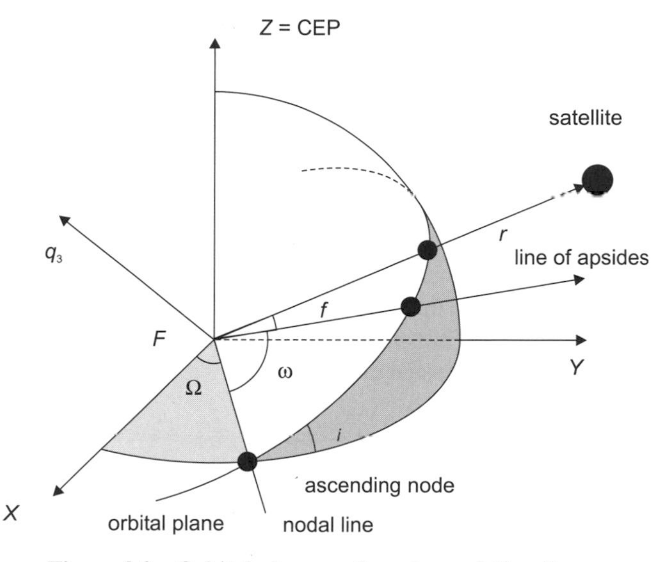

Figure 3.2 Orbital plane on the sphere of direction.

The orbital plane is shown in Figure 3.2, with respect to the true celestial coordinate system. The center of the sphere of directions is located at the focal point F. The X axis is in the direction of the vernal equinox, the Z axis coincides with the celestial ephemeris pole, and Y is located in the equator, thus completing the right-handed coordinate system. The intersection of the orbital plane with the equator is called the nodal line. The point at which the satellite ascends the equator is the ascending node. The right ascension of the ascending node is denoted by Ω. The line of apsides connects the focal point F and the perigee. The angle subtended by the nodal line and the line of apsides is called the argument of perigee, ω. The true anomaly f and the argument of perigee ω lie in the orbital plane. Finally, the angle between the orbital plane and the equator is the inclination i. Figure 3.2 shows that (Ω, i) determines the position of the orbital plane in the true celestial system, (Ω, ω, i) the orbital ellipse in space, and (a, e, f) the position of the satellite within the orbital plane.

The six Kepler elements are $\{\Omega, \omega, i, a, e, f\}$. The true anomaly f is the only Kepler element that is a function of time in the case of normal orbits; the remaining five Kepler elements are constant. For actual satellite orbits, which are not subject to the conditions of normal orbits, all Kepler elements are a function of time. They are called osculating Kepler elements.

3.1.2 Normal Orbital Theory

Normal orbits are particularly useful for understanding and visualizing the spatial motions of satellites. The solutions of the respective equations of motions can be given by simple, analytical expressions. Since normal orbits are a function of the central portion of the earth's gravitational field, which is by far the largest force acting on the satellite, normal orbits are indeed usable for orbital predictions over short periods of time when low accuracy is sufficient. Thus, one of the popular uses of normal orbits is for the construction of satellite visibility charts.

The normal motion of satellites is completely determined by Newton's law of gravitation:

$$F = \frac{k^2 m M}{r^2} \tag{3.1}$$

In (3.1), M and m denote the mass of the earth and the satellite, respectively, k^2 is the universal constant of gravitation, r is the geocentric distance to the satellite, and F is the gravitational force between the two bodies. This force can also be written as

$$F = ma \tag{3.2}$$

where a in this instance denotes the acceleration experienced by the satellite. Combining (3.1) and (3.2) gives

$$a = \frac{k^2 M}{r^2} \tag{3.3}$$

This equation can be written in vector form as

$$\ddot{\mathbf{r}} = -k^2 M \frac{\mathbf{r}}{r^3} = -\mu \frac{\mathbf{r}}{r^3} \tag{3.4}$$

where

$$\mu = k^2 M \tag{3.5}$$

is the earth's gravitational constant. Including the earth's atmosphere, it has the value $\mu = 3{,}986{,}005 \times 10^8\ \mathrm{m^3 s^{-2}}$. The vector $\mathbf{r}$ is directed from the central body (earth) to the satellite. The sign has been chosen such that the acceleration is directed toward the earth. The colinearity of the acceleration and the position vector as in (3.4) is a characteristic of central gravity fields. A particle released from rest would fall along a straight line toward the earth (straight plumb line).

Equation (3.4) is valid for the motion with respect to an inertial origin. In general, one is interested in determining the motion of the satellite with respect to the earth. The modified equation of motion for accomplishing this is given by Escobal (1965, p. 37) as

$$\ddot{\mathbf{r}} = -k^2 (M + m) \frac{\mathbf{r}}{r^3} \tag{3.6}$$

Because $m \ll M$, the second term is often neglected and (3.6) becomes (3.4).

Figure 3.2 gives the position of the satellite in the (q) orbital plane coordinate system $\mathbf{q} = [q_1 \quad q_2 \quad q_3]^{\mathrm{T}}$ as

$$\mathbf{q} = r \begin{bmatrix} \cos f \\ \sin f \\ 0 \end{bmatrix} \tag{3.7}$$

Because the geocentric distance and the true anomaly are functions of time, the derivative with respect to time, denoted by a dot , is

$$\dot{\mathbf{q}} = \dot{r} \begin{bmatrix} \cos f \\ \sin f \\ 0 \end{bmatrix} + r\dot{f} \begin{bmatrix} -\sin f \\ \cos f \\ 0 \end{bmatrix} \tag{3.8}$$

The second derivatives with respect to time are

$$\ddot{\mathbf{q}} = \ddot{r} \begin{bmatrix} \cos f \\ \sin f \\ 0 \end{bmatrix} + 2\dot{r}\dot{f} \begin{bmatrix} -\sin f \\ \cos f \\ 0 \end{bmatrix} + r\ddot{f} \begin{bmatrix} -\sin f \\ \cos f \\ 0 \end{bmatrix} - r(\dot{f})^2 \begin{bmatrix} \cos f \\ \sin f \\ 0 \end{bmatrix} \tag{3.9}$$

The second derivative is written according to (3.4) and (3.7) as

$$\ddot{\mathbf{r}} = \frac{-\mu}{r^2} \begin{bmatrix} \cos f \\ \sin f \\ 0 \end{bmatrix} \tag{3.10}$$

Evaluating (3.9) and (3.10) at $f = 0$ (perigee) and substituting (3.10) for the left-hand side of (3.9) gives

$$\ddot{r} - r(\dot{f})^2 = \frac{-\mu}{r^2} \tag{3.11}$$

$$r\ddot{f} + 2\dot{r}\dot{f} = 0 \tag{3.12}$$

Equation (3.12) is further developed by multiplying by r and integrating

$$\int \left(r^2\ddot{f} + 2r\dot{r}\dot{f}\right)\, dt = C \tag{3.13}$$

The result of the integration is

$$r^2\dot{f} + 2r^2\dot{f} = C \tag{3.14}$$

as can be readily verified through differentiation. Combining both terms yields

$$r^2\dot{f} = h \tag{3.15}$$

where h is a new constant. Equation (3.15) is identified as an angular momentum equation, implying that the angular momentum for the orbiting satellite is conserved.

In order to integrate (3.11), we define a new variable:

$$u \equiv \frac{1}{r} \tag{3.16}$$

By using Equation (3.15) for dt/df, the differential of (3.16) becomes

$$\frac{du}{df} = \frac{du}{dr}\frac{dr}{dt}\frac{dt}{df} = -\frac{\dot{r}}{h} \tag{3.17}$$

Differentiating again gives

$$\frac{d^2u}{df^2} = \frac{d}{dt}\left(-\frac{\dot{r}}{h}\right)\frac{dt}{df} = -\frac{\ddot{r}}{u^2h^2} \tag{3.18}$$

or

$$\ddot{r} = -h^2u^2\frac{d^2u}{df^2} \tag{3.19}$$

By substituting (3.19) in (3.11), substituting $\dot{f}$ from (3.15) in (3.11), and replacing r by u according to (3.16), Equation (3.11) becomes

$$\frac{d^2u}{df^2} + u = \frac{\mu}{h^2} \tag{3.20}$$

which can readily be integrated as

$$\frac{1}{r} \equiv u = C\cos f + \frac{\mu}{h^2} \tag{3.21}$$

where C is a constant.

Equation (3.21) is the equation of an ellipse. This is verified by writing the equation for the orbital ellipse in Figure 3.1 in the principal axis form:

$$\frac{\xi^2}{a^2} + \frac{\eta^2}{b^2} = 1 \tag{3.22}$$

where

$$\xi = ae + r\cos f \tag{3.23}$$

$$\eta = r\sin f \tag{3.24}$$

$$b^2 = a^2(1 - e^2) \tag{3.25}$$

Equation (3.25) is valid for any ellipse. Substituting Equations (3.23) through (3.25) into (3.22) and solving the resulting second-order equation for r gives

$$\frac{1}{r} = \frac{1}{a(1-e^2)} + \frac{e}{a(1-e^2)} \cos f \tag{3.26}$$

With

$$C = \frac{e}{a(1-e^2)} \tag{3.27}$$

and

$$h = \sqrt{\mu a(1-e^2)} \tag{3.28}$$

the identity between the expression for the ellipse (3.26) and Equation (3.21) is established. Thus the motion of a satellite under the condition of a normal orbit is an ellipse. This is the content of Kepler's first law. The focus of the ellipse is at the center of mass. Kepler's second law states that the geocentric vector **r** sweeps equal areas during equal times. Because the area swept for the differential angle df is

$$dA = \frac{1}{2} r^2 \, df \tag{3.29}$$

it follows from (3.15) and (3.28) that

$$\frac{dA}{dt} = \frac{1}{2}\sqrt{\mu a\left(1-e^2\right)} \tag{3.30}$$

which is a constant. The derivation of Kepler's third law requires the introduction of the eccentric anomaly E. From Figure 3.1 we see that

$$q_1 = \xi - ae = a(\cos E - e) \tag{3.31}$$

where

$$\xi = a \cos E \tag{3.32}$$

From Equation (3.22)

$$q_2 \equiv \eta = \sqrt{\left(1 - \frac{\xi^2}{a^2}\right) b^2} \tag{3.33}$$

Substitute (3.32) in (3.33); then

$$q_2 \equiv \eta = b \sin E \tag{3.34}$$

With (3.31), (3.34), and (3.25) the geocentric satellite distance becomes

$$r = \sqrt{q_1^2 + q_2^2} = a(1 - e\cos E) \tag{3.35}$$

Differentiating Equations (3.35) and (3.26) gives

$$dr = ae\sin E\, dE \tag{3.36}$$

$$dr = \frac{r^2 e}{a\left(1 - e^2\right)} \sin f\, df \tag{3.37}$$

Equating (3.37) and (3.36) and using (3.24), (3.25), (3.34), and (3.7) and multiplying the resulting equation by r gives

$$rb\, dE = r^2\, df \tag{3.38}$$

Substituting (3.25) for b and (3.35) for r, replacing df by dt using (3.15), using (3.28) for h, and then integrating, we obtain

$$\int_{E=0}^{E} (1 - e\cos E)\; dE = \int_{t_0}^{t} \sqrt{\frac{\mu}{a^3}}\; dt \tag{3.39}$$

Integrating both sides gives

$$E - e\sin E = M \tag{3.40}$$

$$M = n\,(t - t_0) \tag{3.41}$$

$$n = \sqrt{\frac{\mu}{a^3}} \tag{3.42}$$

Equation (3.42) is Kepler's third law. Equation (3.40) is called the Kepler equation. The symbol n denotes the mean motion, M is the mean anomaly, and t_0 denotes the time of perigee passage of the satellite. The mean anomaly M should not be confused with the same symbol used for the mass of the central body in (3.1). Let P denote the orbital period, i.e., the time required for one complete revolution; then

$$P = \frac{2\pi}{n} \tag{3.43}$$

The mean motion n equals the average angular velocity of the satellite. Equation (3.42) shows that the semimajor axis completely determines the mean motion and thus the period of the orbit.

With the Kepler laws in place, one can identify alternative sets of Kepler elements, such as $\{\Omega, \omega, i, a, e, M\}$ or $\{\Omega, \omega, i, a, e, E\}$. Often the orbit is not specified by the Kepler elements but by the vector $\mathbf{r} = [X \quad Y \quad Z]^{\mathrm{T}} = \mathbf{X}$ and the velocity $\dot{\mathbf{r}} = [\dot{X} \quad \dot{Y} \quad \dot{Z}]^{\mathrm{T}} = \dot{\mathbf{X}}$, expressed in the true celestial coordinate system (X). Figure 3.2 shows that

$$\begin{aligned} \mathbf{q} &= \mathbf{R}_3(\omega)\,\mathbf{R}_1(i)\,\mathbf{R}_3(\Omega)\,\mathbf{X} \\ &= \mathbf{R}_{qX}(\Omega, i, \omega)\,\mathbf{X} \end{aligned} \tag{3.44}$$

where $\mathbf{R}_i$ denotes a rotation around axis i. The inverse transformation is

$$\mathbf{X} = \mathbf{R}_{qX}^{-1}(\Omega, i, \omega)\,\mathbf{q} \tag{3.45}$$

Differentiating (3.45) once gives

$$\dot{\mathbf{X}} = \mathbf{R}_{qX}^{-1}(\Omega, i, \omega)\dot{\mathbf{q}} \tag{3.46}$$

Note that the elements of $\mathbf{R}_{qX}$ are constants, because the orbital ellipse does not change its position in space. Using relations (3.25), (3.31), and (3.34), it follows that

$$\mathbf{q} = \begin{bmatrix} a(\cos E - e) \\ a\sqrt{1 - e^2}\sin E \\ 0 \end{bmatrix} = \begin{bmatrix} r\cos f \\ r\sin f \\ 0 \end{bmatrix} \tag{3.47}$$

The velocity becomes

$$\dot{\mathbf{q}} = \frac{na}{1 - e\cos E}\begin{bmatrix} -\sin E \\ \sqrt{1 - e^2}\cos E \\ 0 \end{bmatrix} = \frac{na}{\sqrt{1 - e^2}}\begin{bmatrix} -\sin f \\ e + \cos f \\ 0 \end{bmatrix} \tag{3.48}$$

The first part of (3.48) follows from (3.39), and the second part can be verified using known relations between the anomalies E and f. Equations (3.45) to (3.48) transform the Kepler elements into Cartesian coordinates and their velocities $(\mathbf{X}, \dot{\mathbf{X}})$.

The transformation from $(\mathbf{X}, \dot{\mathbf{X}})$ to Kepler elements starts with the computation of the magnitude and direction of the angular momentum vector

$$\mathbf{h} = \mathbf{X} \times \dot{\mathbf{X}} = [h_X \quad h_Y \quad h_Z]^{\mathrm{T}} \tag{3.49}$$

which is the vector form of Equation (3.15). The various components of $\mathbf{h}$ are shown in Figure 3.3. The right ascension of the ascending node and the inclination of the orbital plane are, according to Figure 3.3,

$$\Omega = \tan^{-1}\left(\frac{h_X}{-h_Y}\right) \tag{3.50}$$

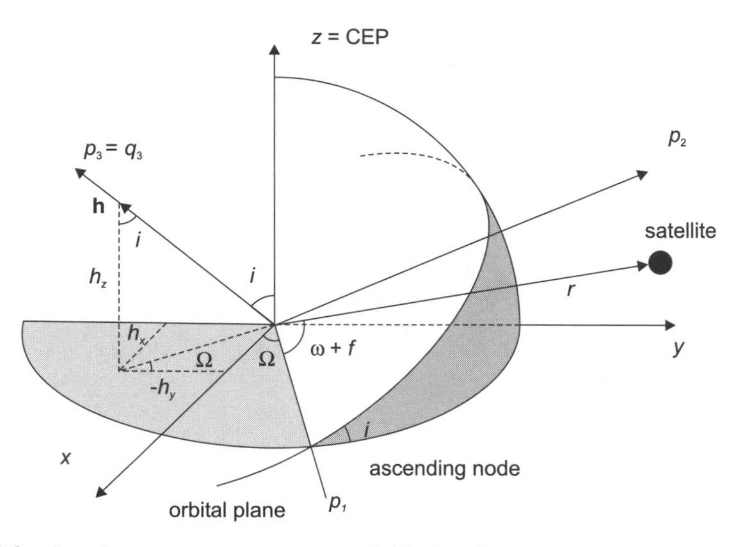

Figure 3.3 Angular momentum vector and Kepler elements. The angular momentum vector is orthogonal to the orbital plane.

$$i = \tan^{-1}\left(\frac{\sqrt{h_X^2 + h_Y^2}}{h_Z}\right) \tag{3.51}$$

By defining the auxiliary coordinate system (p) such that the p_1 axis is along the nodal line, p_3 is along the angular momentum vector, and p_2 completes a right-handed coordinate system, we obtain

$$\mathbf{p} = \mathbf{R}_1(i)\mathbf{R}_3(\Omega)\mathbf{X} \tag{3.52}$$

The sum of the argument of perigee and the true anomaly becomes

$$\omega + f = \tan^{-1}\left(\frac{p_2}{p_1}\right) \tag{3.53}$$

Thus far, the orbital plane and the orientation of the orbital ellipse have been determined. The shape and size of the ellipse depend on the velocity of the satellite. The velocity, geocentric distance, and the magnitude of the angular momentum are

$$v = \|\dot{\mathbf{X}}\| \tag{3.54}$$

$$r = \|\mathbf{X}\| \tag{3.55}$$

$$h = \|\mathbf{h}\| \tag{3.56}$$

The velocity expressed in the (q) coordinate system can be written as follows, using (3.26), (3.42), and (3.48):

$$
\begin{aligned}
v^2 &= \dot{q}_1^2 + \dot{q}_2^2 \\
&= \frac{n^2 a^2}{1 - e^2} \left(\sin^2 f + e^2 + 2e \cos f + \cos^2 f \right) \\
&= \frac{\mu}{a\left(1 - e^2\right)} \left[2 + 2e \cos f - \left(1 - e^2\right) \right] \\
&= \mu \left(\frac{2}{r} - \frac{1}{a} \right)
\end{aligned}
\tag{3.57}
$$

Equation (3.57) yields the expression for the semimajor axis

$$
a = \frac{r}{2 - r v^2 / \mu} \tag{3.58}
$$

From Equation (3.28), it follows that

$$
e = \left(1 - \frac{h^2}{\mu a} \right)^{1/2} \tag{3.59}
$$

and Equations (3.35), (3.47), and (3.48) give an expression for the eccentric anomaly:

$$
\cos E = \frac{a - r}{a\, e} \tag{3.60}
$$

$$
\sin E = \frac{\mathbf{q} \cdot \dot{\mathbf{q}}}{e \sqrt{\mu a}} \tag{3.61}
$$

Equations (3.60) and (3.61) together determine the quadrant of the eccentric anomaly. Having E, the true anomaly follows from (3.47):

$$
f = \tan^{-1} \frac{\sqrt{1 - e^2} \sin E}{\cos E - e} \tag{3.62}
$$

Finally, Kepler's equation yields the mean anomaly:

$$
M = E - e \sin E \tag{3.63}
$$

Equations (3.50) to (3.63) comprise the transformation from $(\mathbf{X}, \dot{\mathbf{X}})$ to the Kepler elements.

Table 3.1 shows six examples of trajectories for which the orbital eccentricity is zero, $e = 0$. The satellites' positions $\mathbf{x}$ in the earth-centered earth-fixed (ECEF)

TABLE 3.1 Trajectories of Normal Orbits

Condition	Trajectory
$\bar{n} = 2 \qquad \delta\lambda > 2\pi$	GPS (1)
$\bar{n} = \dfrac{1}{\cos i} \qquad \delta\lambda > 2\pi$	90 deg at equator (1)
$\dfrac{\pi}{2} + \sin^{-1}\left(\dfrac{\sqrt{1 - \bar{n}\cos i}}{\sin i}\right) - \bar{n}\left\{\dfrac{\pi}{2} + \tan^{-1}\left(\dfrac{\sqrt{(1 - \bar{n}\cos i)\cos i}}{\bar{n} - \cos i}\right)\right\} = 0 \qquad \delta\lambda = 3\,\Delta\lambda$	prograde & touching (3)
$\bar{n} = 1 \quad \delta\lambda = 2\left\{\tan^{-1}\left(\cos i\sqrt{\dfrac{1 - \cos i}{\sin^2 i - 1 + \cos i}}\right) - \sin^{-1}\left(\dfrac{\sqrt{1 - \cos i}}{\sin i}\right)\right\}$	one revolution per day
$\dfrac{3\pi}{2} - \sin^{-1}\left(\dfrac{\sqrt{1 - \bar{n}\cos i}}{\sin i}\right) - \bar{n}\left\{\dfrac{3\pi}{2} - \tan^{-1}\left(\dfrac{\sqrt{(1 - \bar{n}\cos i)\cos i}}{\bar{n} - \cos i}\right)\right\} = 0 \qquad \delta\lambda = 3\,\Delta\lambda$	retrograde & touching (3)
$\bar{n} = \cos i \qquad \delta\lambda = \tan^{-1}\{\cos i \tan(2\pi\bar{n})\} - 2\pi$	retrograde; vertical tangent; (3)

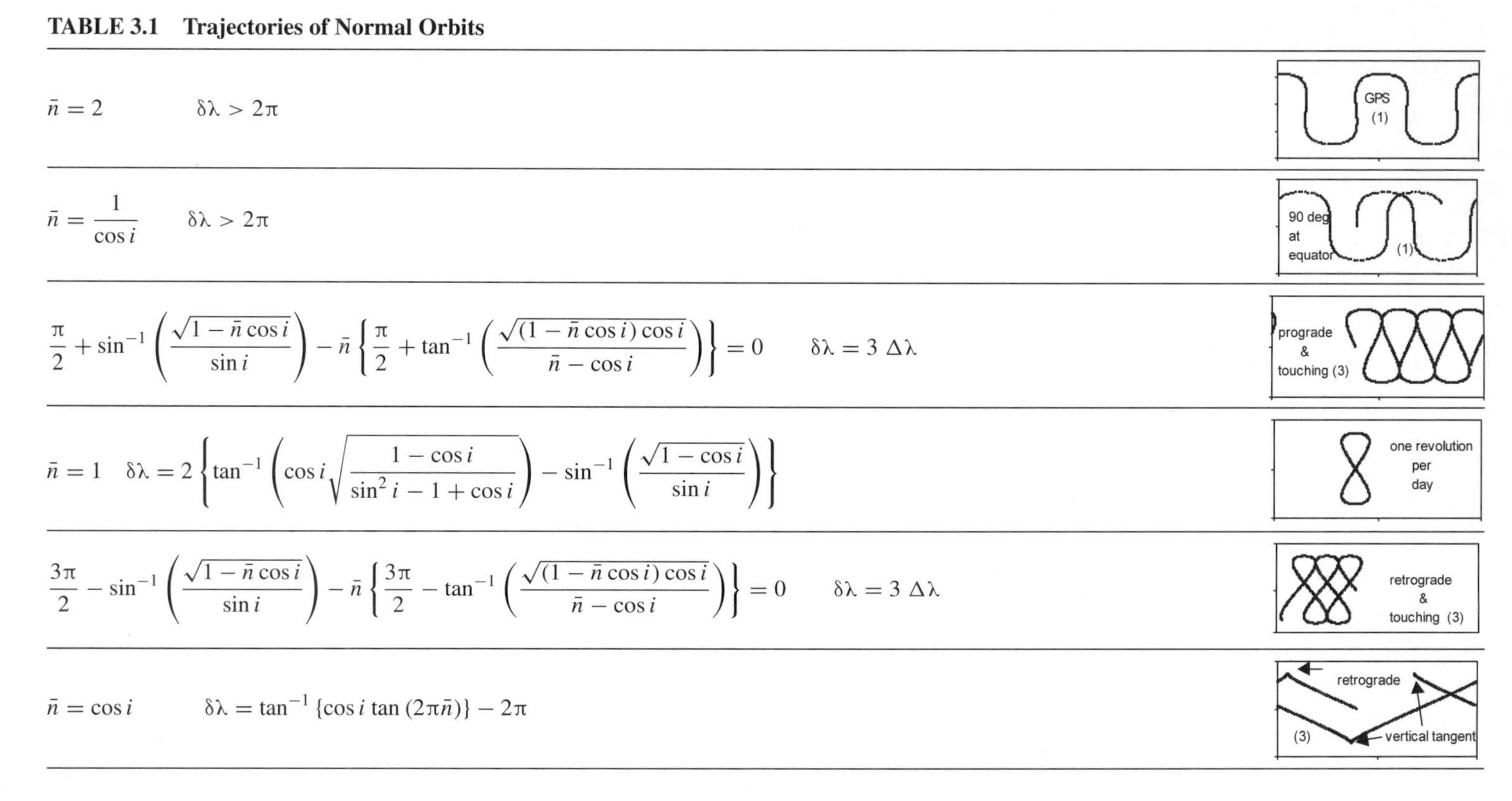

coordinate system can be readily computed from **X** by applying 2.34. We can then readily compute spherical latitude and longitude (ϕ, λ) and the trajectories of the satellites on the sphere. For reasons of convenience, we express the mean motion of the satellites in revolutions per day, $\bar{n} = n/\omega$. The longitude difference between consecutive equator crossings can then be computed from

$$\Delta\lambda = \pi\left(1 - \frac{1}{\bar{n}}\right) \tag{3.64}$$

Table 3.1 also lists the change in longitude of the trajectory over a 24-hour period, denoted by $\delta\lambda$. The number in parentheses on the graphs indicates the number of days plotted. In all cases the inclination is $i = 65°$. The maximum and minimum of the trajectories occur at a latitude of i and $-i$, respectively.

Case 1, specified by $\bar{n} = 2$, applies to GPS because the satellite orbits twice per (sidereal) day. Case 2 has been constructed such that the trajectories intersect the equator at 90°. In case 3 the point at which the trajectory touches, having common vertical tangent, and the point of either maximum or minimum have the same longitude. The mean motion must be computed from a nonlinear equation, but $\bar{n} > 1$ is valid. In case 4 the satellite completes one orbital revolution in exactly one (sidereal) day. Case 5 represents a retrograde motion with $\bar{n} < 1$ but with the same properties as case 3. In case 6 the common tangent at the extrema is vertical. The interested reader may verify that

$$\lambda = \tan^{-1}\left(\cos i \frac{\sin\phi}{\sqrt{\sin^2 i - \sin^2\phi}}\right) - \frac{1}{\bar{n}}\sin^{-1}\left(\frac{\sin\phi}{\sin i}\right) \tag{3.65}$$

and

$$\frac{d\varphi}{d\lambda} = \frac{\cos\phi\sqrt{\sin^2 i - \sin^2\phi}}{\bar{n}\cos i - \cos^2\phi}\,\bar{n} \tag{3.66}$$

is valid for all cases.

3.1.3 Satellite Visibility and Topocentric Motion

The topocentric motion of a satellite as seen by an observer on the surface of the earth can be readily computed from existing expressions. Let $\mathbf{X}_S$ denote the geocentric position of the satellite in the celestial coordinate system (X). These positions could, for example, have been obtained from (3.45) in the case of normal motion or from the integration of perturbed orbits discussed below. The position $\mathbf{X}_S$ can then be readily transformed to crust-fixed coordinate system (x), giving $\mathbf{x}_S$ by applying (2.34). If we further assume that the position of the observer on the ground in the crust-fixed coordinate system is $\mathbf{x}_P$, then the topocentric coordinate difference

$$\Delta\mathbf{x} = \mathbf{x}_S - \mathbf{x}_P \tag{3.67}$$

can be substituted into Equations (2.99) to (2.101) to obtain the topocentric geodetic azimuth, elevation, and distance of the satellite. The geodetic latitude and longitude in these expressions can be computed from $\mathbf{x}_P$, if necessary. For low-accuracy applications such as the creation of visibility charts it is sufficient to use spherical approximations.

3.1.4 Perturbed Satellite Motion

The accurate determination of satellite positions must consider various disturbing forces. Disturbing forces are all those forces causing the satellite to deviate from the simple normal orbit. The disturbances are caused primarily by the nonsphericity of the gravitational potential, the attraction of the sun and the moon, the solar radiation pressure, and other smaller forces acting on the satellites. For example, albedo is a force due to electromagnetic radiation reflected by the earth. There could be thermal reradiation forces caused by anisotropic radiation from the surface of the spacecraft. Additional forces, such as residual atmospheric drag, affect satellites closer to the earth.

Several of the disturbing forces can be readily computed; others, in particular the smaller forces, require detailed modeling and are still subject to further research. Knowing the accurate location of the satellites, i.e., being able to treat satellite position coordinates as known quantities, is important in surveying, in particular for long baseline determination. Most scientific applications of GPS demand the highest orbital accuracy, all the way down to the centimeter level. However, even surveying benefits from such accurate orbits, e.g., in precise point positioning with one receiver. See Section 7.5 for additional detail on this technique. One of the goals of the International GPS Service (IGS) and its contributing agencies and research groups is to refine continuously orbital computation and modeling and to make the most accurate satellite ephemeris available to the users. In this section, we provide only an introductory exposition of orbital determination. The details are found in the extensive literature, going all the way back to the days of the first artificial satellites.

The equations of motion are expressed in an inertial (celestial) coordinate system, corresponding to the epoch of the initial conditions. The initial conditions are either $(\mathbf{X}, \dot{\mathbf{X}})$ or the Kepler elements at a specified epoch. Because of the disturbing forces, all Kepler elements are functions of time. The transformation given above can be used to transform the initial conditions from $(\mathbf{X}, \dot{\mathbf{X}})$ to Kepler elements and vice versa. The equations of motion, as expressed in Cartesian coordinates, are

$$\frac{d\mathbf{X}}{dt} = \dot{\mathbf{X}} \tag{3.68}$$

$$\frac{d\dot{\mathbf{X}}}{dt} = -\frac{\mu \mathbf{X}}{\|\mathbf{X}\|^3} + \ddot{\mathbf{X}}_g + \ddot{\mathbf{X}}_s + \ddot{\mathbf{X}}_m + \ddot{\mathbf{X}}_{\mathrm{SRP}} + \cdots \tag{3.69}$$

These are six first-order differential equations. The symbol μ denotes the geocentric gravitational constant (3.5). The first term in (3.69) represents the acceleration of the central gravity field that generates the normal orbits discussed in the previous section.

Compare (3.69) with (3.4). The remaining accelerations are discussed briefly below. The most simple way to solve (3.68) and (3.69) is to carry out a simultaneous numerical integration. Most of the high-quality engineering or mathematical software packages have such integration routines available. Kaula (1966) expresses the equations of motion in terms of Kepler elements and expresses the disturbing potential in terms of Kepler elements. Kaula (1962) gives similar expressions for the disturbing functions of the sun and the moon.

3.1.4.1 Gravitational Field of the Earth

The acceleration of the noncentral portion of the gravity field of the earth is given by

$$\ddot{\mathbf{X}}_g = \begin{bmatrix} \frac{\partial R}{\partial X} & \frac{\partial R}{\partial Y} & \frac{\partial R}{\partial Z} \end{bmatrix}^{\mathrm{T}} \tag{3.70}$$

The disturbing potential R is

$$R = \sum_{n=2}^{\infty} \sum_{m=0}^{n} \frac{\mu a_e^n}{r^{n+1}} \bar{P}_{nm}(\cos\theta) \left[\bar{C}_{nm} \cos m\lambda + \bar{S}_{nm} \sin m\lambda\right] \tag{3.71}$$

with

$$P_{nm}(\cos\theta) = \frac{\left(1 - \cos^2\theta\right)^{m/2}}{2^n n!} \frac{d^{(n+m)}}{d(\cos\theta)^{(n+m)}} \left(\cos^2\theta - 1\right)^n \tag{3.72}$$

$$\bar{P}_n = \sqrt{2n+1}\; P_n \tag{3.73}$$

$$\bar{P}_{nm} = \left(\frac{(n+m)!}{2(2n+1)(n-m)!}\right)^{-1/2} P_{nm} \tag{3.74}$$

Equation (3.71) expresses the disturbing potential (as used in satellite orbital computations) in terms of a spherical harmonic expansion. The symbol a_e denotes the mean earth radius, r is the geocentric distance to the satellite, and θ and λ are the spherical co-latitude and longitude of the satellite position in the earth-fixed coordinate system, i.e., $\mathbf{x} = \mathbf{x}(r, \theta, \lambda)$. The positions in the celestial system (X) follow from (2.34). $\bar{P}_{nm}$ denotes the associated Legendre functions, which are known mathematical functions of the latitude. $\bar{C}_{nm}$ and $\bar{S}_{nm}$ are the spherical harmonic coefficients of degree n and order m. The bar indicates fully normalized potential coefficients. Note that the summation in (3.71) starts at $n = 2$. The term $n = 0$ equals the central component of the gravitational field. It can be shown that the coefficients for $n = 1$ are zero for coordinate systems whose origin is at the center of mass. Equation (3.71) shows that the disturbing potential decreases exponentially with the power of n. The high-order coefficients represent the detailed structure of the disturbing potential, and, as such, the fine structure of the gravity field of the earth. Only the coefficients of lower degree and order, say, up to degree and order

36, are significant for satellite orbital computations. The higher the altitude of the satellite, the less the impact of higher-order coefficients on orbital disturbances. A set of spherical harmonic coefficients can be found in GSFC (2002).

The largest coefficient in (3.71) is $\bar{C}_{20}$. This coefficient represents the effect of the flattening of the earth on the gravitational field. Its magnitude is about 1000 times larger than any of the other spherical harmonic coefficients.

Useful insight into the orbital disturbance of the flattening of the earth is obtained by considering the effect $\bar{C}_{20}$ only. An analytical expression is obtained if one expresses the equations of motion (3.68) and (3.69) in terms of Kepler elements. The actual derivation of such equations is beyond the scope of this book. The reader is referred to Kaula (1966). Mueller (1964) offers the following result,

$$\dot{\omega} = -\left(\frac{\mu}{a^3}\right)^{1/2} \left(\frac{a_e}{a\left(1-e^2\right)}\right)^2 \frac{3}{2} J_2 \left(1 + \cos^2 i - 1.5 \sin^2 i\right) \tag{3.75}$$

$$\dot{\Omega} = -\left(\frac{\mu}{a^3}\right)^{1/2} \left(\frac{a_e}{a\left(1-e^2\right)}\right)^2 \frac{3}{2} J_2 \cos i \tag{3.76}$$

In these equations we have made the substitution $\bar{C}_{20} = -J_2\sqrt{5}$. The variations of the argument of perigee and the right ascension of the ascending node are shown in Figure 3.4 as a function of the inclination. At the critical inclination of approximately 63.5° the perigee motion is stationary. The perigee and the node regress if $i > 63.5°$. This orbital plane rotation is zero for polar orbits $i = 90°$. Equation (3.76) is also useful for understanding the connection between the earth flattening and precession and the 18.6-year nutation/tidal period.

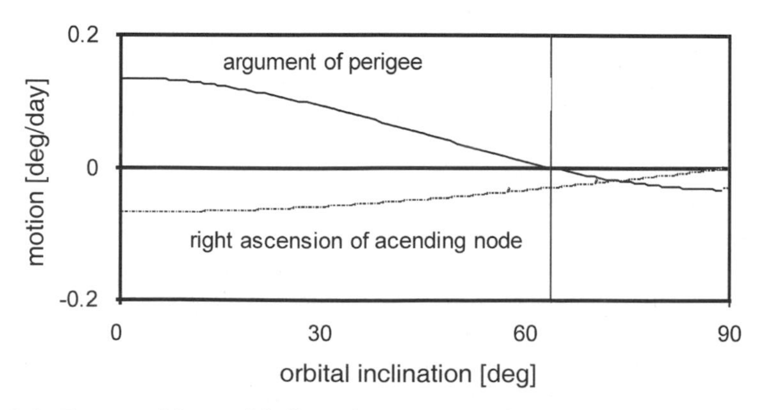

Figure 3.4 Impact of the earth's flattening on the motion of the perigee and the nodal line. The data refer to $a = 26{,}600$ km.

3.1.4.2 *Acceleration due to the Sun and the Moon* The lunar and solar accelerations on the satellites are (Escobal, 1965, p. 37)

$$\ddot{\mathbf{X}}_m = \frac{\mu m_m}{m_e}\left(\frac{\mathbf{X}_m - \mathbf{X}}{\|\mathbf{X}_m - \mathbf{X}\|^3} - \frac{\mathbf{X}_m}{\|\mathbf{X}_m\|^3}\right) \tag{3.77}$$

$$\ddot{\mathbf{X}}_s = \frac{\mu m_s}{m_e}\left(\frac{\mathbf{X}_s - \mathbf{X}}{\|\mathbf{X}_s - \mathbf{X}\|^3} - \frac{\mathbf{X}_s}{\|\mathbf{X}_s\|^3}\right) \tag{3.78}$$

The commonly used values for the mass ratios are $m_m/m_e = 0.0123002$ and $m_s/m_e = 332{,}946$. Mathematical expressions for the geocentric positions of the moon $\mathbf{X}_m$ and the sun $\mathbf{X}_s$ are given, for example, in van Flandern and Pulkkinen (1979).

3.1.4.3 *Solar Radiation Pressure* Solar radiation pressure (SRP) is a result of the impact of light photons emitted from the sun on the satellite's surface. The basic parameters of the SRP are the effective area (surface normal to the incident radiation), the surface reflectivity, thermal state of the surface, luminosity of the sun, and the distance to the sun. Computing SRP requires the evaluation of surface integrals over the illuminated regions, taking shadowed components into account. Even if these regions are known, the evaluation of the surface integrals can still be difficult because of the complex shape of the satellite. The ROCK4 and ROCK42 models represent early attempts to take most of these complex relations and properties into consideration for GPS Block I, Block II, and Block IIa satellites, respectively (Fliegel et al., 1985; Fliegel and Gallini, 1989). Fliegel et al. (1992) describe an SRP force model for geodetic applications. Springer et al. (1999) report on SRP model parameter estimation on a satellite-by-satellite basis, as part of orbital determinations from heavily overdetermined global networks. Ziebart et al. (2002) discuss a pixel array method in connection with finite analysis, in order to delineate even better the illuminated satellite surfaces and surface temperature distribution.

One of the earliest and simplest SRP models uses merely two parameters. Consider the body-fixed coordinate system of Figure 3.5. The z' axis is aligned with the antenna and points toward the center of the earth. The satellite finds this direction and remains locked to it with the help of an earth limb sensor. The x' axis is positive toward the half plane that contains the sun. The y' axis completes the right-handed coordinate system and points along the solar panel axis. The satellites are always oriented such that the y' axis remains perpendicular to the earth-satellite-sun plane. The only motion of the spacecraft in this body-fixed frame is the rotation of the solar panels around the y' axis to make the surface of the solar panels perpendicular to the direction of the sun. The direction of the sun is denoted by $\mathbf{e}$ in Figure 3.5.

In reference to this body-fixed coordinate system, a simple SRP model formulation is

$$\ddot{\mathbf{X}}_{\text{SRP}} = -p\frac{\mathbf{X}_{\text{sun}} - \mathbf{X}}{\|\mathbf{X}_{\text{sun}} - \mathbf{X}\|} + Y\frac{\mathbf{X}_{\text{sun}} \times \mathbf{X}}{\|\mathbf{X}_{\text{sun}} \times \mathbf{X}\|} \tag{3.79}$$

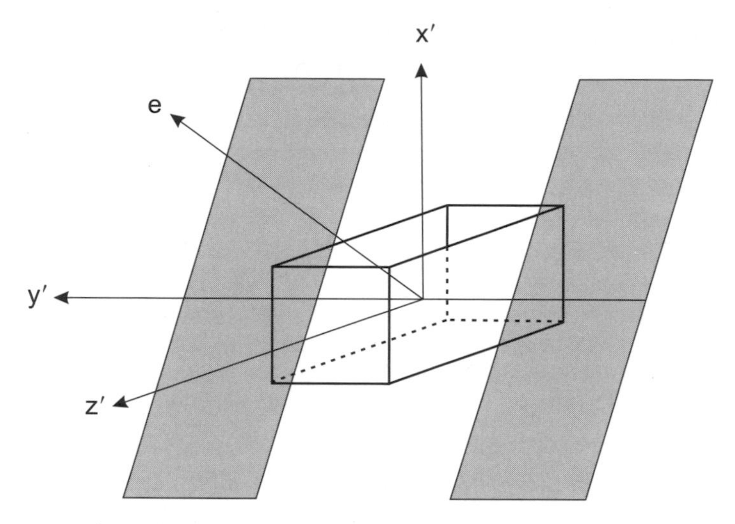

Figure 3.5 The satellite body-fixed coordinate system.

The symbol p denotes the SRP in the direction of the sun. With the sign convention of Equation (3.79), p should be positive. The other parameter is called the Y bias. The reasons for its existence could be structural misalignments, thermal phenomena, or possibly misalignment of the solar panels with the direction of the solar photon flux. The fact that a Y bias exists demonstrates the complexity of accurate solar radiation pressure modeling. Table 3.2 shows the effects of the various perturbations over the period of one day. The table shows the difference between two integrations, one containing the specific orbital perturbation and the others turned off. It is seen that SRP orbital disturbance reaches close to 100 m in a day. This is very significant considering that the goal is centimeter orbital accuracy. Over a period of 1–2 weeks the SRP disturbance can grow to over 1 km.

3.1.4.4 Eclipse Transits and Yaw Maneuvers Orbital determination is further complicated when satellites travel through the earth shadow region (eclipse). The

TABLE 3.2 Effect of Perturbations on GPS Satellites over One Day

Perturbation	Radial	Along	Cross	Total
Earth flattening	1335	12902	6101	14334
Moon	191	1317	361	1379
Sun	83	649	145	670
$\bar{C}_{2,2}$, $\bar{S}_{2,2}$	32	175	9	178
SRP	29	87	3	92
$\bar{C}_{n,m}$, $\bar{S}_{n,m}$ ($n, m = 3 \ldots 8$)	6	46	4	46

Source: Springer et al., 1999.

Note: The units are in meters.

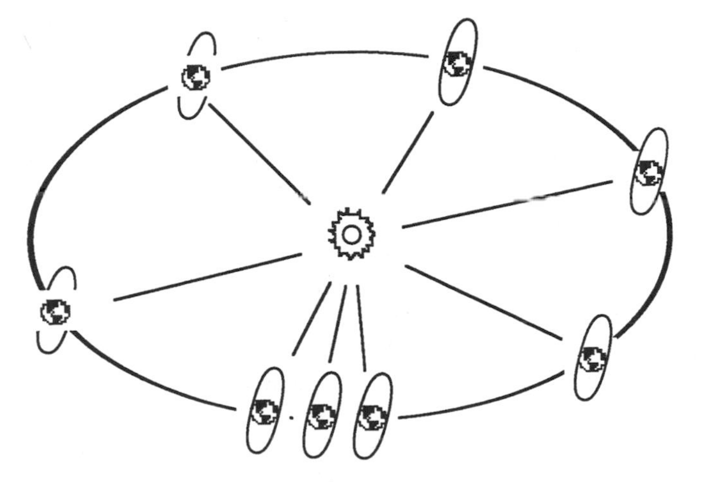

Figure 3.6 Biannual eclipse periods.

satellite travels through the earth's shadow region twice per year. This occurs whenever the sun is in or near the orbital plane. See Figures 3.6 and 3.7 for a graphical presentation. The umbra is that portion of the shadow cone that no light from the sun can reach. The penumbra is the region of partial shadowing; it surrounds the umbra cone. While the satellite transits through the shadow regions, the solar radiation force acting on the satellite is either zero (umbra) or changing (penumbra). These changes in force must be taken into consideration in precise orbital computations. In addition, the thermal reradiation forces change as the temperature of the satellite drops. GPS satellites move through the shadow regions in less than 60 minutes, twice per day.

The shadow regions cause an additional problem for precise orbit determination. The solar panels are orientated toward the sun by the attitude control system (ACS)

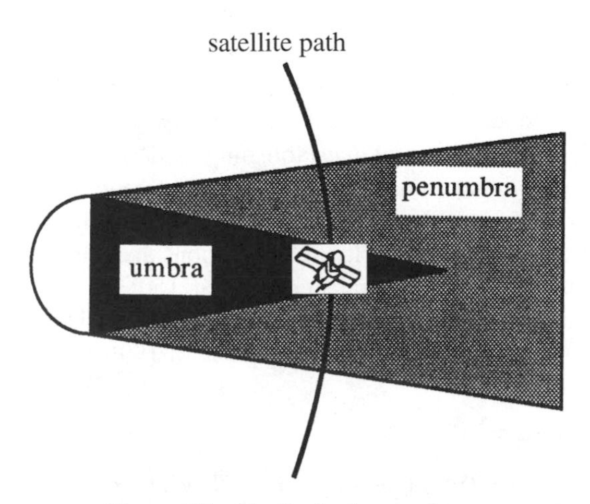

Figure 3.7 Earth shadow regions.

solar sensors mounted on the solar panels. The condition that the z' axis continuously points toward the center of the earth and the solar panels are continuously normal to the satellite-sun direction, the satellite must yaw, i.e., rotate around the z' axis, in addition to rotating the antennas around the y' axis. While the satellite passes through the shadow region, the ACS solar sensors do not receive sunlight and, therefore, cannot maintain the exact alignment of the solar panels. The satellite starts yawing in a somewhat unpredictable way. Errors in yaw cause errors in GPS observations in two ways. First, the range correction from the center of the satellite antenna to the satellite center of mass becomes uncertain. Second, there is an additional but unknown windup error. See Section 7.2.1 for more information on the windup error.

Bar-Sever (1996) has investigated the GPS yaw attitude problem and the compensation method in detail. During shadow, the output of the solar sensors is essentially zero and the ACS is driven by the noise of the system. Even a small amount of noise can trigger a significant yaw change. As a corrective action, a small bias signal is added to the signals of the solar sensors that amounts to a yaw of about 0.5°. As a result, during the time when the sun can be observed, the yaw will be in error by that amount. During eclipse times, the biased noise will yaw the satellite in the direction of the bias, thus avoiding larger and erratic yaw motions. When the satellite leaves the shadow region, the solar sensors provide the information to determine the correct yaw angle. The yaw maneuvers carried out by the satellite from the time it enters the shadow region to the time it leaves it are collectively called "the midnight maneuvers." When the satellite is on the sun-earth axis and between the sun and the earth, the ACS encounters a singularity because any yaw angle represents an optimal orientation of the solar panels for this particular geometry. Any maneuvers that deal with this situation are called "the noon maneuver."

3.2 GPS GLOBAL POSITIONING SYSTEM

Satellite-based positioning has been pursued since the 1960s. An early and very successful satellite positioning system was the Navy Navigation Satellite System (TRANSIT). Since its release for commercial use in 1967, the TRANSIT positioning system was often used to determine widely spaced networks covering large regions—even the globe. It was instrumental in establishing modern geocentric datums and in connecting various national datums to a geocentric reference frame. The TRANSIT satellites were orbiting in polar plane at about 1100 km altitude. The TRANSIT satellites were affected more by gravity field variations than the much higher-orbiting GPS satellites. In addition, their transmissions at 150 and 400 MHz were more susceptible to ionospheric delays and disturbances than the higher GPS frequencies. The TRANSIT system was discontinued at the end of 1996 and replaced by GPS.

3.2.1 General Description

The Navigation Satellite Timing and Ranging (NAVSTAR) GPS provides positioning and timing 24 hours per day, anywhere in the world, and under any weather conditions. The U.S. government operates GPS. It was designed as a dual-use system, with the primary purpose of meeting the military's needs for positioning and timing. Over the past decade, the number of civilian applications has increased significantly, with no end in sight. Because GPS is so well known by now, not just by the experts but by general citizens as well, there is no need to dwell on which innovative application will be next or even attempt to list its numerous current uses.

The buildup of the satellite constellation began with the series Block I satellites. These were concept validation satellites that did not have selective availability (SA) or antispoofing (AS) capability. They were launched into three 63° inclined orbital planes. Their positions within the planes were such that optimal observing geometry was achieved over certain military proving grounds in the continental United States. Eleven Block I satellites were launched between 1978 and 1985 (with one launch failure). The average lifetime was 8–9 years. They were designed to provide 3–4 days of positioning service without contact with the ground control center. The launch of the second generation of GPS satellites, called Block II, began in February 1989. In addition to radiation-hardened electronics, these operational satellites had full SA/AS capability and carried a navigation data message that was valid for fourteen days. Additional modifications resulted in the satellite called Block IIA. These satellites can provide about six weeks of positioning service without contact from the control segment. Twenty-eight Block II/IIA satellites were launched between 1989 and 1997 into six planes, 55° inclined. The first third-generation GPS satellite, called Block IIR (R for replenishment), was successfully launched in 1997. These satellites have the capability to determine their orbits autonomously through UHF cross-link ranging and to generate their own navigation message by onboard processing. They are able to measure ranges between themselves and transmit observations to other satellites as well as to ground control. Currently, GPS is undergoing a major modernization. Most important, GPS satellites will transmit more signals that allow a better delineation of military and civilian uses, and thus increase the performance of GPS even more. Table 3.3 shows the expected progression of the modernization. We anticipate the launch of the IIR-M (M for modified) satellites soon. These satellites will transmit new civil codes on L2 and new military codes on L1 and L2. Given the continued progress in microelectronics, it will then be possible to manufacture inexpensive, compact dual-frequency receivers for civilian uses. Around 2005/2006, we expect the launch of

TABLE 3.3 New GPS Signals

Signal	IIR 1978–2003	IIR-M 2003 (expected)	IIF 2005 (expected)
L1 C/A	X	X	X
L1 P(Y)	X	X	X
L1M		X	X
L2C		X	X
L2 P(Y)	X	X	X
L2 M		X	X
L5C			X

the first IIF (F for follow on) satellites. These satellites will transmit a third civil frequency, called L5.

The U.S. government's current policy is to make GPS available in two services. The precise positioning service (PPS) is available to the military and other authorized users. The standard positioning service (SPS) is available to anyone. See SPS (2001) for a detailed documentation of this service. Without going into detail, let it suffice to say that PPS users have access to the encrypted P(Y)-codes on the L1 and L2 carriers, while SPS users can only observe the public C/A-code on L1. The encryption of the P-codes began January 31, 1994. SPS positioning capability was degraded by SA measures, which entailed an intentional dither of the satellite clocks and falsification of the navigation message. In keeping with the policy, SA was implemented on March 25, 1990, on all Block II satellites. The level of degradation was reduced in September 1990 during the Gulf conflict, but was reactivated to its full level on July 1, 1991, until it was discontinued on May 1, 2000, upon direction of the U.S. president.

Over time, both satellite and receiver technologies have improved significantly. Whereas older receivers could observe the P(Y)-code more accurately than the C/A-codes, this distinction has all but disappeared with modern receiver technology. Dual-frequency P(Y)-code users do have the advantage of being able to correct the effect of the ionosphere on the signals. However, this simple classification of PPS and SPS by no means characterizes how GPS is used today. Researchers have devised various, often patented procedures that make it possible to observe or utilize the encrypted P(Y)-codes effectively, and in doing so, make dual-frequency observations available, at least to high-end receivers. In certain surveying applications where the primary quantity of interest is the vector between nearby stations, intentional degradation of SA could be overcome by differencing the observations between stations and satellites. However, in many applications, positioning with GPS works much better without SA.

The six orbital planes of GPS are evenly spaced in right ascension and are inclined by 55° with respect to the equator. Because of the flattening of the earth, the nodal regression is about −0.04187° per day; an annual orbital adjustment keeps the orbits close to their nominal location. Each orbital plane contains four satellites; however, to optimize global satellite visibility, the satellites are not evenly spaced within the orbital plane. The orbits are nominally circular, with a semimajor axis of about 26,660 km. Using Kepler's third law (3.42), one obtains an orbital period of slightly less than 12 hours. The satellites will complete two orbital revolutions in one sidereal day. This means the satellites will rise about 4 minutes earlier each day. Because the orbital period is an exact multiple of the period of the earth's rotation, the satellite trajectory on the earth (i.e., the trace of the geocentric satellite vector on the earth's surface) repeats itself daily.

Because of their high altitude, the GPS satellites can be viewed simultaneously over large portions of the earth. Usually the satellites are observed only above a certain vertical angle, called the mask angle. Typical values for the mask angle are 10–15°. At low elevation angles the tropospheric effects on the signal can be especially severe and difficult to model accurately. Let ε denote the mask angle, and let α denote the geocentric angle of visibility for a spherical earth; then one can find the relation ($\varepsilon = 0°$, $\alpha = 152°$), ($\varepsilon = 5°$, $\alpha = 142°$), ($\varepsilon = 10°$, $\alpha = 132°$). The viewing angle from the satellite to the limb of the earth is about 27°.

3.2.2 Satellite Transmissions at 2002

The ICD-GPS-200C (2000) is the authoritative source for details on the GPS signal structures, usage of these signals, and other information broadcasts by the satellites. The document can be downloaded from GPS (2002). All satellite transmissions are coherently derived from the fundamental frequency of 10.23 MHz, made available by onboard atomic clocks. This is also true for the new signals discussed further below. Multiplying the fundamental frequency by 154 gives the frequency for the L1 carrier, $f_1 = 1575.42$ MHz, and multiplying by 120 gives the frequency of the L2 carrier, $f_2 = 1227.60$ MHz. The chipping (code) rate of the P(Y)-code is that of the fundamental frequency, i.e., 10.23 MHz, whereas the chipping rate of the C/A-code is 1.023 MHz (one-tenth of the fundamental frequency). The navigation message (telemetry) is modulated on both the L1 and the L2 carriers at a chipping rate of 50 bps. It contains information on the ephemerides of the satellites, GPS time, clock behavior, and system status messages.

The space vehicle time is defined by the onboard atomic clocks of each satellite. The satellite operates on its own time system, i.e., all satellite transmissions such as the C/A-code, the P(Y)-codes, and the navigation message are initiated by satellite time. The data in the navigation message, however, are relative to GPS time. Time is maintained by the control segment and follows UTC(USNO) within specified limits. GPS time is a continuous time scale and is not adjusted for leap seconds. The last common epoch between GPS time and UTC(USNO) was midnight January 5–6, 1980. The navigation message contains the necessary corrections to convert space vehicle time to GPS time. The largest unit of GPS time is one week, defined as 604,800 sec. Additional details on the satellite clock correction are given in Section 5.3.1.

The atomic clocks in the satellites are affected by both special relativity (the satellite's velocity) and general relativity (the difference in the gravitational potential at the satellite's position relative to the potential at the earth's surface). Jorgensen (1986) gives a discussion in lay terms of these effects and identifies two distinct parts in the relativity correction. The predominant portion is common to all satellites and is independent of the orbital eccentricity. The respective relative frequency offset is $\Delta f / f = -4.4647 \times 10^{-10}$. This offset corresponds to an increase in time of 38.3 μs per day; the clocks in orbit appear to run faster. The apparent change in frequency is $\Delta f = 0.0045674$ Hz at the fundamental frequency of 10.23 MHz. The frequency is corrected by adjusting the frequency of the satellite clocks in the factory before launch to 10.22999999543 MHz. The second portion of the relativistic effect is proportional to the eccentricity of the satellite's orbit. For exact circular orbits, this correction is zero. For GPS orbits with an eccentricity of 0.02 this effect can be as large as 45 ns, corresponding to a ranging error of about 14 m. This relativistic effect can be computed from a simple mathematical expression that is a function of the semimajor axis, the eccentricity, and the eccentric anomaly (see Section 5.3.1). In relative positioning as typically carried out in surveying, the relativistic effects cancel for all practical purposes.

The precision P(Y)-code is the principal code used for military navigation. It is a pseudorandom noise (PRN) code which itself is the modulo-2 sum of two other pseudorandom codes. The P(Y)-code does not repeat itself for thirty-seven weeks.

Thus, it is possible to assign weekly portions of this code to the various satellites. As a result, all satellites can transmit on the same carrier frequency and yet can be distinguished because of the mutually exclusive code sequences being transmitted. All codes are initialized once per GPS week at midnight from Saturday to Sunday, thus creating, in effect, the GPS week as a major unit of time. The L1 and L2 carriers are both modulated with the same P(Y)-code.

The period of the coarse/acquisition (C/A) code is merely 1 ms and consists of 1023 bits. Each satellite transmits a different set of C/A-codes. These codes are currently transmitted only on L1. The C/A-codes belong to the family of Gold codes, which characteristically have low cross-correlation between all members. This property makes it possible to distinguish the signals received simultaneously from different satellites rapidly.

One of the satellite identification systems makes use of the PRN weekly number. For example, if one refers to satellite PRN 13, one refers to the satellite that transmits the thirteenth weekly portion of the PRN-code. The short version of PRN 13 is SV 13 (SV = space vehicle). Another identification system uses the space vehicle launch number (SVN). For example, the identification of PRN 13 in terms of launch number is NAVSTAR 9, or SVN 9.

3.2.2.1 Signal Structure The carrier is modulated by several codes and the navigation (data) message. There are at least three commonly used digital modulation methods: amplitude shift keying (ASK), frequency shift keying (FSK), and phase shift keying (PSK). GPS uses PSK. Figure 3.8 briefly demonstrates some of the principles involved. The figure shows an arbitrary digital data stream consisting of binary digits 0 and 1. These binary digits are also called chips, bits, codes, or pulses. In the case of GPS, the digital data stream contains the navigation message or the pseudorandom sequences of the codes. The code sequences look random but actually follow a mathematical formula. ASK corresponds to an on/off operation. The digit 1 might represent turning the carrier on and 0 might mean turning it off. FSK implies transmission on one or the other frequency. The transmitting oscillator is required to switch back and forth between two distinct frequencies. In the case of PSK, the same carrier frequency is used, but the phase changes abruptly. With binary phase shift keying (BPSK), the phase shifts 0° and 180°. The BPSK method is used with GPS signals.

Figure 3.9 shows two data streams. The sequence (a) could represent the navigation data chipped rate of 50 bits per seconds (bps), and (b) could be the C/A-code or the P(Y)-code chipped at the 1.023 MHz or 10.23 MHz, respectively. The times of bit transition are aligned. The navigation message and the code streams have significantly different chipping rates. A chipping rate of 50 bps implies 50 opportunities per second for the digital stream to change from 1 to 0 and vice versa. Within the time of a telemetry chip there are 31,508,400 L1 cycles, 20,460 C/A-code chips, and 204,600 P(Y)-code chips. Looking at this in the distance domain, one telemetry chip is 5950 km long, whereas the lengths of the C/A and P(Y)-codes are 293 m and 29.3 m, respectively. Thus, the P(Y)-code can change the carrier by 180° every 29.3 m, the C/A-code every 293 m, and the telemetry every 5950 km.

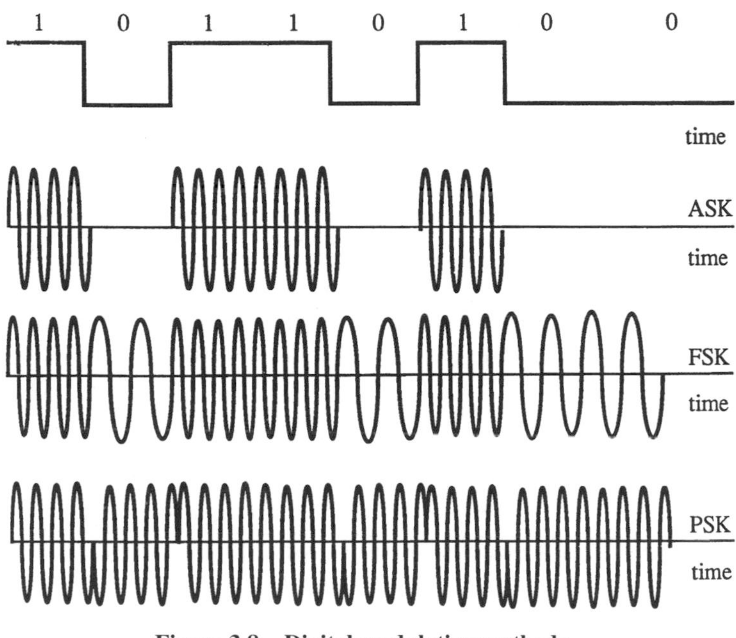

Figure 3.8 Digital modulation methods.

One of the tasks to be accomplished is reading the navigation message at the receiver. We need this information to compute the positions of the satellites. To accomplish this, the data streams (a) and (b) in Figure 3.9 are modulo-2 added before transmission at the satellite. Modulo-2 addition follows the well-known rules: $0+0=0, 1+0=1, 0+1=1$, and $1+1=0$. The result is labeled (c). The figure also shows the phase history of the transmitted carrier. Whenever a binary 1 occurs in the

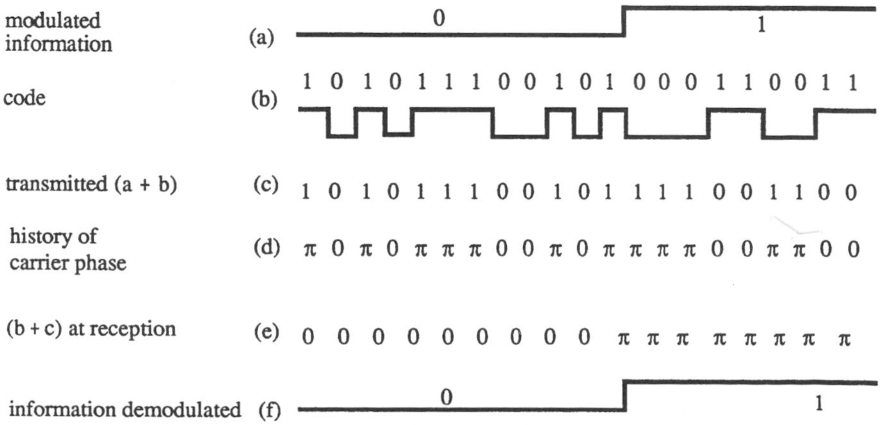

modulated information	(a)	0 1
code	(b)	1 0 1 0 1 1 1 0 0 1 0 1 0 0 0 1 1 0 0 1 1
transmitted (a + b)	(c)	1 0 1 0 1 1 1 0 0 1 0 1 1 1 1 0 0 1 1 0 0
history of carrier phase	(d)	π 0 π 0 π π π 0 0 π 0 π π π π 0 0 π π 0 0
(b + c) at reception	(e)	0 0 0 0 0 0 0 0 0 π π π π π π π π
information demodulated	(f)	0 1

Figure 3.9 Modulo-2 addition of binary data streams.

50 bps navigation data stream, the modulo-2 addition inverts 20,460 adjacent digits of the C/A-code. A binary 1 becomes 0 and vice versa. A binary 0 leaves the next 20,460 C/A-codes unchanged. Let the receiver reproduce the original code sequence that is shifted in time to match the transmitted code. We can then modulo-2 add the receiver-generated code with the received, phase-modulated carrier. The sum is the demodulated 50 bps telemetry data stream.

The modulo-2 addition method must be generalized one additional step because the L1 carrier is modulated by three data streams: the navigation data, the C/A-codes, and P(Y)-codes. Thus, the task of superimposing both code streams and the navigation data stream arises. Two sequential superimpositions are not unique, because the C/A-code and the P(Y)-code have identical bit transition epochs (although their length is different). The solution is called quadrature phase shift keying (QPSK). The carrier is split into two parts, the inphase component (I) and the quadrature component (Q). The latter is shifted by 90°. Each component is then binary phase-modulated, the inphase component is modulated by the P(Y)-code, and the quadrature component is modulated by the C/A-code. Therefore, the C/A-code signal carrier lags the P(Y)-code carrier by 90°. For the L1 and L2 carriers we have

$$S_1^p(t) = A_P P^p(t) D^p(t) \cos(2\pi f_1 t) + A_C G^p(t) D^p(t) \sin(2\pi f_1 t) \tag{3.80}$$

$$S_2^p(t) = B_P P^p(t) D^p(t) \cos(2\pi f_2 t) \tag{3.81}$$

In these equations the symbols denote

p	Superscript identifying the PRN number of the satellite
A_P, A_C, B_P	Amplitudes (power) of P(Y)-codes and C/A-code
$P^p(t)$	Pseudorandom P(Y)-code
$G^p(t)$	C/A-code (Gold code)
$D^p(t)$	Telemetry or navigation data stream

The products $P^p(t)D^p(t)$ and $G^p(t)D^p(t)$ imply modulo-2 addition. The P(Y)-code by itself is a modulo-2 sum of two pseudorandom data streams $X_1(t)$ and $X_2(t - pT)$ as follows:

$$P^p(t) = X_1(t) X_2(t - pT) \tag{3.82}$$

$$0 \leq p \leq 36 \tag{3.83}$$

$$\frac{1}{T} = 10.23 \text{ MHz} \tag{3.84}$$

Expression (3.82) defines the code according to the PRN number p. Using (3.83), one can define thirty-seven mutually exclusive P(Y)-code sequences. At the beginning of

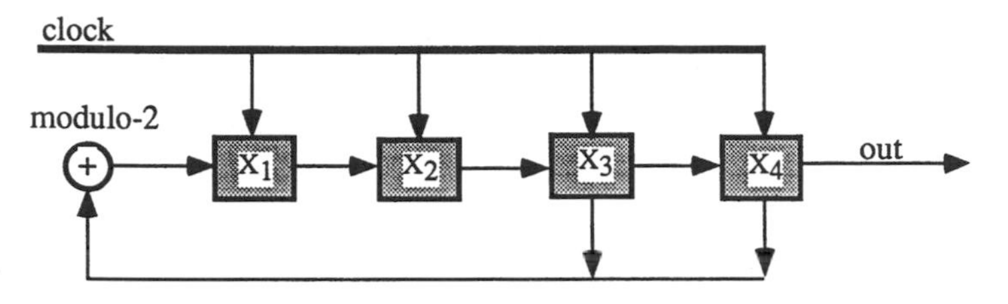

Figure 3.10 Simple FBSR.

the GPS week, the P(Y)-codes are reset. Similarly, the C/A-codes are the modulo-2 sum of two 1023 pseudorandom bit codes as follows:

$$G^p(t) = G_1(t)G_2\left[t - N^p(10T)\right] \tag{3.85}$$

$G^p(t)$ is 1023 bits long or has a 1 ms duration at a 1.023 Mbps bit rate. The $G^p(t)$ chip is ten times as long as the X_1 chip. The G_2-code is selectively delayed by an integer number of chips, expressed by the integer N^p, to produce thirty-six unique Gold codes, one for each of the thirty-six different P(Y)-codes.

The actual generation of the codes X_1, X_2, G_1, and G_2 is accomplished by a feedback shift register (FBSR). Such devices can generate a large variety of pseudorandom codes. These codes look random over a certain interval, but the feedback mechanism causes the codes to repeat after some time. Figure 3.10 shows a very simple register. A block represents a stage register whose content is in either a one or a zero state. When the clock pulse is input to the register, each block has its state shifted one block to the right. In this particular example, the output of the last two stages is modulo-2 added, and the result is fed back into the first stage and modulo-2 added to the old state to create the new state. The successive states of the individual blocks as the FBSR is stepped through a complete cycle are shown in Table 3.4. The elements of the column represent the state of each block, and the successive columns represent the behavior of the shift register as the succession of timing pulses cause it to shift from state to state. In this example, the initial state is (0001). For n blocks, $2^n - 1$ states are possible before repetition occurs. The output corresponds to the state of the last block, and would represent the PRN code if it were generated by such a four-stage FBSR.

The shift registers that are used in GPS code generation are much more complex. They have many more feedback loops and they have many more blocks in

TABLE 3.4 Output of FBSR

x_1	0	1	0	0	...	1	0	0	0
x_2	0	0	1	0	...	1	1	0	0
x_3	0	0	0	1	...	1	1	1	0
x_4	1	0	0	0	...	1	1	1	1
Output	1	0	0	0	...	1	1	1	1

the sequence. The P(Y)-code is derived from two twelve-stage shift registers, $X_1(t)$ and $X_2(t)$, having 15,345,000 and 15,345,037 stages (chips), respectively. Both registers continuously recycle. The modulo-2 sum of both registers has the length of 15,345,000 times 15,345,037 chips. At the chipping rate of 10.23 MHz it takes 266.4 days to complete the whole P(Y)-code cycle. It takes 1.5 s for the X_1 register to go though one cycle. The X_1 cycles (epochs) are known as the Z count.

The bandwidth terminology is often used in connection with pseudorandom noise modulation. Let T denote the duration of the chip (rectangular pulse), then the bandwidth is inverse proportional to T. Therefore, shorter chips (pulses) require greater bandwidth and vice versa. If we subject the rectangular pulse function to a Fourier transform we obtain the well-known sinc (sine-cardinal) function

$$S(\Delta f, f_c) = \frac{1}{f_c}\left(\frac{\sin(\pi\,\Delta f/f_c)}{\pi\,\Delta f/f_c}\right)^2 \tag{3.86}$$

The symbol Δf is the difference with respect to the carrier frequency L1 or L2. The code frequency 10.23 MHz or 1.023 MHz, respectively, is denoted by f_c. The factor $1/f_c$ serves as a normalizing (unit area) scalar. Figure 3.13 shows the power spectral density (3.86) for the P(Y)- and C/A-codes. This symmetric function is zero at multiples of the code rate f_c. The first lobe stretches over the bandwidth, covering the range of $\pm f_c$ with respect to the center frequency. The spectral portion signal beyond one bandwidth is filtered out at the satellite and is not transmitted.

Power ratios in electronics and in connection with signals and antennas are expressed in terms of decibels (dB) on a logarithmic scale. Of course, sound levels are typically also given in units of decibels. One decibel is just detectable by the human ear and a power of 100 watts is perceived to be twice as loud as 10 watts. The latter relationship justifies the preference of using the logarithmic scale in addition to the ability to express very large or very small ratios with a few digits. The power ratio in terms of decibel units is defined as

$$g_{[\mathrm{dB}]} = 10\log_{10}\frac{P_2}{P_1} \tag{3.87}$$

Absolute power can be expressed with respect to a unit power P_1. For example, the units dBW or dBm imply $P_1 = 1$ W or $P_1 = 1$ mW, respectively. Frequently the relation

$$g_{[\mathrm{dB}]} = 20\log_{10}\frac{V_2}{V_1} \tag{3.88}$$

is seen. In (3.88) the symbols V_1 and V_2 denote voltages. Both decibel expressions are related by the fact that the square of the voltage divided by resistance equals power.

The power of the received GPS signals on the ground is lower than the background noise (thermal noise). The specifications call for a minimum power at the user on the earth of -160 dBW for the C/A-code, -163 dBW for the P(Y)-code on L1, and -166 dBW for the P(Y)-code on L2. To track the signal, the receiver correlates the incoming signal by a locally generated replica of the code. This correlation process results in a signal that is well above the noise level.

3.2.2.2 Navigation Message The Master Control Station, located near Colorado Springs, uses data from a network of monitoring stations around the world to monitor the satellite transmissions continuously, compute the broadcast ephemerides, calibrate the satellite clocks, and periodically update the navigation message. This "control segment" ensures that the SPS and PPS are available as specified in SPS (2001).

The satellites transmit a navigation message that contains, among other things, orbital data for computing the positions of all satellites. A complete message consists of 25 frames, each containing 1500 bits. Each frame is subdivided into 5 300-bit subframes, and each subframe consists of 10 words of 30 bits each. At the 50 bps rate it takes 6 seconds to transmit a subframe, 30 seconds to complete a frame, and 12.5 minutes for one complete transmission of the navigation message. The subframes 1, 2, and 3 are transmitted with each frame. Subframes 4 and 5 are each subcommutated 25 times. The 25 versions of subframes 4 and 5 are referred to as pages 1 through 25. Thus, each of these pages repeats every 12.5 minutes.

Each subframe begins with the telemetry word (TLM) and the handover word (HOW). The TLM begins with a preamble and otherwise contains only information that is needed by the authorized user. The HOW is a truncation of the GPS time of week (TOW). HOW, when multiplied by 4, gives the X_1 count at the start of the following subframe. As soon as a receiver has locked to the C/A-code, the HOW word is extracted and is used to identify the X_1 count at the start of the following subframe. In this way, the receiver knows exactly which part of the long P(Y)-code is being transmitted. P(Y)-code tracking can then readily begin, thus the term *handover word*. To lock rapidly to the P(Y)-code, the HOW is included on each subframe (see Figure 3.11).

GPS time is directly related to the X_1 counts of the P(Y)-code. The Z count is a twenty-nine-bit number that contains several pieces of timing information. It can be used to extract the HOW, which relates to the X_1 count as discussed above, and the TOW, which represents the number of seconds since the beginning of the GPS week. A full week has 403,199 X_1 counts. The Z count gives the current GPS week number (modulo-1024). The beginning of the GPS week is offset from midnight UTC by the accumulated number of leap seconds since January 5–6, 1980, the beginning of GPS time.

Subframe 1 contains the GPS week number, space vehicle accuracy and health status, satellite clock correction terms a_{f0}, a_{f1}, a_{f2} and the clock reference time t_{oc} (Section 5.3.1), the differential group delay, T_{GD} (Section 5.4), and the issue of date clock (IODC) term. The latter term is the issue number of the clock data set and can be conveniently used to detect any change in the correction parameters. The messages are updated usually every 4 hours.

Subframes 2 and 3 contain the ephemeris parameters for the transmitting satellite. The various elements are listed in Table 3.5. These elements are a result of least-squares fitting of the predicted ephemeris over a well-specified interval of time. The

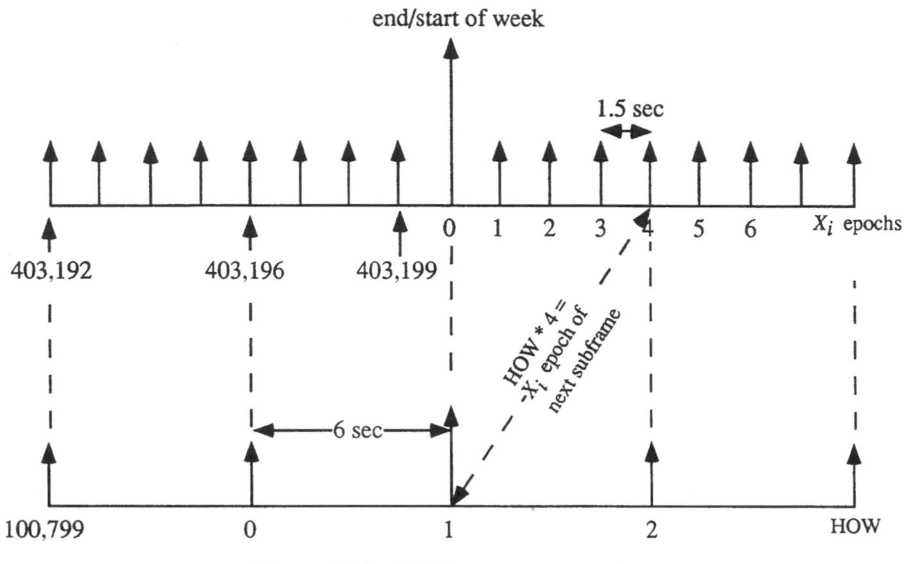

Figure 3.11 HOW versus X_1 epochs.

issue of data ephemeris (IODE) term allows users to detect changes in the ephemeris parameters. For each upload, the control center assigns a new number. The IODE is given in both subframes. During the time of an upload, both IODEs will have different values. Users should download ephemeris data only when both IODEs have the same value. The broadcast elements are used with the algorithm of Table 3.6. The results are coordinates of the phase center of the space vehicle's antennas in the World Geodetic System of 1984 (WGS84). The latter is an ECEF coordinate system that is closely aligned with the international terrestrial reference frame (ITRF). There is no need for an explicit polar motion rotation, since the respective rotations are incorporated in the representation parameters. However, when computing the topocentric distance, the user must account for the rotation of the earth during the signal travel time from satellite to receiver.

Subframes 4 and 5 contain special messages, ionospheric correction terms, coefficients to convert GPS time to universal time coordinated (UTC), and almanac data on pages 2–5 and 7–10 (subframe 4) and 1–24 (subframe 5). The ionospheric terms are the eight coefficients $\{\alpha_n, \beta_n\}$ referenced in Table 6.3. For accurate computation of UTC from GPS time, the message provides a constant offset term, a linear polynomial term, the reference time t_{ot}, and the current value of the leap second. The almanac provides data to compute the positions of satellites other than the transmitting satellite. It is a reduced-precision subset of the clock and ephemeris parameters of subframes 1 to 3. For each satellite, the almanac contains the following: t_{oa}, δ_i, a_{f0}, a_{f1}, e, $\dot{\Omega}$, $a^{1/2}$, Ω_0, ω, and M_0. The almanac reference time is t_{oa}. The correction to the inclination δ_i is given with respect to the fixed value $i_0 = 0.30$ semicircles ($= 54°$). The clock polynomial coefficients a_{f0} and a_{f1} are used to convert space vehicle (SV) time to GPS time, following Equation (5.38). The remaining elements of the almanac are

TABLE 3.5 Elements of Subframes 2 and 3

M_0	Mean anomaly at reference time
Δn	Mean motion difference from computed value
e	Eccentricity
$\sqrt{a}$	Square root of the semimajor axis
Ω_0	Longitude of ascending node of orbit plane at beginning of week
i_0	Inclination angle at reference time
ω	Argument of perigee
$\dot{\Omega}$	Rate of right ascension
IDOT	Rate of inclination angle
$C_{uc}, C_{us}, C_{rc}, C_{rs}, C_{ic}, C_{is}$	Amplitude of second-order harmonic perturbations
t_{oe}	Ephemeris reference time
IODE	Issue of data (ephemeris)

identical to those listed in Table 3.5. The algorithm of Table 3.6 applies, using zero for all elements that are not included in the almanac and replacing the reference time t_{oe} by t_{oa}.

The mean anomaly, the longitude of the ascending node, the inclination, and UTC (if desired) are formulated as polynomials in time; the time argument is GPS time. The polynomial coefficients are, of course, a function of the epoch of expansion. The respective epochs are t_{oc}, t_{oe}, t_{oa}, and t_{ot}.

The navigation message contains other relevant information, such as the user range error (URE). This measure equals the projection of the ephemeris curve fit errors onto the user range and includes effects of satellite timing errors (and possibly SA).

3.2.3 GPS Modernization

GPS modernization becomes possible because of advances in technology as used in the satellite and the receiver. The additional signals transmitted by modernized satellites will improve the antijamming capability, increase protection against anti-spoofing, shorten the time to first fix, and provide a civilian "safety of life" signal (L5) within the protected Aeronautical Radio Navigation Service (ARNS) frequency band. The new L2C signal will increase robustness of the signal, improve resistance to interference, allow for longer integration times in the receiver, thereby reducing tracking noise and increasing accuracy, as well as providing better positioning inside buildings and in wooded areas. The second civil frequency will eliminate the need to use inefficient squaring, cross-correlation, or other patented techniques currently used by civilians in connection with L2. Once the GPS modernization is completed, the dual-frequency or triple-frequency receivers are expected to be in common use and affordable to the mass market.

At the same time, new military codes called the M-codes will be added to L1 and L2, but will be spectrally separated from the civilian codes. There is no military code planned on L5.

TABLE 3.6 GPS Broadcast Ephemeris Algorithm

$\mu = 3.986005 \times 10^{14}\ \mathrm{m}^3/\mathrm{s}^2$	Gravitational constant for WGS84
$\dot{\Omega}_e = 7.2921151467 \times 10^{-5}$ rad/s	Earth's rotation rate for WGS84
$a = \left(\sqrt{a}\right)^2$	Semimajor axis
$n_0 = \sqrt{\dfrac{\mu}{a^3}}$	Computed mean motion—rad/s
$t_k = t - t^*_{oe}$	Time from ephemeris reference epoch
$n = n_0 + \Delta n$	Corrected mean motion
$M_k = M_0 + n t_k$	Mean anomaly
$M_k = E_k - e \sin E_k$	Kepler's equation for eccentric anomaly
$f_k = \tan^{-1}\left\{\dfrac{\sqrt{1-e^2}\sin E_k}{\cos E_k - e}\right\}$	True anomaly
$E_k = \cos^{-1}\left[\dfrac{e + \cos f_k}{1 + e\cos f_k}\right]$	Eccentricity anomaly
$\phi_k = f_k + \omega$	Argument of latitude
$\left.\begin{aligned}\delta u_k &= C_{us}\sin 2\phi_k + C_{uc}\cos 2\phi_k\\ \delta r_k &= C_{rc}\cos 2\phi_k + C_{rs}\sin 2\phi_k\\ \delta i_k &= C_{ic}\cos 2\phi_k + C_{is}\sin 2\phi_k\end{aligned}\right\}$	Second harmonic perturbations
$u_k = \phi_k + \delta u_k$	Corrected argument of latitude
$r_k = a\,(1 - e\cos E_k) + \delta r_k$	Corrected radius
$i_k = i_0 + \delta i_k + (\mathrm{IDOT})\,t_k$	Corrected inclination
$\left.\begin{aligned}x'_k &= r_k \cos u_k\\ y'_k &= r_k \sin u_k\end{aligned}\right\}$	Positions in orbital plane
$\Omega_k = \Omega_0 + \left(\dot{\Omega} - \dot{\Omega}_e\right) t_k - \dot{\Omega}_e t_{oe}$	Corrected longitude of ascending node
$\left.\begin{aligned}x_k &= x'_k \cos\Omega_k - y'_k \cos i_k \sin\Omega_k\\ y_k &= x'_k \sin\Omega_k + y'_k \cos i_k \cos\Omega_k\\ z_k &= y'_k \sin i_k\end{aligned}\right\}$	Earth-fixed coordinates

Note: t is GPS system time at time of transmission, i.e., GPS time corrected for transit time (range/speed of light). Furthermore, t_k shall be the actual total time difference between the time t and the epoch time t_{oe}, and must account for beginning or end of week crossovers. That is, if t_k is greater than 302,400, subtract 604,800 from t_k. If t_k is less than −302,400 sec, add 604,800 sec to t_k.

Fortunately, there is good documentation available on the anticipated GPS signal modernization. GPS (2002) contains several relevant documents and other material on briefings. The material for L2 was extracted from PPIRN-200C-007 (2001). The description of the new L5 signal is found in ICD-GPS-705 (2002). Both documents can be downloaded from GPS (2002). See also Fontana et al. (2001a,b).

3.2.3.1 Civil L2C Codes The new L2 will be shared between civil and military signals. To increase GPS performance for civilian users, the new space vehicles IIR-M and IIF will have two additional civil ranging codes, L2CM (civil moderate length) and L2CL (civil long). As is the case for L1, the new L2 carrier will consist of two BPSK modulated carrier components that are in phase quadrature with each other. The inphase carrier will continue to be BPSK modulated by the bit train that is the modulo-2 sum of the military P(Y)-code and the legacy navigation data $D^p(t)$. There will be three options available for BPSK modulating the quadrature carrier (also called the L2C carrier or the new L2 civil signal):

1. Chip-by-chip time multiplex combinations of bit trains consisting of the modulo-2 sum of the L2CM code and a new navigation message structure $D_C(t)$. The resultant bit trains are then combined with the L2CL code and used to modulate the L2 quadrature carrier. The IIR-M space vehicles will have the option of using the old navigation message $D^p(t)$ instead of $D_C(t)$.
2. Modulo-2 sum of the legacy C/A-code and legacy navigation data $D^p(t)$.
3. C/A-code with no navigation data.

The chipping rate for L2CM and L2CL is 511.5 Kbps. L2CM is 10,230 chips long and lasts 20 ms, whereas L2CL has 767,250 chips and lasts 1.5 s. L2CL is 75 times longer than L2CM. $D_C(t)$ is the new navigation data message and has the same structure as the one adopted for the new L5 civil signal. It is both more compact and more flexible than the legacy message.

3.2.3.2 Civil L5 Code The carrier frequency of L5 is 1176.45 MHz. As is the case for L1, two L5 carriers are in phase quadrature and each is BPSK modulated separately by bit trains. The bit train of the inphase component is a modulo-2 sum of PRN codes and navigation data. The quadraphase code is a separate PRN code but has no navigation data. The chipping rate of the codes is 10.23 MHz. Each code is a modulo-2 sum of two subsequences, whose lengths are 8,190 and 8,191 chips that recycle to generate 10,230 chip codes. Therefore the length of these codes is 1 ms.

3.2.3.3 M-Code For conventional rectangular spreading codes, which are the basis of the P(Y)-codes, the C/A-code heritage signals, and the new L2C and L5 codes, the frequency bandwidth is inversely proportional to the length of the chip. Modulating with faster chipping rates to improve or add additional signals might be impractical because of frequency bandwidth limitations. More advanced modulations have been studied recently that better share existing frequency allocations with each other and with heritage signals by increasing spectral separation, and thus preserve the spectrum. Betz (2002) describes binary-valued modulations, also referred to as binary offset carrier (BOC). Block IIR-M and IIF satellites will transmit a new military M-code signal on L1 and L2 that uses BOC.

If f_c denotes again the chipping (code) rate and if we denote the subcarrier frequency by f_s then (Betz, 2002)

$$f_c = \frac{1}{nT_s} = \frac{2}{n} f_s \tag{3.89}$$

where n is a positive integer, and the normalized power spectral density of the BOC modulation can be written as

$$g(f_s, f_c, \Delta f) = \begin{cases} f_c \left(\dfrac{\tan(\pi\, \Delta f/2f_s)\, \cos(\pi\, \Delta f/f_c)}{\pi\, \Delta f} \right)^2 & \text{if } n \text{ is odd} \\ f_c \left(\dfrac{\tan(\pi\, \Delta f/2f_s)\, \sin(\pi\, \Delta f/f_c)}{\pi\, \Delta f} \right)^2 & \text{if } n \text{ is even} \end{cases} \tag{3.90}$$

We further adopt Betz's abbreviated notation BOC(α, β) to specify the frequencies, i.e., $f_s = 1.023\alpha$ MHz and $f_c = 1.023\beta$ MHz. For example, the modulation BOC(10,5) uses the subcarrier frequency and the spreading code rate of 10.23 MHz and 5.115 MHz, respectively.

A characteristic difference between the BOC and the conventional rectangular spreading code modulation is seen in the power spectral densities of Figure 3.13. The densities for BOC, in this case BOC(10,5), are maximum at the nulls of the P(Y)-codes. Such a property is important for increasing the spectral separation of modulations. The sum of the number of mainlobes and sidelobes between the mainlobes is equal to n, i.e., twice the ratio of the subcarrier frequency to the code rate (3.89). As in conventional PSK the zero crossings of each mainlobe are spaced by twice the code rate, while the zero crossings of each sidelobe are spaced at the code rate. For example, with $n = 5$ the BOC(5,2) modulations have three sidelobes between two mainlobes; with $n = 10$ the BOC(5,1) modulations have eight sidelobes between two mainlobes. In the case of $n = 1$, that is the case of BOC($f_c/2$, f_c), Equations (3.90) and (3.86) give the same power spectral density.

The new military M-codes will use BOC(10,5), which means the subcarrier frequency and the spreading code rate will be 10.23 MHz and 5.115 MHz, respectively, as well as quadraphase modulated, i.e., they share the same carrier with the civilian signals.

3.3 GLONASS

The Russian Global'naya Navigatsionnaya Sputnikkovaya Sistema (GLONASS—global navigation satellite system [GNSS]) system traces its beginnings to 1982, when the first satellite of this navigation satellite system was launched. A time line of the space segment is shown in Figure 3.12. The technical information about GLONASS can be found in the interface control document GLONASS (1998). Additional details on the system and its use, plus many references to relevant publications on the subject, are available in Roßbach (2001).

Like GPS, GLONASS was planned to contain at least twenty-four satellites. The nominal orbits of the satellites are in three orbital planes separated by 120°; the satellites are equally spaced within each plane with nominal inclination of 64.8°. The nominal orbits are circular with each radius being about 25,500 km. This translates into an orbital period of about 11 hours and 15 minutes.

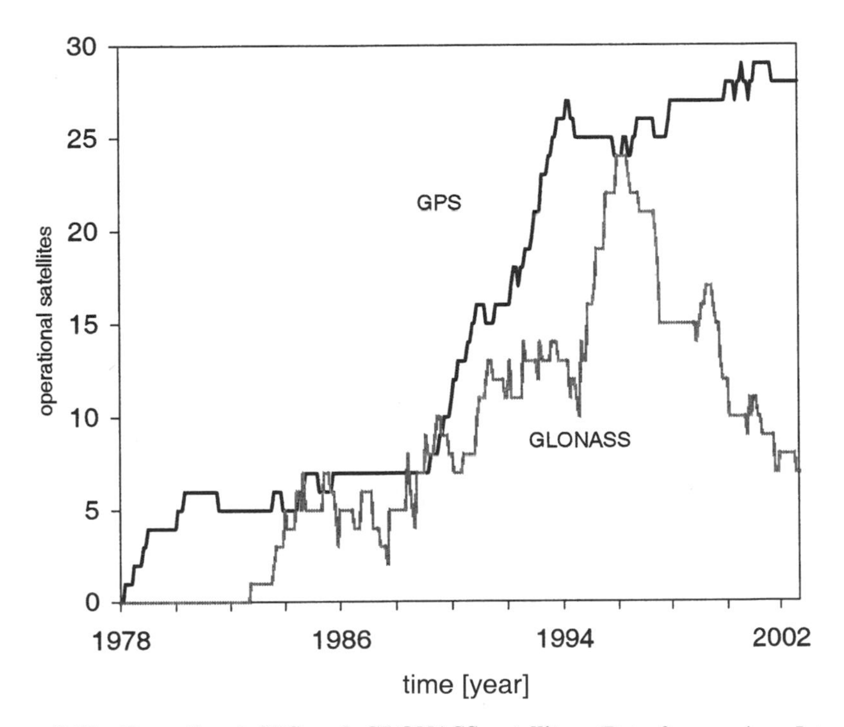

Figure 3.12 Operational GPS and GLONASS satellites. (Data from various Internet documents.)

A major difference between GLONASS and GPS is that each GLONASS satellite transmits at its own carrier frequency. Let p denote the channel number that is specific to the satellite, then

$$f_1^p = 1602 + 0.5625p \text{ MHz} \tag{3.91}$$

$$f_2^p = 1246 + 0.4375p \text{ MHz} \tag{3.92}$$

with

$$\frac{f_1^p}{f_2^p} = \frac{9}{7} \tag{3.93}$$

The original GLONASS signal structure used $1 \leq p \leq 24$, covering a frequency range in L1 from 1602.5625 MHz to 1615.5 MHz. However, receivers have an interference problem in the presence of mobile-satellite terminals that operate at the 1610 to 1621 MHz range. To avoid such interference, it has been suggested that channel numbers will be limited to $-7 \leq p \leq 13$ and satellites located in antipodal slots of the same orbital plane may transmit at the same frequency (GLONASS, 1998).

The L1 and L2 frequencies are coherently derived from common onboard fre-

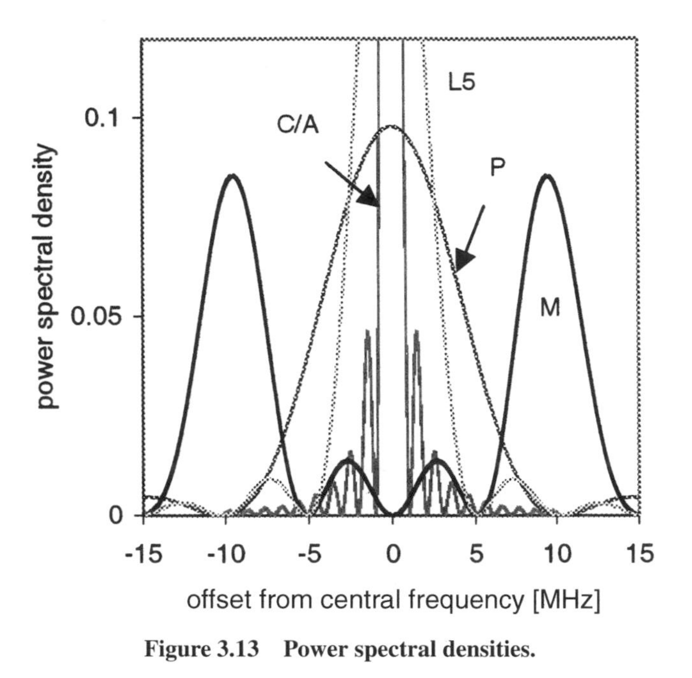

Figure 3.13 Power spectral densities.

quency, standard running at 5.0 MHz. In order to account for relativistic effects, this value is adjusted to 4.99999999782 MHz. As is the case with GPS, there are C/A-codes on L1 and P-codes on L1 and L2, although the code structures differ. The satellite clocks are steered according to UTC(SU). The GLONASS satellite clocks, therefore, are adjusted for leap seconds.

The GLONASS broadcast navigation message contains satellite positions and velocities in the PZ90 ECEF geodetic system and accelerations due to luni-solar attraction at epoch t_0. These data are updated every 30 minutes and serve as initial conditions for orbital integration. The satellite ephemeris at the epoch t_b with $|t_b - t_0| \leq 15$ min is calculated by numerical integration of the differential equations of motion (3.69). Because the integration time is short, it is sufficient to consider a simplified force model for the acceleration of the gravity field of the earth. Since the gravitational potential of the earth is in first approximation rotationally symmetric, the contributions of the tesseral harmonics $m \neq 0$ are neglected in (3.71). Since $\bar{C}_{20} \gg \bar{C}_{n0}$ for $n > 2$, we neglect the higher-order zonal harmonics. With these simplifications the disturbing potential (3.71) becomes

$$
\begin{aligned}
R &= \frac{\mu a_e^2}{r^3} \bar{C}_{20} \bar{P}_n(\cos\theta) = \frac{\mu a_e^2}{r^3} J_2 P_n(\cos\theta) \\
&= \frac{\mu a_e^2}{r^3} J_2 \left(\frac{3}{2}\cos^2\theta - \frac{1}{2} \right)
\end{aligned}
\tag{3.94}
$$

In Expression (3.94) we switched from the fully normalized spherical harmonic coefficients to regular ones and substituted the expression for the Legendre polynomial $P_2(\cos\theta)$. Since $Z = r\cos\theta$, Equation (3.94) can be rewritten as:

$$R = \frac{\mu a_e^2}{r^3} J_2 \left(\frac{3}{2}\frac{Z^2}{r^2} - \frac{1}{2} \right) \tag{3.95}$$

Recognizing that $r = (X^2 + Y^2 + Z^2)^{1/2}$, we can readily differentiate and compute the acceleration $\ddot{\mathbf{X}}_g$ as per (3.70)

$$\ddot{X} = -\frac{\mu}{r^3}X - \frac{3}{2}J_2\frac{\mu a_e^2}{r^5}X\left(1 - 5\frac{Z^2}{r^2}\right) + \ddot{X}_{s+m} \tag{3.96}$$

$$\ddot{Y} = -\frac{\mu}{r^3}Y - \frac{3}{2}J_2\frac{\mu a_e^2}{r^5}Y\left(1 - 5\frac{Z^2}{r^2}\right) + \ddot{Y}_{s+m} \tag{3.97}$$

$$\ddot{Z} = -\frac{\mu}{r^3}Z - \frac{3}{2}J_2\frac{\mu a_e^2}{r^5}Z\left(1 - 5\frac{Z^2}{r^2}\right) + \ddot{Z}_{s+m} \tag{3.98}$$

These equations are valid in the inertial system (X) and could be integrated. The PZ90 reference system, however, is ECEF and rotates with the earth. It is possible to rewrite these equations in the ECEF system (x). Since the integration interval is only ± 15 min, we can neglect the change in precession, nutation, and polar motion and only take the rotation of the earth around the z axis into consideration. The final form of the satellite's equations of motion then becomes:

$$\ddot{x} = -\frac{\mu}{r^3}x - \frac{3}{2}J_2\frac{\mu a_e^2}{r^5}x\left(1 - 5\frac{z^2}{r^2}\right) + \omega_3^2 x + 2\omega_3\dot{y} + \ddot{x}_{s+m} \tag{3.99}$$

$$\ddot{y} = -\frac{\mu}{r^3}y - \frac{3}{2}J_2\frac{\mu a_e^2}{r^5}y\left(1 - 5\frac{z^2}{r^2}\right) + \omega_3^2 y + 2\omega_3\dot{x} + \ddot{y}_{s+m} \tag{3.100}$$

$$\ddot{z} = -\frac{\mu}{r^3}z - \frac{3}{2}J_2\frac{\mu a_e^2}{r^5}z\left(1 - 5\frac{z^2}{r^2}\right) + \ddot{z}_{s+m} \tag{3.101}$$

Note that $(\ddot{x}, \ddot{y}, \ddot{z})_{s+m}$ are the accelerations of the sun and the moon given in the PZ90 frame. These values are assumed constant when integrating over the ± 15 min interval. In order to maintain consistency, the values for μ, a_e, J_2, and ω_3 should be adopted from GLONASS (2002). This document recommends a four-step Runge-Kutta method for integration.

Various international observation campaigns have been conducted to establish accurate transformation parameters between WGS84 and PZ90, with respect to the ITRF. Efforts are continuing to include the precise GLONASS ephemeris into the IGS products.

The most recent launch of three satellites as of this writing (summer 2002) took place on December 1, 2001. The GLONASS program is also undergoing a modernization. The new series of satellites is called GLONASS-M. The interface control documents provide some information on the new (planned) features. The interested reader is advised to consult the current literature to learn about ongoing developments and the status of the GLONASS constellation.

Finally, GLONASS satellites have been used successfully for accurate baseline determination since the mid-1990s (Leick et al., 1995). The additional difficulties encountered in baseline processing because of GLONASS satellites transmitting on different carrier frequencies will be discussed in Chapter 7. GLONASS observations have primarily been used to supplement and strengthen GPS solutions. The improved productivity when including extra GLONASS satellites is clearly noticeable and has heightened the expectations among practitioners for the not-to-distance future when more GLONASS satellites are available again, Galileo becomes available, and, in general, the modernization of the systems has progressed.

3.4 GALILEO

On March 26, 2002, the European Council agreed on the launch of the European Civil Satellite Navigation Program, called Galileo. Civilian European institutions fund this program but complementary financing by public-private partnership is also under consideration. Some of Galileo's services might eventually be subject to a user fee. The space segment is expected to consist of a global constellation of about thirty satellites, distributed over three planes. The nominal orbits are expected to be circular, with semimajor axes being close to those of GPS and GLONASS. Consult Galileo (2002) for up-to-date information on this satellite system, in particular, regarding the details of signal structure and the definition of the various services. The status of Galileo as of fall 2002 can be found in Hein et al. (2002).

As can be seen from Figure 3.14, the Galileo E5A signals share the frequency band with GPS L5. The adjacent region is reserved for Galileo E5B. At the World Radio Conference (WRC) 2000 at Istanbul, Turkey, several decisions were made that deal

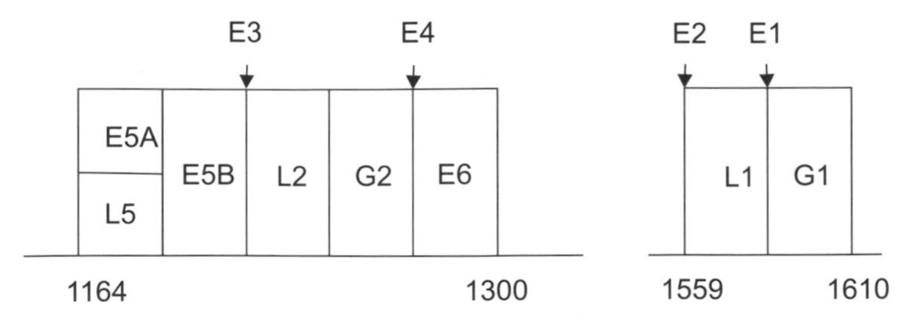

Figure 3.14 Allocation of GPS, GLONASS, and Galileo frequency bands. GPS: L1, L2, L5; GLONASS: G1, G2; Galileo: E1, E2, E3, E4, E5A, E5B, E6. The symbols E1, E2, E3, and E4 indicate the location of very narrow bands.

TABLE 3.7 GPS, GLONASS, and Galileo Carrier Frequencies

Carrier	Multiple of 10.23	Carrier Frequency [MHz]
L5 & E5A	115	1176.45
E5B	117.5	1202.025 (tentatively)
L2	120	1227.60
G 2	see (3.92)	per satellite
E6	125	1278.750
L1, E1-L1-E2	154	1574.42
G 1	see (3.91)	per satellite

with the increasing demand for frequency space. For example, the WRC expanded the bottom end of one of the radio navigation satellite services (RNSS) bands to between 1164 and 1260 MHz, putting E5A, E5B, and L5 under RNSS protection. Galileo has also been assigned the range 1260 to 1300 MHz, labeled E6, at the lower L-band region. At the upper L-band, two narrow bands, labeled E1 and E2, have been reserved for Galileo adjacent to the GPS L1 band. Using BOC modulation techniques, it will be possible to construct a Galileo signal that will have maximum spectral density at E1 and E2 but cover the whole E1-L1-E2 band.

In order to make Galileo and GPS compatible, i.e., allow for the use of common receiver components, the carrier frequency for the Galileo E1-L1-E2 will be 1575.42 MHz, which is the same as GPS L1. Similarly, E5A and L5 will use 1176.45 MHz as the common carrier frequency.

The modulation (inhase and quadrahase) codes and chipping rate for the various carriers must still be finalized. The remaining issues regarding frequency allocation and signal structure are expected to be resolved at a future WRC. Whereas L5, E5, L1, and G1 are within the ARNS bands the middle bands (L2, G2, and E6) currently do not enjoy such a protected status. There is a potential for interference from joint tactical information distribution system (JTIDS), multifunctional information distribution system (MIDS), distance measuring equipment (DME), and tactical air navigation system (TACAN) that requires attention. Table 3.7 summarizes the location of the carrier frequencies.

CHAPTER 4

LEAST-SQUARES ADJUSTMENTS

Least-squares adjustment is a device for carrying out objective quality control of measurements by processing sets of redundant observations according to mathematically well-defined rules. The objectivity of least-squares quality control is especially useful when depositing or exchanging observations. Least-squares solutions require redundant observations, i.e., more observations are required than are necessary to determine a set of unknowns exactly. Details will be given as to what constitutes optimal redundancy. This chapter contains compact but complete derivations of least-squares algorithms.

First, the statistical nature of measurements is analyzed, followed by a discussion of stochastic and mathematical models and the law of variance-covariance propagation of random variables. The mixed adjustment model is derived in detail, and the observation equation and the condition equation models are deduced from the mixed model through appropriate specification. The cases of observed and weighted parameters are presented as well. A special section is devoted to minimal and inner constraint solutions and to those quantities that remain invariant with respect to a change in minimal constraints. Whenever the goal is to perform quality control on the observations, minimal or inner constraint solutions are especially relevant. Statistical testing is important for judging the quality of observations or the outcome of an adjustment. A separate section deals with statistics in least-squares adjustments. The chapter ends with a presentation of additional quality measures, such as internal and external reliability and blunder detection and a brief exposition of Kalman filtering.

4.1 ELEMENTS

Objective quality control of observations is necessary when dealing with any kind of measurements such as angles, distances, pseudoranges, carrier phases, and the geopotential. It is best to separate conceptually quality control of observations and precision or accuracy of parameters. It is unfortunate that least-squares adjustment is most often associated only with high-precision surveying. It may be as important to discover and remove a 10 m blunder in a low-precision survey as a 1 cm blunder in a high-precision survey.

Least-squares adjustment allows the combination of different types of observations (such as angles, distances, and height differences) into one solution and permits simultaneous statistical analysis. For example, there is no need to treat traverses, intersections, and resections separately. Since these geometric figures consist of angle and distance measurements, the least-square rules apply to all of them, regardless of the specific arrangements of the observations or the geometric shape they represent.

Least-squares adjustment simulation is a useful tool to plan a survey and to ensure that accuracy specifications will be met once the actual observations have been made. Simulations allow the observation selection to be optimized when alternatives exist. For example, should one primarily measure angles or rely on distances? Considering the available instrumentation, what is the optimal use of the equipment under the constraints of the project? Experienced surveyors often answer these questions intuitively. Even in these cases, an objective verification using least-squares simulation and the concept of internal and external reliability of networks is a welcome assurance to those who carry responsibility for the project.

4.1.1 Statistical Nature of Surveying Measurement

Assume that a distance of 100 m is measured repeatedly with a tape that has centimeter divisions. A likely outcome of these measurements could be 99.99, 100.02, 100.00, 100.01, etc. Because of the centimeter subdivision of the tape, the surveyor is likely to record the observations to two decimal places. The result therefore is a series of numbers ending with two decimal places. One could wrongly conclude that this measurement process belongs to the realm of discrete statistics yielding discrete outcomes with two decimal places. In reality, however, the series is given two decimal places because of the centimeter division of the tape and the fact that the surveyor did not choose to estimate the millimeters. Imagining a reading device that allows us to read the tape to as many decimal places as desired, we readily see that the process of measuring a distance belongs to the realm of continuous statistics. The same is true for other types of measurements typically used in positioning. A classic textbook case for a discrete statistical process is the throwing of a die in which case the outcome is limited to integers.

When measuring the distance we recognize that any value x_i could be obtained, although experience tells us that values close to 100.00 are most likely. Values such as 99.90 or 100.25 are very unlikely when measured with care. Assume that n

measurements have been made and that they have been grouped into bins of length Δx, with bin i containing n_i observations. Graphing the bins in a coordinate system of relative frequency n_i/n versus x_i gives the histogram. For surveying measurements, the smoothed step function of the rectangular bins typically has a bell-like shape. The maximum occurs around the sample mean. The larger the deviation from the mean, the smaller the relative frequency, i.e., the probability that such a measurement will actually be obtained. A goodness-of-fit test would normally confirm the hypothesis that the observations have a normal distribution. Thus, the typical measurement process in surveying follows the statistical law of normal distribution.

4.1.2 Elementary Statistical Concepts

Several concepts from statistics are required in least-squares adjustment. The following is a partial listing of frequently used concepts and terminology:

- Observation: An observation, or a statistical event, is the outcome of a statistical experiment, e.g., throwing a dice or measuring an angle or a distance.
- Random Variable: A random variable is the outcome of an event. The random variable is denoted by a tilde. Thus, $\tilde{x}$ is a random variable and $\tilde{\mathbf{x}}$ is a vector of random variables. However, we will frequently not use the tilde to simplify the notation when it is unambiguous which symbol represents the random variable.
- Population: The population is the totality of all events. It includes all possible values that the random variable can have. The population is described by a finite set of parameters, called the population parameters. The normal distribution, e.g., describes such a population and is completely specified by the mean and the variance.
- Sample: A sample is a subset of the population. For example, if the same distance is measured ten times, then these ten measurements are a sample of all the possible measurements.
- A statistic represents an estimate of the population parameters or functions of these parameters. It is computed from a sample. For example, the ten measurements of the same distance can be used to estimate the mean and the variance of the normal distribution.
- Probability: Probability is related to the frequency of occurrence of a specific event. Each value of the random variable has an associated probability.
- Probability Density Function: The probability density function relates the probability to the possible values of the random variable. If $f(x)$ denotes the probability density function, then

$$P(a \le \tilde{x} \le b) = \int_a^b f(x)\,dx \tag{4.1}$$

  is the probability that the random variable $\tilde{x}$ assumes a value in the interval $[a, b]$.

4.1.3 Observational Errors

Field observations are not perfect, and neither are the recordings and management of observations. The measurement process suffers from several error sources. Repeated measurements do not yield identical numerical values because of random measurement errors. These errors are usually small, and the probability of a positive or a negative error of a given magnitude is the same (equal frequency of occurrence). Random errors are inherent in the nature of measurements and can never be completely overcome. Random errors are dealt with in least-squares adjustment.

Systematic errors are errors that vary systematically in sign and/or magnitude. Examples are a tape that is 10 cm too short or the failure to correct for vertical or lateral refraction in angular measurement. Systematic errors are particularly dangerous because they tend to accumulate. Adequate instrument calibration, care when observing, such as double centering, and observing under various external conditions help avoid systematic errors. If the errors are known, the observations can be corrected before making the adjustment; otherwise, one might attempt to model and estimate these errors. Discovering and dealing with systematic errors requires a great deal of experience with the data. Success is not at all guaranteed.

Blunders are usually large errors resulting from carelessness. Examples of blunders are counting errors in a whole tape length, transposing digits when recording field observations, continuing measurements after upsetting the tripod, and so on. Blunders can largely be avoided through careful observation, although there can never be absolute certainty that all blunders have been avoided or eliminated. Therefore, an important part of least-squares adjustment is to discover and remove remaining blunders in the observations.

4.1.4 Accuracy and Precision

Accuracy refers to the closeness of the observations (or the quantities derived from the observations) to the true value. Precision refers to the closeness of repeated observations (or quantities derived from repeated sets of observations) to the sample mean. Figure 4.1 shows four density functions that represent four distinctly different measurement processes of the same quantity. Curves 1 and 2 are symmetric with respect to the true value x_T. These measurements have a high accuracy, because the sample mean coincides or is very close to the true value. However, the shapes of both curves are quite different. Curve 1 is tall and narrow, whereas curve 2 is short and broad. The observations of process 1 are clustered closely around the mean (true value), whereas the spread of observations around the mean is larger for process 2. Larger deviations from the true value occur more frequently for process 2 than for process 1. Thus, process 1 is more precise than process 2; however, both processes are equally accurate. Curves 3 and 4 are symmetric with respect to the sample mean x_S, which differs from the true value x_T. Both sequences have equally low accuracy, but the precision of process 3 is higher than that of process 4. The difference $x_T - x_S$ is caused by a systematic error. An increase in the number of observations does not reduce this difference.

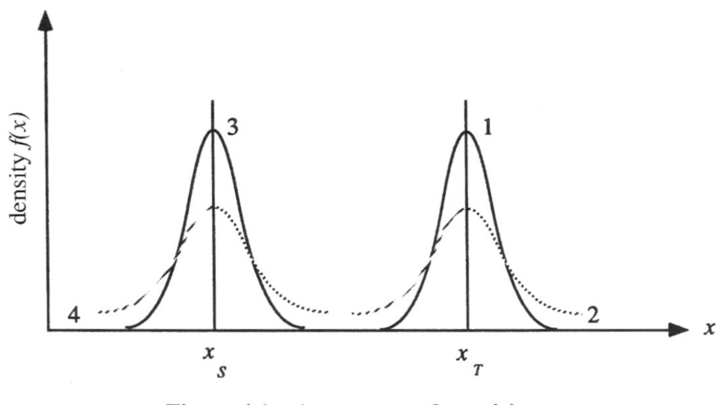

Figure 4.1 Accuracy and precision.

4.2 STOCHASTIC AND MATHEMATICAL MODELS

Least-squares adjustment deals with two equally important components: the stochastic model and the mathematical model. Both components are indispensable and contribute to the adjustment algorithm (see Figure 4.2). We denote the vector of observation with $\boldsymbol{\ell}_b$, and the number of observations by n. The observations are random variables; thus the complete notation for the $n \times 1$ vector of observations is $\tilde{\boldsymbol{\ell}}_b$. To simplify the notation, we do not use the tilde in connection with $\boldsymbol{\ell}_b$. The true value of the observations, i.e., the means of the populations, are estimated from the sample measurements. Since each observation belongs to a different population, the sample size is usually 1. The variances of these distributions comprise the stochastic model. It introduces information about the precision of the observations (or accuracy if only random errors are present). The variance-covariance matrix $\boldsymbol{\Sigma}_{\ell_b}$ expresses the stochastic model. In many cases, the observations are not correlated and the variance-covariance matrix is diagonal. Occasionally, when so-called derived observations are used which are the outcome from a previous adjustment, or when linear combinations of original observations are adjusted, the variance-covariance matrix contains off-diagonal elements. Because in surveying the observations are normal distributed, the vector of observations has a multivariate normal distribution. We use the notation

$$\boldsymbol{\ell}_b \sim N\left(\boldsymbol{\ell}_T, \boldsymbol{\Sigma}_{\ell_b}\right) \tag{4.2}$$

where $\boldsymbol{\ell}_T$ is the vector mean of the population. The cofactor matrix of the observations $\mathbf{Q}_{\ell_b}$ and the weight matrix $\mathbf{P}$ are defined by

$$\mathbf{Q}_{\ell_b} = \frac{1}{\sigma_0^2}\, \boldsymbol{\Sigma}_{\ell_b} \tag{4.3}$$

$$\mathbf{P} = \mathbf{Q}_{\ell_b}^{-1} = \sigma_0^2\, \boldsymbol{\Sigma}_{\ell_b}^{-1} \tag{4.4}$$

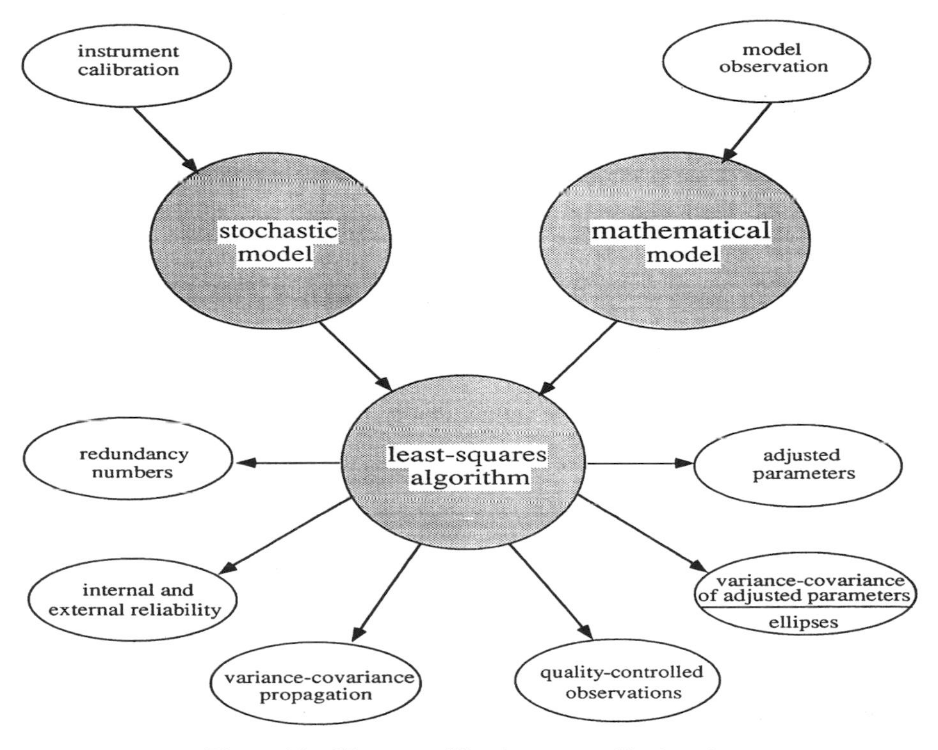

Figure 4.2 Elements of least-squares adjustment.

Typically we do not use a subscript to identify **P** as the weight matrix of the observations. The symbol σ_0^2 denotes the a priori variance of unit weight. It relates the weight matrix and the inverted covariance matrix. An important capability of least-squares adjustment is the estimation of σ_0^2 from observations. We denote that estimate by $\hat{\sigma}_0^2$ and it is the a posteriori variance of unit weight. If the a priori and a posteriori variances of unit weight are statistically equal, the adjustment is said to be correct. More on this fundamental statistical test and its implications will follow in later sections. In general, the a priori variance of unit weight σ_0^2 is set to 1; i.e., the weight matrix is equated with the inverse of the variance-covariance matrix of the observations. The term *variance of unit weight* is derived from the fact that if the variance of an observation equals σ_0^2, then the weight for this observation equals unity. The special cases that **P** equals the identify matrix, $\mathbf{P} = \mathbf{I}$, frequently allow a simple and geometrically intuitive interpretation of the minimization.

The mathematical model expresses a simplification of existing physical reality. It attempts to express mathematically the relations between observations and parameters (unknowns) such as coordinates, heights, and refraction coefficients. Least-squares adjustment is a very general tool that can be used whenever a relationship between observations and parameters has been established. Even though the mathematical model is well known for many routine applications, there are always new cases that require a new mathematical model. Finding the right mathematical model can be a challenge.

Much research has gone into establishing a mathematical formulation that is general enough to deal with all types of globally distributed measurement in a unified model. The collection of observations might include distances, angles, heights, gravity anomalies, gravity gradients, geopotential differences, astronomical observations, and GPS observations. The mathematical models become simpler if one does not deal with all types of observations at the same time but instead uses additional external information. See Chapter 2 for a detailed discussion on the 3D geodetic model.

A popular approach is to reduce (modify) the original observations to be compatible with the mathematical model. These are the model observations. For example, if measured vertical angles are used, the mathematical model must include refraction parameters. On the other hand, the original measurements can be corrected for refraction using an atmospheric refraction model. The thus reduced observations refer to a simpler model that does not require refraction parameters. The more reductions are applied to the original observation, the less general the respective mathematical model is. The final form of the model also depends on the purpose of the adjustment. For example, if the objective is to study refraction, one needs refraction parameters in the model. In surveying applications where the objective typically is to determine location, one prefers not to deal with refraction parameters explicitly. The relation between observations and parameterization is central to the success of estimation and at times requires much attention.

In the most general case, the observations and the parameters are related by an implicit nonlinear function:

$$\mathbf{f}(\mathbf{x}_a, \boldsymbol{\ell}_a) = \mathbf{o} \tag{4.5}$$

This is the mixed adjustment model. The subscript a is to be read as "adjusted." The symbol $\boldsymbol{\ell}_a$ denotes the $n \times 1$ vector of adjusted observations, and the vector $\mathbf{x}_a$ contains u adjusted parameters. There are r nonlinear mathematical functions in $\mathbf{f}$. Often the observations are explicitly related to the parameters, such as in

$$\boldsymbol{\ell}_a = \mathbf{f}(\mathbf{x}_a) \tag{4.6}$$

This is the observation equation model. A further variation is the absence of any parameters as in

$$\mathbf{f}(\boldsymbol{\ell}_a) = \mathbf{o} \tag{4.7}$$

This is the condition equation model.

The application usually dictates which model might be preferred. Selecting another model might require a mathematically more involved formulation. In the case of a leveling network, e.g., the observation equation model and the condition equation model can be applied with equal ease.

The observation equation model has the major advantage in that each observation adds one equation. This allows the observation equation model to be implemented relatively easily and generally in software. One does not have to identify independent loop closures, etc.

Figure 4.2 indicates some of the outcomes from the adjustment. Statistical tests are available to verify the acceptance of the adjustment or aid in discovering and removing blunders. The adjustment provides probability regions for the estimated parameters and allows variance-covariance propagation to determine functions of the estimated parameters and the respective standard deviations. Of particular interest is the ability of the least-squares adjustment to perform internal and external reliability analysis, in order to quantify marginally detectable blunders and to determine their potential influence on the estimated parameters.

Statistical concepts enter the least-squares adjustment in two distinct ways. The actual least-squares solution merely requires the existence of the variance-covariance matrix; there is no need to specify a particular distribution for the observations. If statistical tests are required, then the distribution of the observations must be known. In most cases, one indeed desires to carry out some statistical testing.

4.3 VARIANCE-COVARIANCE PROPAGATION

The purpose of variance-covariance propagation is to compute the variances and covariances of linear functions of random variables. Nonlinear functions must first be linearized. Variance-covariance propagation is applicable to single random variables or to vectors of random variables.

Probability Density and Accumulative Probability For $f(x)$ to be a probability function of the random variable $\tilde{x}$, it has to fulfill certain conditions. First, $f(x)$ must be a nonnegative function, because there is *always* an outcome of an experiment; i.e., the observation can be positive, negative, or even zero. Second, the probability that a sample (observation) is one of all possible outcomes should be 1. Thus the density function $f(x)$ must fulfill the following conditions:

$$f(x) \geq 0 \tag{4.8}$$

$$\int_{-\infty}^{\infty} f(x)\, dx = 1 \tag{4.9}$$

The integration is taken over the whole range (population) of the random variable. Conditions (4.8) and (4.9) imply that the density function is zero at minus infinity and plus infinity. The probability

$$P(\tilde{x} \leq x) = F(x) = \int_{-\infty}^{x} f(t)\, dt \tag{4.10}$$

is called the cumulative distribution function. It is a nondecreasing function because of condition (4.8).

Mean The mean, also called the expected value of a continuously distributed random variable, is defined as

$$\mu_x = E(\tilde{x}) = \int_{-\infty}^{\infty} x\, f(x)\, dx \tag{4.11}$$

The mean is a function of the density function of the random variable. The integration is extended over the whole population. Equation (4.11) is the analogy to the weighted mean in the case of discrete distributions.

Variance The variance is defined by

$$\sigma_{\tilde{x}}^2 = E\left(\tilde{x} - \mu_x\right)^2 = \int_{-\infty}^{\infty} \left(x - \mu_x\right)^2 f(x)\, dx \tag{4.12}$$

The variance measures the spread of the probability density in the sense that it gives the expected value of the squared deviations from the mean. A small variance therefore indicates that most of the probability density is located around the mean.

Multivariate Distribution Any function $f(x_1, x_2, \ldots, x_n)$ of n continuous variables $\tilde{x}_i$ can be a joint density function provided that

$$f(x_1, x_2, \ldots, x_n) \geq 0 \tag{4.13}$$

$$\int_{-\infty}^{\infty} \cdots \int_{-\infty}^{\infty} f(x_1, x_2, \ldots, x_n)\; dx_1 \cdots dx_n = 1 \tag{4.14}$$

It follows as a natural extension from (4.10) that

$$P(\tilde{x}_1 < a_1, \ldots, \tilde{x}_n < a_n) = \int_{-\infty}^{a_1} \cdots \int_{-\infty}^{a_n} f(x_1, x_2, \ldots, x_n)\; dx_1 \cdots dx_n \tag{4.15}$$

The marginal density of a subset of random variables $(x_1, x_2, \ldots, x_p)$ is

$$g\left(x_1, x_2, \ldots, x_p\right) = \int_{-\infty}^{\infty} \cdots \int_{-\infty}^{\infty} f(x_1, x_2, \ldots, x_n)\; dx_{p+1}\, dx_{p+2} \cdots dx_n \tag{4.16}$$

Stochastic Independence The concept of stochastic independence is required when dealing with multivariate distributions. Two sets of random variables, $(\tilde{x}_1, \ldots, \tilde{x}_p)$ and $(\tilde{x}_{p+1}, \ldots, \tilde{x}_n)$, are stochastically independent if the joint density function can be written as a product of the two respective marginal density functions, e.g.,

$$f(x_1, x_2, \ldots, x_n) = g_1\left(x_1, x_2, \ldots, x_p\right)\; g_2\left(x_{p+1}, x_{p+2}, \ldots, x_n\right) \tag{4.17}$$

Vector of Means The expected value for the individual parameter x_i is

$$\mu_{x_i} = E(\tilde{x}_i) = \int_{-\infty}^{\infty} \cdots \int_{-\infty}^{\infty} x_i f(x_1, x_2, \ldots, x_n)\ dx_1\ dx_2 \cdots dx_n \tag{4.18}$$

In vector notation the expected values of all parameters are

$$E(\tilde{\mathbf{x}}) = \begin{bmatrix} E(\tilde{x}_1) & \cdots & E(\tilde{x}_n) \end{bmatrix}^{\mathrm{T}} \tag{4.19}$$

Variance The variance of an individual parameter is given by

$$\sigma_{x_i}^2 = E\left(\tilde{x}_i - \mu_{x_i}\right)^2 = \int_{-\infty}^{\infty} \cdots \int_{-\infty}^{\infty} \left(x_i - \mu_{x_i}\right)^2\ f(x_1, x_2, \ldots, x_n)\ dx_1 \cdots dx_n \tag{4.20}$$

Covariance For multivariate distributions, another quantity called the covariance becomes important. The covariance describes the statistical relationship between two random variables. The covariance is

$$\begin{aligned} \sigma_{x_i,x_j} &= E\left[\left(x_i - \mu_{x_i}\right)\left(x_j - \mu_{x_j}\right)\right] \\ &= \int_{-\infty}^{\infty} \cdots \int_{-\infty}^{\infty} \left(x_i - \mu_{x_i}\right)\left(x_j - \mu_{x_j}\right)\ f(x_1, x_2, \ldots, x_n)\ dx_1 \cdots dx_n \end{aligned} \tag{4.21}$$

Whereas the variance is always larger than or equal to zero, the covariance can be negative, positive, or even zero.

Correlation Coefficients The correlation coefficient of two random variables is defined as

$$\rho_{x_i,x_j} = \frac{E\left[\left(\tilde{x}_i - \mu_{x_i}\right)\left(\tilde{x}_j - \mu_{x_j}\right)\right]}{\sigma_{x_i}\sigma_{x_j}} = \frac{\sigma_{x_i,x_j}}{\sigma_{x_i}\sigma_{x_j}} \tag{4.22}$$

Therefore, the correlation coefficient equals the covariance divided by the respective standard deviations. An important property of the correlation coefficient is that

$$-1 \le \rho_{x_i,x_j} \le 1 \tag{4.23}$$

If two random variables are stochastically independent, then the covariance (and thus the correlation coefficient) is zero. By making use of (4.17) for the density function of stochastically independent random variables, we can write (4.21) as

$$\begin{aligned} \sigma_{x_i,x_j} &= \int_{-\infty}^{\infty} \int_{-\infty}^{\infty} \left(x_i - \mu_{x_i}\right)\left(x_j - \mu_{x_j}\right) g_i(x_i) g_j(x_j)\ dx_i\ dx_j \\ &= \int_{-\infty}^{\infty} \left(x_i - \mu_{x_i}\right) g_i(x_i)\ dx_i \int_{-\infty}^{\infty} \left(x_j - \mu_{x_j}\right) g_j(x_j)\ dx_j \end{aligned} \tag{4.24}$$

These integrals are zero because of the definition of the mean. The converse, i.e., zero correlation, implies stochastic independence is valid only for the multivariate normal distribution.

Variance-Covariance Matrix Equations (4.20) and (4.21) can be used to compute the variances and covariances for all components in the random vector $\tilde{\mathbf{x}}$. Arranging the result in the form of a matrix yields the variance-covariance matrix. Thus, for the random vector

$$\tilde{\mathbf{x}} - \boldsymbol{\mu}_x = \begin{bmatrix} \tilde{x}_1 - \mu_{x_1} & \cdots & \tilde{x}_n - \mu_{x_n} \end{bmatrix}^{\mathrm{T}} \tag{4.25}$$

the $(n \times n)$ variance-covariance matrix becomes

$$\boldsymbol{\Sigma}_x = E\left[\left(\tilde{\mathbf{x}} - \boldsymbol{\mu}_x\right)\left(\tilde{\mathbf{x}} - \boldsymbol{\mu}_x\right)^{\mathrm{T}}\right] = \begin{bmatrix} \sigma_{x_1}^2 & \sigma_{x_1,x_2} & \cdots & \sigma_{x_1,x_n} \\ & \sigma_{x_2}^2 & \cdots & \sigma_{x_2,x_n} \\ & & \ddots & \vdots \\ \text{sym} & & & \sigma_{x_n}^2 \end{bmatrix} \tag{4.26}$$

The variance-covariance matrix is symmetric because of (4.21). The expectation operator E is applied to each matrix element. The variance-covariance matrix is simply called the covariance matrix for the sake of brevity. The correlations are computed according to Equation (4.22) and can be arranged in the same order. Thus, the correlation matrix is

$$\mathbf{C} = \begin{bmatrix} 1 & \rho_{x_1,x_2} & \cdots & \rho_{x_1,x_n} \\ & 1 & \cdots & \rho_{x_2,x_n} \\ & & \ddots & \vdots \\ \text{sym} & & & 1 \end{bmatrix} \tag{4.27}$$

The correlation matrix is symmetric, the diagonal elements equal 1, and the off-diagonal elements are between -1 and $+1$.

Propagation Usually we are more interested in a linear function of the random variables than in the random variables themselves. Typical examples are the adjusted coordinates used to compute distances and angles. From the definition of the mean (4.11), it follows that for a constant c

$$E(c) = c \int_{-\infty}^{\infty} f(x)\, dx = c \tag{4.28}$$

and

$$E(c\tilde{x}) = cE(\tilde{x}) \tag{4.29}$$

The expected value (mean) of a constant equals the constant. Because the mean is a constant, it follows that

$$E\left[E(\tilde{x})\right] = \mu_x \tag{4.30}$$

Relations (4.28) and (4.29) also hold for multivariate density functions, as can be seen from (4.18). Let $\tilde{y} = \tilde{x}_1 + \tilde{x}_2$ be a linear function of random variables, then

$$\begin{aligned} E(\tilde{x}_1 + \tilde{x}_2) &= \int_{-\infty}^{\infty}\int_{-\infty}^{\infty} (x_1 + x_2)\, f(x_1, x_2)\; dx_1\, dx_2 \\ &= \int_{-\infty}^{\infty}\int_{-\infty}^{\infty} x_1 f(x_1, x_2)\; dx_1\, dx_2 + \int_{-\infty}^{\infty}\int_{-\infty}^{\infty} x_2 f(x_1, x_2)\; dx_1\, dx_2 \\ &= E(\tilde{x}_1) + E(\tilde{x}_2) \end{aligned} \tag{4.31}$$

Thus, the expected value of the sum of two random variables equals the sum of the individual expected values. By combining (4.28) and (4.31), we can compute the expected value of a general linear function of random variables. Thus, if the elements of the $n \times u$ matrix $\mathbf{A}$ and the $n \times 1$ vector $\mathbf{a}_0$ are constants and

$$\tilde{\mathbf{y}} = \mathbf{a}_0 + \mathbf{A}\tilde{\mathbf{x}} \tag{4.32}$$

then the expected value is

$$E(\tilde{\mathbf{y}}) = \mathbf{a}_0 + \mathbf{A}E(\tilde{\mathbf{x}}) \tag{4.33}$$

This is the law for propagating the mean. The law of variance-covariance propagation is as follows:

$$\begin{aligned} \boldsymbol{\Sigma}_y &\equiv E\left[(\tilde{\mathbf{y}} - \boldsymbol{\mu}_y)(\tilde{\mathbf{y}} - \boldsymbol{\mu}_y)^{\mathrm{T}}\right] \\ &= E\left\{[\tilde{\mathbf{y}} - E(\tilde{\mathbf{y}})][\tilde{\mathbf{y}} - E(\tilde{\mathbf{y}})]^{\mathrm{T}}\right\} \\ &= E\left\{[\tilde{\mathbf{y}} - \mathbf{a}_0 - \mathbf{A}E(\tilde{\mathbf{x}})][\tilde{\mathbf{y}} - \mathbf{a}_0 - \mathbf{A}E(\tilde{\mathbf{x}})]^{\mathrm{T}}\right\} \\ &= E\left\{[\mathbf{A}\tilde{\mathbf{x}} - \mathbf{A}E(\tilde{\mathbf{x}})][\mathbf{A}\tilde{\mathbf{x}} - \mathbf{A}E(\tilde{\mathbf{x}})]^{\mathrm{T}}\right\} \\ &= \mathbf{A}E\left\{[\tilde{\mathbf{x}} - E(\tilde{\mathbf{x}})][\tilde{\mathbf{x}} - E(\tilde{\mathbf{x}})]^{\mathrm{T}}\right\}\mathbf{A}^{\mathrm{T}} \\ &= \mathbf{A}\boldsymbol{\Sigma}_x\mathbf{A}^{\mathrm{T}} \end{aligned} \tag{4.34}$$

The first line in Expression (4.34) is the general expression for the variance-covariance matrix of the random variable $\tilde{\mathbf{y}}$ according to definition (4.26); $\boldsymbol{\mu}_y$ is the expected value of $\tilde{\mathbf{y}}$. The third line follows by substituting (4.33) for the expected value of $\tilde{\mathbf{y}}$. Equation (4.32) has been substituted in the third line for $\tilde{\mathbf{y}}$, and, finally, the $\mathbf{A}$ matrix has been factored out. Thus the variance-covariance matrix of the random variable $\tilde{\mathbf{y}}$ is obtained by pre- and postmultiplying the variance-covariance matrix of the original random variable $\tilde{\mathbf{x}}$ by the coefficient matrix $\mathbf{A}$ and its transpose. The constant term $\mathbf{a}_0$ cancels. This is the law of variance-covariance propagation for linear functions of random variables. The covariance matrix $\boldsymbol{\Sigma}_y$ is a full matrix in general.

For later reference, the expression for the covariance matrix (4.26) can be rewritten as

$$
\begin{aligned}
\boldsymbol{\Sigma}_x &= E\left[\left(\tilde{\mathbf{x}} - \boldsymbol{\mu}_x\right)\left(\tilde{\mathbf{x}} - \boldsymbol{\mu}_x\right)^{\mathrm{T}}\right] \\
&= E\left[\tilde{\mathbf{x}}\,\tilde{\mathbf{x}}^{\mathrm{T}} - \boldsymbol{\mu}_x \boldsymbol{\mu}_x^T\right]
\end{aligned}
\qquad (4.35)
$$

4.4 MIXED ADJUSTMENT MODEL

To simplify the notation, the tilde will not be used in this section to identify random variables. Observations or functions of observations are always random variables. A caret is used to identify quantities estimated by least-squares, i.e., those quantities that are a solution of a specific minimization. Caret quantities are always random variables because they are functions of observations. To simplify the notation even further, the caret symbol is used consistently only in connection with the parameters $\mathbf{x}$.

In the mixed adjustment model, the observations and the parameters are implicitly related. If $\boldsymbol{\ell}_a$ denotes the vector of n adjusted observations and $\mathbf{x}_a$ denotes u adjusted parameters (unknowns), the mathematical model is given by

$$
\mathbf{f}\left(\boldsymbol{\ell}_a, \mathbf{x}_a\right) = \mathbf{o} \qquad (4.36)
$$

The total number of equations in (4.36) is denoted by r. The stochastic model is

$$
\mathbf{P} = \sigma_0^2\, \boldsymbol{\Sigma}_{\ell_b}^{-1} \qquad (4.37)
$$

where $\mathbf{P}$ denotes the $n \times n$ weight matrix, and $\boldsymbol{\Sigma}_{\ell_b}$ denotes the covariance matrix of the observations. The objective is to estimate the parameters. It should be noted that the observations are stochastic (random) variables and that the parameters are deterministic quantities. The parameters exist, but their values are unknown. The estimated parameters, however, will be functions of the observations and therefore random variables.

4.4.1 Linearization

If we let $\mathbf{x}_0$ denote a vector of known approximate values of the parameters, then the parameter corrections $\mathbf{x}$ are

$$\mathbf{x} = \mathbf{x}_a - \mathbf{x}_0 \tag{4.38}$$

If $\boldsymbol{\ell}_b$ denotes the vector of observations, then the residuals are defined by

$$\mathbf{v} = \boldsymbol{\ell}_a - \boldsymbol{\ell}_b \tag{4.39}$$

With (4.38) and (4.39) the mathematical model can be written as

$$\mathbf{f}\,(\boldsymbol{\ell}_b + \mathbf{v}, \mathbf{x}_0 + \mathbf{x}) = \mathbf{o} \tag{4.40}$$

The nonlinear mathematical model is linearized around the known point of expansion $(\boldsymbol{\ell}_b, \mathbf{x}_0)$, giving

$$_r\mathbf{B}_n\,{}_n\mathbf{v}_1 + {}_r\mathbf{A}_u\,{}_u\mathbf{x}_1 + {}_r\mathbf{w}_1 = {}_r\mathbf{o}_1 \tag{4.41}$$

where

$$\mathbf{B} = \left.\frac{\partial \mathbf{f}}{\partial \boldsymbol{\ell}}\right|_{\mathbf{x}_0, \boldsymbol{\ell}_b} \tag{4.42}$$

$$\mathbf{A} = \left.\frac{\partial \mathbf{f}}{\partial \mathbf{x}}\right|_{\mathbf{x}_0, \boldsymbol{\ell}_b} \tag{4.43}$$

$$\mathbf{w} = \mathbf{f}\,(\boldsymbol{\ell}_b, \mathbf{x}_0) \tag{4.44}$$

See Appendix A for linearization of multivariable functions. The coefficient matrices must be evaluated at the point of expansion, which consists of observations and approximate parameters. The discrepancies $\mathbf{w}$ must be evaluated for the same point of expansion. The better the approximate values $\mathbf{x}_0$, the smaller the parameter corrections $\mathbf{x}$.

4.4.2 Minimization and Solution

The least-squares estimate $\hat{\mathbf{x}}$ is based on the minimization of the function $\mathbf{v}^{\mathrm{T}}\mathbf{P}\mathbf{v}$. A solution is obtained by introducing a vector of Lagrange multipliers, $\mathbf{k}$, and minimizing the function

$$\phi(\mathbf{v}, \mathbf{k}, \mathbf{x}) = \mathbf{v}^{\mathrm{T}}\mathbf{P}\mathbf{v} - 2\mathbf{k}^{\mathrm{T}}(\mathbf{B}\mathbf{v} + \mathbf{A}\mathbf{x} + \mathbf{w}) \tag{4.45}$$

Equation (4.45) is a function of three variables, namely, $\mathbf{v}$, $\mathbf{k}$, and $\mathbf{x}$. A necessary condition for the minimum is that the partial derivatives must be zero. It can be readily shown that this condition is also sufficient. Differentiating (4.45) following the rules of Appendix A and setting the partial derivatives to zero gives

$$\frac{1}{2}\frac{\partial \phi}{\partial \mathbf{v}} = \mathbf{P}\hat{\mathbf{v}} - \mathbf{B}^{\mathrm{T}}\hat{\mathbf{k}} = \mathbf{o} \tag{4.46}$$

$$\frac{1}{2}\frac{\partial \phi}{\partial \mathbf{k}} = \mathbf{B}\hat{\mathbf{v}} + \mathbf{A}\hat{\mathbf{x}} + \mathbf{w} = \mathbf{o} \tag{4.47}$$

$$\frac{1}{2}\frac{\partial \phi}{\partial \mathbf{x}} = -\mathbf{A}^{\mathrm{T}}\hat{\mathbf{k}} = \mathbf{o} \tag{4.48}$$

The solution of (4.46) to (4.48) starts with the recognition that $\mathbf{P}$ is a square matrix and can be inverted. Thus, the expression for the residuals follows from (4.46):

$$\hat{\mathbf{v}} = \mathbf{P}^{-1}\mathbf{B}^{\mathrm{T}}\hat{\mathbf{k}} \tag{4.49}$$

Substituting (4.49) into (4.47), we obtain the solution for the Lagrange multiplier:

$$\hat{\mathbf{k}} = -\mathbf{M}^{-1}(\mathbf{A}\hat{\mathbf{x}} + \mathbf{w}) \tag{4.50}$$

with

$$_{r}\mathbf{M}_{r} = {}_{r}\mathbf{B}_{n}\,{}_{n}\mathbf{P}_{n}^{-1}\,{}_{n}\mathbf{B}_{r}^{\mathrm{T}} \tag{4.51}$$

Finally, the estimate $\hat{\mathbf{x}}$ follows from (4.48) and (4.50)

$$\hat{\mathbf{x}} = -\left(\mathbf{A}^{\mathrm{T}}\mathbf{M}^{-1}\mathbf{A}\right)^{-1}\mathbf{A}^{\mathrm{T}}\mathbf{M}^{-1}\mathbf{w} \tag{4.52}$$

The estimates $\hat{\mathbf{x}}$ and $\hat{\mathbf{v}}$ are independent of the a priori variance of unit weight. The first step is to compute the parameters $\hat{\mathbf{x}}$ from (4.52), then the Lagrange multipliers $\hat{\mathbf{k}}$ from (4.50), followed by the residuals $\hat{\mathbf{v}}$ (4.49). The adjusted parameters and adjusted observations follow from (4.38) and (4.39).

The caret symbol in $\hat{\mathbf{v}}$, $\hat{\mathbf{k}}$, and $\hat{\mathbf{x}}$ indicates that all three estimated values follow from minimizing of $\mathbf{v}^{\mathrm{T}}\mathbf{P}\mathbf{v}$. However, as stated earlier, the caret is only used consistently for the estimated parameters $\hat{\mathbf{x}}$ in order to simplify the notation.

4.4.3 Cofactor Matrices

Equation (4.44) shows that $\mathbf{w}$ is a random variable because it is a function of the observation $\boldsymbol{\ell}_b$. With (4.37), the law of variance-covariance propagation (4.34), and the use of $\mathbf{B}$ in (4.42), the cofactor matrix $\mathbf{Q}_w$ becomes

$$\mathbf{Q}_w = \mathbf{B}\mathbf{P}^{-1}\mathbf{B}^{\mathrm{T}} = \mathbf{M} \tag{4.53}$$

From (4.53) and (4.52) it follows that

$$\mathbf{Q}_x = \left(\mathbf{A}^{\mathrm{T}}\mathbf{M}^{-1}\mathbf{A}\right)^{-1} \tag{4.54}$$

Combining (4.49) through (4.52) the expression for the residuals becomes

$$\mathbf{v} = \left[\mathbf{P}^{-1}\mathbf{B}^{\mathrm{T}}\mathbf{M}^{-1}\mathbf{A}\left(\mathbf{A}^{\mathrm{T}}\mathbf{M}^{-1}\mathbf{A}\right)^{-1}\mathbf{A}^{\mathrm{T}}\mathbf{M}^{-1} - \mathbf{P}^{-1}\mathbf{B}^{\mathrm{T}}\mathbf{M}^{-1}\right]\mathbf{w} \tag{4.55}$$

It follows from the law of variance propagation (4.34) and (4.53) that

$$\mathbf{Q}_v = \mathbf{P}^{-1}\mathbf{B}^{\mathrm{T}}\mathbf{M}^{-1}\left[\mathbf{M} - \mathbf{A}\left(\mathbf{A}^{\mathrm{T}}\mathbf{M}^{-1}\mathbf{A}\right)^{-1}\mathbf{A}^{\mathrm{T}}\right]\mathbf{M}^{-1}\mathbf{B}\mathbf{P}^{-1} \tag{4.56}$$

The adjusted observations are

$$\begin{aligned} \boldsymbol{\ell}_a &= \boldsymbol{\ell}_b + \mathbf{v} \\ &= \boldsymbol{\ell}_b + \left[\mathbf{P}^{-1}\mathbf{B}^{\mathrm{T}}\mathbf{M}^{-1}\mathbf{A}\left(\mathbf{A}^{\mathrm{T}}\mathbf{M}^{-1}\mathbf{A}\right)^{-1}\mathbf{A}^{\mathrm{T}}\mathbf{M}^{-1} - \mathbf{P}^{-1}\mathbf{B}^{\mathrm{T}}\mathbf{M}^{-1}\right]\mathbf{w} \end{aligned} \tag{4.57}$$

Because

$$\frac{\partial \boldsymbol{\ell}_a}{\partial \boldsymbol{\ell}_b} = \mathbf{I} + \mathbf{P}^{-1}\mathbf{B}^{\mathrm{T}}\mathbf{M}^{-1}\mathbf{A}\left(\mathbf{A}^{\mathrm{T}}\mathbf{M}^{-1}\mathbf{A}\right)^{-1}\mathbf{A}^{\mathrm{T}}\mathbf{M}^{-1}\mathbf{B} - \mathbf{P}^{-1}\mathbf{B}^{\mathrm{T}}\mathbf{M}^{-1}\mathbf{B} \tag{4.58}$$

it follows that

$$\mathbf{Q}_{\ell_a} = \mathbf{Q}_{\ell_b} - \mathbf{Q}_v \tag{4.59}$$

where the inverse of $\mathbf{P}$ has been replaced by $\mathbf{Q}_{\ell_b}$ according to (4.4)

4.4.4 A Posteriori Variance of Unit Weight

The minimum of $\mathbf{v}^{\mathrm{T}}\mathbf{P}\mathbf{v}$ follows from (4.49), (4.50), and (4.52) as

$$\mathbf{v}^{\mathrm{T}}\mathbf{P}\mathbf{v} = \mathbf{w}^{\mathrm{T}}\left[\mathbf{M}^{-1} - \mathbf{M}^{-1}\mathbf{A}\left(\mathbf{A}^{\mathrm{T}}\mathbf{M}^{-1}\mathbf{A}\right)^{-1}\mathbf{A}^{\mathrm{T}}\mathbf{M}^{-1}\right]\mathbf{w} \tag{4.60}$$

The expected value of this random variable is

$$\begin{aligned} E\left(\mathbf{v}^{\mathrm{T}}\mathbf{P}\mathbf{v}\right) &= E\left(\mathrm{Tr}\ \mathbf{v}^{\mathrm{T}}\mathbf{P}\mathbf{v}\right) \\ &= E\left\{\mathrm{Tr}\left[\mathbf{w}^{\mathrm{T}}\left(\mathbf{M}^{-1} - \mathbf{M}^{-1}\mathbf{A}\left(\mathbf{A}^{\mathrm{T}}\mathbf{M}^{-1}\mathbf{A}\right)^{-1}\mathbf{A}^{\mathrm{T}}\mathbf{M}^{-1}\right)\mathbf{w}\right]\right\} \\ &= E\left\{\mathrm{Tr}\left[\left(\mathbf{M}^{-1} - \mathbf{M}^{-1}\mathbf{A}\left(\mathbf{A}^{\mathrm{T}}\mathbf{M}^{-1}\mathbf{A}\right)^{-1}\mathbf{A}^{\mathrm{T}}\mathbf{M}^{-1}\right)\mathbf{w}\mathbf{w}^{\mathrm{T}}\right]\right\} \\ &= \mathrm{Tr}\left\{\left[\mathbf{M}^{-1} - \mathbf{M}^{-1}\mathbf{A}\left(\mathbf{A}^{\mathrm{T}}\mathbf{M}^{-1}\mathbf{A}\right)^{-1}\mathbf{A}^{\mathrm{T}}\mathbf{M}^{-1}\right]E\left(\mathbf{w}\mathbf{w}^{\mathrm{T}}\right)\right\} \end{aligned} \tag{4.61}$$

The trace (Tr) of a matrix equals the sum of its diagonal elements. In the first part of (4.61), the property that the trace of a 1×1 matrix equals the matrix element itself is used. Next, the matrix products are switched, leaving the trace invariant. In the last part of the equation, the expected operator and the trace are switched. The expected value $E(\mathbf{w}\mathbf{w}^{\mathrm{T}})$ can be readily computed. Per definition, the expected value of the residuals

$$E(\mathbf{v}) = \mathbf{o} \tag{4.62}$$

is zero because the residuals represent random errors for which positive and negative errors of the same magnitude occur with the same probability. It follows from (4.41) that

$$E(\mathbf{w}) = -\mathbf{A}\mathbf{x} \tag{4.63}$$

Note that $\mathbf{x}$ in (4.63) or (4.41) is not a random variable. In this expression, $\mathbf{x}$ simply denotes the vector of unknown parameters that have fixed values, even though the values are not known. The estimate $\hat{\mathbf{x}}$ is a random variable because it is a function of the observations. By using (4.35) for the definition of the covariance matrix (4.53) and using (4.63), it follows that

$$\begin{aligned} E\left(\mathbf{w}\mathbf{w}^{\mathrm{T}}\right) &= \mathbf{\Sigma}_w + E(\mathbf{w})E(\mathbf{w})^{\mathrm{T}} \\ &= \sigma_0^2\,\mathbf{M} + \mathbf{A}\mathbf{x}\mathbf{x}^{\mathrm{T}}\,\mathbf{A}^{\mathrm{T}} \end{aligned} \tag{4.64}$$

Substituting (4.64) into (4.61) yields the expected value for $\mathbf{v}^{\mathrm{T}}\,\mathbf{P}\mathbf{v}$:

$$\begin{aligned} E\left(\mathbf{v}^{\mathrm{T}}\mathbf{P}\mathbf{v}\right) &= \sigma_0^2\ \mathrm{Tr}\left\{{}_r\mathbf{I}_r - \mathbf{M}^{-1}\mathbf{A}\left(\mathbf{A}^{\mathrm{T}}\mathbf{M}^{-1}\mathbf{A}\right)^{-1}\mathbf{A}^{\mathrm{T}}\right\} \\ &= \sigma_0^2\ (r-u) \end{aligned} \tag{4.65}$$

The difference $r - u$ is called the degree of freedom and equals the number of redundant equations in the model (4.36). Strictly, the degree of freedom is $r - R(\mathbf{A})$ because the second matrix in (4.65) is idempotent. The symbol $R(\mathbf{A})$ denotes the rank of the matrix $\mathbf{A}$. The a posteriori variance of unit weight is computed from

$$\hat{\sigma}_0^2 = \frac{\hat{\mathbf{v}}^{\mathrm{T}}\,\mathbf{P}\hat{\mathbf{v}}}{r-u} \tag{4.66}$$

Using (4.65), we see that

$$E\left(\hat{\sigma}_0^2\right) = \sigma_0^2 \tag{4.67}$$

The expected value of the a posteriori variance of unit weight equals the a priori variance of unit weight.

Finally, the estimated covariance matrices are

$$\mathbf{\Sigma}_x = \hat{\sigma}_0^2\,\mathbf{Q}_x \tag{4.68}$$

$$\mathbf{\Sigma}_v = \hat{\sigma}_0^2\,\mathbf{Q}_v \tag{4.69}$$

$$\mathbf{\Sigma}_{\ell_a} = \hat{\sigma}_0^2\,\mathbf{Q}_{\ell_a} \tag{4.70}$$

With Equation (4.59) it follows that

$$\mathbf{\Sigma}_{\ell_a} = \mathbf{\Sigma}_{\ell_b} - \mathbf{\Sigma}_v \tag{4.71}$$

Because the diagonal elements of all three covariance matrices in (4.71) are positive, it follows that the variances of the adjusted observations are smaller than those of the original observations. The difference is a function of the geometry of the adjustment, as implied by the covariance matrix $\boldsymbol{\Sigma}_v$.

4.4.5 Iterations

Because the mathematical model is generally nonlinear, the least-squares solution must be iterated. Recall that (4.36) is true only for $(\boldsymbol{\ell}_a, \mathbf{x}_a)$. Since neither of these quantities is known before the adjustment, the initial point of expansion is chosen as $(\boldsymbol{\ell}_b, \mathbf{x}_0)$. For the ith iteration, the linearized model can be written

$$\mathbf{B}_{x_{0i},\boldsymbol{\ell}_{0i}}\, \bar{\mathbf{v}}_i + \mathbf{A}_{x_{0i},\boldsymbol{\ell}_{0i}}\, \mathbf{x}_i + \mathbf{w}_{x_{0i},\boldsymbol{\ell}_{0i}} = \mathbf{o} \tag{4.72}$$

where the point of expansion $(\boldsymbol{\ell}_{0i}, \mathbf{x}_{0i})$ represents the previous solution. The symbols $\boldsymbol{\ell}_{ai}$ and $\mathbf{x}_{ai}$ denote the adjusted observations and adjusted parameters for the current (ith) solution. They are computed from

$$\bar{\mathbf{v}}_i = \boldsymbol{\ell}_{ai} - \boldsymbol{\ell}_{0i} \tag{4.73}$$

$$\mathbf{x}_i = \mathbf{x}_{ai} - \mathbf{x}_{0i} \tag{4.74}$$

once the least-squares solution of (4.72) has been obtained. The iteration starts with $\boldsymbol{\ell}_{01} = \boldsymbol{\ell}_b$ and $\mathbf{x}_{01} = \mathbf{x}_0$. If the adjustment converges properly, then both $\bar{\mathbf{v}}_i$ and $\mathbf{x}_i$ converge to zero, or, stated differently, $\boldsymbol{\ell}_{ai}$ and $\mathbf{x}_{ai}$ converge toward $\boldsymbol{\ell}_a$ and $\mathbf{x}_a$, respectively. The quantity $\bar{\mathbf{v}}_i$ does not equal the residuals. The residuals express the random difference between the adjusted observations and the original observations according to Equation (4.39). Defining

$$\mathbf{v}_i = \boldsymbol{\ell}_{ai} - \boldsymbol{\ell}_b \tag{4.75}$$

it follows from (4.73) that

$$\bar{\mathbf{v}}_i = \mathbf{v}_i + (\boldsymbol{\ell}_b - \boldsymbol{\ell}_{0i}) \tag{4.76}$$

Substituting this expression into (4.72) gives

$$\mathbf{B}_{x_{0i},\boldsymbol{\ell}_{01}}\, \mathbf{v}_i + \mathbf{A}_{x_{0i},\boldsymbol{\ell}_{01}} \mathbf{x}_i + \mathbf{w}_{x_{0i},\boldsymbol{\ell}_{01}} + \mathbf{B}_{x_{0i},\boldsymbol{\ell}_{01}}\, (\boldsymbol{\ell}_b - \boldsymbol{\ell}_{0i}) = \mathbf{o} \tag{4.77}$$

The formulation (4.77) assures that the vector $\mathbf{v}_i$ converges toward the vector of residuals $\mathbf{v}$. The last term in (4.77) will be zero for the first iteration when $\boldsymbol{\ell}_{0i} = \boldsymbol{\ell}_b$. The iteration has converged if

$$\left| \mathbf{v}^{\mathrm{T}}\mathbf{P}\mathbf{v}_i - \mathbf{v}^{\mathrm{T}}\mathbf{P}\mathbf{v}_{i-1} \right| < \varepsilon \tag{4.78}$$

where ε is a small positive number.

4.5 OBSERVATION AND CONDITION EQUATION MODELS

Often there is an explicit relationship between the observations and the parameters, such as

$$\boldsymbol{\ell}_a = \mathbf{f}(\mathbf{x}_a) \tag{4.79}$$

This is the observation equation model. Comparing both mathematical models (4.36) and (4.79), and taking the definition of the matrix **B** (4.42) into account, we see that the observation equation model follows from the mixed model using the specification

$$\mathbf{B} \equiv -\mathbf{I} \tag{4.80}$$

$$\boldsymbol{\ell} \equiv \mathbf{w} = \mathbf{f}(\mathbf{x}_0) - \boldsymbol{\ell}_b = \boldsymbol{\ell}_0 - \boldsymbol{\ell}_b \tag{4.81}$$

It is customary to denote the discrepancy by $\boldsymbol{\ell}$ instead of $\mathbf{w}$ when dealing with the observation equation model. The symbol $\boldsymbol{\ell}_0$ equals the value of the observations as computed from the approximate parameters $\mathbf{x}_0$. The point of expansion for the linearization is $\mathbf{x}_0$; the observation vector is not involved in the iteration because of the explicit form of (4.79). The linearized equations

$${}_n\mathbf{v}_1 = {}_n\mathbf{A}_u\, {}_u\mathbf{x}_1 + {}_n\boldsymbol{\ell}_1 \tag{4.82}$$

are the *observation equations*. There is one equation for each observation in (4.82).

If the observations are related by a nonlinear function without use of parameters, we speak of the condition equation model. It is written as

$$\mathbf{f}(\boldsymbol{\ell}_a) = \mathbf{o} \tag{4.83}$$

By comparing this with the mixed model (4.36), and applying the definition of the **A** matrix (4.43) we see that the condition equation model follows upon the specification

$$\mathbf{A} = \mathbf{O} \tag{4.84}$$

The linear equations

$${}_r\mathbf{B}_n\, {}_n\mathbf{v}_1 + {}_r\mathbf{w}_1 = \mathbf{o} \tag{4.85}$$

are called the condition equations. The iteration for the model (4.85) is analogous to a mixed model with the added simplification that there is no **A** matrix and no parameter vector **x**.

The significance of these three models (observation, condition, and mixed) is that a specific adjustment problem can usually be formulated more easily in one of the models. Clearly, that model should be chosen. There are situations in which it is equally easy to use any of the models. A typical example is the adjustment of a level

TABLE 4.1 Three Adjustment Models

	Mixed Model	Observation Model	Condition Model
Nonlinear model	$\mathbf{f}(\boldsymbol{\ell}_a, \mathbf{x}_a) = \mathbf{o}$	$\boldsymbol{\ell}_a = \mathbf{f}(\mathbf{x}_a)$	$\mathbf{f}(\boldsymbol{\ell}_a) = \mathbf{o}$
Specifications		$\mathbf{B} = -\mathbf{I}, \quad \boldsymbol{\ell} = \mathbf{w},$ $r = n$	$\mathbf{A} = \mathbf{O}$
Linear model	$\mathbf{Bv} + \mathbf{Ax} + \mathbf{w} = \mathbf{o}$	$\mathbf{v} = \mathbf{Ax} + \boldsymbol{\ell}$	$\mathbf{Bv} + \mathbf{w} = \mathbf{o}$
Normal equation elements	$\mathbf{M} = \mathbf{BP}^{-1}\mathbf{B}^T$ $\mathbf{N} = \mathbf{A}^T\mathbf{M}^{-1}\mathbf{A}$ $\mathbf{u} = \mathbf{A}^T\mathbf{M}^{-1}\mathbf{w}$	$\mathbf{M} = \mathbf{P}^{-1}$ $\mathbf{N} = \mathbf{A}^T\mathbf{PA}$ $\mathbf{u} = \mathbf{A}^T\mathbf{P}\boldsymbol{\ell}$	$\mathbf{M} = \mathbf{BP}^{-1}\mathbf{B}^T$
Normal equations	$\mathbf{N}\hat{\mathbf{x}} = -\mathbf{u}$	$\mathbf{N}\hat{\mathbf{x}} = -\mathbf{u}$	
Minimum $\mathbf{v}^T\mathbf{Pv}$	$\mathbf{v}^T\mathbf{Pv} = -\mathbf{u}^T\mathbf{N}^{-1}\mathbf{u} + \mathbf{w}^T\mathbf{M}^{-1}\mathbf{w}$	$\mathbf{v}^T\mathbf{Pv} = -\mathbf{u}^T\mathbf{N}^{-1}\mathbf{u} + \boldsymbol{\ell}^T\mathbf{P}\boldsymbol{\ell}$	$\mathbf{v}^T\mathbf{Pv} = \mathbf{w}^T\mathbf{M}^{-1}\mathbf{w}$
Estimated parameters	$\hat{\mathbf{x}} = -\mathbf{N}^{-1}\mathbf{u}$	$\hat{\mathbf{x}} = -\mathbf{N}^{-1}\mathbf{u}$	—
Estimated residuals	$\hat{\mathbf{v}} = \mathbf{P}^{-1}\mathbf{B}^T\hat{\mathbf{k}}$	$\hat{\mathbf{v}} = \mathbf{A}\hat{\mathbf{x}} + \boldsymbol{\ell}$	$\hat{\mathbf{v}} = \mathbf{P}^{-1}\mathbf{B}^T\hat{\mathbf{k}}$
Estimated variance of unit weight	$\hat{\sigma}_0^2 = \dfrac{\hat{\mathbf{v}}^T\mathbf{P}\hat{\mathbf{v}}}{r-u}$	$\hat{\sigma}_0^2 = \dfrac{\hat{\mathbf{v}}^T\mathbf{P}\hat{\mathbf{v}}}{n-u}$	$\hat{\sigma}_0^2 = \dfrac{\hat{\mathbf{v}}^T\mathbf{P}\hat{\mathbf{v}}}{r}$
Estimated parameter cofactor matrix	$\mathbf{Q}_x = \mathbf{N}^{-1}$	$\mathbf{Q}_x = \mathbf{N}^{-1}$	—
Estimated residual cofactor matrix	$\mathbf{Q}_v = \mathbf{P}^{-1}\mathbf{B}^T\mathbf{M}^{-1}(\mathbf{M} - \mathbf{AN}^{-1}\mathbf{A}^T)\mathbf{M}^{-1}\mathbf{BP}^{-1}$	$\mathbf{Q}_v = \mathbf{P}^{-1} - \mathbf{AN}^{-1}\mathbf{A}^T$	$\mathbf{Q}_v = \mathbf{P}^{-1}\mathbf{B}^T\mathbf{M}^{-1}\mathbf{BP}^{-1}$
Adjusted observation cofactor matrix	$\mathbf{Q}_{\ell_a} = \mathbf{Q}_{\ell_b} - \mathbf{Q}_v$	$\mathbf{Q}_{\ell_a} = \mathbf{Q}_{\ell_b} - \mathbf{Q}_v$	$\mathbf{Q}_{\ell_a} = \mathbf{Q}_{\ell_b} - \mathbf{Q}_v$

network. Most of the time, however, the observation equation model is preferred, because the simple rule "one observation, one equation" is suitable for setting up general software. Table 4.1 lists the important expressions for all three models.

4.6 SEQUENTIAL SOLUTION

Assume that observations are made in two groups, with the second group consisting of one or several observations. Both groups have a common set of parameters. The two mixed adjustment models can be written as

$$\mathbf{f}_1\left(\boldsymbol{\ell}_{1a}, \mathbf{x}_a\right) = \mathbf{o} \tag{4.86}$$

$$\mathbf{f}_2\left(\boldsymbol{\ell}_{2a}, \mathbf{x}_a\right) = \mathbf{o} \tag{4.87}$$

Both sets of observations should be uncorrelated, and the a priori variance of unit weight should be the same for both groups; i.e.,

$$\mathbf{P} = \begin{bmatrix} \mathbf{P}_1 & \mathbf{O} \\ \mathbf{O} & \mathbf{P}_2 \end{bmatrix} = \sigma_0^2 \begin{bmatrix} \boldsymbol{\Sigma}_1^{-1} & \mathbf{O} \\ \mathbf{O} & \boldsymbol{\Sigma}_2^{-1} \end{bmatrix} \tag{4.88}$$

The number of observations in $\boldsymbol{\ell}_{1a}$ and $\boldsymbol{\ell}_{2a}$ are n_1 and n_2, respectively; and r_1 and r_2 are the number of equations in the models $\mathbf{f}_1$ and $\mathbf{f}_2$, respectively. The linearization of (4.86) and (4.87) yields

$$\mathbf{B}_1\mathbf{v}_1 + \mathbf{A}_1\mathbf{x} + \mathbf{w}_1 = \mathbf{o} \tag{4.89}$$

$$\mathbf{B}_2\mathbf{v}_2 + \mathbf{A}_2\mathbf{x} + \mathbf{w}_2 = \mathbf{o} \tag{4.90}$$

where

$$\left.\begin{aligned} \mathbf{B}_1 &= \left.\frac{\partial \mathbf{f}_1}{\partial \boldsymbol{\ell}_1}\right|_{\boldsymbol{\ell}_{1b}, x_0} & \mathbf{A}_1 &= \left.\frac{\partial \mathbf{f}_1}{\partial \mathbf{x}}\right|_{\boldsymbol{\ell}_{1b}, x_0} & \mathbf{w}_1 &= \mathbf{f}_1(\boldsymbol{\ell}_{1b}, \mathbf{x}_0) \\ \mathbf{B}_2 &= \left.\frac{\partial \mathbf{f}_2}{\partial \boldsymbol{\ell}_2}\right|_{\boldsymbol{\ell}_{2b}, x_0} & \mathbf{A}_2 &= \left.\frac{\partial \mathbf{f}_2}{\partial \mathbf{x}}\right|_{\boldsymbol{\ell}_{2b}, x_0} & \mathbf{w}_2 &= \mathbf{f}_2(\boldsymbol{\ell}_{2b}, \mathbf{x}_0) \end{aligned}\right\} \tag{4.91}$$

The function to be minimized is

$$\begin{aligned} \phi\left(\mathbf{v}_1, \mathbf{v}_2, \mathbf{k}_1, \mathbf{k}_2, \mathbf{x}\right) = \mathbf{v}_1^{\mathrm{T}}\mathbf{P}_1\mathbf{v}_1 + \mathbf{v}_2^{\mathrm{T}}\mathbf{P}_2\mathbf{v}_2 - 2\mathbf{k}_1^{\mathrm{T}}\left(\mathbf{B}_1\mathbf{v}_1 + \mathbf{A}_1\mathbf{x} + \mathbf{w}_1\right) \\ - 2\mathbf{k}_2^{\mathrm{T}}\left(\mathbf{B}_2\mathbf{v}_2 + \mathbf{A}_2\mathbf{x} + \mathbf{w}_2\right) \end{aligned} \tag{4.92}$$

The solution is obtained by setting the partial derivatives of (4.92) to zero,

$$\frac{1}{2}\frac{\partial\phi}{\partial \mathbf{v}_1} = \mathbf{P}_1\mathbf{v}_1 - \mathbf{B}_1^{\mathrm{T}}\mathbf{k}_1 = \mathbf{o} \tag{4.93}$$

$$\frac{1}{2}\frac{\partial\phi}{\partial \mathbf{v}_2} = \mathbf{P}_2\mathbf{v}_2 - \mathbf{B}_2^{\mathrm{T}}\mathbf{k}_2 = \mathbf{o} \tag{4.94}$$

$$\frac{1}{2}\frac{\partial\phi}{\partial \mathbf{x}} = -\mathbf{A}_1^{\mathrm{T}}\mathbf{k}_1 - \mathbf{A}_2^{\mathrm{T}}\mathbf{k}_2 = \mathbf{o} \tag{4.95}$$

$$\frac{1}{2}\frac{\partial\phi}{\partial \mathbf{k}_1} = \mathbf{B}_1\mathbf{v}_1 + \mathbf{A}_1\hat{\mathbf{x}} + \mathbf{w}_1 = \mathbf{o} \tag{4.96}$$

$$\frac{1}{2}\frac{\partial\phi}{\partial \mathbf{k}_2} = \mathbf{B}_2\mathbf{v}_2 + \mathbf{A}_2\hat{\mathbf{x}} + \mathbf{w}_2 = \mathbf{o} \tag{4.97}$$

and solving for $\mathbf{v}_1$, $\mathbf{v}_2$, $\mathbf{k}_1$, $\mathbf{k}_2$ and $\mathbf{x}$. Equations (4.93) and (4.94) give the residuals

$$\mathbf{v}_1 = \mathbf{P}_1^{-1}\mathbf{B}_1^{\mathrm{T}}\mathbf{k}_1 \tag{4.98}$$

$$\mathbf{v}_2 = \mathbf{P}_2^{-1}\mathbf{B}_2^{\mathrm{T}}\mathbf{k}_2 \tag{4.99}$$

Combining (4.98) and (4.96) yields

$$\mathbf{M}_1\mathbf{k}_1 + \mathbf{A}_1\hat{\mathbf{x}} + \mathbf{w}_1 = \mathbf{o} \tag{4.100}$$

where

$$\mathbf{M}_1 = \mathbf{B}_1\mathbf{P}_1^{-1}\mathbf{B}_1^{\mathrm{T}} \tag{4.101}$$

is an $r_1 \times r_1$ symmetric matrix. The Lagrange multiplier becomes

$$\mathbf{k}_1 = -\mathbf{M}_1^{-1}\mathbf{A}_1\hat{\mathbf{x}} - \mathbf{M}_1^{-1}\mathbf{w}_1 \tag{4.102}$$

Equations (4.95) and (4.97) become, after combination with (4.102) and (4.99),

$$\mathbf{A}_1^{\mathrm{T}}\mathbf{M}_1^{-1}\mathbf{A}_1\hat{\mathbf{x}} + \mathbf{A}_1^{\mathrm{T}}\mathbf{M}_1^{-1}\mathbf{w}_1 - \mathbf{A}_2^{\mathrm{T}}\mathbf{k}_2 = \mathbf{o} \tag{4.103}$$

$$\mathbf{B}_2\mathbf{P}_2^{-1}\mathbf{B}_2^{\mathrm{T}}\mathbf{k}_2 + \mathbf{A}_2\hat{\mathbf{x}} + \mathbf{w}_2 = \mathbf{o} \tag{4.104}$$

By using

$$\mathbf{M}_2 = \mathbf{B}_2\mathbf{P}_2^{-1}\mathbf{B}_2^{\mathrm{T}} \tag{4.105}$$

we can write both Equations (4.103) and (4.104) in matrix form:

$$\begin{bmatrix} \mathbf{A}_1^T \mathbf{M}_1^{-1} \mathbf{A}_1 & \mathbf{A}_2^T \\ \mathbf{A}_2 & -\mathbf{M}_2 \end{bmatrix} \begin{bmatrix} \hat{\mathbf{x}} \\ -\mathbf{k}_2 \end{bmatrix} = \begin{bmatrix} -\mathbf{A}_1^T \mathbf{M}_1^{-1} \mathbf{w}_1 \\ -\mathbf{w}_2 \end{bmatrix} \tag{4.106}$$

Equation (4.106) shows how the normal matrix of the first group must be augmented in order to find the solution of both groups. The whole matrix can be inverted in one step to give the solution for $\hat{\mathbf{x}}$ and $\mathbf{k}_2$. Alternatively, one can compute the inverse using the matrix partitioning techniques of Section A.3.5, giving

$$\hat{\mathbf{x}} = -\mathbf{Q}_{11} \mathbf{A}_1^T \mathbf{M}_1^{-1} \mathbf{w}_1 - \mathbf{Q}_{12} \mathbf{w}_2 \tag{4.107}$$

$$\mathbf{k}_2 = \mathbf{Q}_{21} \mathbf{A}_1^T \mathbf{M}_1^{-1} \mathbf{w}_1 + \mathbf{Q}_{22} \mathbf{w}_2 \tag{4.108}$$

Setting

$$\mathbf{N}_1 = \mathbf{A}_1^T \mathbf{M}_1^{-1} \mathbf{A}_1 \tag{4.109}$$

$$\mathbf{N}_2 = \mathbf{A}_2^T \mathbf{M}_2^{-1} \mathbf{A}_2 \tag{4.110}$$

then

$$\mathbf{Q}_x \equiv \mathbf{Q}_{11} = \left(\mathbf{N}_1 + \mathbf{N}_2\right)^{-1} = \mathbf{N}_1^{-1} - \mathbf{N}_1^{-1} \mathbf{A}_2^T \left[\mathbf{M}_2 + \mathbf{A}_2 \mathbf{N}_1^{-1} \mathbf{A}_2^T\right]^{-1} \mathbf{A}_2 \mathbf{N}_1^{-1} \tag{4.111}$$

$$\mathbf{Q}_{12} = \mathbf{Q}_{21}^T = \mathbf{N}_1^{-1} \mathbf{A}_2^T \left[\mathbf{M}_2 + \mathbf{A}_2 \mathbf{N}_1^{-1} \mathbf{A}_2^T\right]^{-1} \tag{4.112}$$

$$\mathbf{Q}_{22} = -\left[\mathbf{M}_2 + \mathbf{A}_2 \mathbf{N}_1^{-1} \mathbf{A}_2^T\right]^{-1} \tag{4.113}$$

Substituting $\mathbf{Q}_{11}$ and $\mathbf{Q}_{12}$ into (4.107) gives the sequential solution for the parameters. We denote the solution of the first group by an asterisk and the contribution of the second group by Δ. In that notation, the estimated parameters of the first group are denoted by $\hat{\mathbf{x}}^*$, which is simplified to $\mathbf{x}^*$. Thus,

$$\hat{\mathbf{x}} = \mathbf{x}^* + \Delta \mathbf{x} \tag{4.114}$$

Comparing (4.107) and (4.52) the sequential solution becomes

$$\mathbf{x}^* = -\mathbf{N}_1^{-1} \mathbf{A}_1^T \mathbf{M}_1^{-1} \mathbf{w}_1 \tag{4.115}$$

and

$$\Delta \mathbf{x} = -\mathbf{N}_1^{-1} \mathbf{A}_2^T \left[\mathbf{M}_2 + \mathbf{A}_2 \mathbf{N}_1^{-1} \mathbf{A}_2^T\right]^{-1} \left(\mathbf{A}_2 \mathbf{x}^* + \mathbf{w}_2\right) \tag{4.116}$$

Similarly, the expression for the Lagrange multiplier $\mathbf{k}_2$ is

$$\mathbf{k}_2 = -\left[\mathbf{M}_2 + \mathbf{A}_2 \mathbf{N}_1^{-1} \mathbf{A}_2^T\right]^{-1} \left(\mathbf{A}_2 \mathbf{x}^* + \mathbf{w}_2\right) \tag{4.117}$$

A different form for the solution of the augmented system (4.106) is obtained by using alternative relations of the matrix partitioning inverse Expressions (A.82) to (A.89). It follows readily that

$$\begin{aligned}\hat{\mathbf{x}} &= -\left(\mathbf{N}_1 + \mathbf{N}_2\right)^{-1}\left(\mathbf{A}_1^T\mathbf{M}_1^{-1}\mathbf{w}_1 + \mathbf{A}_2^T\mathbf{M}_2^{-1}\mathbf{w}_2\right) \\ &= -\left(\mathbf{N}_1 + \mathbf{N}_2\right)^{-1}\left(-\mathbf{N}_1\mathbf{x}^* + \mathbf{A}_2^T\mathbf{M}_2^{-1}\mathbf{w}_2\right) \\ &= \mathbf{x}^* - \left(\mathbf{N}_1 + \mathbf{N}_2\right)^{-1}\left(\mathbf{N}_2\mathbf{x}^* + \mathbf{A}_2^T\mathbf{M}_2^{-1}\mathbf{w}_2\right)\end{aligned} \tag{4.118}$$

The procedure implied by the first line in (4.118) is called the method of adding normal equations. The contributions of the new observations are simply added appropriately.

The cofactor matrix $\mathbf{Q}_x$ of the parameters can be written in sequential form as

$$\begin{aligned}\mathbf{Q}_x &= \mathbf{Q}_{x^*} - \mathbf{Q}_{x^*}\mathbf{A}_2^T\left[\mathbf{M}_2 + \mathbf{A}_2\mathbf{Q}_{x^*}\mathbf{A}_2^T\right]^{-1}\mathbf{A}_2\mathbf{Q}_{x^*} \\ &= \mathbf{Q}_{x^*} + \Delta\mathbf{Q}_x\end{aligned} \tag{4.119}$$

$\mathbf{Q}_{x^*}$ is the cofactor matrix of the first group of observations and equals $\mathbf{N}_1^{-1}$. The contribution of the second group of observations to the cofactor matrix is

$$\Delta\mathbf{Q}_x = -\mathbf{Q}_{x^*}\mathbf{A}_2^T\left[\mathbf{M}_2 + \mathbf{A}_2\mathbf{Q}_{x^*}\mathbf{A}_2^T\right]^{-1}\mathbf{A}_2\mathbf{Q}_{x^*} \tag{4.120}$$

The change $\Delta\mathbf{Q}_x$ can be computed without having the actual observations of the second group. This is relevant in simulation studies.

The computation of $\mathbf{v}^T\mathbf{P}\mathbf{v}$ proceeds as usual

$$\begin{aligned}\mathbf{v}^T\mathbf{P}\mathbf{v} &= \mathbf{v}_1^T\mathbf{P}_1\mathbf{v}_1 + \mathbf{v}_2^T\mathbf{P}_2\mathbf{v}_2 \\ &= -\mathbf{k}_1^T\mathbf{w}_1 - \mathbf{k}_2^T\mathbf{w}_2\end{aligned} \tag{4.121}$$

The second part of (4.121) follows from (4.95) to (4.99). Using (4.102) for $\mathbf{k}_1$, (4.114) for $\hat{\mathbf{x}}$, (4.116) for $\Delta\mathbf{x}$, and (4.117) for $\mathbf{k}_2$, then the sequential solution becomes

$$\begin{aligned}\mathbf{v}^T\mathbf{P}\mathbf{v} &= \mathbf{v}^T\mathbf{P}\mathbf{v}^* + \Delta\mathbf{v}^T\mathbf{P}\mathbf{v} \\ &= \mathbf{v}^T\mathbf{P}\mathbf{v}^* + \left(\mathbf{A}_2\mathbf{x}^* + \mathbf{w}_2\right)^T\left[\mathbf{M}_2 + \mathbf{A}_2\mathbf{N}_1^{-1}\mathbf{A}_2^T\right]^{-1}\left(\mathbf{A}_2\mathbf{x}^* + \mathbf{w}_2\right)\end{aligned} \tag{4.122}$$

with $\mathbf{v}^T\mathbf{P}\mathbf{v}^*$ being obtained from (4.60) for the first group only.

The a posteriori variance of unit weight is computed in the usual way:

$$\hat{\sigma}_0^2 = \frac{\mathbf{v}^T\mathbf{P}\mathbf{v}}{r_1 + r_2 - u} \tag{4.123}$$

where r_1 and r_2 are the number of equations in (4.86) and (4.87), respectively. The letter u denotes, again, the number of parameters.

The second set of observations contributes to all residuals. From (4.98), (4.102), and (4.114) we obtain

$$\begin{aligned} \mathbf{v}_1 &= \mathbf{v}_1^* + \Delta\mathbf{v}_1 \\ &= -\mathbf{P}_1^{-1}\mathbf{B}_1^T\mathbf{M}_1^{-1}\left(\mathbf{A}_1\mathbf{x}^* + \mathbf{w}_1\right) - \mathbf{P}_1^{-1}\mathbf{B}_1^T\mathbf{M}_1^{-1}\mathbf{A}_1\Delta\mathbf{x} \end{aligned} \tag{4.124}$$

The expression for $\mathbf{v}_2$ follows from Equations (4.99) and (4.117):

$$\mathbf{v}_2 = -\mathbf{P}_2^{-1}\mathbf{B}_2^T\mathbf{T}\left(\mathbf{A}_2\mathbf{x}^* + \mathbf{w}_2\right) \tag{4.125}$$

where

$$\mathbf{T} = \left(\mathbf{M}_2 + \mathbf{A}_2\mathbf{N}_1^{-1}\mathbf{A}_2^T\right)^{-1} \tag{4.126}$$

The cofactor matrices for the residuals follow, again, from the law of variance-covariance propagation. The residuals $\mathbf{v}_1$ are a function of $\mathbf{w}_1$ and $\mathbf{w}_2$, according to (4.124). Substituting the expressions for $\mathbf{x}^*$ and $\Delta\mathbf{x}$, we obtain, from (4.124)

$$\frac{\partial\mathbf{v}_1}{\partial\mathbf{w}_1} = -\mathbf{P}_1^{-1}\mathbf{B}_1^T\mathbf{M}_1^{-1}\left(\mathbf{I} - \mathbf{A}_1\mathbf{N}_1^{-1}\mathbf{A}_1^T\mathbf{M}_1^{-1} + \mathbf{A}_1\mathbf{N}_1^{-1}\mathbf{A}_2^T\mathbf{T}\mathbf{A}_2\mathbf{N}_1^{-1}\mathbf{A}_1^T\mathbf{M}_1^{-1}\right) \tag{4.127}$$

$$\frac{\partial\mathbf{v}_1}{\partial\mathbf{w}_2} = -\mathbf{P}_1^{-1}\mathbf{B}_1^T\mathbf{M}_1^{-1}\mathbf{A}_1\mathbf{N}_1^{-1}\mathbf{A}_2^T\mathbf{T} \tag{4.128}$$

Applying the law of covariance propagation to $\mathbf{w}_1$ and $\mathbf{w}_2$ of (4.91) and knowing that the observations are uncorrelated gives

$$\mathbf{Q}_{w_1,w_2} = \begin{bmatrix} \mathbf{M}_1 & \mathbf{O} \\ \mathbf{O} & \mathbf{M}_2 \end{bmatrix} \tag{4.129}$$

By using the partial derivatives (4.127) and (4.128), Expression (4.129), and the law of variance-covariance propagation, we obtain, after some algebraic computations, the cofactor matrices:

$$\mathbf{Q}_{v_1} = \mathbf{Q}_{v_1^*} + \Delta\mathbf{Q}_{v_1} \tag{4.130}$$

where

$$\mathbf{Q}_{v_1^*} = \mathbf{P}_1^{-1}\mathbf{B}_1^T\mathbf{M}_1^{-1}\left(\mathbf{P}_1^{-1}\mathbf{B}_1^T\right)^T - \left(\mathbf{P}_1^{-1}\mathbf{B}_1^T\mathbf{M}_1^{-1}\mathbf{A}_1\right)\mathbf{N}_1^{-1}\left(\mathbf{P}_1^{-1}\mathbf{B}_1^T\mathbf{M}_1^{-1}\mathbf{A}_1\right)^T \tag{4.131}$$

$$\Delta\mathbf{Q}_{v_1} = \left(\mathbf{P}_1^{-1}\mathbf{B}_1^T\mathbf{M}_1^{-1}\mathbf{A}_1\mathbf{N}_1^{-1}\mathbf{A}_2^T\right)\mathbf{T}\left(\mathbf{P}_1^{-1}\mathbf{B}_1^T\mathbf{M}_1^{-1}\mathbf{A}_1\mathbf{N}_1^{-1}\mathbf{A}_2^T\right)^T \tag{4.132}$$

The partial derivatives of $\mathbf{v}_2$ with respect to $\mathbf{w}_1$ and $\mathbf{w}_2$ follow from (4.125):

$$\frac{\partial\mathbf{v}_2}{\partial\mathbf{w}_1} = \mathbf{P}_2^{-1}\mathbf{B}_2^T\mathbf{T}\mathbf{A}_2\mathbf{N}_1^{-1}\mathbf{A}_1^T\mathbf{M}_1^{-1} \tag{4.133}$$

$$\frac{\partial \mathbf{v}_2}{\partial \mathbf{w}_2} = -\mathbf{P}_2^{-1}\mathbf{B}_2^{\mathrm{T}}\,\mathbf{T} \tag{4.134}$$

By using, again, the law of variance-covariance propagation and (4.129), we obtain the cofactor for $\mathbf{v}_2$:

$$\mathbf{Q}_{v_2} = \mathbf{P}_2^{-1}\mathbf{B}_2^{\mathrm{T}}\,\mathbf{T}\mathbf{B}_2\mathbf{P}_2^{-1} \tag{4.135}$$

The estimated variance-covariance matrix is

$$\hat{\boldsymbol{\Sigma}}_{v_2} = \hat{\sigma}_0^2\mathbf{Q}_{v_2} \tag{4.136}$$

The variance-covariance matrix of the adjusted observations is, as usual,

$$\boldsymbol{\Sigma}_{\ell_a} = \boldsymbol{\Sigma}_{\ell_b} - \boldsymbol{\Sigma}_v \tag{4.137}$$

As for iterations, one has to make sure that all groups are evaluated for the same approximate parameters. If the first system is iterated, the approximate coordinates for the last iteration must be used as expansion points for the second group. Because there are no observations common to both groups, the iteration with respect to the observations can be done individually for each group.

Occasionally, it is desirable to remove a set of observations from an existing solution. Consider again the uncorrelated case in which the set of observations to be removed is not correlated with the other sets. The procedure is readily seen from (4.118), which shows how normal equations are added. When observations are removed, the respective parts of the normal matrix and the right-hand term must be subtracted. Equation (4.118) becomes

$$\begin{aligned}\hat{\mathbf{x}} &= -\left(\mathbf{A}_1^{\mathrm{T}}\,\mathbf{M}_1^{-1}\mathbf{A}_1 - \mathbf{A}_2^{\mathrm{T}}\,\mathbf{M}_2^{-1}\mathbf{A}_2\right)^{-1}\left(\mathbf{A}_1^{\mathrm{T}}\,\mathbf{M}_1^{-1}\mathbf{w}_1 - \mathbf{A}_2^{\mathrm{T}}\,\mathbf{M}_2^{-1}\mathbf{w}_2\right)\\ &= -\left[\mathbf{A}_1^{\mathrm{T}}\,\mathbf{M}_1^{-1}\mathbf{A}_1 + \mathbf{A}_2^{\mathrm{T}}\left(-\mathbf{M}_2^{-1}\right)\mathbf{A}_2\right]^{-1}\left[\mathbf{A}_1^{\mathrm{T}}\,\mathbf{M}_1^{-1}\mathbf{w}_1 + \mathbf{A}_2^{\mathrm{T}}\left(-\mathbf{M}_2^{-1}\right)\mathbf{w}_2\right]\end{aligned} \tag{4.138}$$

One only has to use a negative weight matrix of the group of observations that is being removed, because

$$-\mathbf{M}_2 = \mathbf{B}_2\left(-\mathbf{P}_2^{-1}\right)\mathbf{B}_2^{\mathrm{T}} \tag{4.139}$$

Observations can be removed sequentially following (4.116).

The sequential solution can be used in quite a general manner. One can add or remove any number of groups sequentially. A group may consist of a single observation. Given the solution for $i - 1$ groups, some of the relevant expressions that include all i groups of observations are,

$$\hat{\mathbf{x}}_i = \hat{\mathbf{x}}_{i-1} + \Delta\hat{\mathbf{x}}_i \tag{4.140}$$

$$\Delta\hat{\mathbf{x}}_i = -\mathbf{Q}_{i-1}\mathbf{A}_i^{\mathrm{T}}\left(\mathbf{M}_i + \mathbf{A}_i\mathbf{Q}_{i-1}\mathbf{A}_i^{\mathrm{T}}\right)^{-1}\left(\mathbf{A}_i\hat{\mathbf{x}}_{i-1} + \mathbf{w}_i\right) \tag{4.141}$$

$$\mathbf{v}^{\mathrm{T}}\mathbf{P}\mathbf{v}_i = \mathbf{v}^{\mathrm{T}}\mathbf{P}\mathbf{v}_{i-1} + \Delta\mathbf{v}^{\mathrm{T}}\mathbf{P}\mathbf{v}_i \tag{4.142}$$

$$\Delta\mathbf{v}^{\mathrm{T}}\mathbf{P}\mathbf{v}_i = \left(\mathbf{A}_i\hat{\mathbf{x}}_{i-1} + \mathbf{w}_i\right)^{\mathrm{T}} \left(\mathbf{M}_i + \mathbf{A}_i\mathbf{Q}_{i-1}\mathbf{A}_i^{\mathrm{T}}\right)^{-1} \left(\mathbf{A}_i\hat{\mathbf{x}}_{i-1} + \mathbf{w}_i\right) \tag{4.143}$$

$$\mathbf{Q}_i = \mathbf{Q}_{i-1} - \mathbf{Q}_{i-1}\mathbf{A}_i^{\mathrm{T}} \left(\mathbf{M}_i + \mathbf{A}_i\mathbf{Q}_{i-1}\mathbf{A}_i^{\mathrm{T}}\right)^{-1} \mathbf{A}_i\mathbf{Q}_{i-1} \tag{4.144}$$

Every sequential solution is equivalent to a one-step adjustment that contains the same observations. The sequential solution requires the inverse of the normal matrix. Because computing the inverse of the normal matrix requires many more computations than merely solving the system of normal equations, one might sometimes prefer to use the one-step solution instead of the sequential approach.

4.7 WEIGHTED PARAMETERS AND CONDITIONS

The algorithms developed in the previous section can be used to incorporate exterior information about parameters. This includes weighted functions of parameters, weighted individual parameters, and conditions on parameters. The objective is to incorporate new types of observations that directly refer to the parameters, to specify parameters in order to avoid singularity of the normal equations, or to incorporate the results of prior adjustments. Evaluating conditions between the parameters is the basis for hypothesis testing. These cases are obtained by specifying the coefficient matrices **A** and **B** of the mixed model. For example, the mixed model (4.86) and (4.87) can be specified as

$$\mathbf{f}_1(\boldsymbol{\ell}_{1a}, \mathbf{x}_a) = \mathbf{o} \tag{4.145}$$

$$\boldsymbol{\ell}_{2a} = \mathbf{f}_2(\mathbf{x}_a) \tag{4.146}$$

The linearized form is

$$\mathbf{B}_1\mathbf{v}_1 + \mathbf{A}_1\mathbf{x} + \mathbf{w}_1 = \mathbf{o} \tag{4.147}$$

$$\mathbf{v}_2 = \mathbf{A}_2\mathbf{x} + \boldsymbol{\ell}_2 \tag{4.148}$$

The specifications are $\mathbf{B}_2 = -\mathbf{I}$ and $\boldsymbol{\ell}_2 = \mathbf{w}_2$. For the observation equation model we obtain

$$\boldsymbol{\ell}_{1a} = \mathbf{f}_1(\mathbf{x}_a) \tag{4.149}$$

$$\boldsymbol{\ell}_{2a} = \mathbf{f}_2(\mathbf{x}_a) \tag{4.150}$$

with the linearized form being

$$\mathbf{v}_1 = \mathbf{A}_1\mathbf{x} + \boldsymbol{\ell}_1 \tag{4.151}$$

$$\mathbf{v}_2 = \mathbf{A}_2\mathbf{x} + \boldsymbol{\ell}_2 \tag{4.152}$$

The stochastic model is given by the matrices $\mathbf{P}_1$ and $\mathbf{P}_2$. With proper choice of the elements of $\mathbf{A}_2$ and $\mathbf{P}_2$, it is possible to introduce a variety of relations about the parameters.

As a first case, consider nonlinear relations between parameters. The design matrix $\mathbf{A}_2$ contains the partial derivatives, and $\boldsymbol{\ell}_{2b}$ contains the observed value of the function. This is the case of weighted functions of parameters. Examples are the area or volume of geometric figures as computed from coordinates, angles in geodetic networks, and differences between parameters (coordinates). Each function contributes one equation to (4.148) or (4.152). The respective expressions are identical with those given in Table 4.2 and require no further discussion.

As a second case, consider information about individual parameters. This is a special case of the general method discussed above. Each row of $\mathbf{A}_2$ contains zeros with the exception of one position, which contains a 1. The number of rows in the $\mathbf{A}_2$ matrix corresponds to the number of weighted parameters. The expressions of Table 4.2 are still valid for this case. If information enters into the adjustment in this manner, one speaks of the method of weighted parameters. In the most general case, all parameters are observed and weighted, giving

$$\boldsymbol{\ell}_{2a} = \mathbf{x}_a \tag{4.153}$$

$$\boldsymbol{\ell}_{2b} = \mathbf{x}_b \tag{4.154}$$

$$\mathbf{A}_2 = \mathbf{I} \tag{4.155}$$

$$\boldsymbol{\ell}_2 = \mathbf{f}_2(\mathbf{x}_0) - \boldsymbol{\ell}_{2b} = \mathbf{x}_0 - \mathbf{x}_b \tag{4.156}$$

The symbols $\mathbf{x}_b$ and $\mathbf{x}_0$ denote the observed parameters and approximate parameters. During the iterations, $\mathbf{x}_0$ converges toward the solution, whereas $\mathbf{x}_b$ remains unchanged just as does the vector $\boldsymbol{\ell}_{2b}$. As a special case, the vector $\boldsymbol{\ell}_2$ can be zero, which implies that the current values for the approximate parameters also serve as observations of the parameters. This can generally be done if the intent is to define the coordinate system by assigning weights to the current approximate parameters. Table 4.3 summarizes the solution for weighted parameters for the observation equation model. The parameters are weighted simply by adding the respective weights to the diagonal elements of the normal matrix. The parameters not weighted have zeros in the respective diagonal elements of $\mathbf{P}_2$. This is a convenient way of weighting a subset of parameters. Parameters can be fixed by assigning a large weight.

It is not necessary that the second group of observations represent the observed parameters. Table 4.4 shows the case in which the first group consists of the observed parameters. This approach has the unique feature that all observations can be added to the adjustment in a sequential manner; the first solution is a nonredundant one based solely on the values of the observed parameters. It is important, once again, to distinguish the roles of the observed parameters $\mathbf{x}_b$ and the approximations $\mathbf{x}_0$.

TABLE 4.2 Sequential Adjustment Models

	Mixed Model	Observation Model
Nonlinear model	$\mathbf{f}_1(\boldsymbol{\ell}_{1a}, \mathbf{x}_a) = \mathbf{o}$ $\mathbf{f}_2(\boldsymbol{\ell}_{2a}, \mathbf{x}_a) = \mathbf{o}$ $\mathbf{P} = \begin{bmatrix} \mathbf{P}_1 & \mathbf{O} \\ \mathbf{O} & \mathbf{P}_2 \end{bmatrix}$	$\boldsymbol{\ell}_{1a} = \mathbf{f}_1(\mathbf{x}_a)$ $\boldsymbol{\ell}_{2a} = \mathbf{f}_2(\mathbf{x}_a)$ $\mathbf{P} = \begin{bmatrix} \mathbf{P}_1 & \mathbf{O} \\ \mathbf{O} & \mathbf{P}_2 \end{bmatrix}$
Linear model	$\mathbf{B}_1\mathbf{v}_1 + \mathbf{A}_1\mathbf{x} + \mathbf{w}_1 = \mathbf{o}$ $\mathbf{B}_2\mathbf{v}_2 + \mathbf{A}_2\mathbf{x} + \mathbf{w}_2 = \mathbf{o}$	$\mathbf{v}_1 = \mathbf{A}_1\mathbf{x} + \boldsymbol{\ell}_1$ $\mathbf{v}_2 = \mathbf{A}_2\mathbf{x} + \boldsymbol{\ell}_2$
Normal equation elements	$\mathbf{M}_1 = \mathbf{B}_1\mathbf{P}_1^{-1}\mathbf{B}_1^T \quad \mathbf{M}_2 = \mathbf{B}_2\mathbf{P}_2^{-1}\mathbf{B}_2^T$ $\mathbf{N}_1 = \mathbf{A}_1^T\mathbf{M}_1^{-1}\mathbf{A}_1 \quad \mathbf{N}_2 = \mathbf{A}_2^T\mathbf{M}_2^{-1}\mathbf{A}_2$ $\mathbf{u}_1 = \mathbf{A}_1^T\mathbf{M}_1^{-1}\mathbf{w}_1 \quad \mathbf{u}_2 = \mathbf{A}_2^T\mathbf{M}_2^{-1}\mathbf{w}_2$	$\mathbf{M}_1 = \mathbf{P}_1^{-1} \quad \mathbf{M}_2 = \mathbf{P}_2^{-1}$ $\mathbf{N}_1 = \mathbf{A}_1^T\mathbf{P}_1\mathbf{A}_1 \quad \mathbf{N}_2 = \mathbf{A}_2^T\mathbf{P}_2\mathbf{A}_2$ $\mathbf{u}_1 = \mathbf{A}_1^T\mathbf{P}_1\boldsymbol{\ell}_1 \quad \mathbf{u}_2 = \mathbf{A}_2^T\mathbf{P}_2\boldsymbol{\ell}_2$
Minimum $\mathbf{v}^T\mathbf{P}\mathbf{v}$	$\mathbf{v}^T\mathbf{P}\mathbf{v} = \mathbf{v}^T\mathbf{P}\mathbf{v}^* + \Delta\mathbf{v}^T\mathbf{P}\mathbf{v}$ $\mathbf{v}^T\mathbf{P}\mathbf{v}^* = -\mathbf{u}_1^T\mathbf{N}_1^{-1}\mathbf{u}_1 + \mathbf{w}_1^T\mathbf{M}_1^{-1}\mathbf{w}_1$ $\Delta\mathbf{v}^T\mathbf{P}\mathbf{v} = (\mathbf{A}_2\mathbf{x}^* + \mathbf{w}_2)^T\,\mathbf{T}\,(\mathbf{A}_2\mathbf{x}^* + \mathbf{w}_2)$	$\mathbf{v}^T\mathbf{P}\mathbf{v} = \mathbf{v}^T\mathbf{P}\mathbf{v}^* + \Delta\mathbf{v}^T\mathbf{P}\mathbf{v}$ $\mathbf{v}^T\mathbf{P}\mathbf{v}^* = -\mathbf{u}_1^T\mathbf{N}_1^{-1}\mathbf{u}_1 + \boldsymbol{\ell}_1^T\mathbf{P}_1\boldsymbol{\ell}_1$ $\Delta\mathbf{v}^T\mathbf{P}\mathbf{v} = (\mathbf{A}_2\mathbf{x}^* + \boldsymbol{\ell}_2)^T\,\mathbf{T}\,(\mathbf{A}_2\mathbf{x}^* + \boldsymbol{\ell}_2)$
Estimated parameters	$\hat{\mathbf{x}} = \mathbf{x}^* + \Delta\mathbf{x}$ $\mathbf{x}^* = -\mathbf{N}_1^{-1}\mathbf{u}_1$ $\mathbf{T} = \left(\mathbf{M}_2 + \mathbf{A}_2\mathbf{N}_1^{-1}\mathbf{A}_2^T\right)^{-1}$ $\Delta\mathbf{x} = -\mathbf{N}_1^{-1}\mathbf{A}_2^T\mathbf{T}\left(\mathbf{A}_2\mathbf{x}^* + \mathbf{w}_2\right)$	$\hat{\mathbf{x}} = \mathbf{x}^* + \Delta\mathbf{x}$ $\mathbf{x}^* = -\mathbf{N}_1^{-1}\mathbf{u}_1$ $\mathbf{T} = \left(\mathbf{P}_2^{-1} + \mathbf{A}_2\mathbf{N}_1^{-1}\mathbf{A}_2^T\right)^{-1}$ $\Delta\mathbf{x} = -\mathbf{N}_1^{-1}\mathbf{A}_2^T\mathbf{T}\left(\mathbf{A}_2\mathbf{x}^* + \boldsymbol{\ell}_2\right)$
Estimated residuals	$\mathbf{v}_1 = \mathbf{v}_1^* + \Delta\mathbf{v}_1$ $\mathbf{v}_1^* = -\mathbf{P}_1^{-1}\mathbf{B}_1^T\mathbf{M}_1^{-1}\left(\mathbf{A}_1\mathbf{x}^* + \mathbf{w}_1\right)$ $\Delta\mathbf{v}_1 = -\mathbf{P}_1^{-1}\mathbf{B}_1^T\mathbf{M}_1^{-1}\mathbf{A}_1\,\Delta\mathbf{x}$	$\mathbf{v}_1 = \mathbf{v}_1^* + \Delta\mathbf{v}_1$ $\mathbf{v}_1^* = \mathbf{A}_1\mathbf{x}^* + \boldsymbol{\ell}_1$ $\Delta\mathbf{v}_1 = \mathbf{A}_1\,\Delta\mathbf{x}$
Estimated variance of unit weight	$\hat{\sigma}_0^2 = \dfrac{\mathbf{v}^T\mathbf{P}\mathbf{v}}{r_1 + r_2 - u}$	$\hat{\sigma}_0^2 = \dfrac{\mathbf{v}^T\mathbf{P}\mathbf{v}}{n_1 + n_2 - u}$
Estimated parameter cofactor matrix	$\mathbf{Q}_x = \mathbf{Q}_{x^*} + \Delta\mathbf{Q}$ $\mathbf{Q}_{x^*} = \mathbf{N}_1^{-1}$ $\Delta\mathbf{Q} = -\mathbf{N}_1^{-1}\mathbf{A}_2^T\mathbf{T}\mathbf{A}_2\mathbf{N}_1^{-1}$	$\mathbf{Q}_x = \mathbf{Q}_{x^*} + \Delta\mathbf{Q}$ $\mathbf{Q}_{x^*} = \mathbf{N}_1^{-1}$ $\Delta\mathbf{Q} = -\mathbf{N}_1^{-1}\mathbf{A}_2^T\mathbf{T}\mathbf{A}_2\mathbf{N}_1^{-1}$

Because, in most cases the $\mathbf{P}_1$ matrix will be diagonal, no matrix inverse computation is required. The size of the matrix $\mathbf{T}$ equals the number of observations in the second group. Thus, if one observation is added at a time, only a 1×1 matrix must be inverted. The residuals can be computed directly from the mathematical model as desired.

A third case pertains to the role of the weight matrix of the parameters. The weight matrix expresses the quality of the information known about the observed parameters.

TABLE 4.3 Observed Parameters

$$\boldsymbol{\ell}_{1a} = \mathbf{f}_1(\mathbf{x}_a) \quad \boldsymbol{\ell}_{2a} = \mathbf{x}_a \quad \mathbf{P} = \begin{bmatrix} \mathbf{P}_1 & \mathbf{O} \\ \mathbf{O} & \mathbf{P}_2 \end{bmatrix}$$

$$\mathbf{v}_1 = \mathbf{A}_1\mathbf{x} + \boldsymbol{\ell}_1 \quad \mathbf{v}_2 = \mathbf{x} + \boldsymbol{\ell}_2 \quad \boldsymbol{\ell}_2 = \mathbf{x}_0 - \mathbf{x}_b$$

$$\mathbf{N}_1 = \mathbf{A}_1^T\mathbf{P}_1\mathbf{A}_1 \quad \mathbf{N}_2 = \mathbf{P}_2$$

$$\mathbf{u}_1 = \mathbf{A}_1^T\mathbf{P}_1\boldsymbol{\ell}_1 \quad \mathbf{u}_2 = \mathbf{P}_2\boldsymbol{\ell}_2$$

$$\hat{\mathbf{x}} = -(\mathbf{N}_1 + \mathbf{P}_2)^{-1}(\mathbf{u}_1 + \mathbf{P}_2\boldsymbol{\ell}_2)$$

$$\mathbf{Q}_x = (\mathbf{N}_1 + \mathbf{P}_2)^{-1}$$

Note: Case of observation equation model.

For the adjustment to be meaningful, one must make every attempt to obtain a weight matrix that truly reflects the quality of the additional information. Low weights, or, equivalently, large variances, imply low precision. Even low-weighted parameters can have, occasionally, a positive effect on the quality of the least-squares solution. If the parameters or functions of the parameters are introduced with an infinitely large weight, one speaks of conditions between parameters. The only specifications for implementing conditions are:

$$\mathbf{P}_2^{-1} = \mathbf{O} \tag{4.157}$$

and

$$\mathbf{P}_2 = \infty \tag{4.158}$$

The respective mathematical models are

$$\mathbf{f}(\boldsymbol{\ell}_{1a}, \mathbf{x}_a) = \mathbf{o} \tag{4.159}$$

$$\mathbf{g}(\mathbf{x}_a) = \mathbf{o} \tag{4.160}$$

with

$$\mathbf{B}_1\mathbf{v}_1 + \mathbf{A}_1\mathbf{x} + \mathbf{w}_1 = \mathbf{o} \tag{4.161}$$

$$\mathbf{A}_2\mathbf{x} + \boldsymbol{\ell}_2 = \mathbf{o} \tag{4.162}$$

and

$$\boldsymbol{\ell}_{1a} = \mathbf{f}(\mathbf{x}_a) \tag{4.163}$$

$$\mathbf{g}(\mathbf{x}_a) = \mathbf{o} \tag{4.164}$$

TABLE 4.4 Sequential Solution without Inverting the Normal Matrix

$$\boldsymbol{\ell}_{1a} = \mathbf{x}_a \qquad \mathbf{P} = \begin{bmatrix} \mathbf{P}_1 & \mathbf{O} \\ \mathbf{O} & \mathbf{P}_2 \end{bmatrix}$$
$$\boldsymbol{\ell}_{2a} = \mathbf{f}_{2a}(\mathbf{x}_a)$$

$$\mathbf{v}_1 = \mathbf{x} + \boldsymbol{\ell}_1$$
$$\boldsymbol{\ell}_1 = \mathbf{x}_0 - \mathbf{x}_b$$
$$\mathbf{v}_2 = \mathbf{A}_2\mathbf{x} + \boldsymbol{\ell}_2$$

$$\mathbf{N}_1 = \mathbf{P}_1 \qquad \mathbf{N}_2 = \mathbf{A}_2^T\mathbf{P}_2\mathbf{A}_2$$
$$\mathbf{u}_1 = \mathbf{P}_1\boldsymbol{\ell}_1 \qquad \mathbf{u}_2 = \mathbf{A}_2^T\mathbf{P}_2\boldsymbol{\ell}_2$$

$$\hat{\mathbf{x}}_1 = -(\mathbf{x}_0 - \mathbf{x}_b)$$
$$\mathbf{Q}_1 = \mathbf{P}_1^{-1}$$
$$\mathbf{v}^T\mathbf{P}\mathbf{v}_1 = 0$$

$$\hat{\mathbf{x}}_i = \hat{\mathbf{x}}_{i-1} + \Delta\hat{\mathbf{x}}_{i-1}$$
$$\mathbf{v}^T\mathbf{P}\mathbf{v}_i = \mathbf{v}^T\mathbf{P}\mathbf{v}_{i-1} + \Delta\mathbf{v}^T\mathbf{P}\mathbf{v}_{i-1}$$
$$\mathbf{Q}_i = \mathbf{Q}_{i-1} + \Delta\mathbf{Q}_{i-1}$$

$$\mathbf{T} = \left(\mathbf{P}_i^{-1} + \mathbf{A}_i\mathbf{Q}_{i-1}\mathbf{A}_i^T\right)^{-1}$$
$$\Delta\mathbf{x}_{i-1} = -\mathbf{Q}_{i-1}\mathbf{A}_i^T\mathbf{T}\left(\mathbf{A}_i\hat{\mathbf{x}}_{i-1} + \boldsymbol{\ell}_i\right)$$
$$\Delta\mathbf{v}^T\mathbf{P}\mathbf{v}_{i-1} = (\mathbf{A}_i\hat{\mathbf{x}}_{i-1} + \boldsymbol{\ell}_i)^T\mathbf{T}(\mathbf{A}_i\hat{\mathbf{x}}_{i-1} + \boldsymbol{\ell}_i)$$
$$\Delta\mathbf{Q}_{i-1} = -\mathbf{Q}_{i-1}\mathbf{A}_i^T\mathbf{T}\mathbf{A}_i\mathbf{Q}_{i-1}$$

Note: Case of observation equation model.

with

$$\mathbf{v}_1 = \mathbf{A}_1\mathbf{x} + \boldsymbol{\ell}_1 \tag{4.165}$$

$$\mathbf{A}_2\mathbf{x} + \boldsymbol{\ell}_2 = \mathbf{o} \tag{4.166}$$

Table 4.5 contains the expression of the sequential solution with conditions between parameters. If (4.158) is used to impose the conditions, the largest numbers that can still be represented in the computer should be used. In most situations, it will be readily clear what constitutes a large weight; the weight must simply be large enough so that the respective observations or parameters do not change during the adjustment. For sequential solution, the solution of the first group must exist. Conditions cannot

TABLE 4.5 Conditions on Parameters

	Mixed Model with Conditions	Observation Model with Conditions
Nonlinear model	$\mathbf{f}_1(\boldsymbol{\ell}_{1a}, \mathbf{x}_a) = \mathbf{o}$ $\mathbf{g}(\mathbf{x}_a) = \mathbf{o}$ $\quad \mathbf{P}_1$	$\boldsymbol{\ell}_{1a} = \mathbf{f}_1(\mathbf{x}_a)$ $\mathbf{g}(\mathbf{x}_a) = \mathbf{o}$ $\quad \mathbf{P}_1$
Linear model	$\mathbf{B}_1\mathbf{v}_1 + \mathbf{A}_1\mathbf{x} + \mathbf{w}_1 = \mathbf{o}$ $\mathbf{A}_2\mathbf{x} + \boldsymbol{\ell}_2 = \mathbf{o}$	$\mathbf{v}_1 = \mathbf{A}_1\mathbf{x} + \boldsymbol{\ell}_1$ $\mathbf{A}_2\mathbf{x} + \boldsymbol{\ell}_2 = \mathbf{o}$
Normal equation elements	$\mathbf{M}_1 = \mathbf{B}_1\mathbf{P}_1^{-1}\mathbf{B}_1^{\mathrm{T}}$ $\mathbf{N}_1 = \mathbf{A}_1^{\mathrm{T}}\mathbf{M}_1^{-1}\mathbf{A}_1$ $\mathbf{u}_1 = \mathbf{A}_1^{\mathrm{T}}\mathbf{M}_1^{-1}\mathbf{w}_1$	$\mathbf{M}_1 = \mathbf{P}_1^{-1}$ $\mathbf{N}_1 = \mathbf{A}_1^{\mathrm{T}}\mathbf{P}_1\mathbf{A}_1$ $\mathbf{u}_1 = \mathbf{A}_1^{\mathrm{T}}\mathbf{P}_1\boldsymbol{\ell}_1$
Minimum $\mathbf{v}^{\mathrm{T}}\mathbf{P}\mathbf{v}$	$\mathbf{v}^{\mathrm{T}}\mathbf{P}\mathbf{v} = \mathbf{v}^{\mathrm{T}}\mathbf{P}\mathbf{v}^* + \Delta\mathbf{v}^{\mathrm{T}}\mathbf{P}\mathbf{v}$ $\mathbf{v}^{\mathrm{T}}\mathbf{P}\mathbf{v}^* = -\mathbf{u}_1^{\mathrm{T}}\mathbf{N}_1^{-1}\mathbf{u}_1 + \mathbf{w}_1^{\mathrm{T}}\mathbf{M}_1^{-1}\mathbf{w}_1$ $\Delta\mathbf{v}^{\mathrm{T}}\mathbf{P}\mathbf{v} = \left(\mathbf{A}_2\mathbf{x}^* + \boldsymbol{\ell}_2\right)^{\mathrm{T}}\mathbf{T}\left(\mathbf{A}_2\mathbf{x}^* + \boldsymbol{\ell}_2\right)$	$\mathbf{v}^{\mathrm{T}}\mathbf{P}\mathbf{v} = \mathbf{v}^{\mathrm{T}}\mathbf{P}\mathbf{v}^* + \Delta\mathbf{v}^{\mathrm{T}}\mathbf{P}\mathbf{v}$ $\mathbf{v}^{\mathrm{T}}\mathbf{P}\mathbf{v}^* = -\mathbf{u}_1^{\mathrm{T}}\mathbf{N}_1^{-1}\mathbf{u}_1 + \boldsymbol{\ell}_1^{\mathrm{T}}\mathbf{P}_1\boldsymbol{\ell}_1$ $\Delta\mathbf{v}^{\mathrm{T}}\mathbf{P}\mathbf{v} = \left(\mathbf{A}_2\mathbf{x}^* + \boldsymbol{\ell}_2\right)^{\mathrm{T}}\mathbf{T}\left(\mathbf{A}_2\mathbf{x}^* + \boldsymbol{\ell}_2\right)$
Estimated parameters	$\hat{\mathbf{x}} = \mathbf{x}^* + \Delta\mathbf{x}$ $\mathbf{x}^* = -\mathbf{N}_1^{-1}\mathbf{u}_1$ $\mathbf{T} = \left(\mathbf{A}_2\mathbf{N}_1^{-1}\mathbf{A}_2^{\mathrm{T}}\right)^{-1}$ $\Delta\mathbf{x} = -\mathbf{N}_1^{-1}\mathbf{A}_2^{\mathrm{T}}\mathbf{T}\left(\mathbf{A}_2\mathbf{x}^* + \mathbf{w}_2\right)$	$\hat{\mathbf{x}} = \mathbf{x}^* + \Delta\mathbf{x}$ $\mathbf{x}^* = -\mathbf{N}_1^{-1}\mathbf{u}_1$ $\mathbf{T} = \left(\mathbf{A}_2\mathbf{N}_1^{-1}\mathbf{A}_2^{\mathrm{T}}\right)^{-1}$ $\Delta\mathbf{x} = -\mathbf{N}_1^{-1}\mathbf{A}_2^{\mathrm{T}}\mathbf{T}\left(\mathbf{A}_2\mathbf{x}^* + \boldsymbol{\ell}_2\right)$
Estimated residuals	$\mathbf{v}_1 = \mathbf{v}_1^* + \Delta\mathbf{v}_1$ $\mathbf{v}_1^* = -\mathbf{P}_1^{-1}\mathbf{B}_1^{\mathrm{T}}\mathbf{M}_1^{-1}\left(\mathbf{A}_1\mathbf{x}^* + \mathbf{w}_1\right)$ $\Delta\mathbf{v}_1 = -\mathbf{P}_1^{-1}\mathbf{B}_1^{\mathrm{T}}\mathbf{M}_1^{-1}\mathbf{A}_1\ \Delta\mathbf{x}$	$\mathbf{v}_1 = \mathbf{v}_1^* + \Delta\mathbf{v}_1$ $\mathbf{v}_1^* = \mathbf{A}_1\mathbf{x}^* + \boldsymbol{\ell}_1$ $\Delta\mathbf{v}_1 = \mathbf{A}_1\ \Delta\mathbf{x}$
Estimated variance of unit weight	$\hat{\sigma}_0^2 = \dfrac{\mathbf{v}^{\mathrm{T}}\mathbf{P}\mathbf{v}}{r_1 + r_2 - u}$	$\hat{\sigma}_0^2 = \dfrac{\mathbf{v}^{\mathrm{T}}\mathbf{P}\mathbf{v}}{n_1 + n_2 - u}$
Estimated parameter cofactor matrix	$\mathbf{Q}_x = \mathbf{Q}_{x^*} + \Delta\mathbf{Q}$ $\mathbf{Q}_{x^*} = \mathbf{N}_1^{-1}$ $\Delta\mathbf{Q} = -\mathbf{N}_1^{-1}\mathbf{A}_2^{\mathrm{T}}\mathbf{T}\mathbf{A}_2\mathbf{N}_1^{-1}$	$\mathbf{Q}_x = \mathbf{Q}_{x^*} + \Delta\mathbf{Q}$ $\mathbf{Q}_{x^*} = \mathbf{N}_1^{-1}$ $\Delta\mathbf{Q} = -\mathbf{N}_1^{-1}\mathbf{A}_2^{\mathrm{T}}\mathbf{T}\mathbf{A}_2\mathbf{N}_1^{-1}$

be imposed sequentially to eliminate a singularity in the first group; e.g., conditions should not be used sequentially to define the coordinate system. A one-step solution is given by (4.118).

The a posteriori variance of unit weight is always computed from the final set of residuals. The degree of freedom increases by 1 for every observed parameter function, weighted parameter, or condition. In nonlinear adjustments the linearized condition must always be evaluated for the current point of expansion, i.e., the point of expansion of the last iteration (current solution).

The expressions in Table 4.2 and Table 4.5 are almost identical. The only difference is that the matrix $\mathbf{T}$ contains the matrix $\mathbf{M}_2$ in Table 4.2.

4.8 MINIMAL AND INNER CONSTRAINTS

This section deals with the implementation of minimal and inner constraints to the observation equation model. The symbol r denotes the rank of the design matrix, $R({}_n\mathbf{A}_u) = R(\mathbf{A}^{\mathrm{T}}\mathbf{PA}) = r \leq u$. Note that the use of the symbol r in this context is entirely different from its use in the mixed model, where r denotes the number of equations. The rank deficiency of $u - r$ is generally caused by a lack of coordinate system definition. For example, a network of distances is invariant with respect to translation and rotation, a network of angles is invariant with respect to translation, rotation, and scaling, and a level network (consisting of measured height differences) is invariant with respect to a translation in the vertical. The rank deficiency is dealt with by specifying $u - r$ conditions of the parameters. Much of the theory of inner and minimal constraint solution is discussed by Pope (1971). The main reason for dealing with minimal and inner constraint solutions is that this type of adjustment is important for the quality control of observations. Inner constraint solutions have the additional advantage that the standard ellipses (ellipsoids) represent the geometry as implied by the $\mathbf{A}$ and $\mathbf{P}$ matrices.

The formulation of the least-squares adjustment for the observation equation model in the presence of a rank deficiency is

$$ {}_n\mathbf{v}_1 = {}_n\mathbf{A}_u\,\mathbf{x}_B + {}_n\boldsymbol{\ell}_1 \tag{4.167} $$

$$ \mathbf{P} = \sigma_0^2\,\boldsymbol{\Sigma}_{\ell_b}^{-1} \tag{4.168} $$

$$ {}_{u-r}\mathbf{B}_u\,\mathbf{x}_B = \mathbf{o} \tag{4.169} $$

The subscript B indicates that the solution of the parameters $\mathbf{x}$ depends on the special condition implied by the $\mathbf{B}$ matrix in (4.169). This is the observation equation model with conditions between the parameters that was treated in Section 4.7. The one-step solution is given by (4.106):

$$ \begin{bmatrix} \mathbf{A}^{\mathrm{T}}\mathbf{PA} & \mathbf{B}^{\mathrm{T}} \\ \mathbf{B} & \mathbf{O} \end{bmatrix} \begin{bmatrix} \hat{\mathbf{x}}_B \\ -\hat{\mathbf{k}}_2 \end{bmatrix} = \begin{bmatrix} -\mathbf{A}^{\mathrm{T}}\mathbf{PL} \\ \mathbf{o} \end{bmatrix} \tag{4.170} $$

The matrix on the left side of (4.170) is a nonsingular matrix if the conditions (4.169) are linearly independent; i.e., the $(u-r) \times u$ matrix $\mathbf{B}$ has full row rank, and the rows are linear-independent of the rows of the design matrix $\mathbf{A}$. A general expression for the inverse is obtained from

$$\begin{bmatrix} \mathbf{A}^{\mathrm{T}}\mathbf{PA} & \mathbf{B}^{\mathrm{T}} \\ \mathbf{B} & \mathbf{O} \end{bmatrix} \begin{bmatrix} \mathbf{Q}_B & \mathbf{S}^{\mathrm{T}} \\ \mathbf{S} & \mathbf{R} \end{bmatrix} = \begin{bmatrix} \mathbf{I} & \mathbf{O} \\ \mathbf{O} & \mathbf{I} \end{bmatrix} \tag{4.171}$$

This matrix equation gives the following four equations of submatrices:

$$\mathbf{A}^{\mathrm{T}}\mathbf{PAQ}_B + \mathbf{B}^{\mathrm{T}}\mathbf{S} = \mathbf{I} \tag{4.172}$$

$$\mathbf{A}^{\mathrm{T}}\mathbf{PAS}^{\mathrm{T}} + \mathbf{B}^{\mathrm{T}}\mathbf{R} = \mathbf{O} \tag{4.173}$$

$$\mathbf{BQ}_B = \mathbf{O} \tag{4.174}$$

$$\mathbf{BS}^{\mathrm{T}} = \mathbf{I} \tag{4.175}$$

The solution of these equations requires the introduction of the $(u-r) \times u$ matrix $\mathbf{E}$, whose rows span the null space of the design matrix $\mathbf{A}$ or the null space of the normal matrix. According to (A.53), there is a matrix $\mathbf{E}$ such that

$$\left(\mathbf{A}^{\mathrm{T}}\mathbf{PA}\right)\mathbf{E}^{\mathrm{T}} = \mathbf{O} \tag{4.176}$$

or

$$\mathbf{AE}^{\mathrm{T}} = \mathbf{O} \qquad \text{or} \qquad \mathbf{EA}^{\mathrm{T}} = \mathbf{O} \tag{4.177}$$

Because the rows of $\mathbf{B}$ are linearly independent of the rows of $\mathbf{A}$, the $(u-r) \times (u-r)$ matrix $\mathbf{BE}^{\mathrm{T}}$ has full rank and thus can be inverted. Multiplying (4.172) by $\mathbf{E}$ from the left and using (4.177), we get

$$\mathbf{S} = \left(\mathbf{EB}^{\mathrm{T}}\right)^{-1}\mathbf{E} \tag{4.178}$$

This expression also satisfies (4.175). Substituting $\mathbf{S}$ into (4.173) gives

$$\mathbf{A}^{\mathrm{T}}\mathbf{PAE}^{\mathrm{T}}\left(\mathbf{BE}^{\mathrm{T}}\right)^{-1} + \mathbf{B}^{\mathrm{T}}\mathbf{R} = \mathbf{O} \tag{4.179}$$

Because of (4.176), this expression becomes

$$\mathbf{B}^{\mathrm{T}}\mathbf{R} = \mathbf{O} \tag{4.180}$$

Because $\mathbf{B}$ has full rank, it follows that the matrix $\mathbf{R} = \mathbf{O}$. Thus,

$$\begin{bmatrix} \mathbf{A}^{\mathrm{T}}\mathbf{PA} & \mathbf{B}^{\mathrm{T}} \\ \mathbf{B} & \mathbf{O} \end{bmatrix}^{-1} = \begin{bmatrix} \mathbf{Q}_B & \mathbf{E}^{\mathrm{T}}\left(\mathbf{BE}^{\mathrm{T}}\right)^{-1} \\ \left(\mathbf{EB}^{\mathrm{T}}\right)^{-1}\mathbf{E} & \mathbf{O} \end{bmatrix} \tag{4.181}$$

Substituting Expression (4.178) for $\mathbf{S}$ into (4.172) gives the nonsymmetric matrix

$$\mathbf{T}_B \equiv \mathbf{A}^{\mathrm{T}}\mathbf{P}\mathbf{A}\mathbf{Q}_B = \mathbf{I} - \mathbf{B}^{\mathrm{T}}\left(\mathbf{E}\mathbf{B}^{\mathrm{T}}\right)^{-1}\mathbf{E} \tag{4.182}$$

This expression is modified with the help of (4.174), (4.176), and (4.182):

$$\left(\mathbf{A}^{\mathrm{T}}\mathbf{P}\mathbf{A} + \mathbf{B}^{\mathrm{T}}\mathbf{B}\right)\left[\mathbf{Q}_B + \mathbf{E}^{\mathrm{T}}\left(\mathbf{B}\mathbf{E}^{\mathrm{T}}\right)^{-1}\left(\mathbf{E}\mathbf{B}^{\mathrm{T}}\right)^{-1}\mathbf{E}\right] = \mathbf{I} \tag{4.183}$$

It can be solved for $\mathbf{Q}_B$:

$$\mathbf{Q}_B = \left(\mathbf{A}^{\mathrm{T}}\mathbf{P}\mathbf{A} + \mathbf{B}^{\mathrm{T}}\mathbf{B}\right)^{-1} - \mathbf{E}^{\mathrm{T}}\left(\mathbf{E}\mathbf{B}^{\mathrm{T}}\mathbf{B}\mathbf{E}^{\mathrm{T}}\right)^{-1}\mathbf{E} \tag{4.184}$$

The least-squares solution of $\hat{\mathbf{x}}_B$ subject to condition (4.169) is, according to (4.170), (4.171), and (4.181),

$$\hat{\mathbf{x}}_B = -\mathbf{Q}_B\mathbf{A}^{\mathrm{T}}\mathbf{P}\mathbf{L} \tag{4.185}$$

The cofactor matrix of the parameters follows from the law of variance-covariance propagation

$$\mathbf{Q}_{x_B} = \mathbf{Q}_B\mathbf{A}^{\mathrm{T}}\mathbf{P}\mathbf{A}\mathbf{Q}_B = \mathbf{Q}_B \tag{4.186}$$

The latter part of (4.186) follows from (4.182) upon multiplying from the left by $\mathbf{Q}_B$ and using (4.174). Multiplying (4.182) from the right by $\mathbf{A}^{\mathrm{T}}\mathbf{P}\mathbf{A}$ and using (4.177) gives

$$\mathbf{A}^{\mathrm{T}}\mathbf{P}\mathbf{A} = \mathbf{A}^{\mathrm{T}}\mathbf{P}\mathbf{A}\mathbf{Q}_B\mathbf{A}^{\mathrm{T}}\mathbf{P}\mathbf{A} \tag{4.187}$$

The relation implied in (4.186) is

$$\mathbf{Q}_B\mathbf{A}^{\mathrm{T}}\mathbf{P}\mathbf{A}\mathbf{Q}_B = \mathbf{Q}_B \tag{4.188}$$

$u - r$ conditions are necessary to solve the least-squares problem; i.e., the minimal number of conditions is equal to the rank defect of the design (or normal) matrix. Any solution derived in this manner is called a minimal constraint solution. There are obviously many different sets of minimal constraints possible for the same adjustment. The only prerequisite on the $\mathbf{B}$ matrix is that it have full row rank and that its rows be linearly independent of $\mathbf{A}$. Assume that

$$\mathbf{C}\mathbf{x}_C = \mathbf{o} \tag{4.189}$$

is an alternative set of conditions. The solution $\hat{\mathbf{x}}_C$ follows from the expressions given by simply replacing the matrix $\mathbf{B}$ by $\mathbf{C}$. The pertinent expressions are

$$\hat{\mathbf{x}}_C = -\mathbf{Q}_C\mathbf{A}^{\mathrm{T}}\mathbf{P}\mathbf{L} \tag{4.190}$$

$$\mathbf{Q}_C = \left(\mathbf{A}^{\mathrm{T}}\mathbf{PA} + \mathbf{C}^{\mathrm{T}}\mathbf{C}\right)^{-1} - \mathbf{E}^{\mathrm{T}}\left(\mathbf{EC}^{\mathrm{T}}\mathbf{CE}^{\mathrm{T}}\right)^{-1}\mathbf{E} \tag{4.191}$$

$$\mathbf{T}_C \equiv \mathbf{A}^{\mathrm{T}}\mathbf{PAQ}_C = \mathbf{I} - \mathbf{C}^{\mathrm{T}}\left(\mathbf{EC}^{\mathrm{T}}\right)^{-1}\mathbf{E} \tag{4.192}$$

$$\mathbf{A}^{\mathrm{T}}\mathbf{PAQ}_C\mathbf{A}^{\mathrm{T}}\mathbf{PA} = \mathbf{A}^{\mathrm{T}}\mathbf{PA} \tag{4.193}$$

$$\mathbf{Q}_C\mathbf{A}^{\mathrm{T}}\mathbf{PAQ}_C = \mathbf{Q}_C \tag{4.194}$$

The solutions pertaining to the various alternative sets of conditions are all related. In particular,

$$\hat{\mathbf{x}}_B = \mathbf{T}_B^{\mathrm{T}}\hat{\mathbf{x}}_C \tag{4.195}$$

$$\mathbf{Q}_B = \mathbf{T}_B^{\mathrm{T}}\mathbf{Q}_C\mathbf{T}_B \tag{4.196}$$

$$\hat{\mathbf{x}}_C = \mathbf{T}_C^{\mathrm{T}}\hat{\mathbf{x}}_B \tag{4.197}$$

$$\mathbf{Q}_C = \mathbf{T}_C^{\mathrm{T}}\mathbf{Q}_B\mathbf{T}_C \tag{4.198}$$

Equations (4.195) to (4.198) constitute the transformation of minimal control; i.e., they relate the adjusted parameters and the covariance matrix for different minimal constraints. These transformation expressions are readily proven. For example, by using (4.190), (4.182), (4.192), and (4.177), we obtain

$$\begin{aligned}
\mathbf{T}_B^{\mathrm{T}}\hat{\mathbf{x}}_C &= -\mathbf{T}_B^{\mathrm{T}}\mathbf{Q}_C\mathbf{A}^{\mathrm{T}}\mathbf{PL} \\
&= -\mathbf{Q}_B\mathbf{A}^{\mathrm{T}}\mathbf{PAQ}_C\mathbf{A}^{\mathrm{T}}\mathbf{PL} \\
&= -\mathbf{Q}_B\left[\mathbf{I} - \mathbf{C}^{\mathrm{T}}\left(\mathbf{EC}^{\mathrm{T}}\right)^{-1}\mathbf{E}\right]\mathbf{A}^{\mathrm{T}}\mathbf{PL} \\
&= -\mathbf{Q}_B\mathbf{A}^{\mathrm{T}}\mathbf{PL} \\
&= \hat{\mathbf{x}}_B
\end{aligned} \tag{4.199}$$

With (4.192), (4.187), and (4.194), it follows that

$$\begin{aligned}
\mathbf{T}_C^{\mathrm{T}}\mathbf{Q}_B\mathbf{T}_C &= \mathbf{Q}_C\mathbf{A}^{\mathrm{T}}\mathbf{PAQ}_B\mathbf{A}^{\mathrm{T}}\mathbf{PAQ}_C \\
&= \mathbf{Q}_C\mathbf{A}^{\mathrm{T}}\mathbf{PAQ}_C \\
&= \mathbf{Q}_C
\end{aligned} \tag{4.200}$$

Instead of using the general condition (4.189), we can use the condition

$$\mathbf{Ex}_P = \mathbf{o} \tag{4.201}$$

The rows of $\mathbf{E}$ are linearly independent of $\mathbf{A}$ because of (4.177). Thus, replacing the matrix $\mathbf{C}$ by $\mathbf{E}$ in Equations (4.190) through (4.198) gives this special solution:

$$\hat{\mathbf{x}}_P = -\mathbf{Q}_P \mathbf{A}^{\mathrm{T}} \mathbf{PL} \tag{4.202}$$

$$\mathbf{Q}_P = \left(\mathbf{A}^{\mathrm{T}} \mathbf{PA} + \mathbf{E}^{\mathrm{T}} \mathbf{E}\right)^{-1} - \mathbf{E}^{\mathrm{T}} \left(\mathbf{E}\mathbf{E}^{\mathrm{T}} \mathbf{E}\mathbf{E}^{\mathrm{T}}\right)^{-1} \mathbf{E} \tag{4.203}$$

$$\mathbf{T}_P \equiv \mathbf{A}^{\mathrm{T}} \mathbf{PAQ}_P = \mathbf{I} - \mathbf{E}^{\mathrm{T}} \left(\mathbf{E}\mathbf{E}^{\mathrm{T}}\right)^{-1} \mathbf{E} \tag{4.204}$$

$$\mathbf{A}^{\mathrm{T}} \mathbf{PAQ}_P \mathbf{A}^{\mathrm{T}} \mathbf{PA} = \mathbf{A}^{\mathrm{T}} \mathbf{PA} \tag{4.205}$$

$$\mathbf{Q}_P \mathbf{A}^{\mathrm{T}} \mathbf{PAQ}_P = \mathbf{Q}_P \tag{4.206}$$

$$\hat{\mathbf{x}}_B = \mathbf{T}_B^{\mathrm{T}} \hat{\mathbf{x}}_P \tag{4.207}$$

$$\mathbf{Q}_B = \mathbf{T}_B^{\mathrm{T}} \mathbf{Q}_P \mathbf{T}_B \tag{4.208}$$

$$\hat{\mathbf{x}}_P = \mathbf{T}_P^{\mathrm{T}} \hat{\mathbf{x}}_B \tag{4.209}$$

$$\mathbf{Q}_P = \mathbf{T}_P^{\mathrm{T}} \mathbf{Q}_B \mathbf{T}_P \tag{4.210}$$

The solution (4.202) is called the inner constraint solution. The matrix $\mathbf{T}_P$ in (4.204) is symmetric. The matrix $\mathbf{Q}_P$ is a generalized inverse, called the pseudoinverse of the normal matrix; the following notation is used:

$$\mathbf{Q}_P = \mathbf{N}^{+} = \left(\mathbf{A}^{\mathrm{T}} \mathbf{PA}\right)^{+} \tag{4.211}$$

The pseudoinverse of the normal matrix is computed from available algorithms of generalized matrix inverses or, equivalently, by finding the $\mathbf{E}$ matrix and using Equation (4.203). For typical applications in surveying, the matrix $\mathbf{E}$ can be readily identified. Because of (4.177), the solution (4.202) can also be written as

$$\hat{\mathbf{x}}_P = -\left(\mathbf{A}^{\mathrm{T}} \mathbf{PA} + \mathbf{E}^{\mathrm{T}} \mathbf{E}\right)^{-1} \mathbf{A}^{\mathrm{T}} \mathbf{PL} \tag{4.212}$$

Note that the covariance matrix of the adjusted parameters is

$$\boldsymbol{\Sigma}_x = \hat{\sigma}_0^2 \mathbf{Q}_{B,C,P} \tag{4.213}$$

depending on whether constraint (4.169), (4.189), or (4.201) is used.

The inner constraint solution is yet another minimal constraint solution, although it has some special features. It can be shown that among all possible minimal constraint solutions, the inner constraint solution also minimizes the sum of the squares of the parameters, i.e.,

$$\mathbf{x}^{\mathrm{T}}\mathbf{x} = \text{minimum} \tag{4.214}$$

This property can be used to obtain a geometric interpretation of the inner constraints. For example, it can be shown that the approximate parameters $\mathbf{x}_0$ and the adjusted parameters $\hat{\mathbf{x}}_P$ can be related by a similarity transformation whose least-squares estimates of translation and rotation are zero. For inner constraint solutions, the standard ellipses show the geometry of the network and are not affected by the definition of the coordinate system. It can also be shown that the trace of $\mathbf{Q}_P$ is the smallest compared to the trace of the other cofactor matrices. All minimal constraint solutions yield the same adjusted observations, a posteriori variance of unit weight, covariance matrices for residuals, and adjusted observations and the same values for estimable functions of the parameters and their variances. The next section presents a further explanation of quantities invariant with respect to changes in minimal constraints.

4.9 STATISTICS IN LEAST-SQUARES ADJUSTMENT

Statistics completes the theory of adjustments, because it allows one to make objective statements about the data. The basic requirements, however, are that the mathematical model and the stochastic model be correct and that the observations have a multivariate normal distribution. Statistics cannot guarantee the right decision, but it can be helpful in gaining deeper insight into often unconscious motives that lead to certain decisions.

4.9.1 Multivariate Normal Distribution

This section contains a brief introduction to multivariate normal distribution. A few theorems are given that will be helpful in subsequent derivations. The multivariate normal distribution is especially pleasing, because the marginal distributions derived from multivariate normal distributions are also normally distributed. An extensive treatment of this distribution is found in the standard statistical literature. To simplify notation, the tilde is not used to identify random variables. The random nature of variables can be readily deduced from the context.

Let $\mathbf{x}$ be a vector with n random components with a mean of

$$E(\mathbf{x}) = \boldsymbol{\mu} \tag{4.215}$$

and a covariance matrix of

$$E\left[(\mathbf{x}-\boldsymbol{\mu})(\mathbf{x}-\boldsymbol{\mu})^{\mathrm{T}}\right] = {}_n\boldsymbol{\Sigma}_n \tag{4.216}$$

If $\mathbf{x}$ has a multivariate normal distribution, then the multivariate density function is

$$f\left(x_1, \ldots, x_n\right) = \frac{1}{(2\pi)^{n/2}\,|\boldsymbol{\Sigma}|^{1/2}} e^{-(\mathbf{x}-\boldsymbol{\mu})^{\mathrm{T}}\boldsymbol{\Sigma}^{-1}(\mathbf{x}-\boldsymbol{\mu})/2} \tag{4.217}$$

The mean and the covariance matrix completely describe the multivariate normal distribution. The notation

$$ {}_{n}\mathbf{x}_1 \sim N_n \left({}_{n}\boldsymbol{\mu}_1, {}_{n}\boldsymbol{\Sigma}_n \right) \tag{4.218} $$

is used. The dimension of the distribution is n.

In the following, some theorems on multivariate normal distributions are given without proofs. These theorems are useful in deriving the distribution of $\mathbf{v}^T \mathbf{P} \mathbf{v}$ and some of the basic statistical tests in least-squares adjustments.

Theorem 1 If $\mathbf{x}$ is multivariate normal

$$ \mathbf{x} \sim N(\boldsymbol{\mu}, \boldsymbol{\Sigma}) \tag{4.219} $$

and

$$ \mathbf{z} = {}_{m}\mathbf{D}_{n}\mathbf{x} \tag{4.220} $$

is a linear function of the random variable, where $\mathbf{D}$ is a $m \times n$ matrix of rank $m \le n$, then

$$ \mathbf{z} \sim N_m \left(\mathbf{D}\, \boldsymbol{\mu}, \mathbf{D}\boldsymbol{\Sigma}\mathbf{D}^T \right) \tag{4.221} $$

is a multivariate normal distribution of dimension m. The mean and variance of the random variable $\mathbf{z}$ follow from the laws for propagating the mean (4.33) and variance-covariances (4.34).

Theorem 2 If $\mathbf{x}$ is multivariate normal $\mathbf{x} \sim N(\boldsymbol{\mu}, \boldsymbol{\Sigma})$, the marginal distribution of any set of components of $\mathbf{x}$ is multivariate normal with means, variances, and covariances obtained by taking the proper component of $\boldsymbol{\mu}$ and $\boldsymbol{\Sigma}$. For example, if

$$ \mathbf{x} = \begin{bmatrix} \mathbf{x}_1 \\ \mathbf{x}_2 \end{bmatrix} \sim N \left(\begin{bmatrix} \boldsymbol{\mu}_1 \\ \boldsymbol{\mu}_2 \end{bmatrix}, \begin{bmatrix} \boldsymbol{\Sigma}_{11} & \boldsymbol{\Sigma}_{12} \\ \boldsymbol{\Sigma}_{21} & \boldsymbol{\Sigma}_{22} \end{bmatrix} \right) \tag{4.222} $$

then the marginal distribution of $\mathbf{x}_2$ is

$$ \mathbf{x}_2 \sim N \left(\boldsymbol{\mu}_2, \boldsymbol{\Sigma}_{22} \right) \tag{4.223} $$

The same law holds, of course, if the set contains only one component, say x_i. The marginal distribution of x_i is then

$$ \mathbf{x}_i \sim n \left(\mu_i, \sigma_i^2 \right) \tag{4.224} $$

Theorem 3 If $\mathbf{x}$ is multivariate normal, a necessary and sufficient condition that two subsets of the random variables are stochastically independent is that the covariances be zero. For example, if

$$\begin{bmatrix} \mathbf{x}_1 \\ \mathbf{x}_2 \end{bmatrix} \sim N\left(\begin{bmatrix} \boldsymbol{\mu}_1 \\ \boldsymbol{\mu}_2 \end{bmatrix}, \begin{bmatrix} \boldsymbol{\Sigma}_{11} & \mathbf{O} \\ \mathbf{O} & \boldsymbol{\Sigma}_{22} \end{bmatrix}\right) \tag{4.225}$$

then $\mathbf{x}_1$ and $\mathbf{x}_2$ are stochastically independent. If one set of normally distributed random variables is uncorrelated with the remaining variables, the two sets are independent. The proof of the above theorem follows from the fact that the density function can be written as a product of $f_1(\mathbf{x}_1)$ and $f_2(\mathbf{x}_2)$ because of the special form of the density function (4.217).

4.9.2 Distribution of $\mathbf{v}^T\mathbf{P}\mathbf{v}$

The derivation of the distribution is based on the assumption that the observations have a multivariate normal distribution. The dimension of the distribution equals the number of observations. In the subsequent derivations the observation equation model is used. However, these statistical derivations could just as well have been carried out with the mixed model.

The observation equations are

$$\begin{aligned} \mathbf{v} &= \mathbf{A}\mathbf{x} + \boldsymbol{\ell}_0 - \boldsymbol{\ell}_b \\ &= \mathbf{A}\mathbf{x} + \boldsymbol{\ell} \end{aligned} \tag{4.226}$$

A first assumption is that the residuals are randomly distributed, i.e., the probability for a positive or negative residual of the equal magnitude is the same. From this assumption it follows that

$$E(\mathbf{v}) = \mathbf{o} \tag{4.227}$$

Because $\mathbf{x}$ and $\boldsymbol{\ell}_0$ are constant vectors, it further follows that the mean and variance-covariance matrix, respectively, are

$$E(\boldsymbol{\ell}_b) = \boldsymbol{\ell}_0 + \mathbf{A}\mathbf{x} \tag{4.228}$$

$$E(\mathbf{v}\mathbf{v}^T) = E\{[\boldsymbol{\ell}_b - E(\boldsymbol{\ell}_b)][\boldsymbol{\ell}_b - E(\boldsymbol{\ell}_b)]^T\} = \boldsymbol{\Sigma}_{\ell_b} = \sigma_0^2\mathbf{P}^{-1} \tag{4.229}$$

The second basic assumption refers to the type of distribution of the observations. It is assumed that the distribution is multivariate normal. Using the mean (4.228) and the covariance matrix (4.229), the n-dimensional multivariate normal distribution of $\boldsymbol{\ell}_b$ is written as

$$\boldsymbol{\ell}_b \sim N_n(\boldsymbol{\ell}_0 + \mathbf{A}\mathbf{x}, \boldsymbol{\Sigma}_{\ell_b}) \tag{4.230}$$

Alternative expressions are

$$\boldsymbol{\ell} \sim N_n(-\mathbf{A}\mathbf{x}, \boldsymbol{\Sigma}_{\ell_b}) \tag{4.231}$$

$$\mathbf{v} \sim N_n\left(\mathbf{o}, \boldsymbol{\Sigma}_{\ell_b}\right) = N_n\left(\mathbf{o}, \sigma_0^2\,\mathbf{P}^{-1}\right) \tag{4.232}$$

Applying two orthogonal transformations we can conveniently derive $\mathbf{v}^{\mathrm{T}}\,\mathbf{P}\mathbf{v}$. If $\boldsymbol{\Sigma}_{\ell_b}$ is nondiagonal, one can always find observations that are stochastically independent and have a unit variate normal distribution. As discussed in Appendix A, for a positive definite matrix $\mathbf{P}$ there exists a nonsingular matrix $\mathbf{D}$ such that the following is valid,

$$\mathbf{D} = \mathbf{E}\,\boldsymbol{\Lambda}^{-1/2} \tag{4.233}$$

$$\mathbf{D}^{\mathrm{T}}\,\mathbf{P}^{-1}\,\mathbf{D} = \mathbf{I} \tag{4.234}$$

$$\mathbf{D}^{\mathrm{T}}\,\mathbf{v} = \mathbf{D}^{\mathrm{T}}\,\mathbf{A}\mathbf{x} + \mathbf{D}^{\mathrm{T}}\,\boldsymbol{\ell} \tag{4.235}$$

$$\bar{\mathbf{v}} = \bar{\mathbf{A}}\mathbf{x} + \bar{\boldsymbol{\ell}} \tag{4.236}$$

$$\bar{\boldsymbol{\ell}} = \mathbf{D}^{\mathrm{T}}\,\boldsymbol{\ell}_0 - \mathbf{D}^{\mathrm{T}}\,\boldsymbol{\ell}_b = \bar{\boldsymbol{\ell}}_0 - \bar{\boldsymbol{\ell}}_b \tag{4.237}$$

$$E(\bar{\mathbf{v}}) = \mathbf{D}^{\mathrm{T}}\,E(\mathbf{v}) = \mathbf{o} \tag{4.238}$$

$$\boldsymbol{\Sigma}_{\bar{v}} = \sigma_0^2\,\mathbf{D}^{\mathrm{T}}\mathbf{P}^{-1}\,\mathbf{D} = \sigma_0^2\,\mathbf{I} \tag{4.239}$$

$$\bar{\boldsymbol{\ell}} \sim N_n\left(\mathbf{o}, \sigma_0^2\,\mathbf{I}\right) \tag{4.240}$$

The columns of the orthogonal matrix $\mathbf{E}$ consist of the normalized eigenvectors of $\mathbf{P}^{-1}$; $\boldsymbol{\Lambda}$ is a diagonal matrix having the eigenvalues of $\mathbf{P}^{-1}$ at the diagonal. The quadratic form $\mathbf{v}^{\mathrm{T}}\mathbf{P}\mathbf{v}$ remains invariant under this transformation because

$$R \equiv \mathbf{v}^{\mathrm{T}}\mathbf{P}\mathbf{v} = \bar{\mathbf{v}}^{\mathrm{T}}\boldsymbol{\Lambda}^{1/2}\,\mathbf{E}^{\mathrm{T}}\mathbf{P}\mathbf{E}\boldsymbol{\Lambda}^{1/2}\bar{\mathbf{v}} = \bar{\mathbf{v}}^{\mathrm{T}}\boldsymbol{\Lambda}^{1/2}\boldsymbol{\Lambda}^{-1}\boldsymbol{\Lambda}^{1/2}\bar{\mathbf{v}} = \bar{\mathbf{v}}^{\mathrm{T}}\bar{\mathbf{v}} \tag{4.241}$$

If the covariance matrix $\boldsymbol{\Sigma}_{\ell_b}$ has a rank defect, then one could use matrix $\mathbf{F}$ of (A.52) for the transformation. The dimension of the transformed observations $\bar{\boldsymbol{\ell}}_b$ equals the rank of the covariance matrix.

In the next step, the parameters are transformed to a new set that is stochastically independent. To keep the generality, let the matrix $\bar{\mathbf{A}}$ in (4.236) have less than full column rank, i.e., $R(\bar{\mathbf{A}}) = r < u$. Let the matrix $\mathbf{F}$ be an $n \times r$ matrix whose columns constitute an orthonormal basis for the column space of $\bar{\mathbf{A}}$. One such choice for the columns of $\mathbf{F}$ may be to take the normalized eigenvectors of $\bar{\mathbf{A}}\bar{\mathbf{A}}^{\mathrm{T}}$. Let $\mathbf{G}$ be an $n \times (n-r)$ matrix, such that $[\mathbf{F}\ \mathbf{G}]$ is orthogonal and such that the columns of $\mathbf{G}$ constitute an orthonormal basis to the $n-r$-dimensional null space of $\bar{\mathbf{A}}\bar{\mathbf{A}}^{\mathrm{T}}$. Such a matrix always exists. There is no need to compute this matrix explicitly. With these specifications we obtain

$$\begin{bmatrix} \mathbf{F}^{\mathrm{T}} \\ \mathbf{G}^{\mathrm{T}} \end{bmatrix} [\mathbf{F} \quad \mathbf{G}] = \begin{bmatrix} \mathbf{F}^{\mathrm{T}}\,\mathbf{F} & \mathbf{F}^{\mathrm{T}}\,\mathbf{G} \\ \mathbf{G}^{\mathrm{T}}\,\mathbf{F} & \mathbf{G}^{\mathrm{T}}\,\mathbf{G} \end{bmatrix} = \begin{bmatrix} {}_r\mathbf{I}_r & \mathbf{O} \\ \mathbf{O} & {}_{n-r}\mathbf{I}_{n-r} \end{bmatrix} \tag{4.242}$$

$$[\mathbf{F} \quad \mathbf{G}]\,[\mathbf{F} \quad \mathbf{G}]^{\mathrm{T}} = \mathbf{F}\,\mathbf{F}^{\mathrm{T}} + \mathbf{G}\,\mathbf{G}^{\mathrm{T}} = \mathbf{I} \tag{4.243}$$

$$\bar{\mathbf{A}}^{\mathrm{T}}\,\mathbf{G} = \mathbf{O} \tag{4.244}$$

$$\mathbf{G}^{\mathrm{T}}\,\bar{\mathbf{A}} = \mathbf{O} \tag{4.245}$$

The required transformation is

$$\begin{bmatrix} \mathbf{F}^{\mathrm{T}} \\ \mathbf{G}^{\mathrm{T}} \end{bmatrix} \bar{\mathbf{v}} = \begin{bmatrix} \mathbf{F}^{\mathrm{T}} \\ \mathbf{G}^{\mathrm{T}} \end{bmatrix} \bar{\mathbf{A}}\mathbf{x} + \begin{bmatrix} \mathbf{F}^{\mathrm{T}} \\ \mathbf{G}^{\mathrm{T}} \end{bmatrix} \bar{\boldsymbol{\ell}} \tag{4.246}$$

or, equivalently,

$$\begin{bmatrix} \mathbf{F}^{\mathrm{T}}\,\bar{\mathbf{v}} \\ \mathbf{G}^{\mathrm{T}}\,\bar{\mathbf{v}} \end{bmatrix} = \begin{bmatrix} \mathbf{F}^{\mathrm{T}}\,\bar{\mathbf{A}}\mathbf{x} \\ \mathbf{o} \end{bmatrix} + \begin{bmatrix} \mathbf{F}^{\mathrm{T}}\,\bar{\boldsymbol{\ell}} \\ \mathbf{G}^{\mathrm{T}}\,\bar{\boldsymbol{\ell}} \end{bmatrix} \tag{4.247}$$

Labeling the newly transformed observations by $\mathbf{z}$, i.e.,

$$\mathbf{z} = \begin{bmatrix} \mathbf{z}_1 \\ \mathbf{z}_2 \end{bmatrix} = \begin{bmatrix} \mathbf{F}^{\mathrm{T}}\,\bar{\boldsymbol{\ell}} \\ \mathbf{G}^{\mathrm{T}}\,\bar{\boldsymbol{\ell}} \end{bmatrix} \tag{4.248}$$

we can write (4.247) as

$$\bar{\mathbf{v}}_z = \begin{bmatrix} \bar{\mathbf{v}}_{z_1} \\ \bar{\mathbf{v}}_{z_2} \end{bmatrix} = \begin{bmatrix} \mathbf{F}^{\mathrm{T}}\,\bar{\mathbf{A}}\mathbf{x} \\ \mathbf{o} \end{bmatrix} + \begin{bmatrix} \mathbf{z}_1 \\ \mathbf{z}_2 \end{bmatrix} \tag{4.249}$$

There are r random variables in $\mathbf{z}_1$ and $n - r$ random variables in $\mathbf{z}_2$. The quadratic form again remains invariant under the orthogonal transformation, since

$$\begin{aligned} \bar{\mathbf{v}}_z^{\mathrm{T}}\bar{\mathbf{v}}_z &= \bar{\mathbf{v}}^{\mathrm{T}}\left(\mathbf{F}\mathbf{F}^{\mathrm{T}} + \mathbf{G}\mathbf{G}^{\mathrm{T}}\right)\bar{\mathbf{v}} \\ &= \bar{\mathbf{v}}^{\mathrm{T}}\bar{\mathbf{v}} = R \end{aligned} \tag{4.250}$$

according to (4.243). The actual quadratic form is obtained from (4.249):

$$R = \bar{\mathbf{v}}_z^{\mathrm{T}}\bar{\mathbf{v}}_z = \left(\mathbf{F}^{\mathrm{T}}\bar{\mathbf{A}}\mathbf{x} + \mathbf{z}_1\right)^{\mathrm{T}}\left(\mathbf{F}^{\mathrm{T}}\bar{\mathbf{A}}\mathbf{x} + \mathbf{z}_1\right) + \mathbf{z}_2^{\mathrm{T}}\mathbf{z}_2 \tag{4.251}$$

The least-squares solution requires that R be minimized by variation of the parameters. Generally, equating partial derivatives with respect to $\mathbf{x}$ to zero and solving the resulting equations gives the minimum. The special form of (4.251) permits a much simpler approach. The expressions on the right side of Equation (4.251) consist of the sum of two positive terms (sum of squares). Because only the first term is a function of the parameters $\mathbf{x}$, the minimum is achieved if the first term is zero, i.e.,

$$-\,{}_r\mathbf{F}^{\mathrm{T}}_{n}\,{}_n\bar{\mathbf{A}}_u\,{}_u\hat{\mathbf{x}}_1 = \mathbf{z}_1 \tag{4.252}$$

Note that the caret identifies the estimated parameters. Consequently, the estimate of the quadratic form is

$$\hat{R} = \mathbf{z}_2^{\mathrm{T}} \mathbf{z}_2 \tag{4.253}$$

Because there are $r < u$ equations for the u parameters in (4.252), there always exists a solution for $\hat{\mathbf{x}}$. The simplest approach is to equate $u - r$ parameters to zero. This would be identical to having these $u - r$ parameters treated as constants in the adjustment. They could be left out when setting up the design matrix and thus, the singularity problem would be avoided altogether. Equation (4.252) can be solved subject to $u - r$ general conditions between the parameters. The resulting solution is a minimal constraint solution. If the particular condition (4.201) is applied, one obtains the inner constraint solution. If $\bar{\mathbf{A}}$ has no rank defect, then the system (4.252) consists of u equations for u unknowns.

The estimate for the quadratic form (4.253) does not depend on the parameters $\mathbf{x}$ and, thus, is invariant with respect to the selection of the minimal constraints for finding the least-square estimate of $\mathbf{x}$. Moreover, the residuals themselves are independent of the minimal constraints. Substituting the solution (4.252) into (4.247) gives

$$\begin{bmatrix} \mathbf{F}^{\mathrm{T}} \\ \mathbf{G}^{\mathrm{T}} \end{bmatrix} \hat{\bar{\mathbf{v}}} = \begin{bmatrix} \mathbf{o} \\ \mathbf{G}^{\mathrm{T}} \bar{\boldsymbol{\ell}} \end{bmatrix} \tag{4.254}$$

Since the matrix $[\mathbf{F}\ \mathbf{G}]$ is orthonormal, the expression for the residuals becomes

$$\hat{\bar{\mathbf{v}}} = [\mathbf{F} \quad \mathbf{G}] \begin{bmatrix} \mathbf{o} \\ \mathbf{G}^{\mathrm{T}} \bar{\boldsymbol{\ell}} \end{bmatrix} = \mathbf{G}\mathbf{G}^{\mathrm{T}} \bar{\boldsymbol{\ell}} \tag{4.255}$$

Thus, the residuals are independent of the specific solution for $\hat{\mathbf{x}}$. The matrix $\mathbf{G}$ depends only on the structure of the design matrix $\bar{\mathbf{A}}$. By applying the law of variance-covariance propagation to (4.255), we clearly see that the covariance matrix of the adjusted residuals, and thus the covariance matrix of the adjusted observations, does not depend on the specific set of minimal constraints. Note that the transformation (4.235) does not invalidate these statements, since the $\mathbf{D}$ matrix is not related to the parameters.

Returning to the derivation of the distribution of $\mathbf{v}^{\mathrm{T}}\mathbf{P}\mathbf{v}$, we find from (4.248) that

$$\mathrm{E}(\mathbf{z}) = \begin{bmatrix} -\mathbf{F}^{\mathrm{T}} \bar{\mathbf{A}} \mathbf{x} \\ \mathbf{o} \end{bmatrix} \tag{4.256}$$

using (4.245) and the fact that $E(\bar{\boldsymbol{\ell}}) = -\bar{\mathbf{A}}\mathbf{x}$ according to (4.236). Making use of (4.240) the covariance matrix is

$$\boldsymbol{\Sigma}_z = \sigma_0^2 \begin{bmatrix} \mathbf{F}^{\mathrm{T}} \\ \mathbf{G}^{\mathrm{T}} \end{bmatrix} \mathbf{I}\,[\mathbf{F} \quad \mathbf{G}] = \sigma_0^2 \begin{bmatrix} \mathbf{F}^{\mathrm{T}}\mathbf{F} & \mathbf{F}^{\mathrm{T}}\mathbf{G} \\ \mathbf{G}^{\mathrm{T}}\mathbf{F} & \mathbf{G}^{\mathrm{T}}\mathbf{G} \end{bmatrix} = \sigma_0^2 \begin{bmatrix} \mathbf{I} & \mathbf{O} \\ \mathbf{O} & \mathbf{I} \end{bmatrix} \tag{4.257}$$

Since a linear transformation of a random variable with multivariate normal distribution results in another multivariate normal distribution according to Theorem 1, it follows that $\mathbf{z}$ is distributed as

$$\mathbf{z} \sim N_n \left(\begin{bmatrix} -\mathbf{F}^{\mathrm{T}} \bar{\mathbf{A}} \mathbf{x} \\ \mathbf{o} \end{bmatrix}, \sigma_0^2 \begin{bmatrix} {}_r\mathbf{I}_r & \mathbf{O} \\ \mathbf{O} & {}_{n-r}\mathbf{I}_{n-r} \end{bmatrix} \right) \tag{4.258}$$

The random variables $\mathbf{z}_1$ and $\mathbf{z}_2$ are stochastically independent, as are the individual components according to Theorem 3. From Theorem 2 it follows that

$$\mathbf{z}_2 \sim N_{n-r}(\mathbf{o}, \sigma_0^2 \mathbf{I}) \tag{4.259}$$

Thus

$$z_{2i} \sim n\left(0, \sigma_0^2\right) \tag{4.260}$$

$$\frac{z_{2i}}{\sigma_0} \sim n(0, 1) \tag{4.261}$$

are unit variate normal distributed. As listed in Appendix A5, the square of a standardized normal distributed variable has a chi-square distribution with one degree of freedom. In addition, the sum of chi-square distributed variables is also a chi-square distribution with a degree of freedom equal to the sum of the individual degrees of freedom. Using these functions of random variables, it follows that $\mathbf{v}^{\mathrm{T}}\mathbf{P}\mathbf{v}$

$$\frac{\hat{R}}{\sigma_0^2} = \frac{\mathbf{z}_2^{\mathrm{T}} \mathbf{z}_2}{\sigma_0^2} = \sum_{i=1}^{n-r} \frac{z_{2i}^2}{\sigma_0^2} \sim \chi_{n-r}^2 \tag{4.262}$$

has a chi-square distribution with $n - r$ degrees of freedom.

4.9.3 Testing $\mathbf{v}^{\mathrm{T}}\mathbf{P}\mathbf{v}$ and $\Delta\mathbf{v}^{\mathrm{T}}\mathbf{P}\mathbf{v}$

Combining the result of (4.262) with the expression for the a posteriori variance of unit weight of Table 4.1, we obtain the formulation for a fundamental statistical test in least-squares estimation:

$$\frac{\mathbf{v}^{\mathrm{T}} \mathbf{P} \mathbf{v}}{\sigma_0^2} = \frac{\hat{\sigma}_0^2}{\sigma_0^2}(n - r) \sim \chi_{n-r}^2 \tag{4.263}$$

Note that $n-r$ is the degree of freedom of the adjustment. If there is no rank deficiency in the design matrix, the degree of freedom is $n-u$. Based on the statistics (4.263), the test can be performed to find out whether the adjustment is distorted. The formulation of the hypothesis is as follows:

$$H_0\colon \sigma_0^2 = \hat{\sigma}_0^2 \tag{4.264}$$

$$H_1: \sigma_0^2 \neq \hat{\sigma}_0^2 \tag{4.265}$$

The zero hypothesis states that the a priori variance of unit weight statistically equals the a posteriori variance of unit weight. Recall that the a posteriori variance of unit weight is a random variable; the adjustment makes a sample value available for this quantity on the basis of the observations (the samples). Both variances of unit weight do not have to be numerically equal; but they should be statistically equal in the sense of (4.67). If the zero hypothesis is accepted, the adjustment is judged to be correct. If the numerical value

$$\chi^2 = \frac{\hat{\sigma}_0^2}{\sigma_0^2}(n - r) = \frac{\mathbf{v}^T \mathbf{P} \mathbf{v}}{\sigma_0^2} \tag{4.266}$$

is such that

$$\chi^2 < \chi^2_{n-r,1-\alpha/2} \tag{4.267}$$

$$\chi^2 > \chi^2_{n-r,\alpha/2} \tag{4.268}$$

then the zero hypothesis is rejected. The significance level α, i.e., the probability of a type-I error, or the probability of rejecting the zero hypothesis even though it is true, is generally fixed to 0.05. Here the significance level is the sum of the probabilities in both tails. Table 4.6 lists selected values from the chi-square distribution $\chi^2_{n-r,\alpha}$. Rejection of the zero hypothesis is taken to indicate that something is wrong with the adjustment. The cause for rejection remains to be clarified. Figure 4.3 shows the limits for the posteriori variance of unit weight as a function of the degree of freedom given the significance level $\alpha = 0.05$.

The probability β of the type-II error, i.e., the probability of rejecting the alternative hypothesis and accepting the zero hypothesis even though the alternative hypothesis is true, is generally not computed. Type-II errors are considered in Section 4.10.2 in regards to reliability and in Section 7.8.3 in regards to discernibility of estimated ambiguity sets.

TABLE 4.6 Selected Values for Chi-Square

Degree of Freedom (DF)	Probability α 0.975	0.950	0.050	0.025
1	0.00	0.00	3.84	5.02
5	0.83	1.15	11.07	12.83
10	3.25	3.94	18.31	20.48
20	9.59	10.85	31.41	34.17
50	32.36	34.76	67.50	71.42
100	74.22	77.93	124.34	129.56

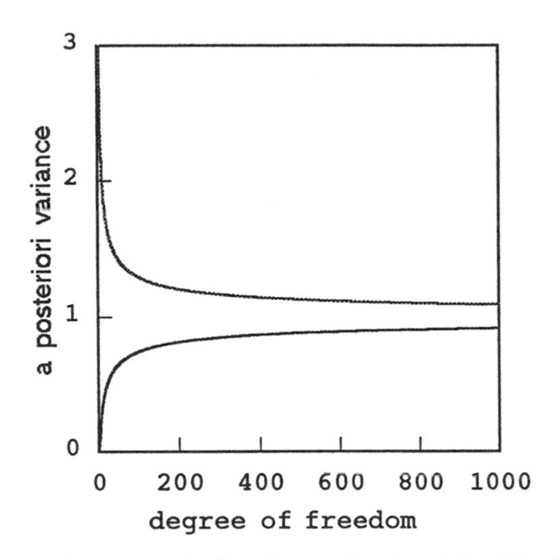

Figure 4.3 Limits on the a posteriori variance of unit weight. The figure refers to $\alpha = 0.05$.

The test statistics for testing groups of observations is based on $\mathbf{v}^{\mathrm{T}}\mathbf{P}\mathbf{v}^*$ and the change $\Delta\mathbf{v}^{\mathrm{T}}\mathbf{P}\mathbf{v}$. According to Table 4.2 we have

$$\begin{aligned}\Delta\mathbf{v}^{\mathrm{T}}\mathbf{P}\mathbf{v} &= \left(\mathbf{A}_2\mathbf{x}^* + \boldsymbol{\ell}_2\right)^{\mathrm{T}} \mathbf{T}\left(\mathbf{A}_2\mathbf{x}^* + \boldsymbol{\ell}_2\right) \\ &= \mathbf{z}_3^{\mathrm{T}}\mathbf{T}\,\mathbf{z}_3\end{aligned} \tag{4.269}$$

The new random variable $\mathbf{z}_3$ is a function of observations $\boldsymbol{\ell}_1$ and $\boldsymbol{\ell}_2$. Applying the laws of propagation of mean and variance, one finds

$$E(\mathbf{z}_3) = \mathbf{A}_2 E(\mathbf{x}^*) + E(\boldsymbol{\ell}_2) = \mathbf{A}_2\mathbf{x} - \mathbf{A}_2\mathbf{x} = \mathbf{o} \tag{2.470}$$

$$\boldsymbol{\Sigma}_{z_3} = \mathbf{T}^{-1} \tag{4.271}$$

$$\mathbf{z}_3 \sim N\left(\mathbf{o}, \sigma_0^2\,\mathbf{T}^{-1}\right) \tag{4.272}$$

Carrying out the orthonormal transformation yields a random vector whose components are stochastically independent and normally distributed. By standardizing these distributions and summing the squares of these random variables, it follows that

$$\frac{\Delta\mathbf{v}^{\mathrm{T}}\mathbf{P}\mathbf{v}}{\sigma_0^2} = \frac{\mathbf{z}_3^{\mathrm{T}}\mathbf{T}\mathbf{z}_3}{\sigma_0^2} \sim \chi^2_{n_2} \tag{4.273}$$

has a chi-square distribution with n_2 degrees of freedom, where n_2 equals the number of observations in the second group. The random variables (4.273) and (4.263) are stochastically independent. To prove this, consider the new random variable $\mathbf{z} =$

$[\mathbf{z}_1\ \mathbf{z}_2\ \mathbf{z}_3]^T$, which is a linear function of the random variables $\boldsymbol{\ell}$ (first group) and $\boldsymbol{\ell}_2$, according to Equations (4.235), (2.248), and (4.269). By using the covariance matrix (4.88) and applying variance-covariance propagation, we find that the covariances between the $\mathbf{z}_i$ are zero. Because the distribution of the $\mathbf{z}$ is multivariate normal, it follows that the random variables $\mathbf{z}_i$ are stochastically independent. Since $\Delta\mathbf{v}^T\mathbf{P}\mathbf{v}$ is a function of $\mathbf{z}_3$ only, it follows that $\mathbf{v}^T\mathbf{P}\mathbf{v}$ in (4.263), which is only a function of $\mathbf{z}_2$, and $\Delta\mathbf{v}^T\mathbf{P}\mathbf{v}$ in (4.273) are stochastically independent. Thus, it is permissible to form the following ratio of random variables:

$$\frac{\Delta\mathbf{v}^T\mathbf{P}\mathbf{v}(n_1 - r)}{\mathbf{v}^T\mathbf{P}\mathbf{v}^*(n_2)} \sim F_{n_2,n_1-r} \tag{4.274}$$

which has an F distribution.

Thus the fundamental test in sequential adjustment is based on the F distribution. The zero hypothesis states that the second group of observations does not distort the adjustment, or that there is no indication that something is wrong with the second group of observations. The alternative hypothesis states that there is an indication that the second group of observations contains errors. The zero hypothesis is rejected, and the alternative hypothesis is accepted if

$$F < F_{n_2,n_1-r,1-\alpha/2} \tag{4.275}$$

$$F > F_{n_2,n_1-r,\alpha/2} \tag{4.276}$$

Table 4.7 lists selected values from the F distribution as a function of the degrees of freedom and probability. The tabulation refers to the parameters as specified in $F_{n1,n2,0.05}$.

4.9.4 General Linear Hypothesis

The general linear hypothesis deals with linear conditions between parameters. Nonlinear conditions are first linearized. The basic idea is to test the change $\Delta\mathbf{v}^T\mathbf{P}\mathbf{v}$ for its statistical significance. Any of the three adjustment models can be used to carry out

TABLE 4.7 Selected Values for F

	n_1			
n_2	1	2	3	4
5	6.61	5.79	5.41	5.19
10	4.96	4.10	3.71	3.48
20	4.35	3.49	3.10	2.87
60	4.00	3.15	2.76	2.53
120	3.92	3.07	2.68	2.45
∞	3.84	3.00	2.60	2.37

the general linear hypothesis test. For the observation equation model with additional conditions between the parameters, one has

$$\mathbf{v}_1 = \mathbf{A}_1\mathbf{x} + \boldsymbol{\ell}_1 \tag{4.277}$$

$$H_0\colon \mathbf{A}_2\mathbf{x} + \boldsymbol{\ell}_2 = \mathbf{o} \tag{4.278}$$

Equation (4.278) expresses the zero hypothesis H_0. The solution of the combined adjustment is found in Table 4.5. Adjusting (4.277) alone results in $\mathbf{v}^{\mathrm{T}}\mathbf{P}\mathbf{v}^*$, which has a chi-square distribution with $n - r$ degrees of freedom according to (4.273). The change $\Delta\mathbf{v}^{\mathrm{T}}\mathbf{P}\mathbf{v}$ resulting from the condition (4.278) is

$$\Delta\mathbf{v}^{\mathrm{T}}\mathbf{P}\mathbf{v} = \left(\mathbf{A}_2\mathbf{x}^* + \boldsymbol{\ell}_2\right)^{\mathrm{T}} \mathbf{T}\left(\mathbf{A}_2\mathbf{x}^* + \boldsymbol{\ell}_2\right) \tag{4.279}$$

The expression in (4.279) differs from (4.269) in two respects. First, the matrix $\mathbf{T}$ differs; i.e., the matrix $\mathbf{T}$ in (4.279) does not contain the $\mathbf{P}_2$ matrix. Second, the quantity $\boldsymbol{\ell}_2$ is not a random variable. These differences, however, do not matter in the proof of stochastic independence of $\mathbf{v}^{\mathrm{T}}\mathbf{P}\mathbf{v}^*$ and $\Delta\mathbf{v}^{\mathrm{T}}\mathbf{P}\mathbf{v}$. Analogously to (4.269), we can express the change $\Delta\mathbf{v}^{\mathrm{T}}\mathbf{P}\mathbf{v}$ in (4.279) as a function of a new random variable $\mathbf{z}_3$. The proof for stochastic independence follows the same lines of thought as given before (for the case of additional observations). Thus, just as (4.274) is the basis for testing two groups of observations, the basic test for the general linear hypothesis (4.278) is

$$\frac{\Delta\mathbf{v}^{\mathrm{T}}\mathbf{P}\mathbf{v}\,(n_1 - r)}{\mathbf{v}^{\mathrm{T}}\mathbf{P}\mathbf{v}^* \quad n_2} \sim F_{n_2,\, n_1-r} \tag{4.280}$$

A small $\Delta\mathbf{v}^{\mathrm{T}}\mathbf{P}\mathbf{v}$ implies that the null hypothesis (4.278) is acceptable; i.e., the conditions are in agreement with the observations. The conditions do not impose any distortions on the adjustment. The rejection criterion is based on the one-tail test at the upper end of the distribution. Thus, reject H_0 at a $100\alpha\%$ significance level if

$$F > F_{n_2,\, n_1-r,\, \alpha} \tag{4.281}$$

The general formulation of the null hypothesis in (4.278) makes it possible to test any hypothesis on the parameters, so long as the hypothesis can be expressed in a mathematical equation. Nonlinear hypotheses must first be linearized. Simple hypotheses could be used to test whether an individual parameter has a certain numerical value, whether two parameters are equal, whether the distance between two stations has a certain length, whether an angle has a certain size, etc. For example, consider the hypothesis

$$H_0\colon \mathbf{x} - \mathbf{x}_T = \mathbf{o} \tag{4.282}$$

$$H_1\colon \mathbf{x} - \mathbf{x}_T \neq \mathbf{o} \tag{4.283}$$

The zero hypothesis states that the parameters equal a certain (true) value $\mathbf{x}_T$. From (4.278) it follows that $\mathbf{A}_2 = \mathbf{I}$ and $\boldsymbol{\ell}_2 = -\mathbf{x}_T$. Using these specifications we can use $\mathbf{T} = \mathbf{N}$ in (4.279), and the statistic (4.280) becomes

$$\frac{\left(\hat{\mathbf{x}}^* - \mathbf{x}_T\right)^{\mathrm{T}} \mathbf{N} \left(\hat{\mathbf{x}}^* - \mathbf{x}_T\right)}{\hat{\sigma}_0^2\, r} \sim F_{r,\, n_1 - r,\, \alpha} \tag{4.284}$$

where the a posteriori variance of unit weight (first group only) has been substituted for $\mathbf{v}^{\mathrm{T}}\mathbf{P}\mathbf{v}^*$. Once the adjustment of the first group (4.277) is completed, the values for the adjusted parameters and the a posteriori variance of unit weight are entered in (4.284), and the fraction is computed and compared with the F value (taking the proper degrees of freedom and the desired significance level into account). Rejection or acceptance of the zero hypothesis follows rule (4.281).

Note that one of the degrees of freedom in (4.284) is $r = R(\mathbf{N}) < u$, instead of u, which equals the number of parameters, even though Equation (4.282) expresses u conditions. Because of the possible rank defect of the normal matrix $\mathbf{N}$, the distribution of $\Delta\mathbf{v}^{\mathrm{T}}\mathbf{P}\mathbf{v}$ in (4.279) is a chi-square distribution with r degrees of freedom. Consider the derivation leading to (4.273). The u components of $\mathbf{z}_3$ are transformed to r stochastically independent unit variate normal distributions that are then squared and summed to yield the distribution of $\Delta\mathbf{v}^{\mathrm{T}}\mathbf{P}\mathbf{v}$. The interpretation is that (4.282) represents one hypothesis on all parameters $\mathbf{x}$, and not u hypotheses on the u components on $\mathbf{x}$.

Expression (4.284) can be used to define the r-dimensional confidence region. Replace the particular $\mathbf{x}_T$ by the unknown parameter $\mathbf{x}$, and drop the asterisk; then

$$P\left[\frac{(\hat{\mathbf{x}} - \mathbf{x})^{\mathrm{T}} \mathbf{N} (\hat{\mathbf{x}} - \mathbf{x})}{\hat{\sigma}_0^2\, r} \le F_{r,n_1-r,\alpha}\right] = \int_0^{F_{r,n_1-r,\alpha}} F_{r,n_1-r}\, dF = 1 - \alpha \tag{4.285}$$

The probability region described by the expression on the left side of Equation (4.285) is an $R(\mathbf{N})$-dimensional ellipsoid. The probability region is an ellipsoid, because the normal matrix $\mathbf{N}$ is positive definite or, at least, semipositive definite. If one identifies the center of the ellipsoid with $\hat{\mathbf{x}}$, then there is $(1 - \alpha)$ probability that the unknown point $\mathbf{x}$ lies within the ellipsoid. The orientation and the size of this ellipsoid are a function of the eigenvectors and eigenvalues of the normal matrix, the rank of the normal matrix, and the degree of freedom. Consider the orthonormal transformation

$$\mathbf{z} = \mathbf{F}^{\mathrm{T}} (\mathbf{x} - \hat{\mathbf{x}}) \tag{4.286}$$

with $\mathbf{F}$ as specified in (A.52) and containing the normalized eigenvectors of $\mathbf{N}$, then

$$\mathbf{F}^{\mathrm{T}}\mathbf{N}\mathbf{F} = \boldsymbol{\Lambda} \tag{4.287}$$

with $\boldsymbol{\Lambda}$ containing the r eigenvalues of $\mathbf{N}$, and

$$(\hat{\mathbf{x}} - \mathbf{x})^{\mathrm{T}} \mathbf{N} (\hat{\mathbf{x}} - \mathbf{x}) = \mathbf{z}^{\mathrm{T}} \boldsymbol{\Lambda} \mathbf{z} = \sum_{i=1}^{r} z_i^2 \lambda_i = \sum_{i=1}^{r} \frac{z_i^2}{\left(1/\sqrt{\lambda_i}\right)^2} \tag{4.288}$$

Combining Equations (4.285) and (4.288), we can write the r-dimensional ellipsoid, or the r-dimensional confidence region, in the principal axes form:

$$\mathrm{P}\left[\frac{z_1^2}{\left(\hat{\sigma}_0\sqrt{r F_{r,n-r,\alpha}/\lambda_1}\right)^2} + \cdots + \frac{z_r^2}{\left(\hat{\sigma}_0\sqrt{r F_{r,n-r,\alpha}/\lambda_r}\right)^2} \leq 1\right] = 1 - \alpha \tag{4.289}$$

The confidence region is centered at $\hat{\mathbf{x}}$. Whenever the zero hypothesis H_0 of (4.282) is accepted, the point $\mathbf{x}_T$ falls within the confidence region. The probability that the ellipsoid contains the true parameters $\mathbf{x}_T$, is $1 - \alpha$. For these reasons, one naturally would like the ellipsoid to be small. Equation (4.289) shows that the semimajor axes are proportional to the inverse of the eigenvalues of the normal matrix. It is exactly this relationship that makes us choose the eigenvalues of $\mathbf{N}$ as large as possible, provided that we have a choice through appropriate network design variation. As an eigenvalue approaches zero, the respective axis of the confidence ellipsoid approaches infinity; this is an undesirable situation, both from a statistical point of view and because of the numerical difficulties encountered during the inversion of the normal matrix.

4.9.5 Ellipses as Confidence Regions

Confidence ellipses are statements of precision. They are frequently used in connection with two-dimensional networks in order to make the directional precision of station location visible. Ellipses of confidence follow from Section 4.9.4 simply by limiting the hypothesis (4.282) to two parameters, i.e., the Cartesian coordinates of a station. Of course, in a three-dimensional network one can compute three-dimensional ellipsoids or several ellipses, e.g., one for the horizontal and others for the vertical. Confidence ellipses or ellipsoids are not limited to the specific application of networks. However, in networks the confidence regions can be referenced with respect to the coordinate system of the network and thus can provide an integrated view of the geometry of the confidence regions and the network.

Consider the following hypothesis:

$$H_0\colon \mathbf{x}_i - \mathbf{x}_{i,T} = \mathbf{o} \tag{4.290}$$

where the notation

$$\mathbf{x}_i = [x_1 \quad x_2]^{\mathrm{T}} \tag{4.291}$$

is used. The symbols x_1 and x_2 denote the Cartesian coordinates of a two-dimensional network station P_i. The test of this hypothesis follows the outline given in the previous section. The $\mathbf{A}_2$ matrix is of size $2 \times u$ because there are two separate equations in the hypothesis and u components in $\mathbf{x}$. The elements of $\mathbf{A}_2$ are zero except those elements of rows 1 and 2, which correspond to the respective positions of x_1 and x_2 in $\mathbf{x}$. With these specifications it follows that

$$\mathbf{Q}_i = \mathbf{A}_2 \mathbf{N}^{-1} \mathbf{A}_2^{\mathrm{T}} = \begin{bmatrix} q_{x_1} & q_{x_1,x_2} \\ q_{x_2,x_1} & q_{x_2} \end{bmatrix} \tag{4.292}$$

$\mathbf{Q}_i$ contains the respective elements of the inverse of the normal matrix. With these specifications $\mathbf{T} = \mathbf{Q}_i^{-1}$ and Expression (4.280) becomes

$$\frac{1}{2\hat{\sigma}_0^2} \left(\hat{\mathbf{x}}_i - \mathbf{x}_{i,T}\right)^{\mathrm{T}} \mathbf{Q}_i^{-1} \left(\hat{\mathbf{x}}_i - \mathbf{x}_{i,T}\right) \sim F_{2,n-r} \tag{4.293}$$

Given the significance level α, the hypothesis test can be carried out. The two-dimensional confidence region is

$$P\left[\frac{(\hat{\mathbf{x}}_i - \mathbf{x}_i)^{\mathrm{T}} \mathbf{Q}_i^{-1} (\hat{\mathbf{x}}_i - \mathbf{x}_i)}{2\hat{\sigma}_0^2} \le F_{2,n-r,\alpha}\right] = \int_0^{F_{2,n-r,\alpha}} F_{2,n-r}\, dF = 1 - \alpha \tag{4.294}$$

The size of the confidence ellipses defined by (4.294) depends on the degree of freedom of the adjustment and the significance level. The ellipses are centered at the adjusted position and delimit the $(1 - \alpha)$ probability area for the true position. The principal axis form of (4.294) is obtained through orthogonal transformation. Let $\mathbf{R}_i$ denote the matrix whose rows are the orthonormal eigenvectors of $\mathbf{Q}_i$, then

$$\mathbf{R}_i^{\mathrm{T}} \mathbf{Q}_i^{-1} \mathbf{R}_i = \mathbf{\Lambda}_i^{-1} \tag{4.295}$$

according to (A.48). The matrix $\mathbf{\Lambda}_i$ is diagonal and contains the eigenvalues λ_i^Q and λ_2^Q of $\mathbf{Q}_1$. With

$$\mathbf{z}_i = \mathbf{R}_i^{\mathrm{T}} \left(\hat{\mathbf{x}}_i - \hat{\mathbf{x}}_i\right) \tag{4.296}$$

Expression (4.294) becomes

$$\mathrm{P}\left\{\left[\frac{z_1^2}{\left(\hat{\sigma}_0\sqrt{\lambda_1^Q 2F_{2,n-r,\alpha}}\right)^2} + \frac{z_2^2}{\left(\hat{\sigma}_0\sqrt{\lambda_2^Q 2F_{2,n-r,\alpha}}\right)^2}\right] \le 1\right\}$$
$$= \int_0^{F_{2,n-r,\alpha}} F_{2,n-r}\, dF = 1 - \alpha \tag{4.297}$$

For $F_{2,n-r,\alpha} = 1/2$, the ellipse is called the *standard ellipse* or the *error ellipse.* Thus, the probability enclosed by the standard ellipse is a function of the degree of freedom $n - r$ and is computed as follows:

$$\mathrm{P(standard\ ellipse)} = \int_0^{1/2} F_{2,n-r}\, dF \tag{4.298}$$

TABLE 4.8 Magnification Factor for Standard Ellipses

	Probability $1 - \alpha$		
$n - r$	95%	98%	99%
1	20.00	50.00	100.00
2	6.16	9.90	14.10
3	4.37	6.14	7.85
4	3.73	4.93	6.00
5	3.40	4.35	5.15
6	3.21	4.01	4.67
8	2.99	3.64	4.16
10	2.86	3.44	3.89
12	2.79	3.32	3.72
15	2.71	3.20	3.57
20	2.64	3.09	3.42
30	2.58	2.99	3.28
50	2.52	2.91	3.18
100	2.49	2.85	3.11
∞	2.45	2.80	3.03

The magnification factor, $\sqrt{2F_{2,n-r,\alpha}}$, as a function of the probability and the degree of freedom, is shown in Table 4.8. The table shows immediately that a small degree of freedom requires a large magnification factor to obtain, e.g., 95% probability. It is seen that in the range of small degrees of freedom, an increase in the degree of freedom rapidly decreases the magnification factor, whereas with a large degree of freedom, any additional observations cause only a minor reduction of the magnification factor. After a degree of freedom of about 8 or 10, the decrease in the magnification factor slows down noticeably. Thus, based on the speed of decreasing magnification factor, a degree of 10 appears optimal, considering the expense of additional observations and the little gain derived from them in the statistical sense. For a degree of freedom of 10, the magnification factor is about 3 to cover 95% probability.

The hypothesis (4.290) can readily be generalized to three dimensions encompassing the Cartesian coordinates of a three-dimensional network station. The magnification factor of the respective *standard ellipsoid* is $\sqrt{3F_{3,n-r,\alpha}}$ for it to contain $(1 - \alpha)$ probability. Similarly, the standard deviation of an individual coordinate is converted to a $(1 - \alpha)$ probability confidence interval by multiplication with $\sqrt{F_{1,n-r,\alpha}}$. These magnification factors are shown in Figure 4.4 for $\alpha = 0.05$. For higher degrees of freedom, the magnification factors converge toward the respective chi-square values because of the relationship $r F_{r,\infty} = \chi^2_r$.

For drawing the confidence ellipse at station P_i, we need the rotation angle φ between the $(\mathrm{x_i})$ and $(\mathrm{z_i})$ coordinate systems as well as the semimajor and semiminor axis of the ellipse. Let (y_i) denotes the translated $(\mathrm{x_i})$ coordinate system through the adjusted point $\hat{\mathbf{x}}_\mathrm{i}$, then Equation (4.296) becomes

$$\mathbf{z}_i = \mathbf{R}_i^{\mathrm{T}} \mathbf{y}_i \tag{4.299}$$

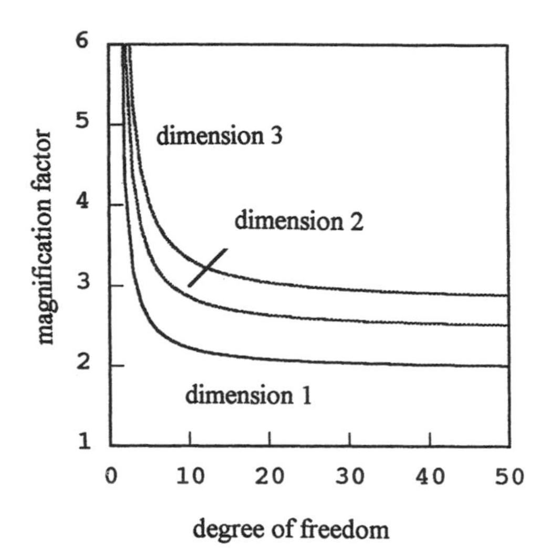

Figure 4.4 Magnification factors for confidence regions. The values refer to $\alpha = 0.05$.

The eigenvectors of $\mathbf{Q}_i$ determine the directions of the semiaxes, and the eigenvalues determine their lengths. Rather than computing the vectors explicitly, we choose to compute the rotation angle φ by comparing coefficients from quadratic forms. Figure 4.5 shows the rotational relation

$$\mathbf{z}_i = \begin{bmatrix} \cos\varphi & \sin\varphi \\ -\sin\varphi & \cos\varphi \end{bmatrix} \mathbf{y}_i \tag{4.300}$$

and Equations (4.295) and (4.299) give the two quadratic forms

$$\mathbf{y}_i^{\mathrm{T}} \mathbf{Q}_i \mathbf{y}_i = \mathbf{z}_i^{\mathrm{T}} \boldsymbol{\Lambda}_i \mathbf{z}_i \tag{4.301}$$

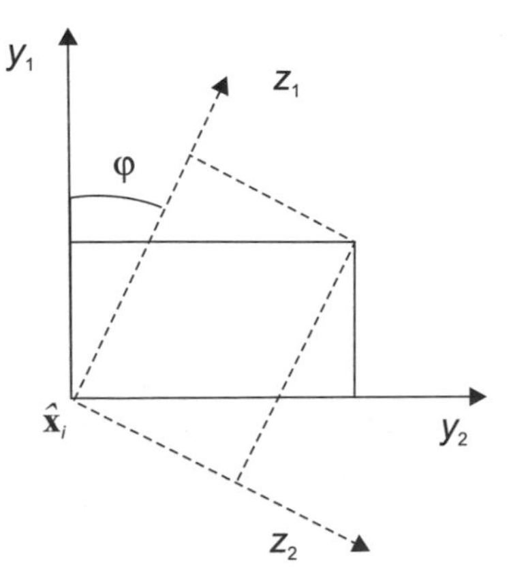

Figure 4.5 Rotation of the principal axis coordinate system.

We substitute (4.300) into the right-hand side of (4.301) and the matrix elements of $\mathbf{Q}_i$ of (4.292) into the left-hand side and compare the coefficient of $y_1 y_2$ on both sides, giving

$$\sin 2\varphi = \frac{2q_{x_1,x_2}}{\lambda_1^Q - \lambda_2^Q} \tag{4.302}$$

The eigenvalues follow directly from the characteristic equation

$$\left|\mathbf{Q}_i - \lambda^Q \mathbf{I}\right| = \begin{vmatrix} q_{x_1} - \lambda^Q & q_{x_1,x_2} \\ q_{x_1,x_2} & q_{x_2} - \lambda^Q \end{vmatrix} = \left(q_{x_1} - \lambda^Q\right)\left(q_{x_2} - \lambda^Q\right) - q_{x_1,x_2}^2 = 0 \tag{4.303}$$

The solution of the quadratic equation is given in (4.304) to (4.308). The terms $\sin 2\varphi$ and $\cos 2\varphi$ determine the quadrant of φ.

$$\lambda_1^Q = \frac{q_{x_1} + q_{x_2}}{2} + \frac{1}{2}W \tag{4.304}$$

$$\lambda_2^Q = \frac{q_{x_1} + q_{x_2}}{2} - \frac{1}{2}W \tag{4.305}$$

$$W = \sqrt{\left(q_{x_1} - q_{x_2}\right)^2 + 4q_{x_1,x_2}^2} \tag{4.306}$$

$$\sin 2\varphi = \frac{2q_{x_1,x_2}}{W} \tag{4.307}$$

$$\cos 2\varphi = \frac{q_{x_1} - q_{x_2}}{W} \tag{4.308}$$

Figure 4.6 shows the defining elements of the standard ellipse. Recall Equation (4.297) regarding the interpretation of the standard ellipses as a confidence region. In any adjustment, any two parameters can comprise $\mathbf{x}_i$, regardless of the geometric meaning of the parameters. Examples are the intercept and slope in the fitting of a straight line or ambiguity parameters in the case of GPS carrier phase solutions. The components $\mathbf{x}_i$ can always be interpreted as Cartesian coordinates for drawing the standard ellipse and thus can give a graphical display of the covariance. In surveying networks, the vectors $\mathbf{x}_i$ contain coordinates of stations in a well-defined coordinate system. If $\mathbf{x}_i$ represents latitude and longitude or northing and easting, the horizontal standard ellipse is computed. If $\mathbf{x}_i$ contains the vertical coordinate and easting, then the standard ellipse in the prime vertical is obtained.

Because the shape of the standard ellipses and ellipsoids depends on the geometry of the network through the design matrix and the weight matrix, the geometric interpretation is enhanced if the network and the standard ellipses are displayed together. Occasionally, users prefer to compute coordinate differences and their covariance matrix and plot relative standard ellipses.

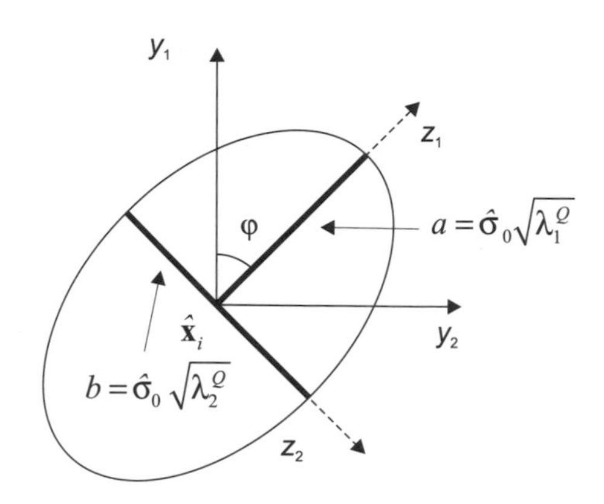

Figure 4.6 Defining elements of standard ellipse.

4.9.6 Properties of Standard Ellipses

The positional error p of a station is directly related to the standard ellipse, as seen in Figure 4.7. The positional error is the standard deviation of a station in a certain direction, say ψ. It is identical with the standard deviation of the distance to a known (fixed) station along the same direction ψ, as computed from the linearized distance equation and variance-covariance propagation. The linear function is

$$r = z_1 \cos\psi + z_2 \sin\psi \tag{4.309}$$

Because of Equations (4.295) and (4.296), the distribution of the random variable $\mathbf{z}_i$ is multivariate normal with

$$\begin{bmatrix} z_1 \\ z_2 \end{bmatrix} \sim N\left(\begin{bmatrix} 0 \\ 0 \end{bmatrix}, \hat{\sigma}_0^2 \begin{bmatrix} \lambda_1^Q & 0 \\ 0 & \lambda_2^Q \end{bmatrix}\right) = N\left(\begin{bmatrix} 0 \\ 0 \end{bmatrix}, \begin{bmatrix} a^2 & 0 \\ 0 & b^2 \end{bmatrix}\right) \tag{4.310}$$

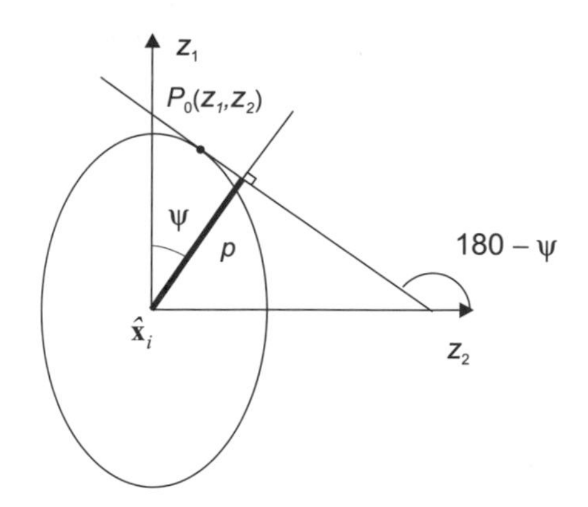

Figure 4.7 Position error.

The variance of the random variable r follows from the law of variance-covariance propagation:

$$\sigma_r^2 = a^2 \cos^2 \psi + b^2 \sin^2 \psi \tag{4.311}$$

The variance (4.311) is geometrically related to the standard ellipse. Let the ellipse be projected onto the direction ψ. The point of tangency is denoted by P_0. Because the equation of the ellipse is

$$\frac{z_1^2}{a^2} + \frac{z_2^2}{b^2} = 1 \tag{4.312}$$

the slope of the tangent is

$$\frac{dz_1}{dz_2} = -\frac{z_2 a^2}{z_1 b^2} = -\tan \psi \tag{4.313}$$

See Figure 4.7 regarding the relation of the slope of the tangent and the angle ψ. The second part of (4.313) yields

$$\frac{z_{01}}{a^2} \sin \psi - \frac{z_{02}}{b^2} \cos \psi = 0 \tag{4.314}$$

This equation relates the coordinates of the point of tangency P_0 to the slope of the tangent. The length p of the projection of the ellipse is, according to Figure 4.7,

$$p = z_{01} \cos \psi + z_{02} \sin \psi \tag{4.315}$$

Next, (4.314) is squared and then multiplied with $a^2 b^2$, and the result is added to the square of (4.315), giving

$$p^2 = a^2 \cos^2 \psi + b^2 \sin^2 \psi \tag{4.316}$$

By comparing this expression with (4.311), it follows that $\hat{\sigma}_r = p$; i.e., the standard deviation in a certain direction is equal to the projection of the standard ellipse onto that direction. Therefore, the standard ellipse is not a standard deviation curve. Figure 4.8 shows the continuous standard deviation curve. We see that for narrow ellipses there are only small segments of the standard deviations that are close to the length of the semiminor axis. The standard deviation increases rapidly as the direction ψ moves away from the minor axis. Therefore, an extremely narrow ellipse is not desirable if the overall accuracy for the station position is important.

As a by-product of the property discussed, we see that the standard deviations of the parameter x_1 and x_2

$$\hat{\sigma}_{x_1} = \hat{\sigma}_0 \sqrt{q_{x_1}} \tag{4.317}$$

$$\hat{\sigma}_{x_2} = \hat{\sigma}_0 \sqrt{q_{x_2}} \tag{4.318}$$

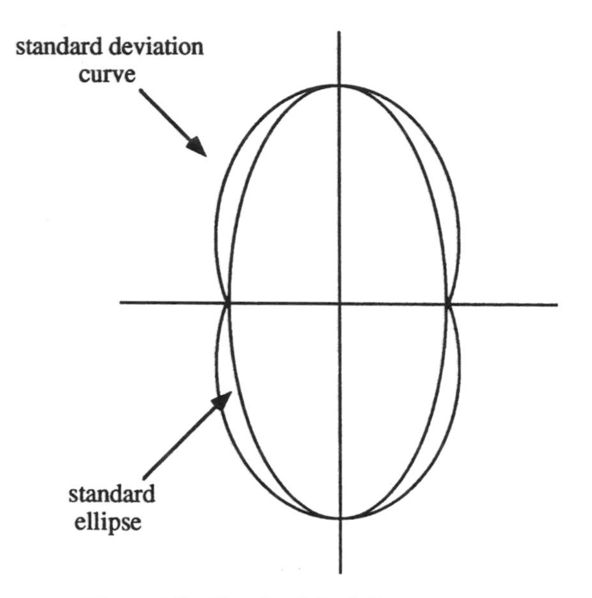

Figure 4.8 Standard deviation curve.

are the projections of the ellipse in the directions of the x_1 and x_2 axes. This is shown in Figure 4.9. Equations (4.317) and (4.318) follow from the fact that the diagonal elements of the covariance matrix are the variances of the respective parameters. Equation (4.316) confirms for $\psi = 0$ and $\psi = 90°$ that the axes a and b equal the maximum and minimum standard deviations, respectively. The rectangle formed by the semisides $\hat{\sigma}_{x_1}$ and $\hat{\sigma}_{x_2}$ encloses the ellipse. This rectangle can be used as an approximation for the ellipses. The diagonal itself is sometimes referred to as the mean position error $\hat{\sigma}$,

$$\hat{\sigma} = \sqrt{\hat{\sigma}_{x_1}^2 + \hat{\sigma}_{x_2}^2} = \hat{\sigma}_0 \sqrt{q_{x_1} + q_{x_2}} \tag{4.319}$$

The points of contact between the ellipse and the rectangle in Figure 4.9 are functions of the correlation coefficients. For these points, the tangent on the ellipse is either horizontal or vertical in the (y_i) coordinate system. The equation of the ellipse in the (y) system is, according to (4.294),

$$[y_1 \ y_2] \begin{bmatrix} q_{x_1} & q_{x_1,x_2} \\ q_{x_1,x_2} & q_{x_2} \end{bmatrix}^{-1} \begin{bmatrix} y_1 \\ y_2 \end{bmatrix} = \hat{\sigma}_0^2 \tag{4.320}$$

By replacing the matrix by its inverse, the expression becomes

$$[y_1 \ y_2] \begin{bmatrix} q_{x_2} & -q_{x_1,x_2} \\ -q_{x_1,x_2} & q_{x_1} \end{bmatrix} \begin{bmatrix} y_1 \\ y_2 \end{bmatrix} = \left(q_{x_1} q_{x_2} - q_{x_1,x_2}^2\right) \hat{\sigma}_0^2 \tag{4.321}$$

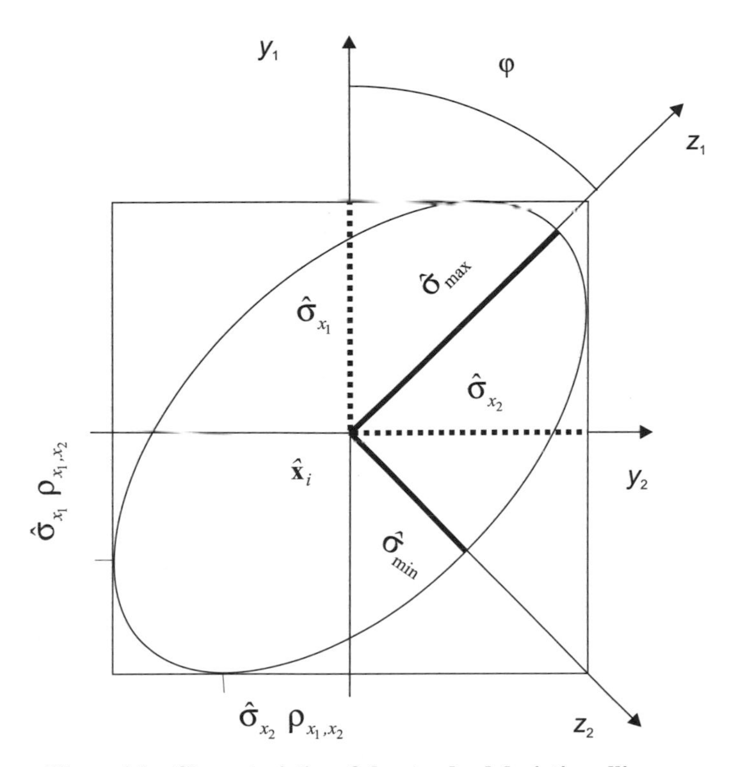

Figure 4.9 Characteristics of the standard deviation ellipse.

Evaluating the left-hand side and dividing both sides by $q_{x_1}q_{x_2}$ gives

$$\frac{y_1^2}{q_{x_1}} + \frac{y_2^2}{q_{x_2}} - \frac{2y_1y_2q_{x_1,x_2}}{q_{x_1}q_{x_2}} = \text{constant} \tag{4.322}$$

from which it follows that

$$\frac{dy_1}{dy_2} = \frac{(2y_2/q_{x_2}) - (2y_1\rho_{x_1,x_2}/\sqrt{q_{x_1}q_{x_2}})}{(2y_2\rho_{x_1,x_2}/\sqrt{q_{x_1}q_{x_2}}) - (2y_1/q_{x_1})} \tag{4.323}$$

Consider the tangent for which the slope is infinity. The equation of this tangent line is

$$y_2 = \hat{\sigma}_0\ \sqrt{q_{x_2}} \tag{4.324}$$

Substituting this expression into the denominator of (4.323) and equating it to zero gives

$$\frac{\hat{\sigma}_0\sqrt{q_{x_2}}\rho_{x_1,x_2}}{\sqrt{q_{x_1}q_{x_2}}} = \frac{y_1}{q_{x_1}} \tag{4.325}$$

which yields the y_1 coordinate for the point of tangency:

$$y_1 = \hat{\sigma}_0 \sqrt{q_{x_1}}\, \rho_{x_1,x_2} = \hat{\sigma}_{x_1} \rho_{x_1,x_2} \tag{4.326}$$

The equation for the horizontal tangent is

$$y_1 = \hat{\sigma}_0 \sqrt{q_{x_1}} \tag{4.327}$$

It follows from the numerator of (4.323) that

$$y_2 = \hat{\sigma}_0 \sqrt{q_{x_2}}\, \rho_{x_1,x_2} = \hat{\sigma}_{x_2} \rho_{x_1,x_2} \tag{4.328}$$

Figure 4.9 shows that the standard ellipse becomes narrower the higher the correlation. For correlation plus or minus 1 (linear dependence), the ellipse degenerates into the diagonal of the rectangle. The ellipse becomes a circle if $a = b$, or $\sigma_{x_1} = \sigma_{x_2}$ and $\rho_{x_1,x_2} = 0$.

4.9.7 Other Measures of Precision

In surveying and geodesy, the most popular measure of precision is the standard deviation. The confidence regions are usually expressed in terms of ellipses and ellipsoids of standard deviation. These figures are often scaled to contain 95% probability or higher. Because GPS is a popular tool for both surveying and navigation, several of the measures of precision used in navigation are becoming increasingly popular in surveying. Examples include the dilution of precision (DOP) numbers. The DOPs are discussed in detail in Section 7.4.1. Other single-number measures refer to circular or spherical confidence regions for which the eigenvalues of the cofactor matrix have the same magnitude. In these cases, the standard deviations of the coordinates and the semiaxes are of the same size. See Equation (4.297). When the standard deviations are not equal, these measures become a function of the ratio of the semiaxes. The derivation of the following measures and additional interpretation are given in Greenwalt and Shultz (1962).

The radius of a circle that contains 50% probability is called the circular error probable (CEP). This function is usually approximated by segments of straight lines. The expression

$$\text{CEP} = 0.5887 \left(\hat{\sigma}_{x_1} + \hat{\sigma}_{x_2}\right) \tag{4.329}$$

is, strictly speaking, valid in the region $\sigma_{\min}/\sigma_{\max} \geq 0.2$, but it is the function used most often. The 90% probability region

$$\text{CMAS} = 1.8227 \times \text{CEP} \tag{4.330}$$

is called the circular map accuracy standard. The mean position error (4.319) is also called the mean square positional error (MSPE), or the distance root mean square (DRMS), i.e.,

$$\mathrm{DRMS} = \sqrt{\hat{\sigma}_{x_1}^2 + \hat{\sigma}_{x_2}^2} \tag{4.331}$$

This measure contains 64% to 77% probability. The related measure

$$2\ \mathrm{DRMS} = 2 \times \mathrm{DRMS} \tag{4.332}$$

contains about 95% to 98% probability.

The three-dimensional equivalent of CEP is the spherical error probable (SEP), defined as

$$\mathrm{SEP} = 0.5127 \left(\hat{\sigma}_{x_1} + \hat{\sigma}_{x_2} + \hat{\sigma}_{x_3}\right) \tag{4.333}$$

Expression (4.333) is, strictly speaking, valid in the region $\sigma_{\min}/\sigma_{\max} \geq 0.35$. The corresponding 90% probability region,

$$\mathrm{SAS} = 1.626 \times \mathrm{SEP} \tag{4.334}$$

is called the spherical accuracy standard (SAS). The mean radial spherical error (MRSE) is defined as

$$\mathrm{MRSE} = \sqrt{\hat{\sigma}_{x_1}^2 + \hat{\sigma}_{x_2}^2 + \hat{\sigma}_{x_3}^2} \tag{4.335}$$

and contains about 61% probability.

These measures of precision are sometimes used to capture the achieved or anticipated precision conveniently using single numbers. However, the geometry of the adjustment seldom produces covariance matrices that yield circular distribution. Consequently, the probability levels contained in these measures of precision inevitably are a function of the correlations between the parameters.

4.10 RELIABILITY

Small residuals are not necessarily an indication of a quality adjustment. Equally important is the knowledge that all blunders in the data have been identified and removed and that remaining small blunders in the observations do not adversely impact the adjusted parameters. Reliability refers to the controllability of observations, i.e., the ability to detect blunders and to estimate the effects that undetected blunders may have on a solution. The theory outlined here follows that of Baarda (1967, 1968), and Kok (1984).

4.10.1 Redundancy Numbers

Following the expressions in Table 4.1 the residuals for the observation equation model are

$$\bar{\mathbf{v}} = \mathbf{Q}_v \mathbf{P} \boldsymbol{\ell} \tag{4.336}$$

with a cofactor matrix for the residuals

$$\mathbf{Q}_v = \mathbf{P}^{-1} - \mathbf{A}\mathbf{N}^{-1}\mathbf{A}^{\mathrm{T}} \tag{4.337}$$

Compute the trace

$$\begin{aligned} \mathrm{Tr}(\mathbf{Q}_v\,\mathbf{P}) &= \mathrm{Tr}\big(\mathbf{I} - \mathbf{A}\mathbf{N}^{-1}\mathbf{A}^{\mathrm{T}}\mathbf{P}\big) \\ &= n - \mathrm{Tr}\big(\mathbf{N}^{-1}\mathbf{A}^{\mathrm{T}}\mathbf{P}\mathbf{A}\big) \\ &= n - u \end{aligned} \tag{4.338}$$

A more general expression is obtained by noting that the matrix $\mathbf{A}\mathbf{N}^{-1}\mathbf{A}^{\mathrm{T}}\,\mathbf{P}$ is idempotent. The trace of an idempotent matrix equals the rank of that matrix. Thus,

$$\mathrm{Tr}\big(\mathbf{A}\mathbf{N}^{-1}\mathbf{A}^{\mathrm{T}}\mathbf{P}\big) = R\big(\mathbf{A}^{\mathrm{T}}\mathbf{P}\mathbf{A}\big) = R(\mathbf{A}) = r \le u \tag{4.339}$$

Thus, from Equations (4.338) and (4.339)

$$\mathrm{Tr}(\mathbf{Q}_v\mathbf{P}) = \mathrm{Tr}(\mathbf{P}\mathbf{Q}_v) = n - R(\mathbf{A}) \tag{4.340}$$

By denoting the diagonal element of the matrix $\mathbf{Q}_v\mathbf{P}$ by r_i, we can write

$$\sum_{i=1}^{n} r_i = n - R(\mathbf{A}) \tag{4.341}$$

The sum of the diagonal elements of $\mathbf{Q}_v\mathbf{P}$ equals the degree of freedom. The element r_i is called the redundancy number for the observation i. It is the contribution of the ith observation to the degree of freedom. If the weight matrix $\mathbf{P}$ is diagonal, which is usually the case when original observations are adjusted, then

$$r_i = q_i\,p_i \tag{4.342}$$

where q_i is the diagonal element of the cofactor matrix $\mathbf{Q}_v$, and p_i denotes the weight of the ith observation. Equation (4.337) implies the inequality

$$0 \le q_i \le \frac{1}{p_i} \tag{4.343}$$

Multiplying by p_i gives the bounds for the redundancy numbers,

$$0 \le r_i \le 1 \tag{4.344}$$

Considering the general relation

$$\mathbf{Q}_{\ell_a} = \mathbf{Q}_{\ell_b} - \mathbf{Q}_v \tag{4.345}$$

given in Table 4.1 and the specification (4.342) for the redundancy number r_i as the diagonal element of $\mathbf{Q}_v\mathbf{P}$, it follows that if the redundancy number is close to 1, then the variance of the residuals is close to the variance of the observations, and the variance of the adjusted observations is close to zero. If the redundancy number is close to zero, then the variance of the residuals is close to zero, and the variance of the adjusted observations is close to the variance of the observations.

Intuitively, it is expected that the variance of the residuals and the variance of the observations are close; for this case, the noise in the residuals equals that of the observations, and the adjusted observations are determined with high precision. Thus the case of r_i close to 1 is preferred, and it is said that the gain of the adjustment is high. If r_i is close to zero, one expects the noise in the residuals to be small. Thus, small residuals as compared to the expected noise of the observations are not necessarily desirable. Because the inequality (4.344) is a result of the geometry as represented by the design matrix $\mathbf{A}$, small residuals can be an indication of a weak part of the network.

Because the weight matrix P is considered diagonal, i.e.,

$$p_i = \frac{\sigma_0^2}{\sigma_i^2} \tag{4.346}$$

it follows that

$$\hat{\sigma}_{v_i} = \hat{\sigma}_0\sqrt{q_i} = \hat{\sigma}_0\sqrt{\frac{r_i}{p_i}} = \hat{\sigma}_0\sqrt{\frac{r_i\sigma_i^2}{\sigma_0^2}} = \frac{\hat{\sigma}_0}{\sigma_0}\sigma_i\sqrt{r_i} \tag{4.347}$$

From (4.341) it follows that the average redundancy number is

$$r_{\mathrm{av}} = \frac{n - R(\mathbf{A})}{n} \tag{4.348}$$

The higher the degree of freedom, the closer the average redundancy number is to 1. However, as seen from Table 4.8, the gain, in terms of probability enclosed by the standard ellipses, reduces noticeably after a certain degree of freedom.

4.10.2 Controlling Type-II Error for a Single Blunder

Baarda's (1967) development of the concept of reliability of networks is based on un-Studentized hypothesis tests, which means that the a priori variance of unit weight is assumed to be known. Consequently, the a priori variance of unit weight (not the a posteriori variance of unit weight) is used in this section. The alternative hypothesis H_a specifies that the observations contain one blunder, that the blunder be located at observation i, and that its magnitude be ∇_i. Thus the adjusted residuals for the case of the alternative hypothesis are

$$\hat{\mathbf{v}}|H_a = \hat{\mathbf{v}} - \mathbf{Q}_v\mathbf{P}\mathbf{e}_i\nabla_i \tag{4.349}$$

where

$$\mathbf{e}_i = [0 \quad \cdots \quad 0 \quad 1 \quad 0 \quad \cdots \quad 0]^{\mathrm{T}} \tag{4.350}$$

denotes an $n \times 1$ vector containing 1 in position i and zeros elsewhere. The expected value and the covariance matrix are

$$E(\hat{\mathbf{v}}|H_a) = -\mathbf{Q}_v \mathbf{P} \mathbf{e}_i \nabla_i \tag{4.351}$$

$$\boldsymbol{\Sigma}_{v|H_a} = \hat{\boldsymbol{\Sigma}}_v = \sigma_0^2 \mathbf{Q}_v \tag{4.352}$$

It follows from Theorem 1 of Section 4.9.1 that

$$\hat{\mathbf{v}}|H_a \sim N\left(-\mathbf{Q}_v \mathbf{P} \mathbf{e}_i \nabla_i, \sigma_0^2 \mathbf{Q}_v\right) \tag{4.353}$$

Since $\mathbf{P}$ is a diagonal matrix, the individual residuals are distributed as

$$\hat{v}_i|H_a \sim n\left(-q_i p_i \nabla_i, \sigma_0^2 q_i\right) \tag{4.354}$$

according to Theorem 2. Standardizing gives

$$\begin{aligned} w_a|H_a &= \frac{\hat{v}_i|H_a}{\sigma_0 \sqrt{q_i}} \sim n\left(\frac{-q_i p_i \nabla_i}{\sigma_0 \sqrt{q_i}}, 1\right) \\ &= n\left(\frac{-\sqrt{q_i}\, p_i \nabla_i}{\sigma_0}, 1\right) \end{aligned} \tag{4.355}$$

or

$$H_a\colon w_a = \frac{\hat{v}_i|H_a}{\sigma_{v_i}} \sim n\left(\frac{-\nabla_i p_i \sqrt{q_i}}{\sigma_0}, 1\right) \tag{4.356}$$

The zero hypothesis, which states that there is no blunder, is

$$H_0\colon w_0 = \frac{\hat{v}_i|H_0}{\sigma_{v_i}} \sim n(0, 1) \tag{4.357}$$

The noncentrality parameter in (4.356), i.e., the mean of the noncentral normal distribution, is denoted by δ_i and is

$$\delta_i = \frac{-\nabla_i p_i \sqrt{q_i}}{\sigma_0} = \frac{-\nabla_i \sqrt{r_i}}{\sigma_i} \tag{4.358}$$

The parameter δ_i is a translation parameter of the normal distribution. The situation is shown in Figure 4.10. The probability of committing an error of the first kind, i.e., of accepting the alternative hypothesis, equals the significance level α of the test

$$P\left(|w_0| \le t_{a/2}\right) = \int_{-t_{a/2}}^{t_{a/2}} n\,(0,1)\;dx = 1 - \alpha \tag{4.359}$$

or

$$P\left(|w_0| \ge t_{a/2}\right) = \int_{-\infty}^{t_{1-a/2}} n\,(0,1)\;dx + \int_{t_{a/2}}^{\infty} n(0,1)\,dx = \alpha \tag{4.360}$$

In 100α% of the cases, the observations are rejected and remeasurement or investigations for error sources are performed, even though the observations are correct (they do not contain a blunder). From Figure 4.10 it is seen that the probability β_i of a type-II error, i.e., the probability of rejecting the alternative hypothesis (and accepting the zero hypothesis) even though the alternative hypothesis is correct, depends on the noncentrality factor δ_i. *Because the blunder ∇_i is not known, the noncentrality factor is not known either.* As a practical matter one can proceed in the reverse: one can assume an acceptable probability β_0 for the error of the second kind and compute the respective noncentrality parameter δ_0. This parameter in turn is used to compute the lower limit for the blunder, which can still be detected. Figure 4.10 shows that

$$P\left(|w_a| \le t_{a/2}\right) = \int_{-t_{a/2}}^{t_{a/2}} n(\delta_i, 1)\;dx \ge \beta_0 \tag{4.361}$$

if

$$\delta_i \le \delta_0 \tag{4.362}$$

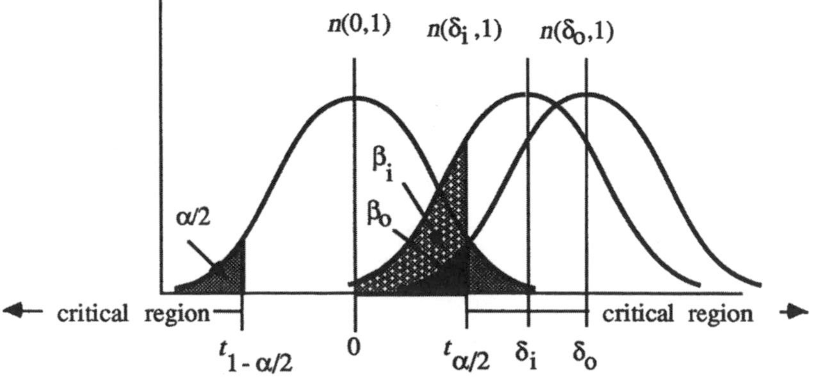

Figure 4.10 Defining the noncentrality.

Substituting Equation (4.358) into (4.362) gives the limit for the marginally detectable blunder, given the probability levels α and β_0:

$$|\nabla_{0i}| \geq \frac{\delta_0}{\sqrt{r_i}} \sigma_i \tag{4.363}$$

Equations (4.361) and (4.363) state that in $100(1 - \beta_0)\%$ of the cases, blunders greater than those given in (4.363) are detected. In $100\beta_0\%$ of the cases, blunders greater than those given in (4.363) remain undetected. The larger the redundancy number, the smaller is the marginally detectable blunder (for the same δ_0 and σ_i). It is important to recognize that the marginally detectable blunders (4.363) are based on adopted probabilities of type-I and type-II errors for the normal distribution. The probability levels α and β_0 refer to the one-dimensional test (4.357) of the individual residual v_i, with the noncentrality being δ_0. The assumption is that only one blunder at a time is present. The geometry is shown in Figure 4.10. It is readily clear that there is a simple functional relationship $\delta_0 = \delta_n(\alpha, \beta_0)$ between two normal distributions. Table 4.9 contains selected probability levels and the respective δ_0 values.

The chi-square test (4.263) of the a posteriori variance of unit weight $\hat{\sigma}_0^2$ is also sensitive to the blunder ∇_i. In fact, the blunder will cause a noncentrality of δ_i for the chi-square distribution of the alternative hypothesis. One can choose the probabilities α_{chi} and β_{chi} for this multidimensional chi-square test such that $\delta_0 = \delta_{\text{chi}}\left(\alpha_{\text{chi}}, \beta_{\text{chi}}, n - u\right)$. The factor δ_0 depends on the degree of freedom because the chi-square distribution depends on it. Baarda's B method suggests equal traceability of errors through one-dimensional tests of individual residuals, v_i, and the multidimensional test of the a posteriori variance of unit weight $\hat{\sigma}_0^2$. This is achieved by requiring that the one-dimensional test and the multidimensional test have the same type-II error, i.e., $\beta_0 = \beta_{\text{chi}}$. Under this condition there exists a relationship between the probability of type-II error, the significance levels, and the degree of freedom expressed symbolically by $\delta_0 = \delta_n\left(\alpha, \beta_0\right) = \delta_{\text{chi}}\left(\alpha_{\text{chi}}, \beta_0, n - r\right)$. The B method assures equal traceability but implies different significance levels for the

TABLE 4.9 Selected Probability Levels in Reliability

α	β_0	δ_0
0.05	0.20	2.80
0.025	0.20	3.1
0.001	0.20	4.12
0.05	0.10	3.24
0.025	0.10	3.52
0.001	0.10	4.57

one-dimensional and multidimensional tests. For details see Baarda (1968, p. 25). In practical applications one chooses the factor δ_0 on the basis of a reasonable value for α and β_0 from Table 4.9.

4.10.3 Internal Reliability

Even though the one-dimensional test is based on the assumption that only one blunder exists in a set of observations, the limit (4.363) is usually computed for all observations. The marginally detectable errors, computed for all observations, are viewed as a measure of the capability of the network to detect blunders with probability $(1 - \beta_0)$. They constitute the internal reliability of the network. Because the marginally detectable errors (4.363) do not depend on the observations or on the residuals, they can be computed as soon as the configuration of the network and the stochastic model are known. If the limits (4.363) are of about the same size, the observations are equally well checked, and the internal reliability is said to be consistent. The emphasis is then on the variability of the marginally detectable blunders rather on their magnitude. A typical value is $\delta_0 = 4$.

4.10.4 Absorption

According to (4.336) the residuals in the presence of one blunder are:

$$\mathbf{v} = \mathbf{Q}_v \mathbf{P} (\boldsymbol{\ell} - \mathbf{e}_i \nabla_i) \tag{4.364}$$

The impact on the residual of observation i is

$$\nabla v_i = -r_i \nabla_i \tag{4.365}$$

Equation (4.365) is used to estimate the blunders that might cause large residuals. Solving for ∇_i gives

$$\nabla_i = -\frac{\nabla v_i}{r_i} \approx -\frac{v_i^* + \nabla v_i}{r_i} \approx -\frac{v_i}{r_i} \tag{4.366}$$

because $v_i^* \ll \nabla v_i$, where v_i^* denotes the residual without the effect of the blunder. The computation (4.366) provides only estimates of possible blunders. Because the matrix $\mathbf{Q}_v \mathbf{P}$ is not a diagonal matrix, a specific blunder has an impact on all residuals. If several blunders are present, their effects overlap and one blunder can mask others; a blunder may cause rejection of a good observation.

Equation (4.365) demonstrates that the residuals in least-squares adjustments are not robust with respect to blunders in the sense that the effect of a blunder on the residuals is smaller than the blunder itself, because r varies between 0 and 1. The absorption, i.e., the portion of the blunder that propagates into the estimated parameters and falsifies the solution, is

$$A_i = (1 - r_i)\,\nabla_i \tag{4.367}$$

The factor $(1-r_i)$ is called the absorption number. The larger the redundancy number, the less is a blunder absorbed, i.e., the less falsification. If $r_i = 1$, the observation is called fully controlled, because the residual completely reflects the blunder. A zero redundancy implies uncontrolled observations in that a blunder enters into the solution with its full size. Observations with small redundancy numbers might have small residuals and instill false security in the analyst. Substituting ∇_i from (4.366) expresses the absorption as a function of the residuals:

$$A_i = -\frac{1 - r_i}{r_i} v_i \tag{4.368}$$

The residuals can be looked on as the visible parts of errors. The factor in (4.368) is required to compute the invisible part from the residuals.

4.10.5 External Reliability

A good and homogeneous internal reliability does not automatically guarantee reliable coordinates. What are the effects of undetectable blunders on the parameters? In deformation analysis, where changes in parameters between adjustments of different epochs indicate existing deformations, it is particularly important that the impact of blunders on the parameters be minimal. The influence of each of the marginally detectable errors on the parameters of the adjustment or on functions of the parameters is called external reliability. The estimated parameters in the presence of a blunder are, for the observation equation model,

$$\hat{\mathbf{x}} = -\mathbf{N}^{-1}\mathbf{A}^{\mathrm{T}}\,\mathbf{P}\,(\boldsymbol{\ell} - \mathbf{e}_i\,\nabla_i) \tag{4.369}$$

The effect of the blunder in observation i is

$$\nabla\mathbf{x} = \mathbf{N}^{-1}\mathbf{A}^{\mathrm{T}}\,\mathbf{P}\mathbf{e}_i\,\nabla_i \tag{4.370}$$

The shifts $\nabla\mathbf{x}$ are sometimes called local external reliability. The blunder affects all parameters. The impact of the marginally detectable blunder ∇_{0i} is

$$\nabla\mathbf{x}_{0i} = \mathbf{N}^{-1}\mathbf{A}^{\mathrm{T}}\,\mathbf{P}\mathbf{e}_i\,\nabla_{0i} \tag{4.371}$$

Because there are n observations, one can compute n vectors (4.371), showing the impact of each marginal detectable blunder on the parameters. Graphical representations of these effects can be very helpful in the analysis. The problem with (4.371) is that the effect on the coordinates depends on the definition (minimal constraints) of the coordinate system. Baarda (1968) suggested the following alternative expression:

$$\lambda_{0i}^2 = \frac{\nabla\mathbf{x}_{0i}^{\mathrm{T}}\;\mathbf{N}\;\nabla\mathbf{x}_{0i}}{\sigma_0^2} \tag{4.372}$$

By substituting (4.371) and (4.363), we can write this equation as

$$\lambda_{0i}^2 = \frac{\nabla_{0i}\mathbf{e}_i^{\mathrm{T}}\,\mathbf{PAN}^{-1}\mathbf{A}^{\mathrm{T}}\,\mathbf{Pe}_i\nabla_{0i}}{\sigma_0^2} = \frac{\nabla_{0i}^2\mathbf{e}_i^{\mathrm{T}}\,\mathbf{P}(\mathbf{I}-\mathbf{Q}_v\mathbf{P})\mathbf{e}_i}{\sigma_0^2} = \frac{\nabla_{0i}^2\,p_i\,(1-r_i)}{\sigma_0^2} \tag{4.373}$$

or

$$\lambda_{0i}^2 = \frac{1-r_i}{r_i}\,\delta_0^2 \tag{4.374}$$

The values λ_{0i} are a measure of global external reliability. There is one such value for each observation. If the λ_{0i} are the same order of magnitude, the network is homogeneous with respect to external reliability. If r_i is small, the external reliability factor becomes large and the global falsification caused by a blunder can be significant. It follows that very small redundancy numbers are not desirable. The global external reliability number (4.374) and the absorption number (4.368) have the same dependency on the redundancy numbers.

4.10.6 Correlated Cases

The derivations for detectable blunders, internal reliability, absorption, and external reliability assumes uncorrelated observations for which the covariance matrix $\boldsymbol{\Sigma}_{\ell_b}$ is diagonal. Correlated observations are decorrelated by the transformation (4.235). It can be readily verified that the redundancy numbers for the decorrelated observations $\bar{\boldsymbol{\ell}}$ are

$$\bar{r}_i = \left(\bar{\mathbf{Q}}_{\bar{v}}\bar{\mathbf{P}}\right)_{ii} = \left(\mathbf{I} - \mathbf{D}^{\mathrm{T}}\,\mathbf{AN}^{-1}\mathbf{A}^{\mathrm{T}}\,\mathbf{D}\right)_{ii} \tag{4.375}$$

In many applications, the covariance matrix $\boldsymbol{\Sigma}_{\ell_b}$ is of block-diagonal form. For example, for GPS vector observations, this matrix consists of 3×3 full block-diagonal matrices if the correlations between the vectors are neglected. In this case, the matrix $\mathbf{D}$ is also block-diagonal and the redundancy numbers can be computed vector by vector from (4.375). The sum of the redundancy numbers for the three vector components varies between 0 and 3. Since, in general, the matrix $\mathbf{D}$ has a full rank, the degree of freedom $(n-r)$ of the adjustment does not change. Once the redundancy numbers $\bar{r}_i$ are available, the marginal detectable blunders $\bar{\nabla}_{0i}$, the absorption numbers $\bar{A}_i$ and other reliability values can be computed for the decorrelated observations. These quantities, in turn, can be transformed back into the physical observation space by premultiplication with the matrix $(\mathbf{D}^{\mathrm{T}})^{-1}$.

4.11 BLUNDER DETECTION

Errors (blunders) made during the recording of field observations, data transfer, the computation, etc., can be costly and time-consuming to find and eliminate. Blunder detection can be carried out before the adjustment or as part of the adjustment. Before the adjustment, the discrepancies (angle and/or distance of simple figures such as triangles and traverses) are analyzed. A priori blunder detection is helpful in detecting extralarge blunders caused by, e.g., erroneous station numbering. Blunder detection in conjunction with the adjustment is based on the analysis of the residuals. The problem with using least-squares adjustments when blunders are present is that the adjustments tend to hide (reduce) their impact and distribute their effects more or less throughout the entire network (see (4.364) and (4.365), noting that the redundancy number varies between zero and 1). The prerequisite for any blunder-detection procedure is the availability of a set of redundant observations. Only observations with redundancy numbers greater than zero can be controlled.

It is important to understand that if a residual does not pass a statistical test, this does not mean that there is a blunder in that observation. The observation is merely flagged so that it can be examined and a decision about its retention or rejection can be made. Blind rejection is never recommended. A blunder in one observation usually affects the residuals in other observations. Therefore, the tests will often flag other observations in addition to the ones containing blunders. If one or more observations are flagged, the search begins to determine if there is a blunder.

The first step is to check the field notes to confirm that no error occurred during the transfer of the observations to the computer file, and that all observations are reasonable "at face value." If a blunder is not located, the network should be broken down into smaller networks, and each one should be adjusted separately. At the extreme, the entire network may be broken down into triangles or other simple geometric entities, such as traverses, and adjusted separately. Alternatively, the observations can be added sequentially, one at a time, until the blunder is found. This procedure starts with weights assigned to all parameters. The observations are then added sequentially. The sum of the normalized residuals squared is then inspected for unusually large variations. When searching for blunders, the coordinate system should be defined by minimal constraints.

Blunder detection in conjunction with the adjustment takes advantage of the total redundancy and the strength provided by the overall geometry of the network, and thus is more sensitive to smaller blunders. Only if the existence of a blunder is indicated does action need to be taken to locate the blunder. The flagged observations are the best hint where to look for errors and thus avoid unnecessary and disorganized searching of the whole observation data set.

4.11.1 The τ Test

The τ test was introduced by Pope (1976). The test belongs to the group of Studentized tests, which make use of the a posteriori variance of unit weight as estimated from the observations. The test statistic is

$$\tau_i = \frac{v_i}{\hat{\sigma}_{v_i}} = \frac{\sigma_0 v_i}{\hat{\sigma}_0 \sigma_i \sqrt{r_i}} \sim \tau_{n-r} \tag{4.376}$$

The symbol τ_{n-r} denotes the τ distribution with $n-r$ degrees of freedom. It is related to Student's t by

$$\tau_{n-r} = \frac{\sqrt{n-r}\; t_{n-r-1}}{\sqrt{n-r-1+t^2_{n-r-1}}} \tag{4.377}$$

For an infinite degree of freedom the τ distribution converges toward the Student distribution or the standardized normal distribution, i.e., $\tau_\infty = t_\infty = n(0, 1)$.

Pope's blunder rejection procedure tests the hypothesis $v_i \sim n(0,\ \hat{\sigma}_{v_i}/\hat{\sigma}_0)$. The hypothesis is rejected, i.e., the observation is flagged for further investigation and possibly rejection, if

$$|\tau_i| \geq c \tag{4.378}$$

The critical value c is based on a preselected significance level. For large systems, the redundancy numbers are often replaced by the average value according to Equation (4.348), in order to reduce computation time; thus

$$\tau_i = \frac{\sigma_0}{\hat{\sigma}_0} \frac{v_i}{\sigma_i \sqrt{(n-r)/n}} \tag{4.379}$$

could be used instead of (4.376).

4.11.2 Data Snooping

Baarda's data snooping applies to the testing of individual residuals as well. The theory assumes that only one blunder be present in the set of observations. Applying a series of one-dimensional tests, i.e., testing consecutively all residuals, is called a data snooping strategy. Baarda's test belongs to the group of un-Studentized tests which assume that the a priori variance of unit weight is known. The zero hypothesis (4.357) is written as

$$n_i = \frac{v_i}{\sigma_0 \sqrt{q_i}} \sim n(0, 1) \tag{4.380}$$

At a significant level of 5%, the critical value is 1.96. The critical value for this test is not a function of the number of observations in the adjustment. The statistic (4.380) uses the a priori value σ_0 and not the a posteriori estimate $\hat{\sigma}_0$.

Both the τ and the data snooping procedures work best for iterative solutions. At each iteration step, the observation with the largest blunder should be removed. Since least-squares attempts to distribute blunders, several correct observations might receive large residuals and might be flagged mistakenly.

4.11.3 Changing Weights of Observations

This method, although not based on rigorous statistical theory, is an automated method whereby blunders are detected and their effects on the adjustment minimized (or even eliminated). The advantage that this method has, compared to previous methods, is that it locates and potentially eliminates the blunders automatically. The method examines the residuals per iteration. If the magnitude of a residual is outside a defined range, the weight of the corresponding observation is reduced. The process of reweighting and readjusting continues until the solution converges, i.e., no weights are being changed. The criteria for judging the residuals and choice for the reweighting function are somewhat arbitrary. For example, a simple strategy for selection of the new weights at iteration $k + 1$ could be

$$p_{k+1,i} = p_{k,i} \begin{cases} e^{-|v_{k,i}|/3\sigma_i} & \text{if} \quad \left|v_{k,i}\right| > 3\sigma_i \\ 1 & \text{if} \quad \left|v_{k,i}\right| \leq 3\sigma_i \end{cases} \tag{4.381}$$

where σ_i denotes the standard deviation of observation i.

The method works efficiently for networks with high redundancy. If the initial approximate parameters are inaccurate, it is possible that correct observations are deweighted after the first iteration because the nonlinearity of the adjustment can cause large residuals. To avoid unnecessary rejection and reweighting, one might not change the weights during the first iteration. Proper use of this method requires some experience. All observations whose weights are changed must be investigated, and the cause for the deweighting must be investigated.

4.12 EXAMPLES

In the following, we use plane two-dimensional networks to demonstrate the geometry of adjustments. As mentioned above, the geometry of a least-squares adjustment is the result of the combined effects of the stochastic model (weight matrix **P**—representing the quality of the observations) and the mathematical model (design matrix **A**—representing the geometry of the network and the spatial distribution of the observations). For the purpose of these examples, it is not necessary to be concerned about the physical realization of two-dimensional networks. The experienced reader might think of such networks as being located on the conformal mapping plane and all that it takes to compute the respective model observations. However, it is entirely sufficient here to stay simply within the area of plane geometry.

We will use the observation equation model summarized in Table 4.1. Assume there is a set of n observations, such as distances and angles that determine the points of a network. For a two-dimensional network of s stations, there could be as many as $u = 2s$ unknown coordinates. Let the parameter vector $\mathbf{x}_a$ consist of coordinates only, i.e., we do not parameterize refraction, centering errors, etc. To be specific, $\mathbf{x}_a$ contains only coordinates that are to be estimated. Coordinates of known stations are constants and not included in $\mathbf{x}_a$. The mathematical model $\boldsymbol{\ell}_a = \mathbf{f}(\mathbf{x}_a)$ is very simple in this case. The n components $\mathbf{f}$ will contain the functions:

$$d_{ij} = \sqrt{(x_i - x_j)^2 + (y_i - y_j)^2} \tag{4.382}$$

$$a_{jik} = \tan^{-1}\frac{x_k - x_i}{y_k - y_i} - \tan^{-1}\frac{x_j - x_i}{y_j - y_i} \tag{4.383}$$

In these expressions the subscripts i, j, and k identify the network points. The notation a_{jik} implies that the angle is measured at station i, from j to k in a clockwise sense. The ordering of the components in $\mathbf{f}$ does not matter, as long as the same order is maintained with respect to the rows of $\mathbf{A}$ and diagonal elements of $\mathbf{P}$.

Although the $\mathbf{f}(\mathbf{x}_a)$ have been expressed in terms of $\mathbf{x}_a$, the components typically depend only on a subset of the coordinates. The relevant partial derivatives in a row of $\mathbf{A}$ are for distances and angles:

$$\left\{\frac{-(y_k - y_i)}{d_{ik}}, \frac{-(x_k - x_i)}{d_{ik}}, \frac{y_k - y_i}{d_{ik}}, \frac{x_k - x_i}{d_{ik}}\right\} \tag{4.384}$$

$$\left\{\frac{x_i - x_j}{d_{ij}^2}, -\frac{y_i - y_j}{d_{ij}^2}, \frac{x_k - x_j}{d_{kj}^2} - \frac{x_i - x_j}{d_{ij}^2}, -\frac{y_k - y_j}{d_{kj}^2} + \frac{y_i - y_j}{d_{ij}^2}, -\frac{x_k - x_j}{d_{kj}^2}, \frac{y_k - y_j}{d_{kj}^2}\right\} \tag{4.385}$$

Other elements are zero. The column location for these partials depends on the sequence in $\mathbf{x}_a$. In general, if α is the α-th component of $\boldsymbol{\ell}_b$ and β the β-th component of $\mathbf{x}_a$, then the element $a_{\alpha,\beta}$ of $\mathbf{A}$ is

$$a_{\alpha,\beta} = \frac{\partial \ell_\alpha}{\partial x_\beta} \tag{4.386}$$

The partial derivatives and the discrepancy $\boldsymbol{\ell}_0$ must be evaluated for the approximate coordinates $\mathbf{x}_0$.

Example 1: This example demonstrates the impact of changes in the stochastic model. Figure 4.11 shows a traverse connecting two known stations. Three solutions

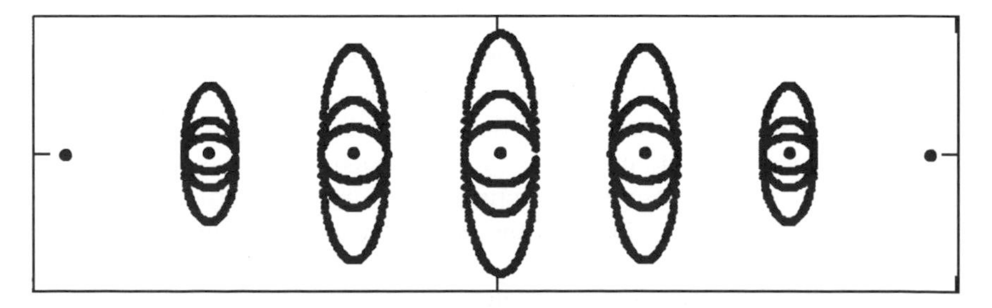

Figure 4.11 Impact of changing the stochastic model.

are given. In all cases, the distances are of the same length and observed with the same accuracy. The angle observations are 180° and measured with the same accuracy but are changed by a common factor for each solution. If we declare the solutions with the smallest ellipses in Figure 4.11 as the base solutions with observational standard deviation of σ_a then the other solutions use $2\sigma_a$ and $4\sigma_a$, respectively. The shape of the ellipses elongates as the standard deviation of the angles increases.

Example 2: This example demonstrates the impact of changing network geometry using a resection. Four known stations lie exactly on an imaginary circle with radius r. The coordinates of the new station are determined by angle measurements, i.e., no distances are involved. For the first solution, the unknown station is located at the center of the circle. In subsequent solutions it location moves to $0.5\ r$, $0.9\ r$, $1.1\ r$, and $1.5\ r$ from the center while retaining the same standard deviation for the angle observations in each case. Figure 4.12 shows that the ellipses become more elongated the closer the unknown station moves to the circle. The solution is singular if the new station is located exactly on the circle.

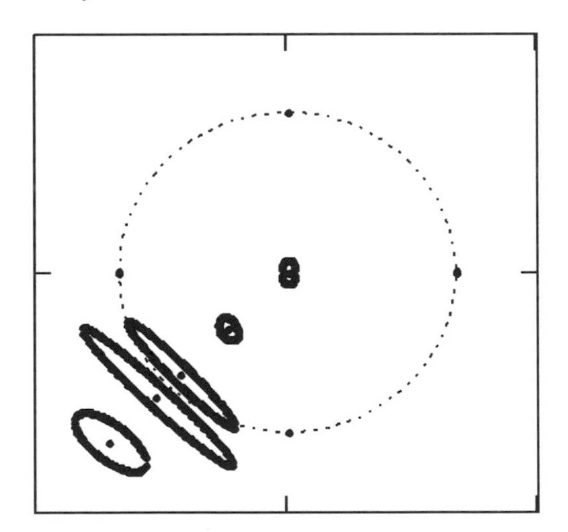

Figure 4.12 Impact of changing network geometry.

Example 3: Three cases are given that demonstrate how different definitions of the coordinate system affect the ellipses of standard deviation. All cases refer to the same plane network using the same observed angles and distances and the same respective standard deviations of the observations. A plane network that contains angle and distance observations requires three minimal constraints. Simply holding three coordinates fixed imposes such minimal constraints. The particular coordinates are constants and are not included in the parameter vector $\mathbf{x}_a$, and, consequently, there are no columns in the $\mathbf{A}$ matrix that pertain to these three coordinates. Inner constraints offer another possibility of defining the coordinate system.

Figure 4.13 shows the results of two different minimal constraints. The coordinates of station 2 are fixed in both cases. In the first case, we hold one of the coordinates of

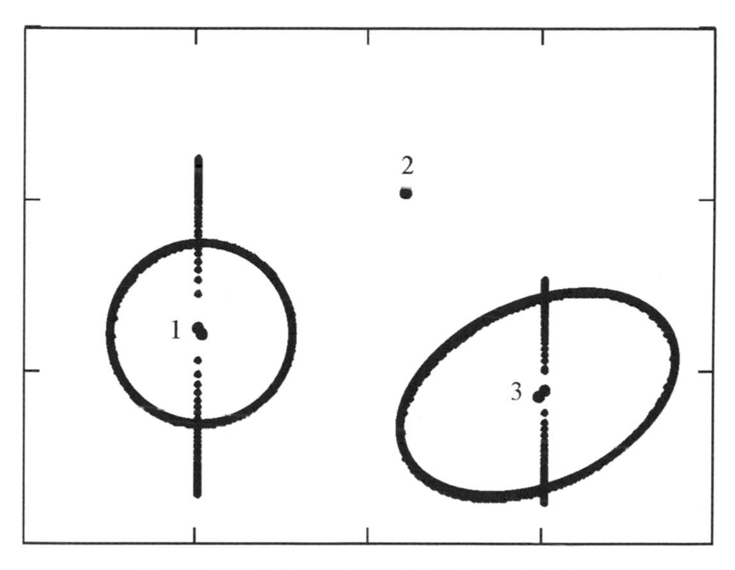

Figure 4.13 Changing minimal constraints.

station 1 fixed. This results in a degenerated ellipse (straight line) at station 1 and a regular ellipse at station 3. In the second case, we hold one of the coordinates of station 3 fixed. The result is a degenerated ellipse at station 3 and a regular ellipse at station 1. The ellipses of standard deviation change significantly due to the change in minimal constraints. Clearly, if one were to specify the quality of a survey in terms of ellipses of standard deviation, one must also consider the underlying minimal constraints. Figure 4.13 also shows that the adjusted coordinates for stations 1 and 2 differ in both cases, although the internal shape of the adjusted network 1-2-3 is the same.

The inner constraint solution, which is a special case of the minimal constraint solutions, has the property that no individual coordinates need to be held fixed. All coordinates become adjustable; for s stations of a plane network, the vector $\mathbf{x}_a$ contains $2s$ coordinate parameters. The ellipses reflect the geometry of the network, the distribution of the observations, and their standard deviations. Section 4.8 contains the theory of inner constraints. The elements for drawing the ellipses are taken from the cofactor matrix (4.203) and Equation (4.212) gives the adjusted parameters. A first step is to find a matrix $\mathbf{E}$ that fulfills $\mathbf{AE}^{\mathrm{T}} = \mathbf{O}$ according to (4.177). The number of rows of $\mathbf{E}$ equals the rank defect of $\mathbf{A}$. For trilateration networks with distances and angles we have

$$\mathbf{E} = \begin{bmatrix} \cdots & 1 & 0 & \cdots & 1 & 0 & \cdots \\ \cdots & 0 & 1 & \cdots & 0 & 1 & \cdots \\ \cdots & -y_i & x_i & \cdots & -y_k & x_k & \cdots \end{bmatrix} \tag{4.387}$$

Four constraints are required for triangulation networks that contain only angle observations. In addition to fixing translation and rotation, triangulation networks also require scaling information. The $\mathbf{E}$ matrix for such networks is

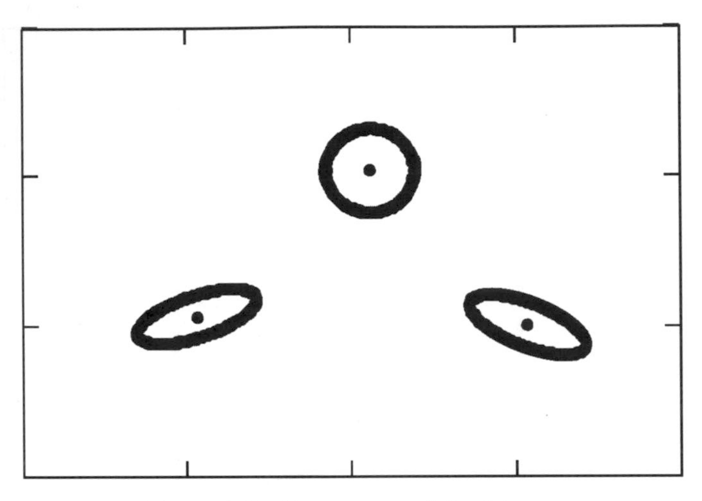

Figure 4.14 Inner constraint solution.

$$\mathbf{E} = \begin{bmatrix} \cdots & 1 & 0 & \cdots & 1 & 0 & \cdots & 1 & 0 & \cdots \\ \cdots & 0 & 1 & \cdots & 0 & 1 & \cdots & 0 & 1 & \cdots \\ \cdots & -y_i & x_i & \cdots & -y_j & x_j & \cdots & -y_k & x_j & \cdots \\ \cdots & x_i & y_i & \cdots & x_j & y_j & \cdots & x_k & y_k & \cdots \end{bmatrix} \tag{4.388}$$

The inner constraint solution is shown in Figure 4.14. Every station has an ellipse. The minimal constraint solutions and the inner constraint solution give the same estimates for residuals, a posteriori variance of unit weight, and redundancy numbers. While the estimated parameters (station coordinates) and their covariance matrix differ for these solutions, the same result is obtained when using these quantities in covariance propagation to compute other observables and their standard deviations.

Example 4: Weighting all approximate coordinates can also provide the coordinate system definition. Table 4.3 contains expressions that include a priori weights on the parameters. If the purpose of the adjustment is to control the quality of the observations, it is important that the weights of the approximate coordinate are small enough to allow observations to adjust freely. For example, if the approximate coordinates are accurate to 1 m, one can use a standard deviation of, say, 1–2 m, or even larger. Ideally, of course, the weight should reflect our knowledge of the approximate coordinates by using meaningful standard deviation. One may prefer to use large standard deviations just to make sure that the internal geometry of the network solution is not affected.

Figure 4.15 shows all ellipses for the case when each approximate station coordinate is assigned a standard deviation of 10 m. The ellipse at each network point is approximately circular. The size of the ellipses is in the range of the a priori coordinate standard deviations. The ellipses in Figure 4.15 imply a scale factor of 10^6 when compared to those in Figures 4.13 and 4.14, which roughly corresponds to the

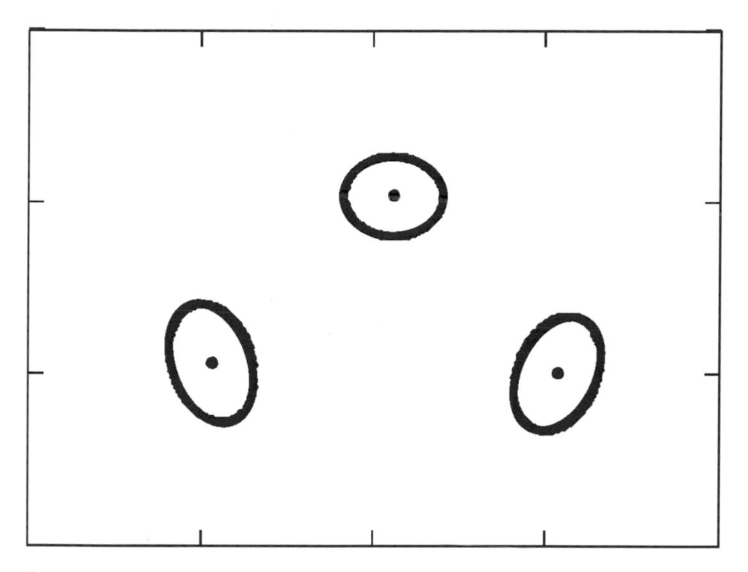

Figure 4.15 Weighting approximate coordinates to define the coordinate system.

ratio of the variances of the approximate coordinates over the average variance of the observations.

The weighted parameter approach is also a convenient way of imposing minimal constraints. Only a subset of three approximate coordinates needs to be weighted in case of a plane angle and distance network.

4.13 KALMAN FILTERING

Least-squares solutions are often applied to surveying networks whose network points refer to monuments that are fixed to the ground. When using the sequential least-squares approach (4.140) to (4.144), the parameters $\mathbf{x}$ are typically treated as a time invariant. The subscript i in these expressions identifies the set of additional observations added to the previous solution that contains the sets $1 \leq i \leq i-1$. Each set of observations merely updates $\mathbf{x}$, resulting in a more accurate determination of the fixed monuments.

We generalize the sequential least-squares formulation by allowing the parameter vector $\mathbf{x}$ to change with time. For example, the vector $\mathbf{x}$ might now contain the three-dimensional coordinates of a moving receiver, the coordinates of satellites, tropospheric delay of signals, or other time-varying parameters. We assume that the dynamic model between parameters of adjacent epochs follows the system of linear equations

$$\mathbf{x}_k(-) = \mathbf{\Phi}_{k-1}\mathbf{x}_{k-1} + \mathbf{w}_k \tag{4.389}$$

We have used the subscript k, instead of i, to emphasize that it now indicates the epoch. The matrix $\mathbf{\Phi}_{k-1}$ is called the parameter transition matrix. The random vector

$\mathbf{w}_k$ is the system process noise and is distributed as $\mathbf{w}_k \sim N(\mathbf{o}, \mathbf{Q}_{w_k})$. The notation $(-)$ indicates the predicted value. Thus,

$$\hat{\mathbf{x}}_k(-) = \mathbf{\Phi}_{k-1}\hat{\mathbf{x}}_{k-1} + \mathbf{w}_k \tag{4.390}$$

$\hat{\mathbf{x}}_k(-)$ is the predicted parameter vector at epoch k, based on the estimated parameter $\hat{\mathbf{x}}_{k-1}(-)$ from the previous epoch and the dynamic model. The solution that generated $\hat{\mathbf{x}}_{k-1}$ also generated the respective cofactor matrix $\mathbf{Q}_{k-1}$. The observation equations for epoch k are given in the familiar form

$$\mathbf{v}_k = \mathbf{A}_k\mathbf{x}_k + \boldsymbol{\ell}_k \tag{4.391}$$

with $\mathbf{v}_k \sim N(\mathbf{o}, \mathbf{Q}_{\ell_k})$.

The first step in arriving at the Kalman filter formulation is to apply variance-covariance propagation to (4.389) to predict the parameter cofactor matrix at the next epoch,

$$\mathbf{Q}_k(-) = \mathbf{\Phi}_{k-1}\mathbf{Q}_{k-1}\mathbf{\Phi}_{k-1}^{\mathrm{T}} + \mathbf{Q}_{w_k} \tag{4.392}$$

Expression (4.392) assumes that the random variables $\boldsymbol{\ell}_k$ and $\mathbf{w}_k$ are uncorrelated. The various observation sets $\boldsymbol{\ell}_k$ are also uncorrelated, as implied by (4.88). The second step involves updating the predicted parameters $\hat{\mathbf{x}}_k(-)$, based on the observations $\boldsymbol{\ell}_k$. Following the sequential least-squares formulation (4.140) to (4.144), we obtain

$$\mathbf{T}_k = \left[\mathbf{Q}_{\ell_k} + \mathbf{A}_k\mathbf{Q}_k(-)\mathbf{A}_k^{\mathrm{T}}\right]^{-1} \tag{4.393}$$

$$\hat{\mathbf{x}}_k = \hat{\mathbf{x}}_k(-) - \mathbf{K}_k\left[\mathbf{A}_k\hat{\mathbf{x}}_k(-) + \boldsymbol{\ell}_k\right] \tag{4.394}$$

$$\mathbf{Q}_k = \left[\mathbf{I} - \mathbf{K}_k\mathbf{A}_k\right]\mathbf{Q}_k(-) \tag{4.395}$$

$$\mathbf{v}^{\mathrm{T}}\mathbf{P}\mathbf{v}_k = \mathbf{v}^{\mathrm{T}}\mathbf{P}\mathbf{v}_{k-1} + \left[\mathbf{A}_k\hat{\mathbf{x}}_k(-) + \boldsymbol{\ell}_k\right]^{\mathrm{T}}\mathbf{T}_k\left[\mathbf{A}_k\hat{\mathbf{x}}_k(-) + \boldsymbol{\ell}_k\right] \tag{4.396}$$

where the matrix

$$\mathbf{K}_k = \mathbf{Q}_k(-)\mathbf{A}_k^{\mathrm{T}}\mathbf{T}_k \tag{4.397}$$

is called the Kalman gain matrix.

If the parameter x_{k+1} depends only on the past (previous) solution x_k, we speak of a first-order Markov process. If noise w_k has a normal distribution, we talk about a first-order Gauss-Markov process,

$$x_{k+1} = \varphi x_k + w_k \tag{4.398}$$

with $w_k \sim n(0, q_{w_k})$. In many applications a useful choice for φ is

$$\varphi = e^{-T/\tau} \tag{4.399}$$

which implies that the variable x is exponentially correlated, i.e., the autocorrelation function is decreasing exponentially (Gelb, 1974, p. 81). The symbol τ denotes the correlation time, and T denotes the time difference between epochs $k+1$ and k. The variance of the process noise for correlation time τ is

$$q_{w_k} = E(w_k w_k) = \frac{\tau}{2}\left[1 - e^{-2T/\tau}\right] q_k \tag{4.400}$$

with q_k being the variance of the process noise (Gelb, 1974, p. 82). The quantities (τ, q_k) could be initially determined from data by fitting a sample mean and sample autocorrelation function.

As τ approaches zero, then $\varphi = 0$. This describes the pure white noise model with no correlation from epoch to epoch. x can be thought of as a random constant that is a nondynamic quantity.

As τ approaches infinity, we obtain the pure random walk. Applying l'Hospital rule for computing the limit or using series expansion, we obtain $\varphi = 1$ and $q_{w_k} = T q_k$. The random noises w_k are uncorrelated.

In general, both the dynamic model (4.389) and the observation model (4.391) are nonlinear. The extended Kalman filter formulation (Gelb, 1974, p. 187) applies to this general case. The reader is urged to consult that reference or other specialized literature for additional details on Kalman filtering.

CHAPTER 5

PSEUDORANGE AND CARRIER PHASE OBSERVABLES

Pseudoranges and carrier phases are the most important GPS observations (observables) used for positioning. Solutions are available that use pseudoranges only, carrier phases only, or both types of observations. The early solutions for navigation relied on pseudoranges. More recently, even point positioning often includes the carrier phase observable. Carrier phases are always required for accurate surveying at the centimeter level. Processing algorithms exist that use the (undifferenced) observations directly. However, one often uses certain linear combinations. Popular examples are the double differences and the triple differences.

Measuring pseudoranges and carrier phases involves advanced techniques in electronics and digital signal processing. This chapter deals with the equations that directly apply to the pseudoranges and carrier phases as downloaded from the receiver. The goal is to determine geocentric positions (point positioning) or relative positions between co-observing stations (differential or relative positioning). These equations are also the basis for estimating ionospheric and tropospheric parameters with GPS, or the transfer of time.

In addition to deriving and discussing the basic pseudorange and carrier phase equations and the double- and triple-difference functions, we address frequently asked questions of novice GPS users. Examples include simultaneity of observations, singularities, and a priori knowledge of initial station and ephemeris. The implications of relativity for GPS observables have widely been addressed in the literature, e.g., Grafarend (1992), Hatch (1992), Ashby (1993), and Schwarze et al. (1993). In this chapter, remarks on relativity are limited to Section 5.3.1, where a clock correction term is given to account for satellite orbital eccentricity. The correction and the adjustment to the fundamental frequency of 10.23 MHz mentioned in Section 3.2.2 are the only references to the applicability of relativity to GPS in this book. In relative positioning, most of the relativistic effects cancel or become negligible.

Throughout this and subsequent chapters, a superscript identifies the satellite, and a subscript identifies the receiver. Lowercase letters p and q generally label the satellites, whereas k and m refer to receivers. In relative positioning when two or more receivers observe at the same time, we occasionally refer to the concept of a base satellite and a base station. Usually the letter p will denote the base satellite and k the base station. One may think of p and q as the pseudorandom noise (PRN) numbers of the satellite or simply as sequential numbers identifying the satellites. The L1 and L2 carrier frequencies are indicated by the subscripts 1 and 2, respectively. In this notation we write the L1-pseudorange observation at station k to satellite p as $P_{k,1}^{p}$. The respective notation for the L1-carrier phase in cycles is $\varphi_{k,1}^{p}$. The symbol $\Phi_{k,1}^{p}$ refers to carrier phases scaled to distance. Occasionally we identify other terms that relate to pseudoranges or carrier phases with subscripts P, φ, and Φ. For example, $I_{k,2,P}^{p}$ identifies the ionospheric delay of the L2-pseudorange measurement from station k to satellite p.

In many cases, the superscripts and subscripts also indicate specific functions of the observables. For example, if we label the epoch of the observations by t, the single difference (SD) and double difference (DD) are defined as

$$\varphi_{km}^{p}(t) = \varphi_{k}^{p}(t) - \varphi_{m}^{p}(t) \tag{5.1}$$

$$\varphi_{km}^{pq}(t) = \varphi_{km}^{p}(t) - \varphi_{km}^{q}(t) \tag{5.2}$$

This notation does not use a comma between subscripts k and m and superscripts p and q. Notice that the subscript combinations km and pq as used here imply a differencing of "k-m" and "p-q" respectively. The triple difference (TD) refers to the difference over time,

$$\Delta\varphi_{km}^{pq}\left(t_2, t_1\right) = \varphi_{km}^{pq}\left(t_2\right) - \varphi_{km}^{pq}\left(t_1\right) \tag{5.3}$$

The between satellite difference (BSD) is identified by

$$\varphi_{k}^{pq}(t) = \varphi_{k}^{p}(t) - \varphi_{k}^{q}(t) \tag{5.4}$$

The receiver sets the epochs of observation internally. Without going into the inner workings of receivers, suffice it to note from the users' point of view that receivers make measurements of several satellites (usually all satellites in view) at the same epoch. The output is available typically at the full second, half second, etc. The user usually can set the rate of output. With this understanding, we will be able to simplify notation at times by simply dropping the time label t.

5.1 PSEUDORANGES AND CARRIER PHASES

The pseudorange is related to the distance between the satellite and the receiver's antenna, implied by the epochs of emission and reception of the codes. The transmission (travel) time of the codes is measured by correlating identical PRN codes generated by the satellite with those generated internally by the receiver. The code-tracking loop within the receiver shifts the internal replica of the PRN code in time until maximum correlation occurs. The codes generated by the receiver are based on the receiver's own clock, and the codes of the satellite transmissions are generated by the satellite clock. Unavoidable timing errors at the satellite and the receiver will cause the measured pseudorange to differ from the geometric distance corresponding to the instants of emission and reception. Pseudoranging is applicable to P(Y)-codes and C/A-codes.

The equation for the pseudorange observable can easily be built by considering first the spatial distance in vacuum,

$$\rho_k^p(\widehat{t}^{\,p}) = \left(\widehat{t}_k - \widehat{t}^{\,p}\right)c = \left(t_k + d\underline{t}_{\,k} - t^p - d\bar{t}^p\right)c \tag{5.5}$$

$\rho_k^p(\widehat{t}^{\,p})$ Geometric distance (vacuum distance) traveled by the code from transmission at satellite p to reception at the receiver antenna k. This distance will eventually have to be computed as part of the receiver position computations. See details below.

$\widehat{t}_k$ True time at the receiver at the epoch the code entered the antenna. The nominal time, i.e., the receiver clock reading, is denoted by t_k. This nominal receiver time is in error by $d\underline{t}_{\,k}$.

$\widehat{t}^{\,p}$ True time at the epoch of code transmission. The nominal satellite time, i.e., the satellite clock reading, is denoted by t^p. This nominal satellite clock time is in error by $d\bar{t}^p$.

c Velocity of light.

There is a direct linear relationship between codes and nominal clock time. The code generation sequence is specified by time as a parameter. Therefore, the nominal satellite time determines which code leaves the satellite and when. The same is true regarding the nominal receiver time and the generation of the receiver's code sequence. The measure pseudorange is, therefore, a function of the nominal times. For the vacuum we have

$$P_k^p(t_k) \equiv \left(t_k - t^p\right)c = \rho_k^p(\widehat{t}^{\,p}) - c\,d\underline{t}_{\,k} + c\,d\bar{t}^p \tag{5.6}$$

where $P_k^p(t_k)$ denotes the pseudorange.

The mathematical expression for the pseudorange observable must take into account the effects of the ionosphere and the troposphere, as well as hardware delays at the satellite and at the receiver. Adding the subscript to identify the frequency, the actual expression for the pseudorange observable becomes

$$P_{k,1}^p(t_k) = \rho_k^p(\widehat{t}^{\,p}) - c\,d\underline{t}_{\,k} + c\,d\bar{t}^p + I_{k,1,P}^p(t_k) + T_k^p(t_k) + \delta_{k,1,P}^p(t_k) + \varepsilon_{1,P} \tag{5.7}$$

with

$$\rho_k^p(\hat{t}^p) = \left\|\mathbf{x}^p - \mathbf{x}_k\right\| = \sqrt{(\mathbf{x}^p - \mathbf{x}_k) \cdot (\mathbf{x}^p - \mathbf{x}_k)}$$
$$= \sqrt{(x^p - x_k)^2 + (y^p - y_k)^2 + (z^p - z_k)^2} \tag{5.8}$$

$$\delta_{k,1,P}^p(t_k) = d_{k,1,P}(t_k) + d_{k,1,P}^p(t_k) + d_{1,P}^p(t_k) \tag{5.9}$$

$\rho_k^p(\hat{t}^p)$ This is again the geometric vacuum distance that is often computed from the ECEF receiver coordinates $\mathbf{x}_k$ and satellite coordinates $\mathbf{x}^p$, taking the earth's rotation during signal travel time into account. Given the nominal time t^p, we add the satellite clock correction given in the broadcast message to compute and estimate the true time $\hat{t}^p$. Because the satellites carry atomic clocks that are carefully monitored and are fairly stable, one can safely assume that the residual error in $\hat{t}^p$ is less than 1 microsecond. For a topocentric range rate of $\left|\dot{\rho}_k^p(t)\right| < 800$ m/s, the computation error in distance is $d\rho < 1$ mm because of dt^p. This error is negligible.

$I_{k,1,P}^p$ Ionospheric P(Y)-code delay at L1. This delay is always positive. It depends on the ionospheric condition along the path and on the frequency. Details are provided in Chapter 6.

T_k^p Tropospheric delay. This delay is always positive. It depends upon the tropospheric condition along the path but is independent of the carrier frequency. Therefore, there is no need to identify the frequency.

$d_{k,1,P}$ Receiver hardware delay. This delay does not depend on the satellite being observed.

$d_{k,1,P}^p$ Multipath delay. This delay depends on the direction of the satellite.

$d_{1,P}^p$ Satellite hardware delay.

$\varepsilon_{1,P}$ Pseudorange measurement noise (approximately 30 cm for P(Y) code pseudoranges and worse for C/A-code pseudoranges, depending on the technology used).

The pseudorange (5.7) would equal the geometric distance from the satellite at epoch of transmission to reception at the receiver if the propagation medium were a vacuum and if there were no clock errors and no other biases.

The phase observable is the sum of the fractional carrier phase at nominal frequency f_1, which arrives at the antenna at the nominal time t_k, and an unknown integer constant representing full waves. In units of cycles the equation for the carrier phase L1 is

$$\varphi_{k,1}^p(t_k) = \frac{f_1}{c}\rho_k^p(\hat{t}^p) + N_k^p(1) - f_1\, d\underline{t}_k + f_1\, d\bar{t}^p + I_{k,1,\varphi}^p(t_k) + \frac{f_1}{c}T_k^p(t_k)$$
$$+ \delta_{k,1,\varphi}^p(t_k) + \varepsilon_{1,\varphi} \tag{5.10}$$

$$\delta_{k,1,\varphi}^p(t_k) = d_{k,1,\varphi}(t_k) + d_{k,1,\varphi}^p(t_k) + d_{1,\varphi}^p(t_k) \tag{5.11}$$

This expression differs from the pseudorange (5.7) as follows:

$N_k^p(1)$ Integer ambiguity: This integer refers to the first epoch of observations and remains constant during the period of observation. During this period, the receiver accumulates the phase differences between arriving phases and internally generated receiver phases. The receiver, therefore, effectively generates an accumulated carrier phase observable that reflects the changes in distance to the satellite. The observation series is continuous until a cycle slip occurs, which introduces an integer jump. After the cycle slip has occurred, the observation series continues with a new integer constant N_k^p.

$I_{k,1,\varphi}^p$ Ionospheric L1 carrier phase advance. This value is negative. The numerical value is a function of the frequency and the ionospheric condition along the path. See Chapter 6 for details of this delay and its relation to the corresponding ionospheric pseudorange delay.

$\delta_{k,1,\varphi}^p$ Hardware delays and multipath effects on the L1 carrier phase.

$\varepsilon_{1,\varphi}$ L1 phase measurement noise (< 0.01 cycles)

The carrier phase can be scaled to unit of length by multiplying with $\lambda_1 = c/f_1$. Thus (5.10) becomes

$$\Phi_{k,1}^p(t_k) = \rho_k^p(\hat{t}^p) + \lambda_1 N_k^p(1) - c\, d\underline{t}_k + c\, d\bar{t}^p + I_{k,1,\Phi}^p(t_k) + T_k^p(t_k) + \delta_{k,1,\Phi}^p(t_k) + \varepsilon_{1,\Phi} \tag{5.12}$$

The subscript Φ implies that the respective quantities are in units of length, e.g., $I_{k,1,\Phi}^p = \lambda_1\, I_{k,1,\varphi}^p$.

Receivers observe pseudoranges and carrier phases to several satellites at the same time. Today's all-in-view receivers generate these observables for all visible satellites at the same nominal time t_k, for L1 and L2 frequencies. Therefore, the receiver clock error and the hardware delays are the same for all observations at the same epoch. Since GLONASS satellites transmit at different frequencies, special attention must be given to the implications of the receiver clock errors when expressing the carrier phases. For details on GLONASS phase processing, see Section 7.7.6.

The following simple functions of the L1 and L2 frequencies are useful for future references,

$$\alpha_f = (f_1/f_2)^2 = (77/60)^2 \approx 1.647 \tag{5.13}$$

$$\beta_f = f_1^2/\left(f_1^2 - f_2^2\right) = \alpha_f/\left(\alpha_f - 1\right) = 77^2/\left(77^2 - 60^2\right) \approx 2.546 \tag{5.14}$$

$$\gamma_f = f_2^2/\left(f_1^2 - f_2^2\right) = 1/\left(\alpha_f - 1\right) = 60^2/\left(77^2 - 60^2\right) \approx 1.546 \tag{5.15}$$

$$\delta_f = f_1 f_2/\left(f_1^2 - f_2^2\right) = \sqrt{\alpha_f}/(\alpha_f - 1) = 77 * 60/\left(77^2 - 60^2\right) \approx 1.984 \tag{5.16}$$

5.2 DIFFERENCING

For receivers at stations k and m, observing the same satellite p at the nominal times t_k and t_m, one can write two pseudorange equations (5.7) and two carrier phase equations (5.10). Let t_k be approximately equal to t_m. Some manufacturers time-shift the observations to make these nominal times equal, say, to the full second. Even if the nominal times are the same, the respective signals leave the satellite p at slightly different times. This is so because the distances between receivers and satellite differ. Because the satellite clocks are highly stable, we assume that the satellite clock errors are the same for these near-simultaneous transmissions. The same assumption is made in regard to the internal satellite hardware delays. Under such a condition of near-simultaneity, the single-difference phase observable (5.1) becomes

$$\begin{aligned} \varphi^p_{km,1}(t) = \frac{f}{c}\rho^p_{km}(\hat{t}^p) + N^p_{km,1}(1) - f(d\underline{t}_k - d\underline{t}_m) \\ + I^p_{km,1,\varphi}(t) + \frac{f}{c}T^p_{km}(t) + d_{km,1,\varphi}(t) + d^p_{km,1,\varphi}(t) + \varepsilon^p_{km,1,\varphi} \end{aligned} \tag{5.17}$$

To simplify the notation, we have used the symbol t to denote the time of observations in those terms where the distinction between t_k and t^p is not necessary. Computing ρ^p_{km} requires two different emission times, one with respect to the observation from receiver k and one with respect to receiver m. We introduce no additional notation to label these two different emission times because it is clear from the context which times must be used for ephemeris interpolation. Following the subscript convention for differencing, we have

$$\rho^p_{km}(\hat{t}^p) = \rho^p_k(\hat{t}^p) - \rho^p_m(\hat{t}^p) \tag{5.18}$$

$$N^p_{km}(1) = N^p_k(1) - N^p_m(1) \tag{5.19}$$

$$I^p_{km,1,\varphi}(t) = I^p_{k,1,\varphi}(t) - I^p_{m,1,\varphi}(t) \tag{5.20}$$

$$T^p_{km}(t) = T^p_k(t) - T^p_m(t) \tag{5.21}$$

$$d_{km,1,\varphi}(t) = d_{k,1,\varphi}(t) - d_{m,1,\varphi}(t) \tag{5.22}$$

$$d^p_{km,1,\varphi}(t) = d^p_{k,1,\varphi}(t) - d^p_{m,1,\varphi}(t) \tag{5.23}$$

$$\varepsilon^p_{km,1,\varphi}(t) = \varepsilon^p_{k,1,\varphi}(t) - \varepsilon^p_{m,1,\varphi}(t) \tag{5.24}$$

We notice that the satellite clock error and the satellite hardware delay have canceled in the single differences. However, the single-difference observations remain sensitive to both receiver clock errors $d\underline{t}_k$ and $d\underline{t}_m$, and to signal multipath at the receiver.

If two receivers k and m observe two satellites p and q at the same nominal time, the double-difference phase observable (5.2) is

$$\varphi_{km,1}^{pq}(t) = \frac{f}{c}\rho_{km}^{pq}(\hat{t}^p) + N_{km,1}^{pq}(1) + I_{km,1,\varphi}^{pq}(t) + \frac{f}{c}T_{km}^{pq}(t) + d_{km,1,\varphi}^{pq}(t) + \varepsilon_{km,1,\varphi}^{pq} \tag{5.25}$$

where

$$\rho_{km}^{pq}(\hat{t}^p) = \rho_{km}^{p}(\hat{t}^p) - \rho_{km}^{q}(\hat{t}^p) \tag{5.26}$$

$$N_{km}^{pq}(1) = N_{km}^{p}(1) - N_{km}^{q}(1) \tag{5.27}$$

$$I_{km,\varphi}^{pq}(t) = I_{km,\varphi}^{p}(t) - I_{km,\varphi}^{q}(t) \tag{5.28}$$

$$T_{km}^{pq}(t) = T_{km}^{p}(t) - T_{km}^{q}(t) \tag{5.29}$$

$$d_{km,\varphi}^{pq}(t) = d_{km,\varphi}^{p}(t) - d_{km,\varphi}^{q}(t) \tag{5.30}$$

$$\varepsilon_{km,\varphi}^{pq} = \varepsilon_{km,\varphi}^{p} - \varepsilon_{km,\varphi}^{q} \tag{5.31}$$

The most important feature of the double-difference observation is the cancellation of the large receiver clock errors $d\underline{t}_k$ and $d\underline{t}_m$ (in addition to the cancellation of the satellite clock errors and the satellite hardware delays). The receiver hardware delays at a given receiver also cancel, as long as they are the same for every satellite observed. Because multipath is a function of the geometry between receiver, satellite, and reflector, the term (5.30) does not cancel in the double-difference observable.

The double-difference integer ambiguity N_{km}^{pq} plays an important role in accurate relative positioning using double differences. Estimating the ambiguity together with the other parameters as a real number, one gets the so-called *float solution*. If the estimated ambiguities $\hat{N}_{km}^{pq}$ can be successfully constrained to integer, one gets the *ambiguity fixed solution*. Because of residual model errors the estimated ambiguities will, at best, be close to integers. Imposing integer constraints adds strength to the solution, because the number of parameters is reduced and the correlations between parameters reduce as well. Much effort has gone into extending the baseline length over which ambiguities can be fixed. At the same time, much research has been carried out to develop algorithms that allow the ambiguities to be fixed from short observation spans over short baselines. Having the possibility of imposing the integer constraint on the estimated ambiguity is a major strength of the double-differencing approach. For details on this topic, see Chapter 7.

The triple difference (5.3) is the difference of two double differences over time

$$\begin{aligned}\Delta\varphi_{km,1}^{pq}(t_2, t_1) = {} & \frac{f_1}{c}\left[\Delta\rho_{km}^{pq}(\hat{t}^p, \hat{t}^q)\right] + \\ & + \Delta I_{km,1,\varphi}^{pq}(t_2, t_1) + \frac{f}{c}\Delta T_{km}^{pq}(t_2, t_1) + \Delta d_{km,1,\varphi}^{pq}(t_2, t_1) + \Delta\varepsilon_{km,1,\varphi}^{pq}\end{aligned} \tag{5.32}$$

$$\Delta\rho_{km}^{pq}(\widehat{t}^{\,p}, \widehat{t}^{\,q}) = \rho_{km}^{pq}(\widehat{t}_2^{\,p}, \widehat{t}_2^{\,q}) - \rho_{km}^{pq}(\widehat{t}_1^{\,p}, \widehat{t}_1^{\,q}) \tag{5.33}$$

The initial integer ambiguity $N_{km}^{pq}(1)$ cancels in (5.32). The triple-difference observable is probably the easiest to deal with because of this cancellation. Often the triple-difference solution serves as a preprocessor to get good initial positions for the double-difference solution. The triple differences have the advantage in that cycle slips are mapped as individual outliers in the computed residuals. Individual outliers can usually be detected and removed. The resulting cycle slip free observations can then be used in the double-difference solution.

A delta range is the difference in time of observables at the same station. For example,

$$\begin{aligned}\Delta\varphi_{k,1}^{p}(t_2, t_1) = {} & \frac{f_1}{c}\Delta\rho_k^p(\widehat{t}_2^{\,p}, \widehat{t}_1^{\,p}) - f_1\, \Delta d\underline{t}_k + f_1\, \Delta d\bar{t}^p + \Delta I_{k,1,\varphi}^p(t) \\ & + \frac{f_1}{c}\Delta T_k^p(t) + \Delta\delta_{k,1,\varphi}^p(t) + \Delta\varepsilon_{1,\varphi}\end{aligned} \tag{5.34}$$

$$\Delta\rho_k^p(\widehat{t}_2^{\,p}, \widehat{t}_1^{\,p}) = \rho_k^p(\widehat{t}_2^{\,p}) - \rho_k^p(\widehat{t}_1^{\,p}) \tag{5.35}$$

These delta ranges are a function of the change in topocentric distance between the station and the satellite, provided there is no cycle slip between the epochs t_1 and t_2. They do not depend on the initial ambiguity because of the differencing over time. The delta range (5.34) depends on the change of the receiver and satellite clock errors from epoch t_1 to epoch t_2.

The between-satellite difference (5.4)

$$\begin{aligned}\varphi_{k,1}^{pq}(t) = {} & \frac{f_1}{c}\rho_k^{pq}(\widehat{t}^{\,p}, \widehat{t}^{\,q}) + N_k^{pq}(1) + f_1\, d\underline{t}^{pq} + I_{k,1,\varphi}^{pq}(t) + \frac{f_1}{c}T_k^{pq}(t) \\ & + \delta_{k,1,\varphi}^{pq}(t) + \varepsilon_{1,\varphi}\end{aligned} \tag{5.36}$$

$$\rho_k^{pq}(\widehat{t}^{\,p}, \widehat{t}^{\,q}) = \rho_k^p(\widehat{t}^{\,p}) - \rho_k^q(\widehat{t}^{\,q}) \tag{5.37}$$

does not depend on the receiver clock error but, instead, contains again an integer ambiguity.

5.3 INITIAL EVALUATION

5.3.1 Satellite Clock Corrections

The control segment maintains GPS time to within 1 μs of UTC(USNO) according to the Interface Control Document (ICD-GPS-200C, 2000), but GPS time does not follow the UTC leap-second jumps. The full second offset is readily available on the Internet and from various data services, if needed. The user needs GPS time and not UTC because the observations are time-tagged with GPS time; it is also the time argument for the broadcast and precise ephemerides. Because the satellite transmissions are steered by the nominal time of the individual satellite (satellite time), it is important to know the differences between GPS time and the individual satellite time. In the notation and sign convention as used by the interface control document, the time correction to the nominal space vehicle time t_{SV} is

$$\Delta t_{\mathrm{SV}} = a_{f0} + a_{f1}\left(t_{\mathrm{SV}} - t_{\mathrm{oc}}\right) + a_{f2}\left(t_{\mathrm{SV}} - t_{\mathrm{oc}}\right)^2 + \Delta t_R \tag{5.38}$$

with

$$t_{\mathrm{GPS}} = t_{\mathrm{SV}} - \Delta t_{\mathrm{SV}} \tag{5.39}$$

and

$$\Delta t_R = -\frac{2}{c^2}\sqrt{a\mu}\; e \sin E = -\frac{2}{c^2}\,\mathbf{X}\cdot\dot{\mathbf{X}} \tag{5.40}$$

The polynomial coefficients are transmitted in units of sec, sec/sec, and sec/sec^2; the clock data reference time t_{oc} is also broadcast in seconds in subframe 1 of the navigation message. As is required when using the ephemeris expressions, the value of t_{SV} must account for the beginning or end-of-week crossovers. That is, if $(t_{\mathrm{SV}} - t_{\mathrm{oc}})$ is greater than 302,400, subtract 604,800 from t_{SV}. If $(t_{\mathrm{SV}} - t_{\mathrm{oc}})$ is less than $-302{,}400$, add 604,800 to t_{SV}.

The second part of (5.40) follows from (3.61). Δt_R is a small relativistic clock correction caused by the orbital eccentricity e. The symbol μ denotes the gravitational constant, a is the semimajor axis of the orbit, and E is the eccentric anomaly. See Chapter 3 for details on these elements. Using $a \approx 26{,}600$ km we have

$$\Delta t_{R[\mu\mathrm{sec}]} \approx -2e \sin E \tag{5.41}$$

5.3.2 Topocentric Range

The pseudorange equation (5.7) and the carrier phase equation (5.10) require that the topocentric distance ρ_k^p be computed. In the inertial coordinate system (X), this is simply accomplished by

$$\rho_k^p = \left\| \mathbf{X}_k(\widehat{t}_k) - \mathbf{X}^p(\widehat{t}^{\,p}) \right\| \tag{5.42}$$

In the inertial coordinate system, the receiver coordinates are a function of time due to the earth's rotation. If the receiver antenna and satellite ephemeris are available in the terrestrial coordinate system, we must take the earth's rotation explicitly into account. Neglecting polar motion, the Greenwich apparent sidereal time relates the terrestrial coordinate system (x) to the inertial system (X) by (2.34). Let θ^p and θ_k denote the Greenwich apparent sidereal times for transmission and reception of the signal, then

$$\rho_k^p = \left\| \mathbf{R}_3(-\theta_k)\mathbf{x}_k - \mathbf{R}_3(-\theta^p)\mathbf{x}^p(\hat{t}^p) \right\| \tag{5.43}$$

If τ denotes the travel time for the signal, then the earth rotates during that time by

$$\theta = \theta_k - \theta^p = \dot{\Omega}_e \left(\hat{t}_k - \hat{t}^p \right) = \dot{\Omega}_e \tau \tag{5.44}$$

with $\dot{\Omega}_e$ being the earth rotation rate and τ the travel time of the signal. The topocentric distance becomes

$$\begin{aligned} \rho_k^p &= \left\| \mathbf{R}_3(-\theta_k)\mathbf{x}_k - \mathbf{R}_3(-\theta_k + \theta)\mathbf{x}^p(\hat{t}^p) \right\| \\ &= \left\| \mathbf{R}_3(-\theta^p)\left\{ \mathbf{R}_3(-\theta)\mathbf{x}_k - \mathbf{x}^p(\hat{t}^p) \right\} \right\| \\ &= \left\| \mathbf{R}_3(-\theta)\mathbf{x}_k - \mathbf{x}^p(\hat{t}^p) \right\| \\ &= \left\| \mathbf{x}_k - \mathbf{R}_3(\theta)\mathbf{x}^p(\hat{t}^p) \right\| \end{aligned} \tag{5.45}$$

In modifying (5.45), we used the facts that a distance is invariant with respect to the rotation of the coordinate system and that the rotation matrix $\mathbf{R}_3$ is orthonormal.

Because θ is a function of τ, Equation (5.45) must be iterated. A good initial estimate is $\tau_0 = 0.75$ sec. Computing θ_1 from (5.44) and using this value in (5.45) gives the initial value ρ_1 for the distance. The second estimate of the travel time follows from $\tau_2 = \rho_1/c$. This value is used in (5.44) to continue the iteration loop.

5.3.3 Cycle Slips

A cycle slip is a sudden jump in the carrier phase observable by an integer number of cycles. The fractional portion of the phase is not affected by this discontinuity in the observation sequence. Cycle slips are caused by the loss of lock of the phase lock loops. Loss of lock may occur briefly between two epochs or may last several minutes or more, if the satellite signals cannot reach the antenna. If receiver software would not attempt to correct for cycle slips, it would be a characteristic of a cycle slip that all observations after the cycle slip would be shifted by the same integer. This is demonstrated in Table 5.1, where a cycle slip is assumed to have occurred for

TABLE 5.1 Effect of Cycle Slips on Carrier Phase Differences

Carrier Phase				Double Difference	Triple Difference
$\varphi_k^p(i-2)$	$\varphi_m^p(i-2)$	$\varphi_k^q(i-2)$	$\varphi_m^q(i-2)$	$\varphi_{km}^{pq}(i-2)$	$\Delta\varphi_{km}^{pq}(i-1, i-2)$
$\varphi_k^p(i-1)$	$\varphi_m^p(i-1)$	$\varphi_k^q(i-1)$	$\varphi_m^q(i-1)$	$\varphi_{km}^{pq}(i-1)$	$\Delta\varphi_{km}^{pq}(i, i-1) - \Delta$
$\varphi_k^p(i)$	$\varphi_m^p(i)$	$\varphi_k^q(i) + \Delta$	$\varphi_m^q(i)$	$\varphi_{km}^{pq}(i) - \Delta$	$\Delta\varphi_{km}^{pq}(i+1, i)$
$\varphi_k^p(i+1)$	$\varphi_m^p(i+1)$	$\varphi_k^q(i+1) + \Delta$	$\varphi_m^q(i+1)$	$\varphi_{km}^{pq}(i+1) - \Delta$	$\Delta\varphi_{km}^{pq}(i+2, i+1)$
$\varphi_k^p(i+2)$	$\varphi_m^p(i+2)$	$\varphi_k^q(i+2) + \Delta$	$\varphi_m^q(i+2)$	$\varphi_{km}^{pq}(i+2) - \Delta$	

receiver k while observing satellite q between the epochs $i-1$ and i. The cycle slip is denoted by Δ. Because the double differences are a function of observations at one epoch, all double differences starting with epoch i are offset by the amount Δ. Only one of the triple-differences is affected by the cycle slip, because triple differences are differences over time. For each additional slip there is one additional triple-difference outlier and one additional step in the double-difference sequence. A cycle slip may be limited to just one cycle or could be millions of cycles.

This simple relation can break down if the receiver software attempts to fix the slips internally. Assume the receiver successfully corrects for a slip immediately following the epoch of occurrence. The result is an outlier (not a step function) for double differences and two outliers for the triple differences.

There is probably no best method for cycle slips removal, leaving lots of space for optimization and innovation. For example, in the case of simple static applications, one could fit polynomials, generate and analyze higher-order differences, visually inspect the observation sequences using graphical tools, or introduce new ambiguity parameters to be estimated whenever a slip might have occurred. The latter option is very attractive in kinematic positioning.

It is best to inspect the discrepancies rather than the actual observations. The observed double and triple differences show a large time variation that depends on the length of the baseline and the satellites selected. These variations can mask small slips. The discrepancies are the difference between the computed observations and the actual observed values. If good approximate station coordinates are used then the discrepancies are rather flat and make even small slips easily detectable.

For static positioning, one could begin with the triple-difference solution. The affected triple-difference observations can be treated as observations with blunders and dealt with using the blunder detection techniques given in Chapter 4. A simple method is to change the weights of those triple-difference observations that have particularly large residuals. Once the least-squares solution has converged, the residuals will indicate the size of the cycle slips. Not only is triple-difference processing a robust technique for cycle slip detection, it also provides good station coordinates, which, in turn, can be used as approximations in a subsequent double-difference solution.

Before computing the double-difference solution, the double-difference observations should be corrected for cycle slips identified from the triple-difference solution. If only two receivers observe, it is not possible to identify the specific undifferenced phase sequence where the cycle slip occurred from analysis of the double difference. Consider the double differences

$$\varphi_{12}^{1p} = \left(\varphi_1^1 - \varphi_2^1\right) - \left(\varphi_1^p - \varphi_2^p\right) \tag{5.46}$$

for stations 1 and 2 and satellites 1 and p. The superscript p denoting the satellites varies from 2 to S, the total number of satellites. Equation (5.46) shows that a cycle slip in φ_1^1 or φ_2^1 will affect all double differences for all satellites and cannot be separately identified. The slips Δ_1^1 and $-\Delta_2^1$ cause the same jump in the double-difference observation. The same is true for slips in the phase from station 1 to satellite

p and station 2 to satellite p. However, a slip in the latter phase sequences affects only the double differences containing satellite p. Other double-difference sequences are not affected.

For a session network, the double-difference observation is

$$\varphi_{1m}^{1p} = \left(\varphi_1^1 - \varphi_m^1\right) - \left(\varphi_1^p - \varphi_m^p\right) \tag{5.47}$$

The superscript p goes from 2 to S, and the subscript m runs from 2 to R. It is readily seen that a cycle slip in φ_1^1 affects all double-difference observations, an error in φ_m^1 affects all double differences pertaining to the baseline 1 to m, an error in φ_1^p affects all double differences containing satellite p, and an error in φ_m^p affects only one series of double differences, namely, the one that contains station m and satellite p. Thus, by analyzing the distribution of a blunder in all double differences at the same epoch, we can identify the undifferenced phase observation sequence that contains the blunder. This identification gets more complicated if several slips occur at the same epoch. In session network processing, it is always necessary to carry out cross-checks. The same cycle slip must be verified in all relevant double differences before it can be declared an actual cycle slip. Whenever a cycle slip occurs in the undifferenced phase observations from the base station or to the base satellite, the cycle slip enters several double-difference sequences. Actually it is not necessary that the undifferenced phase observations be corrected; it is sufficient to limit the correction to the double-difference phase observations if the final position computation is based on double differences.

It is also possible to use the geometry-free functions of the observables to detect cycle slips. The geometry-free functions are discussed in Chapter 7.

5.3.4 Singularities

A case of a critical configuration for terrestrial observations is discussed in Chapter 4. For example, Figure 4.12 shows how ellipses of standard deviation display the change in the geometry as the critical configuration (singularity) is reached for the plane resections. The dilution of precision (DOP) introduced in Section 7.4.1 is a one-number indicator for the geometry of the point positioning solutions. At the critical configurations the columns of the design matrix become linearly dependent. When the satellite constellation approaches a critical configuration, the resulting positioning solution can be ill conditioned.

Linearizing the pseudorange equation (5.7) around the receiver location $\mathbf{x}_k$ gives

$$\begin{aligned} dP_k^p &= -\begin{bmatrix} \dfrac{x^p - x_k}{\rho_k^p} & \dfrac{y^p - y_k}{\rho_k^p} & \dfrac{z^p - z_k}{\rho_k^p} \end{bmatrix} \begin{bmatrix} dx_k \\ dy_k \\ dz_k \end{bmatrix} = -\mathbf{e}_k^p \cdot d\mathbf{x}_k \\ &= -\frac{1}{\rho_k^p} \boldsymbol{\rho}_k^p \cdot d\mathbf{x}_k \end{aligned} \tag{5.48}$$

where $\mathbf{e}_k^p$ is the unit vector pointing from the station to the satellite and $\boldsymbol{\rho}_k^p$ is the respective unscaled topocentric vector.

Figure 5.1 shows a situation where all satellites are located on a circular cone. This is obviously a special situation. The vertex of the cone is at the receiver. The unit vector $\mathbf{e}_{\text{axis}}$ specifies the axis of the cone. For all satellites that are located on the cone, the dot product

$$\mathbf{e}_k^i \cdot \mathbf{e}_{\text{axis}} = \cos\theta \tag{5.49}$$

is constant. $\mathbf{e}_k^i$ represents the first three elements of row i of the design matrix. Therefore, (5.49) expresses a perfect linear dependency of the four columns. Another critical configuration occurs when the satellites and the receiver are located in the same plane. In this case, the first three columns of the design matrix fulfill the cross-product vector function

$$\mathbf{e}_k^i \cdot \mathbf{e}_k^j = \mathbf{n} \tag{5.50}$$

where $\mathbf{n}$ is a constant vector.

Critical configurations usually do not last long because of the continuous motion of the satellites. The critical configurations present a problem only in continuous kinematic or very short rapid static applications. The more satellites are available, the less likely it is that a critical configuration will ever occur.

In relative positioning, one can encounter critical configurations as well. Clearly, the satellites cannot be located on a perfectly circular cone as viewed from each of the

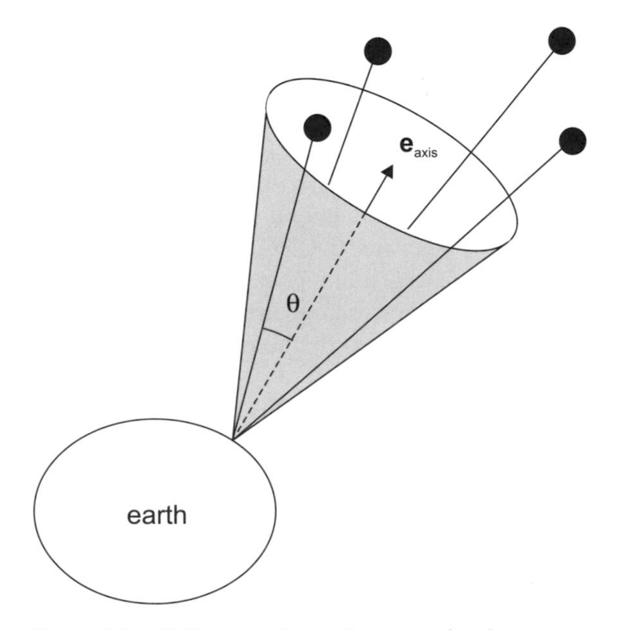

Figure 5.1 Critical configuration on a circular cone.

stations. However, for short baselines, the satellites could be located approximately on circular cones. Consider the relevant portion of the double-difference phase equation (5.25) scaled to distances,

$$P_{km}^{pq} = \rho_k^p - \rho_m^p - \left[\rho_k^q - \rho_m^q\right] + \cdots \tag{5.51}$$

The total differential

$$\begin{aligned} dP_{km}^{pq} &= -\mathbf{e}_k^p \cdot d\mathbf{x}_k + \mathbf{e}_m^p \cdot d\mathbf{x}_m + \mathbf{e}_k^q \cdot d\mathbf{x}_k - \mathbf{e}_m^q \cdot d\mathbf{x}_m \\ &= \left[\mathbf{e}_k^q - \mathbf{e}_k^p\right] \cdot d\mathbf{x}_k + \left[\mathbf{e}_m^p - \mathbf{e}_m^q\right] \cdot d\mathbf{x}_m \end{aligned} \tag{5.52}$$

expresses the change in the double-difference observable in terms of differential changes in station coordinates. The coefficients in the brackets represent the differences in the direction cosines from one station and two satellites. For short baselines these differences approach zero. It can readily be seen that the direction vectors $\mathbf{e}_k^p$ are related to the vector of directions from the center of the baseline to the satellite $\mathbf{e}_c^p$ as

$$\mathbf{e}_k^p = \mathbf{e}_c^p + \boldsymbol{\varepsilon}_k^p \tag{5.53}$$

where the components of the vector $\boldsymbol{\varepsilon}_k^p$ are of the order $O(b/\rho_k^p)$. The symbol b denotes the length of the baseline. Referencing the other vectors also to the center of the baseline, Equation (5.52) becomes

$$dP_{km}^{pq} = \left[\mathbf{e}_c^q - \mathbf{e}_c^p + \boldsymbol{\varepsilon}_k^q - \boldsymbol{\varepsilon}_k^p\right] \cdot d\mathbf{x}_k + \left[\mathbf{e}_c^p - \mathbf{e}_c^q + \boldsymbol{\varepsilon}_m^p - \boldsymbol{\varepsilon}_m^q\right] \cdot d\mathbf{x}_m \tag{5.54}$$

For the special case that the vertex of the circular cone is at the center of the baseline, the condition

$$\mathbf{e}_c^i \cdot \mathbf{e}_{\text{axis}} = \cos\theta \tag{5.55}$$

is valid for all satellites on the cone. This means that the dot products

$$\left[\mathbf{e}_c^q - \mathbf{e}_c^p + \boldsymbol{\varepsilon}_k^q - \boldsymbol{\varepsilon}_k^p\right] \cdot \mathbf{e}_{\text{axis}} = \left[\boldsymbol{\varepsilon}_k^q - \boldsymbol{\varepsilon}_k^p\right] \cdot \mathbf{e}_{\text{axis}} = O\left(b/\rho_k^p\right) \tag{5.56}$$

in (5.54) are of the order $O(b/\rho_k^p)$. These products become smaller the shorter the baseline. A product like (5.56) applies to every double-difference observation. Therefore, we are dealing with a near-singular situation since the columns of the double-difference design matrix are nearly dependent. The shorter the baseline, the more likely it is that the near-singularity damages the baseline solution.

5.3.5 Impact of a Priori Position Errors

A frequent concern is the need for a priori knowledge of geocentric station positions and the effects of ephemeris errors on the relative positions. The answer to these concerns lies again in the linearized double-difference equations. Without loss of generality, it is sufficient to investigate the difference between one satellite and two ground stations. Scaled to distances, the relevant portion of the double-difference equation is

$$P_{km}^{pq}(t) = \rho_k^p(t) - \rho_m^p(t) + \cdots \tag{5.57}$$

The linearized form is

$$dP_{km}^{pq} = -\mathbf{e}_k^p \cdot d\mathbf{x}_k + \mathbf{e}_m^p \cdot d\mathbf{x}_m + \left[\mathbf{e}_k^p - \mathbf{e}_m^p\right] \cdot d\mathbf{x}^p \tag{5.58}$$

Next, we transform the coordinate corrections into their differences and sums. This is accomplished by

$$d\mathbf{x}_k - d\mathbf{x}_m = d(\mathbf{x}_k - \mathbf{x}_m) = d\mathbf{b} \tag{5.59}$$

$$\frac{d\mathbf{x}_k + d\mathbf{x}_m}{2} = d\left(\frac{\mathbf{x}_k + \mathbf{x}_m}{2}\right) = d\mathbf{x}_c \tag{5.60}$$

The difference (5.59) represents the change in the baseline vector, i.e., the change in length and orientation of the baseline, and (5.60) represents the change in the geocentric location of the baseline center. The latter can be interpreted as the translatory uncertainty of the baseline, or the uncertainty of the fixed baseline station. Transforming (5.58) to the difference and sum gives

$$dP_{km}^{pq} = -\frac{1}{2}\left[\mathbf{e}_k^p + \mathbf{e}_m^p\right] \cdot d\mathbf{b} - \left[\mathbf{e}_k^p - \mathbf{e}_m^p\right] \cdot d\mathbf{x}_c + \left[\mathbf{e}_k^p - \mathbf{e}_m^p\right] \cdot d\mathbf{x}^p \tag{5.61}$$

There is a characteristic difference in magnitude between the first bracket and the others. Allowing an error of the order $O(b/\rho_k^p)$, the first bracket simplifies to $2\mathbf{e}_m^p$ or $2\mathbf{e}_k^p$. The second and the third brackets are of opposite signs but the same magnitude. It is readily verified that the terms in the latter two brackets are of the order $O(b/\rho_k^p)$. When the baseline vector is defined by

$$\mathbf{b} \equiv \boldsymbol{\rho}_m^p - \boldsymbol{\rho}_k^p \tag{5.62}$$

Equation (5.61) becomes, after neglecting the usual small terms,

$$dP_{km}^{pq} = -\mathbf{e}_m^p \cdot d\mathbf{b} + \frac{\mathbf{b}}{\rho_m^p} \cdot d\mathbf{x}_c - \frac{\mathbf{b}}{\rho_m^p} \cdot d\mathbf{x}^p \tag{5.63}$$

The orders of magnitude for the coefficients in this equation will not change, even if double-difference expressions are fully considered. Equating the first two terms in (5.63), we get the relative impact of changes in the baseline and the translatory position of the baseline from

$$\rho_m^p \cdot d\mathbf{b} = \mathbf{b} \cdot d\mathbf{x}_c \tag{5.64}$$

Similarly, changes in the baseline vector and ephemeris position are related by

$$\boldsymbol{\rho}_m^p \cdot d\mathbf{b} = \mathbf{b} \cdot d\mathbf{x}^p \tag{5.65}$$

These relations are usually quoted in terms of absolute values, thereby neglecting the cosine terms of the dot product. In this sense, a rule of thumb for relating baseline accuracy, a priori geocentric position accuracy, and ephemeris accuracy is

$$\frac{\|d\mathbf{b}\|}{b} = \frac{\|d\mathbf{x}_c\|}{\rho_m^p} = \frac{\|d\mathbf{x}^P\|}{\rho_m^p} \tag{5.66}$$

Equation (5.66) shows that the accuracy requirements for the a priori geocentric station coordinates and the satellite orbital positions are the same. The accuracy requirement is proportional to the baseline length. This means that for short baselines an accurate position of the reference station might not be required and that the simple point positioning might be sufficient. A 1000 km line can be measured to 1 cm if the ephemeris errors and the geocentric location error can be reduced to 0.2 m, according to the rule of thumb given above.

The simplified derivation given in this section neglects the impact of the satellite constellation on the geometry of the solution. The only elements that enter the derivations are the baseline length and the receiver-satellite distance.

5.3.6 Cancellation of Common Mode Errors

GPS positioning benefits considerably from the fact that common-mode errors can be combined with other parameters or canceled at times. It has been pointed out in detail how single and double differences reduce the effects of clock errors. Additional detail is provided here for both point and relative positioning.

5.3.6.1 Point Positioning Generally, the propagation media affects satellite signals as a function of azimuth and elevation angle. For example, in the case of the ionosphere we split the total effect into a station average component $I_{k,P}$ and one that is a function of the direction of the satellite $\delta I_{k,P}^p$,

$$I_{k,P}^p = I_{k,P} + \delta I_{k,P}^p \tag{5.67}$$

The tropospheric delay can be split in a similar manner. The receiver hardware delay can also be a common source of errors. It is even possible that the satellite clocks contain a common offset, e.g., an incomplete correction due to relativity. The common components are combined with the receiver clock error into a new epoch parameter ξ_k, giving

$$\xi_k = dt_{\underline{\ }k} - \frac{I_{k,P}}{c} - \frac{T_k}{c} - \frac{d_{k,P}}{c} \tag{5.68}$$

The symbols for the ionosphere and the troposphere have no superscript p in (5.68) to indicate the common component. The symbol ξ_k represents a new unknown, which, in addition to the station clock error, contains the common components of all the other errors. The relevant portion of the pseudorange equations can now be written as

$$P_k^p = \rho_k^p - c\,\xi_k + cdt^p + \delta I_{k,P}^p + \delta T_k^p \tag{5.69}$$

The linearly dependent common components ξ_k in (5.68) cannot be estimated separately from the epoch point positioning solution but rather are absorbed by the estimate of the receiver clock error $d\underline{t}_k$. Unmodeled errors that are common to all observations at a particular station do not affect the estimated position. Thus, modeling of the ionosphere and troposphere, e.g., is useful only if it reduces the variability with respect to the common portion.

Equation (5.68) also demonstrates how the requirements for positioning and timing with GPS are quite different. If the goal is to determine time, then modeling or controlling the common station errors is of critical importance.

5.3.6.2 Relative Positioning

In relative positioning, the errors common to both stations tend to cancel during double differencing. For example, the tropospheric correction can be decomposed into the common station parts T_k and T_m and the satellite-dependent part as follows:

$$\begin{aligned} T_{km}^{pq} &= \left[T_{km} + \left(\delta T_k^p - \delta T_m^p\right)\right] - \left[T_{km} + \left(\delta T_k^q - \delta T_m^q\right)\right] \\ &= \left(\delta T_k^p - \delta T_m^p\right) - \left(\delta T_k^q - \delta T_m^q\right) \end{aligned} \tag{5.70}$$

It is useful to apply tropospheric and ionospheric corrections if the differential correction between the stations can be determined accurately. If this is not the case, because, say, the meteorological data are not representative of the actual tropospheric conditions, it might be better not to apply the correction at all and to rely on the common-mode elimination. Because the ionosphere and the troposphere are highly correlated over short distances, most of their delays are common to both stations. In terms of the tropospheric effect, an exception to this rule might apply to nearby stations that are located at significantly different elevations.

Because of the cancellation of most of the effects of the propagation media, the clock errors, and hardware delays, relative positioning has become especially popular and useful in surveying. Although the presence of the ambiguity parameters in the double differences might initially be perceived as a nuisance, they provide a unique vehicle to improve the solution if they can be successfully constrained to integers.

5.4 SATELLITE CODE OFFSETS

According to the ICD-GPS-200C (2000), the P1 and P2 codes are offset by T_{GD}^p. This offset is also referred to as the differential group delay (DGD) or more generally the differential code bias (DCB) for P1 and P2. The purpose of this delay is to allow dual-frequency users conveniently to eliminate the ionospheric effect on pseudorange observations when computing positions. This aspect is treated in detail in Chapter 7.

The satellite manufacturer initially determines the offset T_{GD}^{p} for each satellite during factory testing. The offset is available to users through the broadcast message. The code emission times on both frequencies are related as follows:

$$t_{L1}^{p} - t_{L2}^{p} = T_{\mathrm{GD}}^{p}\left(1 - \alpha_f\right) \tag{5.71}$$

The symbol α_f is given in (5.13). The offset can be treated as an additional clock correction for code phase observations,

$$\begin{aligned} P_{k,1}^{p}(t) = {} & \rho_k^p(\bar{t}^p) - c\, d\underline{t}_k + c\left(d\bar{t}^p + T_{\mathrm{GD}}^{p}\right) + I_{k,1,P}^{p}(t) + T_k^p(t) \\ & + \delta_{k,1,P}^{p}(t) + \varepsilon_{1,P} \end{aligned} \tag{5.72}$$

$$\begin{aligned} P_{k,2}^{p}(t) = {} & \rho_k^p(\bar{t}^p) - c\, d\underline{t}_k + c\left(dt^p + \alpha_f T_{\mathrm{GD}}^{p}\right) + I_{k,2,P}^{p}(t) + T_k^p(t) \\ & + \delta_{k,2,P}^{p}(t) + \varepsilon_{2,P} \end{aligned} \tag{5.73}$$

Following the sign convention of (5.39), L1 users can modify the computed satellite clock correction by subtracting T_{GD}^{p}; L2-only users must subtract $\alpha_f T_{\mathrm{GD}}^{p}$. The offset T_{GD}^{p} can be combined with and is indeed inseparable from the satellite hardware delay denoted by $d_{1,P}^{p}$ above. The same is true for L2.

The T_{GD}^{p} offsets change with time. For example, changes in satellite configuration due to use of backup hardware might impact the offsets. The offsets are routinely estimated from ground observations. See Sardón et al. (1994) and Sardón and Zarraoa (1997) for details on the observation and processing techniques. As an example, the results of a recent determination (Wilson et al., 1999) show a spread in the offset between the various satellites from close to zero to 12 ns, corresponding to about 4 m. The broadcast values are updated approximately four times per year. The actual biases are monitored daily to identify any abrupt changes (B. D. Wilson, JPL, private communication).

In addition to the P1-P2 interfrequency P-code bias, there is a bias between C/A and P-code pseudoranges. The C/A-P1 biases can also reach several nanoseconds. Current value estimates of these biases are available on the Internet. For example, JPL estimates these biases as part of the real-time Internet-based global differential GPS solution.

CHAPTER 6

TROPOSPHERE AND IONOSPHERE

This chapter begins with a general overview of the troposphere and ionosphere and a brief discussion of the relevancy of the atmosphere to GPS surveying. In Section 6.2 the tropospheric refraction is derived starting with the commonly used equation that expresses the refractivity as a function of partial pressure of dry air, partial water vapor pressure, and temperature. The equation for the zenith hydrostatic delay (ZHD) by Saastamoinen (1972), the expression for the zenith wet delay (ZWD) by Mendes and Langley (1999), and Niell's (1996) function for mapping the slant delays to the zenith delays are given without derivation. The horizontal gradient method is briefly discussed as a means to incorporate azimuth dependency of the refractivity. We then establish the relationship between the zenith wet delay and precipitable water vapor (PWV). Section 6.3 deals with tropospheric absorption and water vapor radiometers (WVR) that measure the tropospheric wet delay. We present and discuss the radiative transfer equation and the concept of brightness temperature. To demonstrate further the principles of the water vapor radiometer, we discuss the relevant absorption line profiles for water vapor, oxygen, and liquid water. This is followed by a brief discussion of retrieval techniques to compute the wet delay and radiometer calibration using tipping curves.

In Section 6.4 the causes of ionization are briefly discussed. The derivation of the ionospheric refraction is sketched beginning with the Appleton-Hartree formula. Section 6.5 gives expressions for the ionospheric delay of codes and ionospheric advances of carrier phases. Section 6.6 centers around the ionospheric-free and ionospheric functions for pseudoranges and carrier phases. The chapter concludes with brief remarks on the global ionospheric model (GIM).

6.1 OVERVIEW

The propagation media affect electromagnetic wave propagation at all frequencies, resulting in a bending of the signal path, time delays of arriving modulations, advances of carrier phases, scintillation, and other changes. In GPS positioning one is primarily concerned with the arriving times of carrier modulations and carrier phases. Geometric bending of the signal path causes a small delay that is negligible for elevation angles above 5°. The propagation of electromagnetic waves through the various atmospheric regions varies with location and time in a complex manner and is still the subject of active research. The relevant propagation regions are the troposphere and the ionosphere. Whereas positioning with GPS requires careful consideration of the impacts of the propagation media, GPS, in turn, has become a tool for studying the atmosphere. The subject of propagation of electromagnetic signals in the GPS frequency range, which is approximately the microwave region, is discussed but only to the extent required for GPS positioning.

Most of the mass of the atmosphere is located in the troposphere. We are concerned with the tropospheric delay of pseudoranges and carrier phases. For frequencies below 30 GHz, the troposphere behaves essentially like a nondispersive medium; i.e., the refraction is independent of the frequency of the signals passing through it. This tropospheric refraction includes the effect of the neutral, gaseous atmosphere. The effective height of the troposphere is about 40 km. The density in higher regions is too small to have a measurable effect. Mendes (1999) and Schüler (2001) recently studied the details of tropospheric refractions. Typically, tropospheric refraction is treated in two parts. The first part is the hydrostatic component that follows the laws of ideal gases. It is responsible for a zenith delay of about 240 cm at sea level locations. It can be computed accurately from pressure measured at the receiver antenna. The more variable second part is the wet component, or more precisely labeled the nonhydrostatic wet component, which is responsible for up to 40 cm of delay in the zenith direction. Computing the wet delay accurately is a difficult task because of the spatial and temporal variation of water vapor. Figure 6.1 shows the ZWD every 5 minutes for eleven consecutive days, beginning on July 10, 1999, at Lamont, Oklahoma, as determined by GPS, and the difference between the GPS and WVR determination. Both determinations agree within 1 cm. The gaps indicate times when suitable observations were not available.

Figure 6.2 demonstrates the impact water vapor variation can have over a 43 km baseline. The observations were taken over eleven days and processed with the precise ephemeris. Essentially, two cases are compared: (a) measuring the ZWD with the WVR and reducing the measured value to the slant delay using a mapping function that has no azimuth dependency; and (b) measuring the slant wet delay (SWD) with the WVR pointed in the direction of the satellite. In both cases, the hydrostatic delay was computed from the Saastamoinen model using barometric pressure. The largest ellipse in Figure 6.2 shows the repeatability over eleven days using the zenith radiometer corrections; the second largest (closest to spherical shape) shows the repeatability over eleven days using the pointed radiometer corrections. The next to

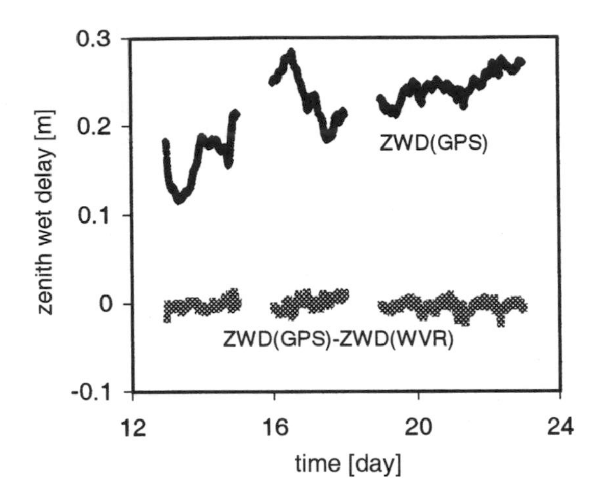

Figure 6.1 PWV from GPS and WVR. (Data from Bar-Sever, JPL.)

smallest are daily repeatability using zenith corrections, and the smallest ellipses are daily repeatability using pointed corrections.

The ionosphere covers the region between approximately 50 and 1500 km above the earth and is characterized by the presence of free (negatively charged) electrons and positively charged atoms and molecules called ions. The total electron content

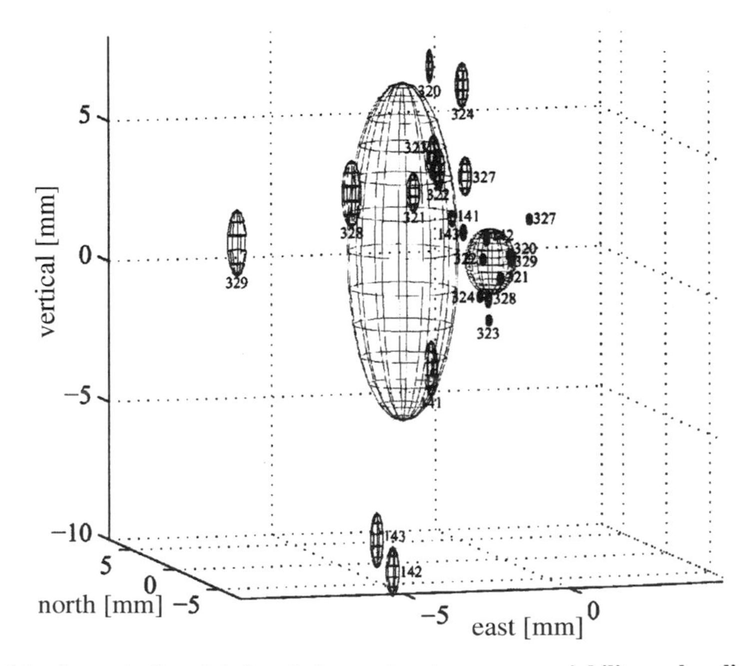

Figure 6.2 Impact of modeled and observed water vapor variability on baseline. (Permission of the American Geophysical Union.)

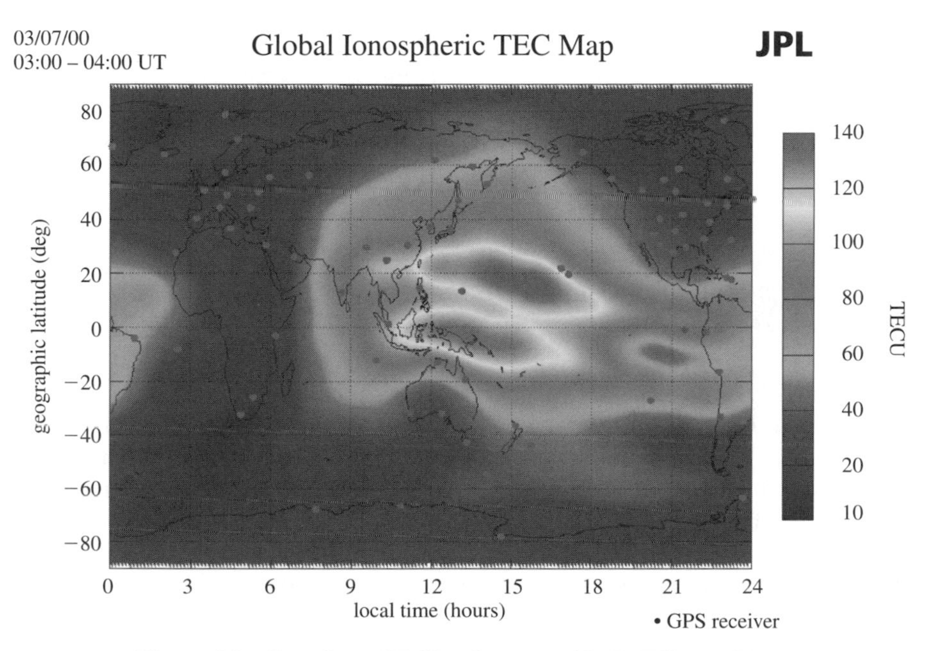

Figure 6.3 Snapshot of TEC. (Courtesy of B. D. Wilson, JPL.)

(TEC) equals the number of free electrons in the column of unit area along which the signal travels between receiver and satellite. Figure 6.3 shows a snapshot of the TEC. The free electrons delay the pseudoranges and advance the carrier phases by equal amounts. The size depends on the TEC and the carrier frequency; i.e., the ionosphere is a dispersive medium. For GPS frequency the delays or advances can amount to the tens of meters. Transmissions below 30 MHz are reflected. The texts by Hargreaves (1992) and Davies (1990) are recommended for in-depth studies of the physics of the ionosphere.

This GIM of the TEC on March 7, 2000, at 03 UT shows the typical global morphology of the ionosphere when the Appleton (equatorial) anomaly is well developed. There are two very strong peaks of ionization that lie on either side of the geomagnetic equator. The peaks begin in the afternoon and stretch into the nighttime region. Also, notice that the ionosphere is large since it is near solar maximum, although vertical TECs can be larger than the 140 to 150 TECU on this day (1 TECU $= 10^{16}$ el/m^2). The peak of the ionosphere is typically in the equatorial region at 14:00 local time. Each dot is the location of a GPS receiver that was used in the GIM model run. GIM's time resolution is 15 minutes, but images of the vertical TEC are typically only made every hour for animations, which is why the image is labeled with the time range 03–04 UT (B. D. Wilson, JPL, private communication).

The atmospheric parameters must be known with sufficient accuracy when applying respective corrections to observations. We typically use temperature, pressure, and humidity at the receiver antenna, as well as the TEC. Mapping the spatial and temporal distribution of these atmospheric parameters is also an area to which GPS

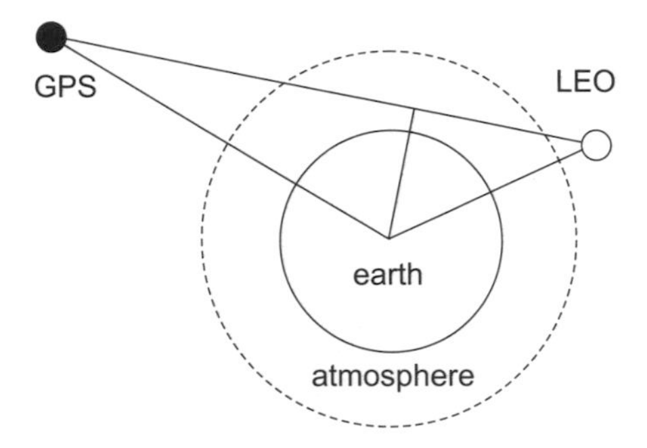

Figure 6.4 Schematic view of an LEO satellite and a GPS satellite configuration.

contributes. Figure 6.4 shows a schematic view of a low earth orbiter (LEO) and a GPS satellite. As viewed from the LEO, an occultation takes place when the GPS satellite rises or sets behind the earth's ionosphere and troposphere. When the signals pass through the media they experience tropospheric delays, ionospheric code delays, and phase advances. If the accurate position of the LEO is known and if the LEO carries a GPS receiver, one can estimate atmospheric parameters by comparing the travel time of the signal and the geometric distance between both satellites. Since the modeling associated with GPS occultations is still evolving, as is accurate orbit determination of LEOs, the reader should consult the current literature for details; a recommended start is Kursinski et al. (1997). One often assumes in these computations that the travel path through the media is symmetric and considers the tangent point as the point of measurement.

We present two figures showing typical products that can be derived from GPS occultation. Figures 6.5 and 6.6 indicate results from the GPS/MET experiment that was managed by the University Cooperation of Atmospheric Research (UCAR) and lasted from April 1995 to March 1997. A 2 kg TurboRogue receiver modified for use in space was piggybacked on a LEO with a 730 km circular orbit and 60° inclination. Figure 6.5 shows a temperature profile as determined by GPS occultation, direct radiosonde measurements, and an atmospheric weather model. The occultation occurred at 1:33 UT on May 5, 1995, over Hall Beach, Northwest Territory, Canada. The radiosonde at 0:00 UT was 85 km from the occultation location and spatially interpolated to the occultation location. The surface temperature was below freezing with a sharply defined tropopause near 8 km. The good agreement with the radiosonde in resolving the sharp tropopause and the change below 3 km illustrates the high sensitivity and vertical resolution of the occultation technique. Figure 6.6 shows an electron density profile of the ionosphere as a function of height derived from GPS/MET occultation, May 5, 1995, 3:20 UT. The figure also shows another independent determination of the electron density at 3:40 UT using incoherent scatter radar with a 320 μs pulse mode located at Millstone Hill (Massachusetts). Some of the discrepancies seen in the figure are a result of spatial and temporal mismatch of the observations and the spherical symmetry assumption for the signal path. The latter assumption can be a problem because the signal travels through a large portion of the ionosphere, in particular, the upper ionospheric region. Such a profile, of course, changes with time and location.

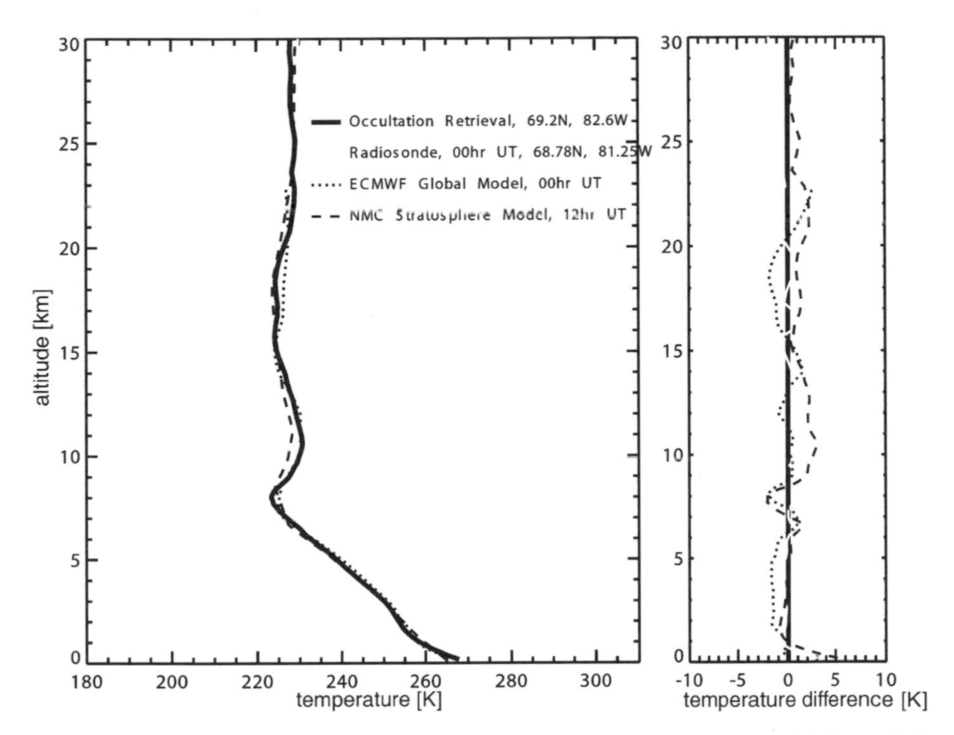

Figure 6.5 Comparison between occultation, radiosonde, and atmospheric model. *Source:* Kursinski et al., 1996. Permission by American Geophysical Union.

The sections below provide details on the index of refraction and on absorption. The general form of the index of refraction for electromagnetic wave can be written as a complex number

$$\bar{n} = \mu - i\chi \tag{6.1}$$

where μ and χ are related to refraction and absorption, respectively. Let A_0 denote the amplitude, we can write the equation of a wave as

$$A = A_0 e^{i(\omega t - \bar{n}\omega x/c)} = A_0 e^{i(\omega t - \mu\omega x/c)} e^{-\chi\omega x/c} \tag{6.2}$$

The wave propagates at speed c/μ, where c denotes the speed of light. The absorption in the medium is given by the exponential attenuation $e^{-\chi\omega x/c}$. The absorption coefficient is $\kappa = \omega\chi/c$. It is readily seen that the amplitude of the wave will reduce by factor e at distance $1/\kappa$.

For GPS frequencies and for frequencies in the microwave region, the index of refraction can be written as

$$\bar{n} = n + n'(f) + in''(f) \tag{6.3}$$

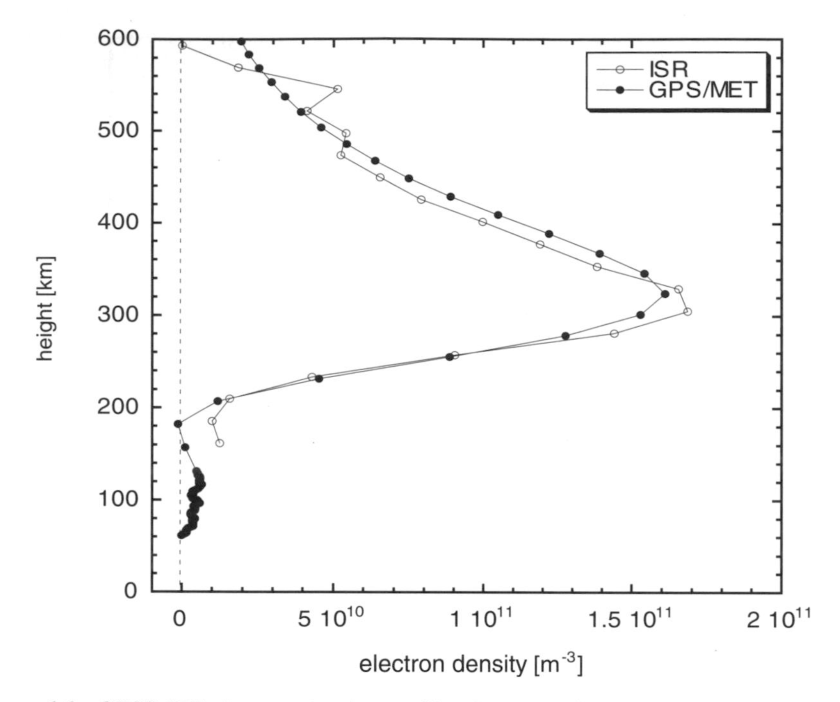

Figure 6.6 GPS/MET electron density profile. *Source:* Hajj and Romans, 1998. Permission by American Geophysical Union.

The medium is called dispersive if $\bar{n}$ is a function of the frequency. When applying (6.3) to the troposphere the real parts n and $n'(f)$ determine refraction that causes the delays in pseudoranges and carrier phases. The nondispersive part of the index of refraction is n. For frequencies in the microwave range the frequency-dependent real term $n'(f)$ is negligible. The latter term causes delays around the millimeter level at 60 GHz and centimeter level at 300 GHz (Janssen, 1993, p. 218). In general, $n'(f)$ and $n''(f)$ are due to interactions with line resonances of molecules in the vicinity of the carrier frequency. The GPS frequencies are far from atmospheric resonance lines. The imaginary part $n''(f)$, however, quantifies absorption (emission) and is important to the WVR observable. When applying (6.3) to the ionosphere the term $n'(f)$ is very important.

6.2 TROPOSPHERIC REFRACTION AND DELAY

The index of refraction is a function of the actual tropospheric path through which the ray passes, starting at the receiver antenna and continuing up to the end of the effective troposphere. Let s denote the distance; the delay due to refraction is

$$v = \int n(s)\, ds - \int ds = \int (n(s) - 1)\, ds \tag{6.4}$$

The first integral refers to the curved propagation path. The path is curved due to the decreasing index of refraction with height above the earth. The second integral is the geometric straight-line distance the wave would take if the atmosphere were a vacuum. The integration begins at the height of the receiver antenna.

Because the index of refraction $n(s)$ is numerically close to unity, it is convenient to introduce a separate symbol for the difference,

$$n(s) - 1 = N(s) \cdot 10^{-6} \tag{6.5}$$

$N(s)$ is called the refractivity. Great efforts have been made during the second part of the last century to determine the refractivity for microwaves. Examples of relevant literature are Thayer (1974) and Askne and Nordius (1987). The refractivity is usually given in the form

$$N = k_1 \frac{p_d}{T} Z_d^{-1} + k_2 \frac{p_{wv}}{T} Z_{wv}^{-1} + k_3 \frac{p_{wv}}{T^2} Z_{wv}^{-1} \tag{6.6}$$

p_d — Partial pressure of dry air (mbar). The dry gases of the atmosphere are, in decreasing percentage of the total volume: N_2, O_2, Ar, CO_2, Ne, He, Kr, Xe, CH_4, H_2, and N_2O. These gases represent 99.96% of the volume.

p_{wv} — Partial pressure of water vapor (mbar). Water vapor is highly variable but hardly exceeds 1% of the mass of the atmosphere. Most of the water in the air is from water vapor. Even inside clouds, precipitation and turbulence ensure that water droplet density remains low. This variability presents a challenge to accurate GPS applications over long distances on one hand, but on the other hand opens up a new field of activity, i.e., remotely sensing the atmosphere for water vapor.

T — Absolute temperature in degrees Kelvin [K].

Z_d, Z_{wv} — Compressibility factors that take into account small departures in behavior of moist atmosphere and ideal gas. Spilker (1996, p. 528) lists the expressions. These factors are often set to unity.

k_1, k_2, k_3 — Physical constants that are based in part on theory and in part on experimental observations. Bevis et al. (1994) lists: $k_1 = 77.60$ K/mbar, $k_2 = 69.5$ K/mbar, $k_3 = 370100$ K^2/mbar.

The partial water vapor pressure and the relative humidity R_h are related by the well-known expression, e.g., WMO (1961),

$$p_{wv[\text{mbar}]} = 0.01\ R_{h[\%]}\ e^{-37.2465+0.213166T-0.000256908T^2} \tag{6.7}$$

The two partial pressures are related to the total pressure p, which is measured directly, by

$$p = p_d + p_{wv} \tag{6.8}$$

The first term of (6.6) expresses the sum of distortions of electron charges of the dry-gas molecules under the influence of an applied magnetic field. The second term of (6.6) refers to the same effect but for water vapor. The third term is caused by the permanent dipole moment of the water vapor molecule; it is a direct result of the geometry of the water vapor molecular structure. Within the GPS frequency range the third term is practically independent of frequency. This is not necessarily true for higher frequencies that are close to the major water vapor resonance lines. Equation (6.6) is further developed by splitting the first term into two terms, one that gives refractivity of an ideal gas in hydrostatic equilibrium and another term that is a function of the partial water vapor pressure. The large hydrostatic constituent can then be accurately computed from ground-based total pressure. The smaller and more variable water vapor contribution must be dealt with separately.

The modification of the first term (6.6) begins by applying the equation of state for the gas constituent i, $(i = d, i = wv)$,

$$p_i = Z_i \rho_i R_i T \tag{6.9}$$

where ρ_i is the mass density and R_i is the specific gas constant ($R_i = R/M_i$, where R is the universal gas constant and M_i is the molar mass). Substituting p_d in (6.9) for the first term in (6.6), replacing the ρ_d by the total density ρ and ρ_{wv}, and applying (6.9) for ρ_{wv} gives for the first term

$$k_1 \frac{p_d}{T} Z_d^{-1} = k_1 R_d \rho_d = k_1 R_d \rho - k_1 R_d \rho_{wv} = k_1 R_d \rho - k_1 \frac{R_d}{R_{wv}} \frac{p_{wv}}{T} Z_{wv}^{-1} \tag{6.10}$$

Substituting (6.10) in (6.6) and combining it with the second term of that equation gives

$$N = k_1 R_d \rho + k_2' \frac{p_{wv}}{T} Z_{wv}^{-1} + k_3 \frac{p_{wv}}{T^2} Z_{wv}^{-1} \tag{6.11}$$

The new constant k_2' is

$$k_2' = k_2 - k_1 \frac{R_d}{R_{wv}} = k_2 - k_1 \frac{M_{wv}}{M_d} \tag{6.12}$$

Bevis et al. (1994) gives $k_2' = 22.1$ K/mbar.

We can now define the hydrostatic and wet (nonhydrostatic) refractivity as

$$N_d = k_1 R_d \rho = k_1 \frac{p}{T} \tag{6.13}$$

$$N_{wv} = k_2' \frac{p_{wv}}{T} Z_{wv}^{-1} + k_3 \frac{p_{wv}}{T^2} Z_{wv}^{-1} \tag{6.14}$$

If we integrate (6.6) along the zenith direction using (6.13) and (6.14), we obtain the ZHD and ZWD, respectively,

$$\text{ZHD} = 10^{-6} \int N_d(h)\, dh \tag{6.15}$$

$$\text{ZWD} = 10^{-6} \int N_{wv}(h)\, dh \tag{6.16}$$

The hydrostatic refractivity N_d depends on total density ρ or the total pressure p. When integrating N_d along the ray path the hydrostatic equilibrium condition to ideal gases is applied. The integration of N_{wv} is complicated by the temporal and spatial variation of the partial water vapor pressure p_{wv} along the path.

6.2.1 Model Zenith Delay Functions

Even though the hydrostatic refractivity is based on the laws of ideal gases, the integration (6.15) still requires assumptions about the variation of temperature and gravity along the path. Examples of solutions for the ZHD are Hopfield (1969) and Saastamoinen (1972). Saastamoinen's solution is given in Davis et al. (1985) in the form

$$\text{ZHD}_{[\text{m}]} = \frac{0.0022768\, p_{0[\text{mbar}]}}{1 - 0.00266 \cos 2\varphi - 0.00028 H_{[\text{km}]}} \tag{6.17}$$

The symbol p_0 denotes the total pressure at the site whose orthometric height is H and latitude is φ.

The model assumptions regarding the wet refractivity are more problematic because of temporal and spatial variability of water vapor. Mendes and Langley (1999) analyzed radiosonde data and explored the correlation between the ZWD and the surface partial water vapor pressure $p_{wv,0}$. Their model is

$$\text{ZWD}_{[\text{m}]} = 0.0122 + 0.00943\, p_{wv,0[\text{mbar}]} \tag{6.18}$$

Surface meteorological data should be used with caution in the estimation of the ZWD. Typical field observations can be influenced by "surface layer biases" introduced by micro-meteorological effects. The measurements at the earth's surface are not necessarily representative of adjacent layers along the line of sight to the satellites. Temperature inversion can occur during nighttime when the air layers close to the ground are cooler than the higher air layers, due to ground surface radiation loss. Convection can occur during noontime when the sun heats the air layers near the ground.

Expressions exist that do not explicitly separate between ZHD and ZWD. In some cases, the models are independent of direct meteorological measurements. The latter typically derive their input from model atmospheres.

6.2.2 Model Mapping Functions

Tropospheric delay is shortest in the zenith direction and increases with the zenith angle ϑ as the air mass traversed by the signal increases. The exact functional relationship is again complicated by temporal and spatial variability of the troposphere. The mapping function models this dependency. We relate the slant hydrostatic and wet delays, SHD and SWD, to the respective zenith delays by

$$\text{SHD} = \text{ZHD} \cdot m_h(\vartheta) \tag{6.19}$$

$$\text{SWD} = \text{ZWD} \cdot m_{wv}(\vartheta) \tag{6.20}$$

The slant total delay (STD) is

$$\text{STD} = \text{ZHD} \cdot m_h(\vartheta) + \text{ZWD} \cdot m_{wv}(\vartheta) \tag{6.21}$$

The literature contains many models for the mapping functions m_h and m_{wv}. The one in common use is Niell's (1996) function,

$$m(\vartheta) = \frac{1 + \cfrac{a}{1 + \cfrac{b}{1 + c}}}{\cos\vartheta + \cfrac{a}{\cos\vartheta + \cfrac{b}{\cos\vartheta + c}}} + h_{[\text{km}]} \left(\frac{1}{\cos\vartheta} - \frac{1 + \cfrac{a_h}{1 + \cfrac{b_h}{1 + c_h}}}{\cos\vartheta + \cfrac{a_h}{\cos\vartheta + \cfrac{b_h}{\cos\vartheta + c_h}}} \right) \tag{6.22}$$

The coefficients for this expression are listed in Table 6.1 (for m_h) and Table 6.2 (for m_{wv}) as a function of the latitude φ of the station. If $\varphi < 15°$ one should use the tabulated values for $\varphi = 15°$; if $\varphi > 75°$ then use the values for $\varphi = 75°$; if $15° \leq \varphi \leq 75°$, linear interpolation applies. Expression (6.22) gives the hydrostatic mapping functions if the coefficients of Table 6.1 are used. Before substitution, however, the coefficients a, b, and c must be corrected for periodic terms following the general formula

$$a(\varphi, \text{DOY}) = \tilde{a} - a_p \cos\left(2\pi \frac{\text{DOY} - \text{DOY}_0}{365.25}\right) \tag{6.23}$$

TABLE 6.1 Coefficients for Niell's Hydrostatic Mapping Function

φ	$\tilde{a} \cdot 10^3$	$\tilde{b} \cdot 10^3$	$\tilde{c} \cdot 10^3$	$a_p \cdot 10^5$	$b_p \cdot 10^5$	$c_p \cdot 10^5$
15	1.2769934	2.9153695	62.610505	0	0	0
30	1.2683230	209152299	62.837393	1.2709626	2.1414979	9.0128400
45	102465397	209288445	63.721774	2.6523662	3.0160779	4.3497037
60	102196049	209022565	63.824265	3.4000452	7.2562722	84.795348
75	102045996	2.9024912	64.258455	4.1202191	11.723375	170.37206
	$a_h \cdot 10^5$	$b_h \cdot 10^3$	$c_h \cdot 10^3$			
	2.53	5.49	1.14			

TABLE 6.2 Coefficients for Niell's Wet Mapping Function

φ	$a \cdot 10^4$	$b \cdot 10^3$	$c \cdot 10^2$
15	5.8021897	1.4275268	4.3472961
30	5.6794847	1.5138625	4.6729510
45	5.8118019	1.4572752	4.3908931
60	5.9727542	1.5007428	4.4626982
75	6.1641693	1.7599082	5.4736038

where DOY denotes the day of year and DOY_0 is 28 or 211 for stations in the Southern or Northern Hemisphere, respectively. When computing the wet mapping function, the height-dependent second term in (6.22) is dropped and the coefficients of Table 6.2 apply.

The Niell function enjoys such popularity because it is accurate, is independent of surface meteorology, and requires only site location and time of year as input. The Niell model assumes azimuthal symmetry. However, efforts have been reported in Niell (2000) and Rocken et al. (2001) to improve the mapping function for low elevation angles by incorporating temperature, pressure, and humidity profiles for a specific location and time period.

6.2.3 Horizontal Gradient Model

As has been mentioned, the variability of the water vapor is of much concern in accurate GPS application. The water vapor exists mostly in the lower 5 km of the troposphere. Its distribution may show an azimuthal dependency primarily due to terrain and wind effects. One could attempt to model the lateral water vapor refractivity by the gradient method.

Assume that a point is parameterized in the local geodetic coordinate system specified by the northing, easting, and up coordinates, $\mathbf{w} = [n \quad e \quad u]^T$. See Section 2.3.5 for the exact definition of this coordinate system. The refractive index at height u above the station can be expanded as

$$N_{wv}(\mathbf{w}) = N_{wv}(w = 0) + \frac{\partial N_{wv}}{\partial n} n + \frac{\partial N_{wv}}{\partial e} e + \frac{\partial N_{wv}}{\partial u} u \tag{6.24}$$

Next we solve (2.91) for the distance s and substitute it in (2.89) and (2.90) and then substitute the resulting expressions for northing n and easting e into (6.24), giving

$$N_{wv}(\mathbf{w}) = N_{wv,0}(w = 0) + \frac{\partial N_{wv}}{\partial u} u + \frac{1}{\tan \beta} \left(u \frac{\partial N_{wv}}{\partial n} \cos \alpha + u \frac{\partial N_{wv}}{\partial e} \sin \alpha \right) \tag{6.25}$$

The zenith delay is obtained by integrating along the vertical from the station to the end of the effective troposphere,

$$\begin{aligned}\text{ZWD}(\alpha,\beta) &= 10^{-6}\left(N_{wv}(w=0) + \int \frac{\partial}{\partial u} N_{wv}\, du\right) \\ &\quad + \frac{10^{-6}}{\tan\beta}\left(\int \frac{\partial N_{wv}}{\partial n}\cos\alpha\, du + \int \frac{\partial N_{wv}}{\partial e}\sin\alpha\, du\right) \\ &= \text{ZWD} + \frac{1}{\tan\beta}\left(G_n\cos\alpha + G_e \sin a\right)\end{aligned} \tag{6.26}$$

One may attempt to estimate the model coefficients G_n and G_e from observations. Depending on the application and weather conditions, a possibly piecewise linear modeling might be appropriate. Applications of the horizontal gradient method are reported, e.g., by Bar-Sever et al. (1998) and Liu (1999).

6.2.4 Precipitable Water Vapor

The GPS observables directly depend on the STD. This quantity, therefore, can be estimated from GPS observations. One might envision the scenario where widely spaced receivers are located at known stations and that the precise ephemeris is also available. If all other errors are taken into consideration, then the residual misclosures of the observations are the STD. We compute the ZHD from surface pressure measurements and a hydrostatic delay model. Using appropriate mapping functions, we could then compute ZWD from (6.21) using the estimated STD. Input to weather models typically requires that the ZWD be converted to precipitable water.

The integrated water vapor (IWV) along the vertical and the precipitable water vapor (PWV) are defined as

$$\text{IWV} \equiv \int \rho_{wv}\, dh \tag{6.27}$$

$$\text{PWV} \equiv \frac{\text{IWV}}{\rho_w} \tag{6.28}$$

where ρ_w is the density of liquid water. To relate the ZWD to these measures, it is convenient to introduce the mean temperature T_m,

$$T_m \equiv \frac{\int \frac{p_{wv}}{T} Z_{wv}^{-1}\, dh}{\int \frac{p_{wv}}{T^2} Z_{wv}^{-1}\, dh} \tag{6.29}$$

The ZWD follows then from (6.16), using (6.14),

$$\text{ZWD} = 10^{-6}\left(k_2' + \frac{k_3}{T_m}\right)\int \frac{p_{wv}}{T} Z_{wv}^{-1}\, dh \tag{6.30}$$

To be precise let us recall that (6.30) represents the nonhydrostatic zenith delay. Using the state equation of water vapor gas,

$$\frac{p_{wv}}{T} Z_{wv}^{-1} = R_{wv} \rho_{wv} \tag{6.31}$$

in the integrant gives

$$\text{ZWD} = 10^{-6} \left(k_2' + \frac{k_3}{T_m} \right) R_{wv} \int \rho_{wv} \, dh \tag{6.32}$$

We replace the integrant in (6.32) by IWV according to (6.27) and then replace the specific gas constant R_{wv} by the universal gas constant R and the molar mass M_{wv}. The conversion factor Q that relates the zenith nonhydrostatic wet delay to the precipitable water becomes

$$Q \equiv \frac{\text{ZWD}}{\text{PWV}} = \rho_w \frac{R}{M_{wv}} \left(k_2' + \frac{k_3}{T_m} \right) 10^{-6} \tag{6.33}$$

The constants needed in (6.33) are known with sufficient accuracy. The largest error contribution comes from T_m, which varies with location, height, season, and weather. The Q value varies between 5.9 and 6.5, depending on the air temperature. For warmer conditions, when the air can hold more water vapor, the ratio is toward the low end. Bevis et al. (1992) correlate T_m with the surface temperature T_0 and offer the model

$$T_{m[\text{K}]} = 70.2 + 0.72 T_{0[\text{K}]} \tag{6.34}$$

The following models for Q are based on radiosonde observations (Keihm, JPL, private communication).

$$Q = 6.135 - 0.01294\,(T_0 - 300) \tag{6.35}$$

$$Q = 6.517 - 0.1686\ \text{PWV} + 0.0181\ \text{PWV}^2 \tag{6.36}$$

$$Q = 6.524 - 0.02797\ \text{ZWD} + 0.00049\ \text{ZWD}^2 \tag{6.37}$$

If no surface temperatures are available, one can use (6.36) and (6.37), which take advantage of the fact that Q correlates with PWV (since higher PWV values are generally associated with higher tropospheric temperatures).

6.3 TROPOSPHERIC ABSORPTION

This section deals briefly with some elements of remote sensing by microwaves. The interested reader may consult general texts on remote sensing. We recommend the book by Janssen (1993) because it is dedicated to atmospheric remote sensing by microwave radiometry. The material presented below very much depends on that source. Solheim's (1993) dissertation is also highly recommended for additional reading.

6.3.1 The Radiative Transfer Equation

The energy emission and absorption of molecules are due to transitions between allowed energy states. Several fundamental laws of physics relate to the emissions and absorptions of gaseous molecules. Bohr's frequency condition relates the frequency f of a photon emission or absorption to the energy levels E_a and E_b of the molecule and to Planck's constant h. Einstein's law of emission and absorption specifies that if $E_a > E_b$, the probability of stimulated emission of a photon by a transition from state a to state b is equal to the probability of absorption of a photon by a transition from b to a. These two probabilities are proportional to the incident energy at frequency f. Dirac's perturbation theory gives the conditions that must be fulfilled, in order to enable the electromagnetic field to introduce transitions between states. For wavelengths that are very long compared to molecular dimensions, this operator is the dipole moment. This is the case in microwave radiometry. We typically observe the rotation spectra, corresponding to radiation emitted in transition between rotational states of a molecule having an electric dipole moment. The rotational motion of a diatomic molecule can be visualized as a rotation of a rigid body about its center of mass. Other types of transitions of molecular quantum states that emit at the ultraviolet, gamma, or infrared range are not relevant to sensing of water vapor. Although the atmosphere contains other polar gases, only water vapor and oxygen are present in enough quantity to emit significantly at microwave range.

Let $I(f)$ denote the instantaneous radiant power that flows at a point in a medium, over a unit area, per unit-frequency interval at a specified frequency f, and in a given direction per unit solid angle. As the signal travels along the path s, the power changes when it encounters sources and sinks of radiation. This change is described by the differential equation

$$\frac{dI(f)}{ds} = -I(f)\,\alpha + S \tag{6.38}$$

The symbol α denotes the absorption (describing the loss) and S is the source (describing the gain) into the given direction.

Scattering from other directions can lead to losses and gains to the intensity. In the following we will ignore scattering. We assume thermodynamic equilibrium, which means that each point along the path s the source can be characterized by temperature T. The law of conservation of energy for absorbed and emitted energy relates the source and absorption as

$$S = \alpha\, B(f, T) \tag{6.39}$$

where

$$B(f, T) = \frac{2\pi h f^3}{c^2\left(e^{hf/kT} - 1\right)} \tag{6.40}$$

$B(f, T)$ is the Planck function, h is the Planck constant, k is the Boltzmann constant, T is the physical temperature, and c denotes the speed of light. Please consult the specialized literature for details on (6.40).

With stated assumptions, Equation (6.38) becomes a standard differential equation with all terms depending only on the intensity along the path of propagation. The solution can be written as

$$I(f,0) = I(f,s_0)\, e^{-\tau(s_0)} + \int_0^{s_0} B(f,T)\, e^{-\tau(s)}\, \alpha\, ds \tag{6.41}$$

$$\tau(s) = \int_0^{s} \alpha(s')\, ds' \tag{6.42}$$

Equation (6.41) is called the radiative transfer equation. $I(f,0)$ is the intensity at the measurement location $s = 0$, and $I(f, s_0)$ is the intensity at some boundary location $s = s_0$. The symbol $\tau(s)$ denotes the optical depth or the opacity.

If $hf \ll kT$, as is the case for microwaves and longer waves, the denominator in (6.40) can be expanded in terms of hf/kT. After truncating the expansion, the Planck function becomes the Rayleigh-Jeans approximation

$$B(\lambda, T) \approx \frac{2f^2kT}{c^2} = \frac{2kT}{\lambda^2} \tag{6.43}$$

The symbol λ denotes the wavelength. Expression (6.43) expresses a linear relationship between Planck function and temperature T. For a given opacity (6.42) the intensity (6.41) is proportional to the temperature of the field of view of the radiometer antenna given (6.43).

The Rayleigh-Jeans brightness temperature $T_b(f)$ is defined by

$$T_b(f) \equiv \frac{\lambda^2}{2k} I(f) \tag{6.44}$$

$T_b(f)$ is measured in degrees Kelvin; it is a simple function of the intensity of the radiation at the measurement location. If we declare the space beyond the boundary s_0 as the background space, we can write the Rayleigh-Jeans background brightness temperature as

$$T_{b0}(f) \equiv \frac{\lambda^2}{2k} I(f, s_0) \tag{6.45}$$

Using definitions (6.44) and (6.45), the approximation (6.43), and $T = T_b$, the radiative transfer equation (6.41) becomes

$$T_b = T_{b0}\, e^{-\tau(s_0)} + \int_0^{s_0} T(s)\, \alpha\, e^{-\tau(s)}\, ds \tag{6.46}$$

This is Chandrasekhar's equation of radiative transfer as used in microwave remote sensing. For ground-based GPS applications, the sensor (radiometer) is on the ground ($s = 0$) and senses all the way to $s = \infty$. T_{b0} becomes the cosmic background temperature T_{cosmic}, which results from the residual cosmic radiation of outer space that is left from the Big Bang. Thus

$$T_b = T_{\text{cosmic}}\, e^{-\tau(\infty)} + \int_0^\infty T(s)\, \alpha\, e^{-\tau(s)}\, ds \tag{6.47}$$

$$\tau(\infty) = \int_0^\infty \alpha(s)\, ds \tag{6.48}$$

$$T_{\text{cosmic}} = 2.7K \tag{6.49}$$

The brightness temperature (6.47) depends on the atmospheric profiles of physical temperature T and absorption α. For the atmosphere the latter is a function of pressure, temperature, and humidity. Equation (6.47) represents the forward problem, i.e., given temperature and absorption profiles along the path one can compute brightness temperature. The inverse solution of (6.47) is of much practical interest. It potentially allows the determination of atmospheric properties such as T and α, as well as their spatial distribution from brightness temperature measurements.

Consider the following special cases. Assume that the temperature T is constant. Neglecting the cosmic term, using $d\tau = \alpha\, ds$, the radiative transfer equation (6.47) becomes

$$T_b = T \int_0^{\tau(a)} e^{-\tau}\, d\tau = T\left(1 - e^{-\tau(a)}\right) \tag{6.50}$$

For a large optical depth $\tau(a) \gg 1$ we get $T_b = T$ and the radiometer acts like a thermometer. For a small optical path $\tau(a) \ll 1$ we get $T_b = T\, \tau(a)$. If the temperature is known, then $\tau(a)$ can be determined. If we also know the absorption properties of the constituencies, it might be possible to estimate the concentration of a particular constituent of the atmosphere.

For the sake of clarity, we reiterate that (6.44) defines the Rayleigh-Jeans brightness temperature. The thermodynamic brightness temperature is defined as the temperature of a blackbody radiator that produces the same intensity as the source being observed. The latter definition refers to the physical temperature, whereas the Rayleigh-Jeans definition directly relates to the radiated intensity. The difference between both definitions can be traced back to the approximation implied in (6.43). A graphical representation of the differences is found in Janssen (1993, p. 10).

6.3.2 Absorption Line Profiles

Microwave radiometers measure the brightness temperature. In ground-based radiometry, the relevant molecules are water vapor, diatomic oxygen (O_2), and liquid water. Mathematical models have been developed for the absorption. For isolated molecules, the quantum mechanic transitions occur at well-defined resonance frequencies (line spectrum). Collision with other molecules broadens these spectral lines. When gas molecules interact the potential energy changes, due to changed relative positions and orientations of the molecules. As a result, the gas is able to absorb photons at frequencies well removed from the resonance lines. Pressure broadening converts the line spectrum into a continuous absorption spectrum, called the line profile. The interactions and thus the broadening increase with pressure. Given the structure of molecules it is possible to derive mathematical functions for the line profiles. Because of the complexities of these computations and the presence of collisions, these functions typically require refinement with laboratory observations. The results are line profile models.

Figures 6.7 and 6.8 show line profiles for water vapor, oxygen, and liquid water computed with Fortran routines provided by Rosenkranz. (See also Rosenkranz, 1998). All computations refer to a temperature of 15°C. The top three lines in Figure 6.7 show the line profiles for water vapor for pressures of 700 mbar, 850 mbar, and 1013 mbar, and a water vapor density of 10 g/m^3. The maximum absorption occurs at the resonance frequency of 22.235 GHz. The effect of pressure broadening on the absorption curve is readily visible. Between about 20.4 GHz and 23.8 GHz the absorption is less, the higher the pressure. The reverse is true in wings of the line profile. In the vicinity of these two particular frequencies, the absorption is relatively independent of pressure. Most WVRs use at least one of these frequencies to minimize the sensitivity of brightness temperature to the vertical distribution of water vapor. The water vapor absorption is fairly stable in regard to changes in frequency around 31.4 GHz. Dual-frequency WVRs for ground-based sensing of water vapor typically also use the 31.4 GHz frequency to separate the effects of water vapor from cloud liquid. The 31.4 GHz channel is approximately twice as sensitive to cloud liquid emissions as the channel near 20.4 GHz. The opposite is true for water vapor, allowing separate retrievals of the two most variable atmospheric constituents. The absorption line of oxygen in Figure 6.7 refers to a water vapor density of 10 g/m^3 and a pressure

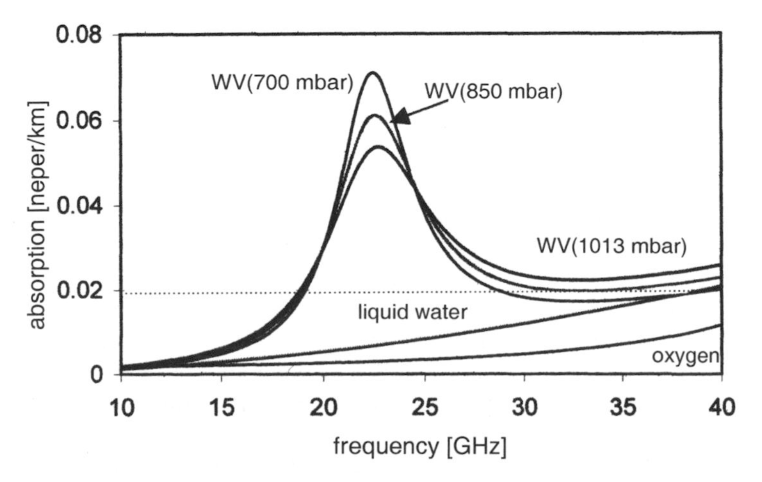

Figure 6.7 Absorption of water vapor, liquid water, and oxygen between 10 and 40 GHz.

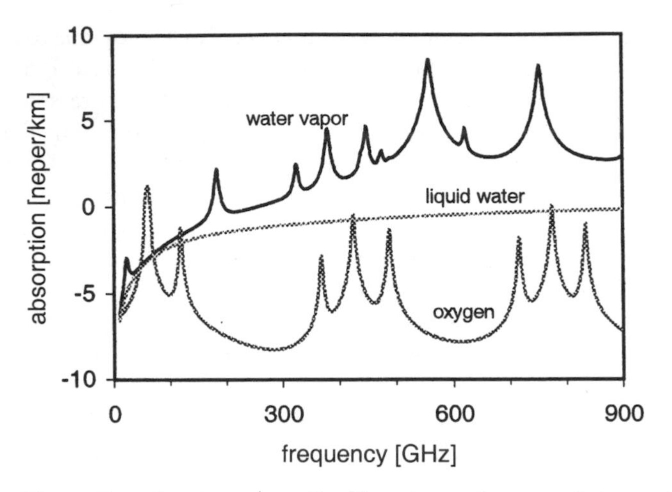

Figure 6.8 Absorption of water vapor, liquid water, and oxygen between 10 and 900 GHz.

of 1013 mbar. The line of liquid water (suspended water droplets) is based on a water density of 0.1 g/m^3. The absorption used in the radiative transfer equation (6.47) is the sum of the absorption of the individual molecular constituencies, i.e.,

$$\alpha = \alpha_{wv}(f, T, p, \rho_{wv}) + \alpha_{lw}\big(f, T, \rho_{lw}\big) + \alpha_{ox}(f, T, p, \rho_{wv}) \tag{6.51}$$

The absorption units are typically referred to as neper per kilometer. The absorption unit refers to the fractional loss of intensity per unit distance (km) traveled in a logarithmic sense. That is, an absorption value of 1 neper/km would imply that the power would be attenuated by $1/e$ fractional amount over 1 km given that the absorption properties remained constant over that kilometer. A neper is the natural logarithm of a voltage ratio and is related to the *dB* unit as follows:

$$dB = \frac{20}{\ln(10)}\ \text{neper} \approx 8.686\ \text{neper} \tag{6.52}$$

The line profiles contain other maxima, as seen in Figure 6.8. A large maximum for water vapor at 183.310 GHz is relevant to water vapor sensing in airborne radiometry. The liquid water absorption increases monotonically with frequency in the microwave range. Oxygen has a band of resonance near 60 GHz. The oxygen absorption is well modeled with pressure and temperature measurements on the ground; the absorption is small compared to that of water vapor and nearly constant for a specific site because oxygen is mixed well in the air. The profiles of Figure 6.8 refer to a temperature of 15°C, a water vapor density of 10 g/m^3, a pressure of 1013 mbar, and a liquid water density 0.1 g/m^3.

Since the absorption of oxygen can be computed from the model and ground-based observations, it is possible to separate its known contribution in (6.47) and invert the radiative transfer equation to determine integrated water vapor and liquid water as a function of the observed brightness temperatures. Westwater (1978) provides a thorough error analysis for this standard dual-frequency case. The fact that at 23.8 GHz the absorption of water vapor is significantly higher than at 31.4 GHz (while the absorption of liquid water changes monotonically over that region) can be used to retrieve separately integrated water vapor and liquid water from the inversion of the radiative transfer equation. With more channels distributed appropriately over the frequency, one can roughly infer the water vapor profiles as well as integrated water vapor and liquid water, or even temperature, vapor, and liquid profiles (Ware, 2002, private communication).

6.3.3 General Statistical Retrieval

Consider the following experiment. Use a radiosonde to measure the temperature and water vapor density profile along the vertical and use equations (6.27) and (6.28) to compute IWV and PWV. Compute the brightness temperature T_b from the radiative transfer equation (6.47) for each radiometer frequency using the frequency-dependent absorption model for water vapor $\alpha_{wv}(f, T, p, p_w)$ and oxygen absorption.

Figure 6.9 shows the result of such an experiment. The plot shows the observed T_b for WVR channels at 20.7 and 31.4 GHz. The data refer to a Bermuda radiosonde station and were collected over a three-year period. The Bermuda site experiences nearly the full range of global humidity and cloud cover conditions. The scatter about the heavily populated "clear" lines is due to the occurrence of cloudy cases. The slopes of $T_b(20.7)$ are approximately 2.2 times the slopes of the $T_b(31.4)$. The scatter about the $T_b(31.4)$ "clear" line is approximately twice as large as the scatter about the $T_b(20.7)$ "clear" line. These results are indicative of the facts that (1) the sensitivity of $T_b(20.7)$ to PWV is approximately 2.2 times greater than that of $T_b(31.4)$ and (2) that the sensitivity of $T_b(31.4)$ to liquid water is approximately 2 times greater than that of $T_b(20.7)$. The sensitivity to liquid water is also illustrated in Figure 6.10, which

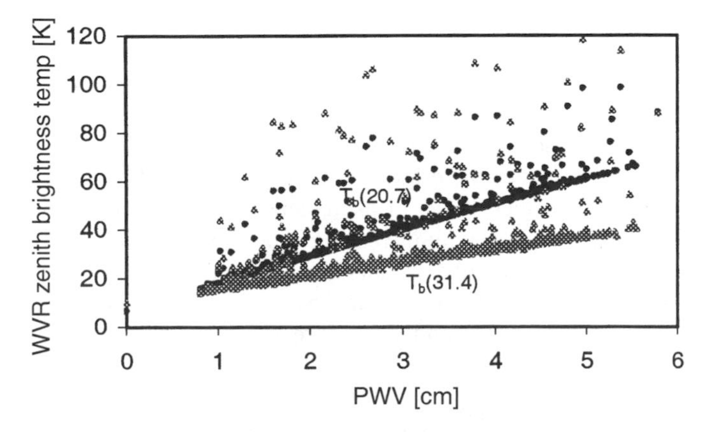

Figure 6.9 Brightness temperature versus precipitable water vapor. (Date source: Keihm, JPL.)

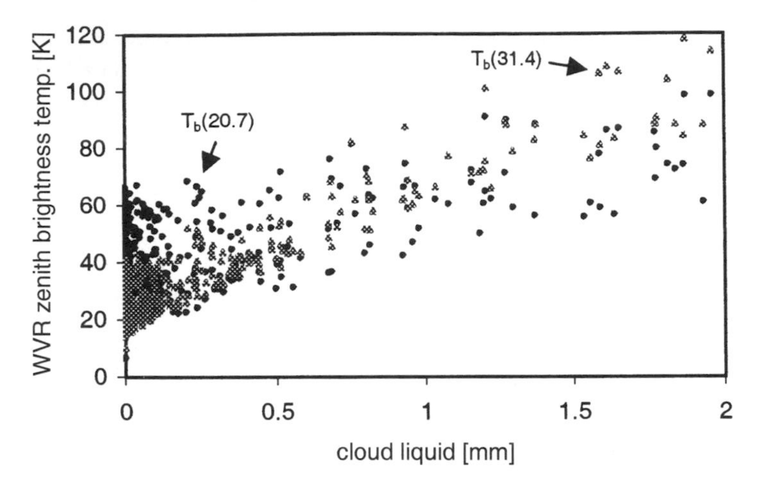

Figure 6.10 Brightness temperature versus cloud liquid. (Date source: Keihm, JPL.)

shows T_b variations versus cloud liquid. Despite the large scatter (due to variable PWV), one can see that the slope of the $T_b(31.4)$ data is approximately twice as large as the slope of the $T_b(20.7)$ data.

Because of the relationships between ZWD, IWV, and PWV as seen by (6.32), (6.27), and (6.28), the strong correlation seen in Figure 6.9 between PWV and the brightness temperature makes a simple statistical retrieval procedure for the ZWD possible. Assume a radiosonde reference station is available to determine ZWD and that a WVR measures zenith $T_{20.7}$ and $T_{31.4}$. Using the model

$$\mathrm{ZWD} = c_0 + c_{20.7} T_{20.7} + c_{31.4} T_{31.4} \tag{6.53}$$

we can estimate accurate retrieval coefficients $\hat{c}_0$, $\hat{c}_{20.7}$, and $\hat{c}_{31.4}$. When users operate a WVR in the same climatological region, they can then readily compute the ZWD at their location from the observed brightness temperature and the estimated regression coefficients. This statistical retrieval procedure can be generalized by using an expanded regression model in (6.53) and by incorporating brightness temperature measurements from several radiosonde references distributed over a region.

The opacity may also be used in this regression. In fact, opacity varies more linearly with PWV than does the brightness temperature T_b. At high levels of water vapor, or low elevation angles, the T_b measurements will eventually begin to saturate, i.e., the rate of the T_b increase with increasing vapor will start to fall off. This is not true for opacity, which essentially remains linear with the in-path vapor abundance. Opacity is available from (6.48) but also can be conveniently related to the brightness temperature. Define mean radiation temperature T_{mr} as

$$T_{\mathrm{mr}} \equiv \frac{\int_0^{\infty} T(s)\,\alpha(s)\,e^{-\tau(s)}\,ds}{\int_0^{\infty} \alpha(s)\,e^{-\tau(s)}\,ds} \tag{6.54}$$

This auxiliary quantity can be accurately estimated from climatologic data. Corrections with surface temperature permit T_{mr} estimates to be computed to a typical accuracy of ~3 K. Using the relationship

$$\int_0^{\tau(\infty)} \alpha e^{-\tau}\, ds = 1 - e^{-\tau(\infty)} \tag{6.55}$$

where we used again $d\tau = \alpha\, ds$, the radiative transfer equation (6.47) can be written as

$$T_b = T_{\text{cosmic}}\, e^{-\tau(\infty)} + T_{mr}\left(1 - e^{-\tau(\infty)}\right) \tag{6.56}$$

which, in turn, can be rewritten as

$$\tau(\infty) = \ln\left(\frac{T_{mr} - T_{\text{cosmic}}}{T_{mr} - T_b}\right) \tag{6.57}$$

The opacities and brightness temperature show similarly high correlations with the wet delay. In fact, at low elevation angles the opacities correlate even better with the wet delay than do brightness temperatures.

If the user measures the brightness temperatures along the slant path rather than the zenith direction, the observed T_b must be converted to the vertical to estimate ZWD using (6.53). Given the slant T_b measurement at zenith angle ϑ, and an estimate of T_{mr}, the slant opacity can be computed and converted to the zenith opacity using the simple $1/\cos(\vartheta)$ mapping function. The equivalent zenith T_b follows from (6.56). For elevation angles above 15° this conversion is very accurate.

T_{mr} for a specific site is computed from (6.54) using radiosonde data that typify the site. The variation of T_{mr} with slant angle is minimal for elevations down to about 20°. The value used for WVR calibration and water vapor retrievals can be a site-average (standard deviation typically about 10 K), or can be adjusted for season to reduce the uncertainty. If surface temperatures T are available, then T_{mr} correlations with T can reduce the T_{mr} uncertainty to about 3 K.

6.3.4 Calibration of WVR

Because the intensity of the atmospheric microwave emission is very low, the WVR calibration is important. Microwave radiometers receive roughly a billionth of a watt in microwave energy from the atmosphere. The calibration establishes a relationship between the radiometer reading and the brightness temperature. Here we briefly discuss the calibration with tipping curves. This technique provides accurate brightness temperatures and the instrument gain without any prior knowledge of either.

Under the assumption that the atmosphere is horizontally homogeneous and that the sky is clear, the opacity is proportional to the thickness of the atmosphere. Clearly the amount of atmosphere sensed increases with the zenith angle. For zenith angles less than about 60° one might consider adopting the following model for the mapping function for the opacity:

$$m_\tau(\vartheta) \equiv \frac{\tau(\vartheta)}{\tau(\vartheta = 0)} = \frac{1}{\cos(\vartheta)} \tag{6.58}$$

Figure 6.11 shows an example of radiometer calibration using tipping. The opacity is plotted versus air mass. Looking straight up, the opacity of one air mass is observed. Looking at 30°, the opacity of two air masses is observed, etc. Since opacity is linear, we can extrapolate to zero air mass. At zero air mass, we have $m_\tau(\vartheta) = 0$ because there is no opacity for a zero atmosphere.

The calibration starts with a radiometer voltage (noise diode, labeled ND in Figure 6.11) reading N_{bb} of an internal reference object, which one might think of as a black body. The physical temperature of that object is T_{bb}. Let G denote the initial estimate of the gain factor (change in radiometer count reading over change in temperature). The observed brightness temperature at various zenith angles, measured by tipping the antenna, is then computed by

$$T(\vartheta) = T_{bb} - \frac{1}{G}\left(N_{bb} - N(\vartheta)\right) \tag{6.59}$$

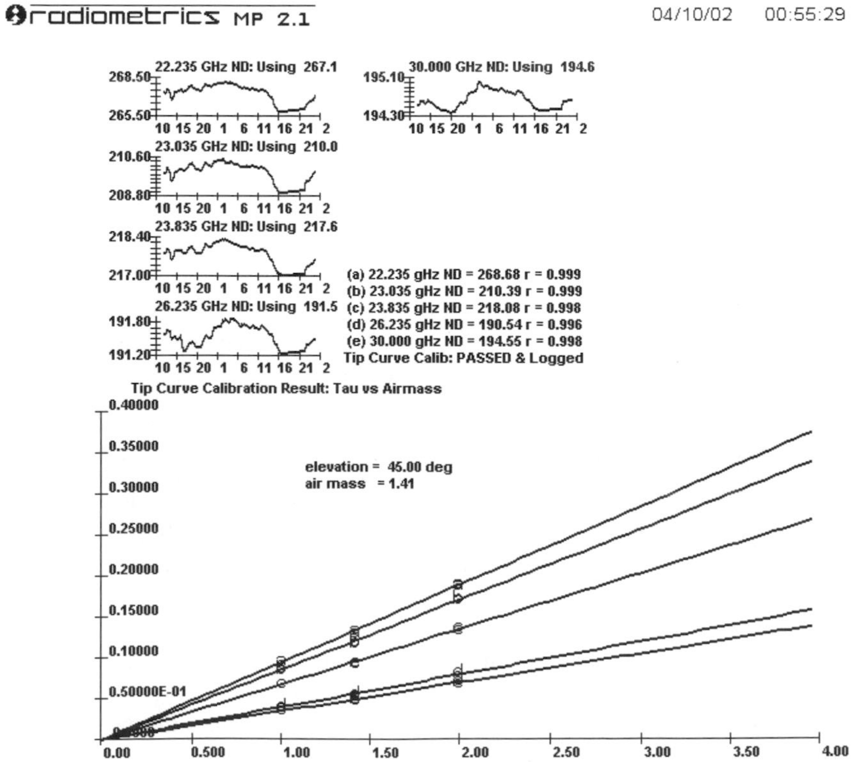

Figure 6.11 Tipping curve example. (Courtesy of R. Ware, Radiometrics Corporation, Boulder)

The brightness temperatures are substituted into (6.57) to get the opacity. If the linear regression line through the computed opacities does not pass through the origin, the gain factor G is adjusted until it passes though the origin. If the regression coefficient of the linear fit is better than a threshold value, typically $r = 0.99$, the tip curve calibration is accepted. The time series in Figure 6.11 show the history of passed tip curve calibrations at the various microwave frequencies. Additional details on radiometer calibration are best obtained from manufacturers.

The tipping curve calibration assumes that we know the microwave cosmic background brightness temperature $T_{\text{cosmic}} = 2.7$ K. Arno Penzias and Robert Wilson received the Nobel Prize for physics in 1978 for their discovery of the cosmic background radiation. Conducting their radio astronomy experiments, they realized a residual radiation that was characteristically independent of the orientation of the antenna.

6.4 IONOSPHERIC REFRACTION

Coronal mass ejections (CMEs) and extreme ultraviolet (EUV) solar radiation (solar flux) are the primary cause of the ionization (Webb and Howard, 1994). A CME is a major solar eruption. When passing the earth it causes at times sudden and large geomagnetic storms, which generate convection motions within the ionosphere, as well as enhanced localized currents. The phenomena can produce large spatial and temporal variation in the TEC and increased scintillation in phase and amplitude. Complicating matters are coronal holes, which are pathways of low density through which high-speed solar wind can escape the sun. Coronal holes and CMS are the two major drivers of magnetic activities on the earth. Larger magnetic storms are rare but may occur at any time.

Solar flux originates high in the sun's chromosphere and low in its corona. Even a quiet sun emits radio energy across a broad frequency spectrum, with slowly varying intensity. EUV radiation is absorbed by the neutral atmosphere and therefore cannot be measured accurately from ground-based instrumentation. Accurate determination of the EUV flux requires observations from space-based platforms above the ionosphere. A popular surrogate measure to the EUV radiation is the widely observed flux at 2800 MHz (10.7 cm). The 10.7 cm flux is useful for studying the ozone layer and global warming. However, Doherty et al. (2000) point out that predicting the TEC by using the daily values of solar 10.7 cm radio flux is not useful due to the irregular, and sometimes very poor, correlation between the TEC and the flux. The TEC at any given place and time is not a simple function of the amount of solar ionizing flux.

The transition from a gas to an ionized gas, i.e., plasma, occurs gradually. During the process, a molecular gas dissociates first into an atomic gas that, with increasing temperature, ionizes as the collisions between atoms break up the outermost orbital electrons. The resulting plasma consists of a mixture of neutral particles, positive ions (atoms or molecules that have lost one or more electrons), and negative electrons. Once produced, the free electron and the ions tend to recombine, and a balance is established between the electron-ion production and loss. The net concentration of free electrons is what impacts electromagnetic waves passing through the ionosphere. In order for gases to be ionized, a certain amount of radiated energy must be absorbed. Hargreaves (1992, p. 223) gives maximum wavelengths for radiation needed to ionize various gases. The average wavelength is about 900 Å (1 Å equals 0.1 nm). The primary gases available at the upper atmosphere for ionization are oxygen, ozone, nitrogen, and nitrous oxide.

Because the ionosphere contains particles that are electrically charged and capable of creating and interacting with electromagnetic fields, there are many phenomena in the ionosphere that are not present in ordinary fluids and solids. For example, the degree of ionization does not uniformly increase with the distance from the earth's surface. Instead, there are regions of ionization, historically labeled D, E, and F, that have special characteristics as a result of variation in the EUV absorption, the predominant type of ions present, or pathways generated by the electromagnetic field. The electron density is not constant within such a region and the transition to another region is continuous. Whereas the TEC determines the amount of pseudorange delays and carrier phase advances, it is the layering that is relevant to radio communication in terms of signal reflection and distance that can be bridged at a given time of the day. In the lowest D region, approximately 60–90 km above the earth, the atmosphere is still dense and atoms that have been broken up into ions recombine quickly. The level of ionization is directly related to radiation that begins at sunrise, disappears at sunset, and generally varies with the sun's elevation angle. There is still some residual ionization left at local midnight. The E region extends from about 90–150 km and peaks around 105–110 km. In the F region, the electrons and ions recombine slowly due to low pressure. The observable effect of the solar radiation develops more slowly and peaks after noon. During daytime this region separates into the $F1$ and $F2$ layers. The $F2$ layer (upper layer) is the region of highest electron density. The top part of the ionosphere reaches up to 1000 to 1500 km. There is no real boundary between the ionosphere and the outer magnetosphere.

Ionospheric convection is the main result of the coupling between the magnetosphere and ionosphere. While in low altitudes the ionospheric plasma co-rotates with the earth, at higher latitudes it is convecting under the influence of the large-scale magnetospheric electric field. Electrons and protons that speed along the magnetic field lines until they strike the atmosphere not only generate the spectacular lights of the aurora in higher latitudes, but they also cause additional ionization. Peaks of electron densities are also found at lower latitudes on both sides of the magnetic equator. The electric field and the horizontal magnetic field interact at the magnetic equator to raise ionization from the magnetic equator to greater heights, where it diffuses along magnetic field lines to latitudes approximately $\pm15°$ to $20°$ on either side of the magnetic equator. The largest TEC values in the world typically occur at these so-called equatorial anomaly latitudes.

There are local disturbances of electron density in the ionosphere. On a small scale, irregularities of a few hundred meters in size can cause amplitude fading and phase scintillation of GPS signals. Larger disturbances of the size of a few kilometers can significantly impact the TEC. Amplitude fading and scintillation can cause receivers to lose lock, or receivers may not be able to maintain lock for a prolonged period of time. Scintillation on GPS frequencies is rare in the midlatitudes, and amplitude scintillation, even under geomagnetically disturbed conditions, is normally not large in the auroral regions. However, rapid phase scintillation can be a problem in both the equatorial and the auroral regions, especially for semicodeless L2 GPS receivers, as the bandwidth of such receivers might be too narrow to follow rapid phase scintillation effects. Strong scintillation in the equatorial region generally occurs in the postsunset to local midnight time period, during geomagnetically quiet periods, mostly during equinoctial months in years having high solar activity. Even during times of strong amplitude scintillation the likelihood of simultaneous deep amplitude fading to occur on more than one GPS satellite is small. Thus, a modern GPS receiver observing all satellites in view should be able to operate continuously through strong scintillation albeit with a continuously changing geometric dilution of precision (GDOP) due to the continually changing "mix" of GPS satellites in lock.

Sunspots are seen as dark areas in the solar disk. At the dark centers the temperature drops to about 3700 K from 5700 K for the surrounding photosphere. They are magnetic regions with field strengths thousands of times stronger than the earth's magnetic field. Sunspots often appear in groups with sets of two spots, one with positive (north) magnetic fields, and one with negative (south) magnetic fields. Sunspots have an approximate lifetime of a few days to a month. The systematic recording of these events began in 1849 when the Swiss astronomer Johann Wolf introduced the sunspot number. This number captures the total number of spots seen, the number of disturbed regions, and the sensitivity of the observing instrument. Wolf searched observatory records to tabulate past sunspot activities. He apparently traced the activities to 1610, the year Galileo Galilei first observed sunspots through his telescope (McKinnon, 1987). Sunspot activities follow a periodic variation, with a principal period of eleven years, as seen in Figure 6.12. The cycles are usually not symmetric. The time from minimum to maximum is shorter than the time from maximum to minimum.

Sunspots are good indicators of solar activities. Even though sunspots have a high correlation with CME and solar flux, there is no strict mathematical relationship

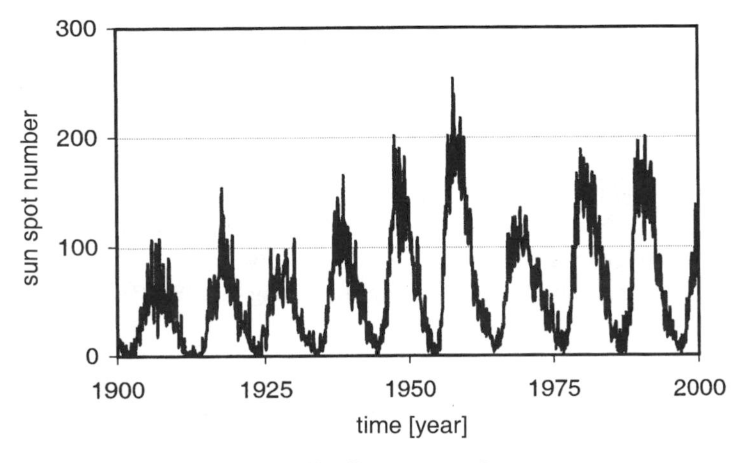

Figure 6.12 Sunspot numbers.

between them. It can happen that GPS is adversely affected even when daily sunspot numbers are actually low. Kunches and Klobuchar (2000) point out that GPS operations are more problematic during certain years of the solar cycle and during certain months of those years. The years at or just after the solar maximum will be stormy, and the months near the equinoxes will contain the greatest number of storm days. Sunspots are good for long-term prediction of ionospheric states.

The Appleton-Hartree formula is usually taken in the literature as the start for developing the ionospheric index of refraction that is applicable to the range of GPS frequencies. The formula is valid for a homogeneous plasma that consists of electrons and heavy positive ions, a uniform magnetic field, and a given electron collision frequency. Following Davies (1990, p. 72), the Appleton-Hartree formula is

$$n^2 = 1 - \frac{X}{1 - iZ - \dfrac{Y_T^2}{2\,(1 - X - iZ)} \pm \sqrt{\dfrac{Y_T^4}{4\,(1 - X - iZ)^2} + Y_L^2}} \tag{6.60}$$

Since the goal is to find the ionospheric index of refraction that applies to the GPS frequency f, several simplifications are permissible. The element $Z = \nu/f$ is the ratio of the electron collision frequency ν and the satellite frequency. This term quantifies the absorption. We simply set $Z = 0$. The index n now becomes a real number; in the notation of (6.3) we have $n''(f) = 0$. The symbols Y_T and Y_L relate to the magnetic field with reference to the direction of the wave normal, i.e., phase propagation. The commonly used first-order ionospheric delay expression is obtained by setting $Y_T = Y_L = 0$. An excellent summary of the higher-order ionospheric terms and their effects on the GPS observables is given in Odijk (2002). With these simplifications we obtain

$$n^2 = 1 - X = 1 - \frac{f_N^2}{f^2} \tag{6.61}$$

The plasma frequency f_N is a measure of the electron motion (oscillation) around the heavy ions. It is a basic constant of plasma. Davies (1990, pp. 21, 73) gives

$$f_N^2 = \frac{N_e\, e^2}{4\pi^2 \varepsilon_0 m_e} = 80.6 N_{e[\mathrm{el/m^3}]} \tag{6.62}$$

In (6.62) the symbol $e = 1.60218 \cdot 10^{-19}$ *coulombs* denotes the electron charge with mass $m_e = 9.10939 \cdot 10^{-31}$ kg; $\varepsilon_0 = 8.854119 \cdot 10^{-12}$ faradays/m is the permittivity of free space. The relevant term is the electron density N_e, which is typically given in units of electrons per cubic meter [el/m^3]. Substituting (6.62) in (6.61) and developing a series gives

$$n = \sqrt{1 - \frac{f_N^2}{f^2}} = 1 - \frac{f_N^2}{2f^2} + \cdots = 1 - N_I \tag{6.63}$$

The ionospheric refractivity is

$$N_I = \frac{f_N^2}{2f^2} + \cdots = \frac{40.30}{f^2} N_e \ll 1 \tag{6.64}$$

The total electron content (TEC) along the path from receiver to the end of the effective ionosphere is

$$\text{TEC} = \int N_e \, ds \tag{6.65}$$

The TEC represents the number of free electrons in a 1-square-meter column along the path and is given in units of [el/m^2].

6.5 IONOSPHERIC CODE ADVANCES AND PHASE DELAYS

We need to deal with the phenomena of carrier phase advancement and group delay of the codes due to the ionosphere. As an introduction to the propagation in a dispersive medium, we consider the simplified situation of wave propagation in a homogeneous and isotropic medium. In a homogeneous medium, the index of refraction is constant and the isotropic property implies that the propagation velocity at any given point in the medium is independent of the direction of the propagation. In such medium a harmonic wave with unit amplitude is described by

$$\varphi = \cos \omega \left(t - \frac{x}{c_\varphi} \right) \tag{6.66}$$

The symbol t denotes the time, c_φ[m/sec] is the phase velocity (propagation speed of the wave), and x is the distance from the transmitting source. The angular frequency ω[rad/sec], the frequency f[Hz], the wavelength λ_φ[m], and the wave number k[rad/m] (phase propagation constant), are related by

$$\omega = 2\pi f \tag{6.67}$$

$$\lambda_\varphi = \frac{c_\varphi}{f} \tag{6.68}$$

$$k = \frac{2\pi}{\lambda_\varphi} \tag{6.69}$$

Using the relations (6.67) to (6.69) the wave equation (6.66) can be written as

$$\varphi_1 = \cos(\omega \, t - k \, x) \tag{6.70}$$

Let us consider another wave that has a slightly different frequency and wave number,

$$\varphi_2 = \cos\left[(\omega + \Delta\omega)\, t - (k + \Delta k)\, x\right] \tag{6.71}$$

These two harmonic waves can be superimposed by addition,

$$\varphi_s = \varphi_1 + \varphi_2 = 2\cos\frac{\Delta\omega\, t - \Delta k\, x}{2}\,\cos\left[\left(\omega + \frac{\Delta\omega}{2}\right) t - \left(k + \frac{\Delta k}{2}\right) x\right] \tag{6.72}$$

This resultant wave is displayed in Figure 6.13. The combined signal shows two component waves of significantly different frequency. The slowly varying amplitude modulation represented by the envelope wave is

$$\Psi = 2\cos\tfrac{1}{2}(\Delta\omega\, t - \Delta k\, x) \tag{6.73}$$

having a propagation velocity wave of $\Delta\omega/\Delta k$. At the limit, $\Delta\omega \to 0$ and $\Delta k \to 0$ we obtain

$$c_g = \frac{d\omega}{dk} \tag{6.74}$$

The quantity c_g is the velocity of the modulation and called the group velocity. In the context of GPS signals c_g is the velocity of the P-code or C/A-codes. The second wave component in (6.72) can be viewed as representing the carrier.

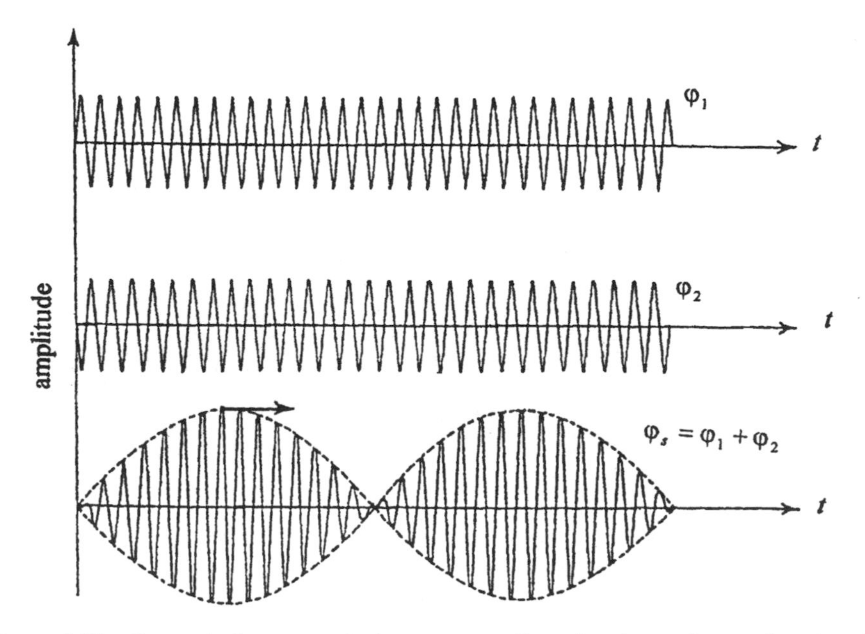

Figure 6.13 Concept of group and phase propagation. A point on the envelope travels with group velocity c_g, whereas the waveform within the envelope travels with phase velocity c_φ.

The products of phase index of refraction n_φ and velocity c_φ, and group index of refraction n_g and velocity c_g equal the speed of light in vacuum, i.e.,

$$n_\varphi \, c_\varphi = c \tag{6.75}$$

$$n_g \, c_g = c \tag{6.76}$$

Using the various relationships identified above, the group index of refraction can be expressed as

$$n_g = n_\varphi + f \frac{dn_\varphi}{df} \tag{6.77}$$

If the group index of refraction depends on the frequency f, i.e., the derivative dn_φ / df in (6.77) is not zero, then $n_\varphi \neq n_g$ and we call the medium dispersive. It follows that the phase velocity and the group velocity are not the same in a dispersive medium whereas in a nondispersive medium we have $c_\varphi = c_g$ and the wave envelope moves with the same velocity as the wave.

Expression (6.77) is applicable in quantifying the impact of the ionosphere on the GPS signals. Substituting the phase index of refraction (6.63) into (6.77) and carrying out the differentiation gives the expression for the group index of refraction,

$$n_g = n_\varphi + f \frac{dn_\varphi}{df} = 1 + N_I \tag{6.78}$$

Neglecting terms of the order N_I squared and higher, the expressions for the phase and group velocities become

$$c_\varphi = \frac{c}{n_\varphi} = \frac{c}{1 - N_I} = c\,(1 + N_I) \tag{6.79}$$

$$c_g = \frac{c}{n_g} = \frac{c}{1 + N_I} = c\,(1 - N_I) \tag{6.80}$$

Since N_I is a positive number, the phase velocity is larger than vacuum speed and the group velocity is smaller than vacuum speed by the same amount $\Delta\, c$; i.e.,

$$\Delta\, c = c\, N_I = \frac{40.30\, c}{f^2} N_e \tag{6.81}$$

The time of a code delay or the phase advancement that is registered at the receiver is directly related to the velocity difference Δc and its variations along the path. Integrating (6.81) over time and realizing that $ds = c\,dt$ gives the ionospheric delay in units of distance

$$I_{f,P} \equiv I^p_{k,f,P} = \frac{40.30}{f^2} \int N_e \, ds \tag{6.82}$$

The subscript f in (6.82) denotes the frequency of the carrier and the next subscript identifies the sign and the unit. If that subscript is P, as in case of (6.82), the distance is given in units of meters. The integral in (6.82) equals the total electron content (TEC) according to (6.65). Using short notation, i.e., neglecting the subscript k (receiver) and superscript p (satellite), we obtain

$$I_{f,P} = \frac{40.30}{f^2} \text{TEC} \tag{6.83}$$

The corresponding ionospheric time delay (codes) or time advance (phases) follows as

$$\nu_f = \frac{I_{f,P}}{c} = \frac{40.30 \text{ TEC}}{cf^2} \tag{6.84}$$

The time delay is proportional to the inverse of the frequency squared. Consequently, the ionosphere affects transmissions at higher frequencies less.

The unit for the ionospheric code delay is in meters, whereas it is typically expressed in cycles for carrier phases, unless the carrier phases have explicitly been scaled to distance. The following notation convention applies to identify sign and units,

$$I_{1,P} = -I_{1,\Phi} = -\frac{c}{f_1} I_{1,\varphi} \tag{6.85}$$

$$I_{2,P} = -I_{2,\Phi} = -\frac{c}{f_2} I_{2,\varphi} \tag{6.86}$$

$$\frac{I_{1,P}}{I_{2,P}} = \frac{f_2^2}{f_1^2} \tag{6.87}$$

$$\frac{I_{1,\varphi}}{I_{2,\varphi}} = \frac{f_2}{f_1} \tag{6.88}$$

Figure 6.14 shows the ionospheric delays for GPS frequencies as a function of TEC. Typically the TEC values range from 10^{16} to 10^{18}. Often the total electron content is expressed in terms of TEC units (TECU), with one TECU being 10^{16} electrons per 1-square-meter column.

Even though they are very important, phase advancement and group delay are not the only manifestations of the ionosphere on the signal propagation. Some of the phase variations are converted to amplitude variation by means of diffraction. The result can be an irregular but rapid variation in amplitude and phase, called scintillation. The signal can experience short-term fading by losing strength. Scintillations might occasionally cause phase-lock problems to occur in receivers. A receiver's bandwidth must be sufficiently wide not only to accommodate the normal rate of change of the geometric Doppler shift, (up to 1 Hz), but also the phase fluctuations due to strong

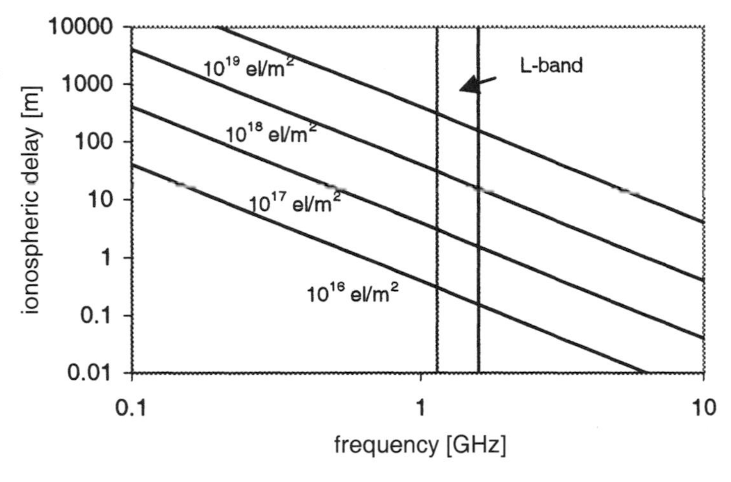

Figure 6.14 Ionospheric range correction.

amplitude and phase scintillation. These scintillation effects generally require a minimum receiver bandwidth of at least 3 Hz under severe fading and phase jitter conditions. Semicodeless L2 receivers generally do not perform well under conditions of severe phase scintillation due to the required narrow bandwidth of such receivers. If the receiver bandwidth is set to 1 Hz to deal with the rate of change of the geometric Doppler shift, and if the ionosphere causes an additional 1 Hz shift, the receiver might lose phase lock. Assuming a maximum TEC of 10^{18} el/m^2, a change of 1.12% TEC causes a single-cycle change in L1.

6.6 IONOSPHERIC SOLUTIONS

The primary purpose of multiple frequencies is to neutralize the effect of the ionosphere on position determination. Functions of the dual-frequency observables are readily available that do not depend on the ionosphere.

6.6.1 Single Frequencies and the Broadcast Ionospheric Model

To support point positioning for single-frequency users, the broadcast message contains eight ionospheric model coefficients for computing the ionospheric group delay along the signal path. The algorithm was developed by Klobuchar (1987) and is listed in Table 6.1. In addition to the broadcast coefficients, other input parameters are the geodetic latitude and longitude of the receiver, the azimuth and elevation angle of the satellite as viewed from the receiver, and the time. Note that several angular arguments are expressed in semicircles (SC). All auxiliary quantities in the middle portion of the table can be computed one at a time starting from the top. The function in the third part of the table has been multiplied with the velocity of light, in order to yield the slant group delay directly in meters. The algorithm presented here compensates for 50–60% of the actual group delay.

The Klobuchar algorithm is based on the "shell model" or single-layer model of the ionosphere. The implicit assumption is that the TEC is concentrated in an infinitesimally thin spherical layer at a certain height, e.g., 350 km. The model further assumes that the maximum ionospheric disturbance occurs at 14:00 local time. The factor F in Table 6.3 is the mapping function that converts the vertical ionospheric delay at the ionospheric pierce point to the slant delay at the receiver location. The ionospheric pierce point is the intersection of the line of sight and the ionospheric layer. The geomagnetic latitude of the ionospheric pierce point is ϕ, and its geodetic latitude and longitude is φ_{IP} and λ_{IP}. The angle ψ denotes the earth's central angle between the user and the ionospheric pierce point, t is the local time, P is the period in seconds, x is the phase in radians, and A denotes the amplitude in seconds.

TABLE 6.3 The Broadcast Ionospheric Model

φ, λ geodetic latitude and longitude of receiver [SC] — $\mathrm{T} = $ GPS time [s]

α_k^p, β_k^p azimuth and altitude of satellite [SC] — α_n, γ_n broadcast coefficients

$$F = 1 + 16\left(0.53 - \beta_k^p\right)^3 \quad \text{(a)} \qquad \psi = \frac{0.0137}{\alpha_k^p + 0.11} - 0.022 \quad \text{(b)}$$

$$\varphi_{IP} = \begin{cases} \varphi + \psi \cos \alpha_k^p & \text{if } |\varphi_{IP}| \le 0.416 \\ 0.416 & \text{if } \varphi_{IP} > 0.416 \\ -0.416 & \text{if } \varphi_{IP} < -0.416 \end{cases} \quad \text{(c)} \qquad \lambda_{IP} = \lambda + \frac{\psi \sin \alpha_k^p}{\cos \varphi_{IP}} \quad \text{(d)}$$

$$\phi = \varphi_{IP} + 0.064 \cos\left(\lambda_{IP} - 1.617\right) \quad \text{(e)}$$

$$t = \begin{cases} \lambda_{IP}\, 4.32 \times 10^4 + \mathrm{T} & \text{if } 0 \le t < 86400 \\ \lambda_{IP}\, 4.32 \times 10^4 + \mathrm{T} - 86400 & \text{if } t \ge 86400 \\ \lambda_{IP}\, 4.32 \times 10^4 + \mathrm{T} + 86400 & \text{if } t < 0 \end{cases} \quad \text{(f)}$$

$$x = \frac{2\pi(t - 50400)}{P} \quad \text{(g)}$$

$$P = \begin{cases} \sum_{n=0}^{3} \gamma_n\, \phi^n & \text{if } P \ge 72000 \\ 72000 & \text{if } P < 72000 \end{cases} \quad \text{(h)} \qquad A = \begin{cases} \sum_{n=0}^{3} \alpha_n \phi^n & \text{if } A \ge 0 \\ A = 0 & \text{if } A < 0 \end{cases} \quad \text{(i)}$$

$$I_{k,1,P}^p = \begin{cases} c\, F\left[5 \times 10^{-9} + A\left(1 - \frac{x^2}{2} + \frac{x^4}{24}\right)\right] & \text{if } |x| < 1.57 \\ c\, F\left(5 \times 10^{-9}\right) & \text{if } |x| > 1.57 \end{cases} \quad \text{(j)}$$

Conversion of SC unit: 1 SC = 180°

For accurate relative positioning with carrier phase observations, single-frequency users still depend on the elimination of ionospheric effects through single or double differencing, as discussed in connection with Equations (5.20) and (5.28). Better ionospheric corrections will be available to single-frequency users in the near future from services provided by the IGS or other entities.

6.6.2 Ionospheric-Free Functions

Since the ionospheric code delay and the phase advance are dependent on frequency, it is possible to eliminate the ionospheric effects for dual-frequency observation. Using simplified notation, the pseudoranges of Equation (5.7) for L1 and L2 can be expressed as

$$P_1 = \rho - c\,d\underline{t} + c\left(d\bar{t} + T_{\mathrm{GD}}\right) + I_{1,P} + T + \delta_{1,P} + \varepsilon_{1,P} \tag{6.89}$$

$$P_2 = \rho - c\,d\underline{t} + c\left(d\bar{t} + \alpha_f T_{\mathrm{GD}}\right) + I_{2,P} + T + \delta_{2,P} + \varepsilon_{2,P} \tag{6.90}$$

The objective is to find functions that do not depend on the ionosphere. Using the coefficients α_f, β_f, γ_f and δ_f defined in (5.13) to (5.16), the ionospheric-free pseudorange function P_{IF},

$$P_{\mathrm{IF}} \equiv \beta_f P_1 - \gamma_f P_2 = \frac{1}{\left(1-\alpha_f\right)}\left\{P_2 - \alpha_f P_1\right\} = \rho - c\,d\underline{t} + c\,d\bar{t} + T + \delta_{P,\mathrm{IF}} + \varepsilon_{P,\mathrm{IF}} \tag{6.91}$$

serves this purpose. In Equation (6.91) the ionospheric terms cancel. The symbols $\delta_{P,\mathrm{IF}}$ and $\varepsilon_{P,\mathrm{IF}}$ are functions of $\delta_{1,P}$, $\delta_{2,P}$, $\varepsilon_{1,P}$ and $\varepsilon_{2,P}$. The satellite code phase offset T_{GD} has also canceled, whereas the other hardware delays and multipath terms do not cancel (but are not listed explicitly in (6.91)).

The dual-frequency carrier phase equations (5.10) in units of cycles are in simplified notation

$$\varphi_1 = \frac{f_1}{c}\rho + N_1 - f_1\,d\underline{t} + f_1\,d\bar{t} - \frac{f_1}{c}I_{1,P} + \frac{f_1}{c}T + \delta_{1,\varphi} + \varepsilon_{1,\varphi} \tag{6.92}$$

$$\varphi_2 = \frac{f_2}{c}\rho + N_2 - f_2\,d\underline{t} + f_2\,d\bar{t} - \frac{f_2}{c}I_{2,P} + \frac{f_2}{c}T + \delta_{2,\varphi} + \varepsilon_{2,\varphi} \tag{6.93}$$

The ionospheric-free carrier phase function φ_{IF} is

$$\varphi_{\mathrm{IF}} \equiv \beta_f\,\varphi_1 - \delta_f\,\varphi_2 = \frac{f_1}{c}\rho - f_1\,d\underline{t} + f_1\,d\bar{t} + \beta_f N_1 - \delta_f N_2 + \frac{f_1}{c}T + \delta_{\varphi,\mathrm{IF}} + \varepsilon_{\varphi,\mathrm{IF}} \tag{6.94}$$

where $\delta_{\varphi,\mathrm{IF}}$ and $\varepsilon_{\varphi,\mathrm{IF}}$ are functions of $\delta_{1,\varphi}$, $\delta_{2,\varphi}$, $\varepsilon_{1,\varphi}$, and $\varepsilon_{2,\varphi}$. The ionospheric-free phase function (6.94) does not contain the ionospheric term. Unfortunately, the integer-nature of the ambiguities has been lost, because the multipliers β_f and δ_f are not integers. Using (6.94) alone, in either undifferenced or double-differenced form, only the N_1 and N_2 linear ambiguity combination is estimable. Any hardware delays in receiver or satellite that are constant in time will also be absorbed by the estimated ambiguities.

6.6.3 Ionospheric Functions

Because the ionosphere delays the codes by the same amount as it advances the carrier phases, the difference of both observations depends on twice the ionosphereic delay while some other terms cancel. Differencing (6.89) and (6.92), and recalling that $I_{1,P} = -I_{1,\Phi}$, one obtains for a single frequency

$$R_{1,I} \equiv P_1 - \Phi_1 = 2I_{1,P} - \frac{c}{f_1} N_1 + cT_{\mathrm{GD}} + \delta_{1,R,I} + \varepsilon_{1,R,I} \tag{6.95}$$

where $\delta_{1,R,I}$ and $\varepsilon_{1,R,I}$ are functions of $\delta_{1,P}$, $\delta_{1,\Phi}$, $\varepsilon_{1,P}$, and $\varepsilon_{1,\Phi}$. The multipath of the pseudorange measurement typically sets the accuracy limit for this function. Since the ambiguity is not known, this function does not give the absolute ionospheric delay. The initial ambiguity and the code phase offsets cancel when differencing over time,

$$R_1\,(t_1, t_2) = 2\left[I_{1,P}(t_2) - I_{1,P}(t_1)\right] = \frac{2 \times 40.30}{f_1}\,[\mathrm{TEC}\,(t_2) - \mathrm{TEC}\,(t_1)] \tag{6.96}$$

as long as the general hardware and multipath terms are constant.

Differencing the dual-frequency pseudoranges (6.89) and (6.90) gives

$$P_I \equiv P_1 - P_2 = \left(1 - \alpha_f\right) I_{1,P} + c\left(1 - \alpha_f\right) T_{\mathrm{GD}} + \delta_{P,I} + \varepsilon_{P,I} \tag{6.97}$$

where $\delta_{P,I}$ and $\varepsilon_{P,I}$ are functions of $\delta_{1,P}$, $\delta_{2,P}$, $\varepsilon_{1,P}$, and $\varepsilon_{2,P}$. This function readily shows the difficulties encountered when measuring the total ionosphere, or the TEC, with dual-frequency receivers. The system specification for the stability of the satellite offset T_{GD} is ± 3 ns (2-sigma) level. This poses a limitation on determining the TEC, because 3 ns of differential delay between L2 minus L1 corresponds to 0.9 m delay or 8.5 TECU and has resulted in efforts to determine these delays more accurately. The separation of the hardware delays and the TEC estimates becomes possible because the impact of the ionosphere depends on the elevation angle, whereas that of the satellite hardware delay does not.

The ionospheric function for the carrier phases follows readily from (6.92) and (6.93),

$$\varphi_I \equiv \varphi_1 - \frac{f_1}{f_2}\varphi_2 = N_1 - \frac{f_1}{f_2} N_2 - \frac{f_1}{c}\left(1 - \alpha_f\right) I_{1,P} + \delta_{\varphi,I} + \varepsilon_{\varphi,I} \tag{6.98}$$

where $\delta_{\varphi,I}$ and $\varepsilon_{\varphi,I}$ are the respective functions of the carrier phase hardware and multipath. The hardware delays are not listed explicitly. The ionospheric function (6.98) reflects the time variation of the TEC. This variation can be measured accurately because of high carrier phase resolution and the small multipath (as compared to the one for code measurements). Unfortunately, this function alone does not permit the estimation of the absolute TEC, because the initial ambiguities are not known. Expressing (6.92) and (6.93) in units of cycles, we obtain the function

$$\Phi_I \equiv \Phi_2 - \Phi_1 = \left(1 - \alpha_f\right) I_{1,P} + \frac{c}{f_2} N_2 - \frac{c}{f_1} N_1 + \delta_{\Phi,I} + \varepsilon_{\Phi,I} \tag{6.99}$$

where $\delta_{\Phi,I}$ and $\varepsilon_{\Phi,I}$ are the respective functions of the carrier phase hardware and multipath. In Expression (6.99) we have made used of Equations (6.85) to (6.88) to convert the ionospheric phase delays to respective pseudorange delays. The functions P_I and Φ_I are affected by the ionosphere by the same amount.

6.6.4 Discriminating Small Cycle Slips

Analysis of dual-frequency carrier phase functions requires some extra attention because certain combinations of slips in L1 and L2 phases generate almost identical effects. For example, consider the ionosphere-free phase observable (6.94). Unfortunately, the ambiguities enter this function not as integers but in the combination of $\alpha_f N_1 - \delta_f N_2$, necessitating a search for a noninteger fraction in the residuals of the ionosphere-free phase. Table 6.4 lists in columns 1 and 2 small changes in the ambiguities and illustrates in columns 3 and 4 their effects on the ionospheric-free and the ionospheric phase functions, respectively. Certain combinations of both integers produce almost identical changes in the ionosphere-free phase function. For example, a change of (−7, −9) causes a small change of 0.033 cycles, whereas (1, 1) causes a change of 0.562 cycles, which is almost identical to the one caused by (8, 10). If pseudorange positioning is accurate enough to resolve the ambiguities within three to four cycles, then these additional difficulties in identifying slip combination can be resolved.

TABLE 6.4 Small Cycle Slips and Phase Functions

ΔN_1	ΔN_2	$\alpha_f\ \Delta N_1 - \delta_f\ \Delta N_2$	$\Delta N_1 - \sqrt{\alpha_f}\ \Delta N_2$
±1	±1	±0.562	∓0.283
±2	±2	±1.124	∓0.567
±1	±2	∓1.422	∓1.567
±2	±3	±0.860	∓1.850
±3	±4	∓0.298	∓2.133
±4	±5	±0.264	∓2.417
±5	±6	±0.827	∓2.700
±6	±7	±1.389	∓2.983
±5	±7	∓1.157	∓3.983
±6	±8	∓0.595	∓4.267
±7	±9	∓0.033	∓4.550
±8	±10	±0.529	∓4.833

TABLE 6.5 Effects of Selected Slips on the Ionospheric Phase Function

ΔN_1	ΔN_2	$\Delta N_1 - \sqrt{\alpha_f}\,\Delta N_2$	ΔN_1	ΔN_2	$\Delta N_1 - \sqrt{\alpha_f}\,\Delta N_2$
−2	−7	6.983	7	0	7.000
−2	−6	5.700	7	1	5.717
−2	−5	4.417	7	2	4.433
−2	−4	3.133	7	3	3.150
−2	−3	1.850	7	4	1.867
−2	−2	0.567	7	5	0.583
−2	−1	−0.718	7	6	−0.700
−2	0	−2.000	7	7	−1.983
2	0	2.000	−7	−7	1.983
2	1	0.717	−7	−6	0.700
2	2	−0.567	−7	−5	−0.583
2	3	−1.850	−7	−4	−1.867
2	4	−3.133	−7	−3	−3.150
2	5	−4.417	−7	−2	−4.433
2	6	−5.700	−7	−1	−5.717
2	7	−6.983	−7	0	−7.000

Table 6.5 shows an arrangement of integers that have a practically undistinguishable effect on the ionospheric function. It is seen that, e.g., the impact of the combinations $(-2, -7)$ and $(7, 0)$ differs by only 0.02 cycle. This amount is too small to be discovered reliably in an observation sequence. Unfortunately, there is no unique combination of small $(\Delta N_1, \Delta N_2)$ that smooths the ionospheric function if slips are present.

6.6.5 Multipath Equations

The multipath equations relate a pseudorange and carrier phases of both frequencies as follows,

$$M1 \equiv P_1 - \Phi_1 + \frac{2}{1-\alpha_f}(\Phi_1 - \Phi_2) = -\lambda_1 N_1 + \frac{2}{1-\alpha_f}(\lambda_1 N_1 - \lambda_2 N_2) + cT_{GD} + \delta_{M1} \tag{6.100}$$

$$M2 \equiv P_2 - \Phi_2 + \frac{2\alpha_f}{1-\alpha_f}(\Phi_1 - \Phi_2) = -\lambda_2 N_2 + \frac{2\alpha_f}{1-\alpha_f}(\lambda_1 N_1 - \lambda_2 N_2) + c\alpha_f T_{GD} + \delta_{M2} \tag{6.101}$$

These expressions can be readily verified. Analyzing these expressions over time is useful for initial cycle slip scanning. While these multipath functions should theoretically be constant in time, the actual variation is dominated by measurement accuracy and multipath of the pseudoranges.

6.6.6 Generalizing the Dual-Frequency Phase Function

The general linear combination of dual-frequency carrier phase observations at a given station is,

$$
\begin{aligned}
\varphi_{m,n} &\equiv m\varphi_1 + n\varphi_2 \\
&= \lambda_{m,n}^{-1}\,\rho + N_{m,n} - f_{m,n}\,d\underline{t} + f_{m,n}\,d\bar{t} + I_{m,n,\varphi} + \lambda_{m,n}^{-1}\,T + \delta_{m,n,\varphi} + \varepsilon_{m,n,\varphi}
\end{aligned} \tag{6.102}
$$

The frequency, wavelength, ambiguity, and the ionospheric terms for the general carrier phase function $\varphi_{m,n}$ are

$$f_{m,n} = mf_1 + nf_2 \tag{6.103}$$

$$\lambda_{m,n} = \frac{c}{f_{m,n}} = \frac{c}{mf_1 + nf_2} \tag{6.104}$$

$$N_{m,n} = mN_1 + nN_2 \tag{6.105}$$

$$I_{m,n,\varphi} = mI_{1,\varphi} + nI_{2,\varphi} = \frac{mf_2 + nf_1}{f_2} I_{1,\varphi} \tag{6.106}$$

$$I_{m,n,\Phi} = \lambda_{m,n}\, I_{m,n,\varphi} \tag{6.107}$$

Because the GPS L1 and L2 frequencies are related as $f_1/f_2 = 77/60$, the $m = 70$ and $n = -60$ combination does not depend in the ionosphere. Expressed in units of length, the function (6.102) becomes

$$\Phi_{m,n} = \rho + \lambda_{m,n} N_{m,n} - c\,d\underline{t} + c\,d\bar{t} + I_{m,n,\Phi} + T + \delta_{m,n,\Phi} + \varepsilon_{m,n,\Phi} \tag{6.108}$$

with $\delta_{m,n,\Phi}$ and $\varepsilon_{m,n,\Phi}$ being the respective functions of the hardware delays, multipath, and measurement noise. The ionospheric ratio with respect to the L1 carrier can be written as

$$\frac{I_{k,m,n,\Phi}^{p}}{I_{k,1,\Phi}^{p}} = \frac{f_1}{f_2}\left(\frac{mf_2 + nf_1}{mf_1 + nf_2}\right) \tag{6.109}$$

In (6.102) the distances are expressed in units of the wavelength. A change in $\varphi_{m,n}$ by one cycle, or a change of the ambiguity $N_{m,n}$ by one cycle, represents a distance change along the station-satellite direction by one wavelength of $\lambda_{m,n}$. The distance corresponding to one wavelength is frequently called a lane. Determination of the ambiguity $N_{m,n}$ thus implies that the topocentric range has been resolved within the unit of $\lambda_{m,n}$. One might, therefore, prefer transformations that give large wavelengths and solve the respective ambiguities. The assumption is that the unmodeled errors are small enough to allow a unique determination of these ambiguities. In a subsequent solution, when estimating N_1 and N_2, one could constrain the $N_{m,n}$. Unfortunately, the m and n factors that generate long wavelengths according to (6.104), might also increase the impact of multipath errors and other disturbances according to (6.102).

Combinations for which m and n have different signs are called the wide-lane observables. Because the specific observable $(1, -1)$ is the most important of all the wide-lane observables, it is usually referred to simply as the widelane (without explicitly mentioning the m and n); the subscript w is also used to identify this combination. If the m and n have the same sign, we speak of narrow-lane observables. The particular combination (1, 1) is simply the narrowlane (without explicitly mentioning the m and n). The subscript n identifies the narrowlane. For example,

$$\varphi_n = \varphi_1 + \varphi_2 \tag{6.110}$$

$$\varphi_w = \varphi_1 - \varphi_2 \tag{6.111}$$

$$\lambda_n = \frac{c}{f_n} = \frac{c}{f_1 + f_2} \approx 0.11\,\text{m} \tag{6.112}$$

$$\lambda_w = \frac{c}{f_w} = \frac{c}{f_1 - f_2} \approx 0.86\,\text{m} \tag{6.113}$$

It is important to note that for any linear combination of the carrier phase observations, the respective variance-covariance preparation must be carried out properly. Finding the optimal combination has at times generated considerable interest. However, that is no longer the case because of the optimal performance of LAMBDA (Teunissen, 1999). LAMBDA automatically includes widelaning but is even more general.

6.6.7 Global Ionospheric Models

The ionosphere can be estimated from (6.97) and (6.99), given dual-frequency observations. Although multipath of the GPS signals is a limiting factor in all GPS applications, we neglect the multipath terms in these equations assuming that their effect averages out or has been corrected computationally using multipath models. Adding the subscript k and superscript p for clarity, we can write

$$P_{k,I}^{p} = \left(1 - \alpha_f\right)\, I_{k,1,P}^{p} + c\left(1 - \alpha_f\right)\, T_{\text{GD}}^{p} + d_{1,P} - d_{2,P} \tag{6.114}$$

$$\Phi_{k,I}^{p} = \left(1 - \alpha_f\right)\, I_{k,1,P}^{p} + \lambda_2 N_{k,2}^{p} - \lambda_1 N_{k,1}^{p} + d_{2,\Phi} - d_{1,\Phi} \tag{6.115}$$

The first step in estimating the ionosphere is to correct all cycle slips, using, e.g., the "phase-connected" arc method (Blewitt, 1990) or any other suitable technique. In the second step, we assume that the receiver hardware delays $d_{1,P} - d_{2,P}$ and $d_{2,\Phi} - d_{1,\Phi}$ are constant over the time of the arc and compute the offset for the arc

$$\Delta_k^p = \frac{1}{n} \sum_{i=1}^{n} \left(P_{k,I}^p - \Phi_{k,I}^p \right)_i \tag{6.116}$$

The summation goes over the n epochs in the arc. Perhaps one might adopt an elevation-depended weighting scheme in (6.116) to take into account the decrease of measurement accuracy with elevation angle. The computed offset is added to (6.115) which can then be modeled as

$$\Phi_{k,I}^p(t) - \Delta_k^p = \left(1 - \alpha_f\right) I_{k,1,P}^p(t) + d^p - d_k \tag{6.117}$$

over the arc. The term d^k is the residual interfrequency satellite delay, which is essentially an estimate of T_{GD}^p, and d_k is a residual interfrequency receiver delay. As with the broadcast ionospheric model, we also relate the slant and vertical ionospheric delays by the mapping function $F(\beta)$ such that

$$I_{k,1,P}^p(\lambda, \varphi, t) = F(\beta) I_{k,1,P}(\lambda_{\mathrm{IP}}, \varphi_{\mathrm{IP}}, t) \tag{6.118}$$

One could use the simple mapping function of Table 6.3 or one that is based on a realistic electron density profile model, such as the extended slab density model by Coster et al. (1992). The symbols φ_{IP} and λ_{IP} denote the latitude and longitude of the ionospheric pierce point, whereas λ and φ identify the receiver location. Since the ionospheric disturbances follow the motion of the sun (the maximum disturbances occur around 14:00 local time) and tend to follow geomagnetic field lines, it is advantageous to parameterize the model for the vertical ionospheric delay $I_{k,1,P}$ in a solar-fixed coordinate system whose third axis coincides with the geomagnetic pole rather the geographic pole. One might model $I_{k,1,P}$ by a spherical harmonic series and estimate the spherical harmonic coefficients for global ionospheric models. The Kalman filter implementation of Mannucci et al. (1998) divides the surface of the earth into tiles (triangles) and estimates the vertical TEC for the vertices. Only observations that fall within the triangle are used to estimate the TEC at the vertices of that triangle. They assume that the TEC varies linearly within the triangle. Because the instrumental biases d^p and d_k are geometry-independent but the ionospheric delay depends on the azimuth and elevation of the satellite the biases and the ionospheric effect are estimable. The biases are fairly stable and need to be estimated less often than the rapidly varying ionospheric parameters. See Sardón et al. (1994) for additional details on the parameterization of TEC and satellite and receiver biases.

CHAPTER 7

PROCESSING PSEUDORANGES AND CARRIER PHASES

In keeping with the objectives of this book, this chapter does not contain the depth required to determine global positioning system (GPS) orbits or orbits of near-earth satellites. Rather, the position is taken that precise GPS satellite ephemerides will be available from the International GPS Service (IGS) and participating agencies and individuals who will continue to fine-tune their models as part of their research agenda. Therefore, the first section deals with the IGS and its products.

A separate section is devoted to antennas. Because of the increasing popularity of precise point positioning (PPP), material is included on the phase windup correction that results from the fact that the GPS satellite transmissions are right circularly polarized. The separation between the center of the satellite antenna and the satellite's center of mass must be properly dealt with in precise PPP application and respective corrections must be applied. The two types of corrections are little known among users performing relative positioning over short distances, because they cancel. All users however, must be concerned with receiver antenna phase center offsets and variations, and, certainly, signal multipath. These phenomena are treated in some detail.

The various GPS positioning techniques are subdivided into geometry-free solutions, point positioning (navigation solution), precise point positioning, real-time precise point positioning, relative positioning (differential positioning), and real-time relative positioning (real-time kinematic positioning [RTK] and network-aided RTK). PPP, in either the static or kinematic mode, is becoming increasingly important because of the availability of very precise postprocessed satellite ephemerides within a short delay time or even predicted precise ephemerides, and the discontinuation of selective availability (SA). PPP becomes even more attractive because of computational services that conveniently are available over the Internet. An example is the service provided by the Jet Propulsion Laboratory (JPL). Relative positioning has thus far been the backbone of most positioning with GPS and will still remain very important in the future. The burden of processing for relative positioning has been lessened because of Internet processing services provided by, e.g., the National Geodetic Survey (NGS). Ambiguity fixing, with emphasis on least-squares ambiguity decorrelation adjustment (LAMBDA), is dealt with in a separate section because of its importance to achieving centimeter-accurate relative positioning. The nonlinear pseudorange position solutions are given for point positioning and relative positioning.

7.1 THE IGS AND ITS PRODUCTS

The IGS is a response to a call by international users for an organizational structure that helps maximize the potential of GPS. It is a globally decentralized organization that is self-governed by its members and is without a central resource of funding. The support comes from various member organizations and agencies around the world, called contributing organizations. The IGS was formerly established by the International Association of Geodesy (IAG) in 1993 and officially began its operations on January 1, 1994.

A governing board sets the IGS policies and exercises broad oversight of all IGS functions. The executive arm of the board is the central bureau, which is located at the JPL and is sponsored by NASA. There are nearly 300 globally distributed permanent GPS tracking sites (Figure 7.1). These stations operate continuously and deliver data hourly or daily to data centers. There are currently three global data centers, five regional data centers, and twenty-three operational data centers. This scheme of data centers provides for efficient access and storage of data, data redundancy,

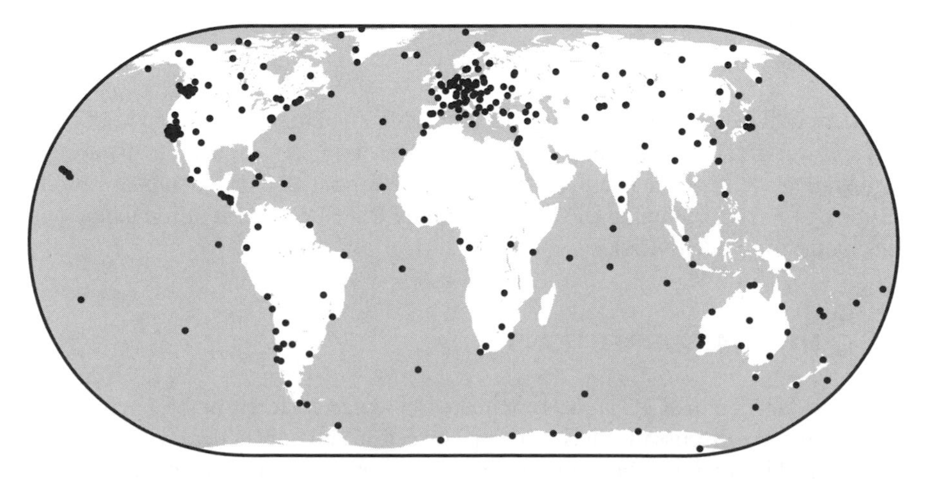

Figure 7.1 IGS permanent tracking network in 2002. (Courtesy NASA/JPL/Caltech.)

and security at the same time. There are eight analysis centers. These centers are the scientific backbone of the IGS that use the global data sets to produce products of the highest quality. The analysis centers cooperate with an analysis center coordinator, whose main task is to combine the products of the centers into single product, which becomes the official IGS product. The global data centers and central bureau make it available. The analysis centers are: (1) Astronomical Institute University of Bern, Center for Orbit Determination in Europe, Switzerland; (2) European Space Agency/European Space Operations Center, Germany; (3) GeoForschungsZentrum, Potsdam, Germany; (4) NASA Jet Propulsion Laboratory, California Institute of Technology, United States; (5) National Geodetic Survey, National Oceanic and Atmospheric Administration, United States; (6) Natural Resource, Canada; (7) Scripps Orbit and Permanent Array Center, Scripps Institution of Oceanography, United States; and (8) U.S. Naval Observatory, United States.

Detailed information about the IGS is available at the website (IGS, 2002). A strategic plan for the years 2002–2007 found at that website lists the long-term goals and objectives of the IGS:

- Provide the highest-quality, reliable global navigation satellite system (GNSS) data and products openly and readily available to all user communities.
- Promote universal acceptance of IGS products and conventions as the world standard.
- Continuously innovate by attracting leading-edge expertise and pursuing challenging projects and ideas.
- Seek to implement new growth opportunities while responding to changing user needs.
- Sustain and nurture the IGS culture of collegiality, openness, inclusiveness, and cooperation.
- Maintain a voluntary organization with effective leadership, governance, and management.

The various IGS products are summarized in Table 7.1. These products have become a de facto world standard for many GPS applications. Examples of universally accepted formats include receiver independent EXchange format (RINEX), standard product #3 for ECEF (earth centered earth fixed) orbital files (SP3), and solution independent EXchange format (SINEX).

7.2 ANTENNA CORRECTIONS

It is important that the GPS signals are treated/modeled correctly at the satellite and at the receiver. We discuss the phase windup correction and how to deal with the separation of satellite antenna phase and satellite center of mass. The receiver phase center offset and variation is generally dealt with in terms of relative and absolute antenna calibration. Several examples are given to shed light on the nature of signal multipath.

TABLE 7.1 IGS Products in 2002

Product	Accuracy	Latency	Updates
	GPS Satellite Ephemeris and Satellite Clocks		
Predicted (Ultra Rapid)	~25 cm; ~5 ns	real time	Twice daily
Rapid	5 cm; 0.2 ns	17 hours	daily
Final	< 5 cm; 0.1 ns	~13 days	weekly
	Geocentric Coordinates of IGS Tracking Stations		
Final horizontal and vertical positions	3 mm & 6 mm	12 days	weekly
Final horizontal and vertical velocities	2 mm/yr & 3 mm/yr	12 days	weekly
	Earth Rotation Parameters		
Rapid polar motion	0.2 mas		
Polar motion rates	0.4 mas/day	17 hours	daily
Length-of-day	0.030 ms		
Final polar motion	0.1 mas		
Polar motion rates	0.2 mas/day	~13 days	weekly
Length-of-day	0.020 ms		
	Atmospheric Parameters		
Final tropospheric	4 mm zenith path delay	< 4 weeks	weekly
Ionospheric TEC grid	Under development		

Source: IGS (2002).

7.2.1 Phase Windup Correction

One must go back to the electromagnetic nature of GPS transmissions in order to understand this correction. In short, the GPS carrier waves are right circularly polarized (RCP). The electromagnetic wave may be visualized as a rotating electric vector field that propagates from the satellite antenna to the receiver antenna. The vector rotates 360° every spatial wavelength or every temporal cycle of the wave. The observed carrier phase can be viewed as the geometric angle between the instantaneous electric field vector at the receiving antenna and some reference direction on the antenna. As the receiving antenna rotates in azimuth, this measured phase changes. The same is true if the transmitting antenna changes its orientation with respect to the receiver antenna. Since the phase is measured in the plane of the receiving antenna, its value depends on the direction of the line of sight to the satellite, in addition to the orientation of the antenna.

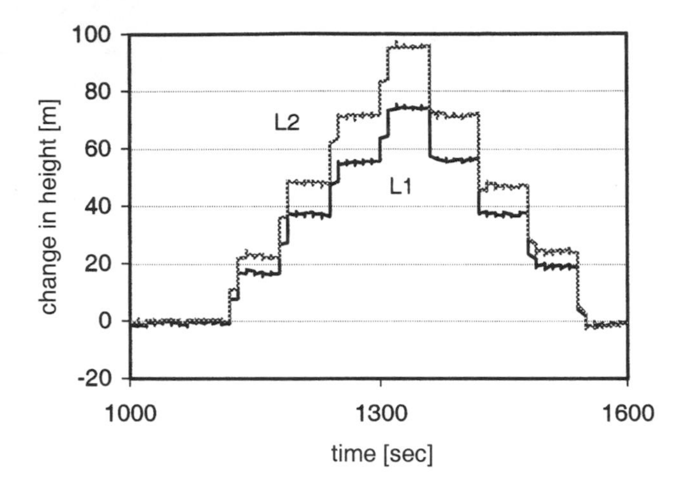

Figure 7.2 Antenna rotation test. (Data from R. J. Muellerschoen, JPL.)

Figure 7.2 shows the results of a simple test to demonstrate RCP of GPS signals. Two antennas, about 5 m apart, were connected to the same receiver and oscillator. Observations were recorded once per second for half an hour. One of the antennas was rotated 360° in azimuth four times clockwise (as viewed looking down on the ground plate), with 1 minute between the rotations, and then four times rotated counterclockwise, again with 1 minute between the rotations. The carrier phase observations were differenced and a linear trend was removed to account for the phase biases and a differential rate (caused by the separation of the antennas). Figure 7.2 shows the change in the single differences for both L1 and L2. Each complete antenna rotation in azimuth causes a change of 1 wavelength.

An introductory discussion of the carrier phase windup correction for rotating GPS antennas is found in Tetewsky and Mullen (1997). Wu et al. (1993) derived the phase windup correction expressions for a crossed dipole antenna, but their results are applicable to cases that are more general. Following their derivations, at a given instant the windup correction is expressed as a function of the directions of the dipoles and of the line of sight to the satellite.

Let $\hat{\mathbf{x}}$ and $\hat{\mathbf{y}}$ denote the unit vectors in the direction of the two-dipole elements in the receiving antenna in which the signal from the y-dipole element is delayed by 90° relative to that from the x-dipole element. $\mathbf{k}$ is the unit vector pointing from the satellite to the receiver. We consider a similar definition for $\hat{\mathbf{x}}'$ and $\hat{\mathbf{y}}'$ at the satellite, i.e., the current in the y'-dipole lags that in the x'-dipole by 90°. They define the effective dipole that represents the resultant of a crossed dipole antenna for the receiver and the transmitter, respectively,

$$\mathbf{d} = \hat{\mathbf{x}} - \mathbf{k}\,(\mathbf{k} \cdot \hat{\mathbf{x}}) + \mathbf{k} \times \hat{\mathbf{y}} \tag{7.1}$$

$$\mathbf{d}' = \hat{\mathbf{x}}' - \mathbf{k}\left(\mathbf{k} \cdot \hat{\mathbf{x}}'\right) - \mathbf{k} \times \hat{\mathbf{y}}' \tag{7.2}$$

The windup correction is (Wu et al., 1993, p. 95)

$$\delta\varphi = \text{sign}\left[\mathbf{k} \cdot \left(\mathbf{d}' \times \mathbf{d}\right)\right] \cos^{-1}\left(\frac{\mathbf{d}' \cdot \mathbf{d}}{\|\mathbf{d}'\| \; \|\mathbf{d}\|}\right) \tag{7.3}$$

At a given instant in time, the windup correction $\delta\varphi$ cannot be separated from the undifferenced ambiguities, nor is it absorbed by the receiver clock error because it is a function of the receiver and the satellite. In practical applications it is therefore sufficient to interpret $\hat{\mathbf{x}}$ and $\hat{\mathbf{y}}$ as unit vectors along northing and easting and $\hat{\mathbf{x}}'$ and $\hat{\mathbf{y}}'$ as unit vectors in the satellite body coordinate system. Any additional windup error resulting from this redefinition of the coordinate system will also be absorbed by the undifferenced ambiguities. Taken over time, however, the values of $\delta\varphi$ reflect the change in orientation of receiver and satellite antennas.

The value of the windup correction for single and double differences has an interesting connection to spherical trigonometry. Consider a spherical triangle whose vertices are given by the latitudes and longitudes of the receivers k and m, and the satellite. In addition, we assume that GPS transmitting antennas are pointing toward the center of the earth and that the ground receiver antennas are pointing upward. This assumption is usually met in the real world. It can be shown that single difference windup correction $\delta\varphi_{km}^{p} = \delta\varphi_{k}^{p} - \delta\varphi_{m}^{p}$ is equal to the spherical excess if the satellite appears on the left as viewed from station k to station m, and it equals the negative spherical excess if the satellite appears to the right. The double-differencing windup correction $\delta\varphi_{km}^{pq}$ equals the spherical excess of the respective quadrilateral. The sign of the correction depends on orientation of the satellite with respect to the baseline. For details, refer to Wu et al. (1993).

The windup correction can be neglected for short baselines because the spherical excess of the respective triangles is small. Neglecting the windup correction might cause problems when fixing the double-difference ambiguities, in particular for longer lines. The float ambiguities absorb the constant part of the windup correction. The variation of the windup correction over time might not be negligible in float solutions of long baselines.

There is no windup-type correction for the pseudoranges. Consider the simple case of a rotating antenna that is at a constant distance from the transmitting source and the antenna plane perpendicular to the direction of the transmitting source. Although the measured phase would change due to the rotation of the antenna the pseudorange will not change because the distance is constant.

7.2.2 Satellite Antenna Phase Center Offset

The satellite antenna phase center offsets are usually given in the satellite-fixed coordinate system (x′) that is also used to express solar radiation pressure (see Section 3.1.4.3). The origin of this coordinate system is at the satellite's center of mass. If $\mathbf{e}$ denotes the unit vector pointing to the sun, expressed in the ECEF coordinate system (x), then the axes of (x′) are defined by the unit vector $\mathbf{k}$ (pointing from the satellite toward the earth's center), the vector $\mathbf{j} = \mathbf{k} \times \mathbf{e}$ (pointing along the solar panel axis), and the unit vector $\mathbf{i}$ that completes the right-handed coordinate system (also located in the sun-satellite-earth plane). For example, the offsets adopted for Block II/IIA satellites are $\mathbf{x}' = [0.279 \quad 0 \quad 1.023]^T$ meters. It can readily be verified that

$$\mathbf{x}_{sa} = \mathbf{x}_{sc} + [\,\mathbf{i} \quad \mathbf{j} \quad \mathbf{k}\,]^{-1}\,\mathbf{x}' \tag{7.4}$$

where $\mathbf{x}_{sa}$ is the position of the satellite antenna and $\mathbf{x}_{sc}$ denotes the position of the satellite's center of mass.

The satellite phase center offsets must be determined for each satellite type. When estimating the offsets from observations while the satellite is in orbit, the effect of the offsets might be absorbed, at least in part, by other parameters. An example is the offset in direction $\mathbf{k}$ and the receiver clock error. Mader and Czopek (2001) report on an effort to calibrate the phase center of the satellite antenna for a Block IIA antenna using ground measurements.

7.2.3 Receiver Antenna Phase Center Offset and Variation

The immediate reference point in positioning with GPS is the phase center of the receiver antenna. Since the phase center cannot be accessed directly with tape we need to know the relationship between the phase center and an external antenna reference point (ARP) in order to relate the GPS-determined positions to a surveying monument. Unfortunately, the phase center is not well defined. Its location varies with the elevation angle of the arriving signal, i.e., the direction of the satellite. For some antennas it also depends, although slightly, on the azimuth. The relationship between the ARP and the phase center, which is the object of antenna calibration, is usually parameterized in term of phase offset (PO) and phase center variation (PCV). The largest offset is in height, which can be as much as 10 cm. The PO and the PCV also depend on the frequency.

Imagine a perfect antenna that has an ARP and a phase center offset that is well known. Imagine further that you connect a "phase meter" to the antenna and that you move a transmitter along the surface of a sphere that is centered on the phase center. In this ideal case, since the distance from the transmitter to the phase center never changes, the output phase will always read a constant amount. In actuality, there is no perfect antenna, and that situation can never be realized. Instead, one effectively moves a source along a sphere centered on a point that one selects as an average phase center. Now instead of recording a constant phase, one detects phase variations, primarily as a function of elevation. Since the distance from source to antenna is constant, these phase variations must be removed so that constant geometric distance is represented by constant phase measurements. Had we picked another phase center, we would get another set of phase variations. That is why the PO and PCVs must be used together and why different POs and PCVs sets will lead one back to the same ARP.

For a long observation series one might hope that the average location of the PCV is well defined and that the position refers to the average phase center. For RTK applications there is certainly no such averaging possible. For short baselines where the antennas at the ends of the line see a satellite at approximately the same elevation angle, orienting both antennas in the same direction can largely eliminate the PO and PCV. This elimination procedure works only for the same antenna types. For large baselines or when mixing antenna types, the antenna calibration is necessary and corrections must be applied. Antenna calibration is also important when estimating tropospheric parameters, since both the PCV and the tropospheric delay depend on the elevation angle.

The NGS (Mader, 1999) has developed procedures for relative antenna calibration using field observations. All test antennas are calibrated with respect to the same reference antenna, which happens to be an AOAD/M_T choke ring antenna. The basic idea is that if the same reference antenna is always used for all calibrations, the PO and PCV of the reference antenna cancel when double-differencing observations of a new baseline and applying the calibrated PO and PCV to both antennas. This technique is accurate as long as the elevation difference of a satellite, as seen from both antennas, is negligible in terms of the PCV (which is parameterized as a function of the elevation angle). Since the PCV amounts to about only 1–2 cm and varies smoothly with elevation angle, relative phase calibration is applicable to baselines of several thousand kilometers in length. NGS uses a calibration baseline of 5 m. The reference antenna and the test antenna are connected to the same type of receiver. Both receivers use the same rubidium oscillator as an external frequency standard. Because the test baseline is known, a common frequency standard is used and because the tropospheric and ionospheric effects cancel over such a short baseline, the single-difference discrepancies over time are very flat and can be modeled as

$$\left(\varphi_{12,b}^{p} - \varphi_{12,0}^{p}\right)_i = \tau_i + \alpha_1 \beta_i^p + \alpha_2 \left(\beta_i^p\right)^2 + \alpha_3 \left(\beta_i^p\right)^3 + \alpha_4 \left(\beta_i^p\right)^4 \tag{7.5}$$

The subscript i denotes the epoch, the superscript p identifies the satellite having elevation angle β_i, and τ_i is the remaining relative time delay (receiver clock error). The coefficients α_1 to α_4 and τ_i are estimated by observing all satellites from rising to setting. The result of the relative calibration of the test antenna is then given by

$$\hat{\varphi}_{\text{antenna,PCV}}(\beta) = \hat{\alpha}_1 \beta + \hat{\alpha}_2 \beta^2 + \hat{\alpha}_3 \beta^3 + \hat{\alpha}_4 \beta^4 + \xi \tag{7.6}$$

The symbol ξ denotes a translation such that $\hat{\varphi}_{\text{antenna,PCV}}(90°) = 0$. The remaining clock difference estimate $\hat{\tau}$ is not included in (7.6), Both $\hat{\tau}$ and ξ cancel in double differencing. Recall that the NGS calibration is relative and therefore (7.6) must be applied in the double-differencing mode. An example of relative PCV is seen in Figure 7.3. The vertical axis shows the difference $\left(\varphi_{12,b}^{p} - \varphi_{12,0}^{p}\right)_i - \hat{\tau}_i$ for all satellites. The multipath is clearly visible in this figure; it has a much higher frequency than the PCV and therefore does not affect the polynomial estimation. Equation (7.6) is used together with the PO that NGS derives from 24-hour data sets. Again, it is sufficient to define the PO of the reference antenna because the calibrated POs are used in the relative mode.

Automated absolute field calibration of GPS antennas in real time is discussed in Wübbena et al. (2000), Schmitz et al. (2002), and references listed therein. They use a robotic arm to determine the absolute PO and PCV as a function of elevation

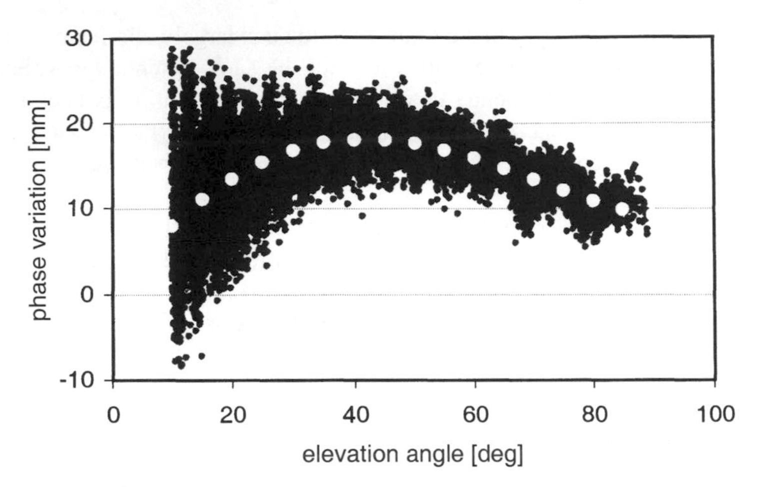

Figure 7.3 Relative vertical antenna phase variation calibration at NGS. (Data Source: Gerry Mader, NGS.)

and azimuth. This real-time calibration uses undifferenced observations from the test antenna that are differenced over very short time intervals. The intervals are sufficiently short so that multipath is eliminated in the differencing. This calibration technique therefore becomes site-independent. Rapid changes of orientation of the calibration robot allow the separation of PCV and any residual multipath effects. Several thousand observations are taken at different robot positions. The calibration takes only a few hours. Figure 7.4 shows the results of the absolute calibration for the AOAD/M_T choke ring antenna.

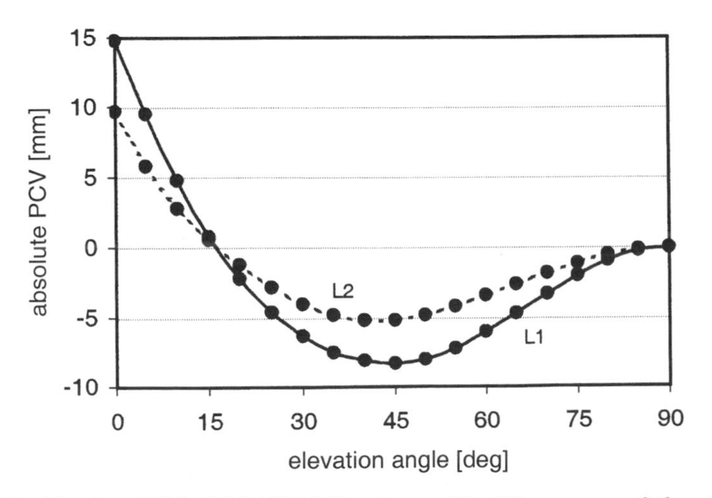

Figure 7.4 Absolute PCV of AOAD/M_T antenna. The PO are $n_{PO} = 0.6$, $e_{PO} = -0.5$, $u_{PO} = 91.2$, and $(-0.1, -0.6, 120.1)$ millimeters for L1 and L2, respectively. (Data from NGS, 2002.)

There are other approaches available for absolute antenna calibration. For example, Schupler and Clark (2001) mount the antenna on a platform that allows it to be rotated in elevation and azimuth and then place the whole device in an anechoic chamber. The interior of the chamber is lined with radiofrequency absorbent material that reduces signal reflections or "echoes" to a minimum. A signal source antenna generates the signals. Since the source antenna can transmit at different frequencies, these anechoic chamber techniques are suitable for studying the frequency dependency of PO and PCV for L1, L2, and other frequencies.

The PO can be dealt with like an eccentricity offset at the station in order to reference it to the surveying monument. Since the up component u_{PO} is the largest one it might be sufficient to correct the carrier phase by

$$\Delta\varphi_{\mathrm{PO}} = \lambda^{-1} u_{\mathrm{PO}} \sin\theta \tag{7.7}$$

where λ denotes the wavelength.

7.2.4 Multipath

Once the satellite signals reach the earth's surface, ideally they enter the antenna directly. However, objects in the receiver's vicinity may reflect some signals before they enter the antenna, causing unwanted signatures in pseudorange and carrier phase observations. Although the direct and reflected signals have a common emission time at the satellite, the reflected signals are always delayed relative to the line-of-sight signals because they travel longer paths. The amplitude (voltage) of the reflected signal is always reduced because of attenuation. The attenuation depends on the properties of the reflector material, the incident angle of the reflection, and the polarization. In general, reflections with very low incident angle have little attenuation. In addition, the impact of multipath on the GPS observables depends on the sensitivity of the antenna in terms of sensing signals from different directions, and the receiver's internal processing to mitigate multipath effects. Multipath is still one of the dominating, if not the dominant, sources of error in GPS positioning.

Signals can be reflected at the satellite (satellite multipath) or in the surroundings of the receiver (receiver multipath). Satellite multipath is likely to cancel in the single-difference observables for short baselines. Reflective objects for receivers on the ground can be the earth's surface itself (ground and water), buildings, trees, hills, etc. Rooftops are known to be bad multipath environments because there are often many vents and other reflective objects within the antenna's field of view.

The impact of multipath on the carrier phases can be demonstrated using a planar vertical reflection surface at distance d from the antenna (Georgiadou and Kleusberg, 1988; Bishop et al., 1985). The geometry is shown in Figure 7.5. We write the direct line-of-sight carrier phase observable for receiver k and satellite p as

$$S_D = A\cos\varphi \tag{7.8}$$

In Equation (7.8) we do not use the subscript k and superscript p in order to simplify the notation. The symbols A and φ denote the amplitude (signal voltage) and the phase, respectively. The reflected signal is written as

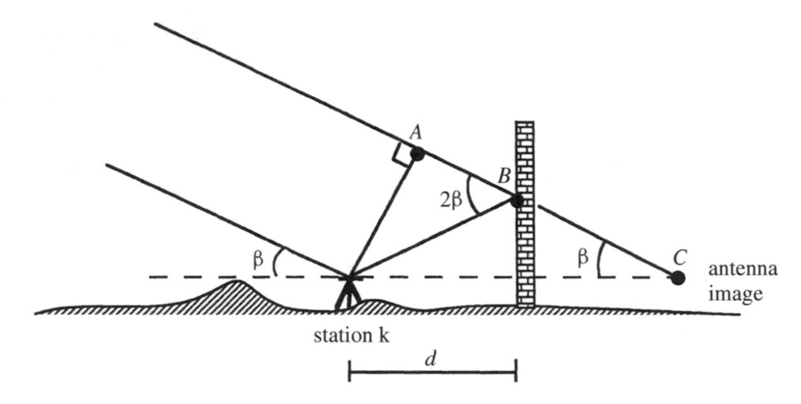

Figure 7.5 Geometry for reflection on a vertical planar plane.

$$S_R = \alpha\, A \cos(\varphi + \theta), \qquad 0 \le \alpha \le 1 \tag{7.9}$$

The amplitude reduction factor (attenuation) is $\alpha = A'/A$, where A' is the amplitude of the reflected signal. The total multipath phase shift is

$$\theta = f\ \Delta\tau + \phi \tag{7.10}$$

where f is the frequency, $\Delta\tau$ is the time delay, and ϕ is the fractional shift. The multipath delay shown in Figure 7.5 is the sum of the distances *AB* and *BC*, which equals $2d \cos\beta$. Converting this distance into cycles and then to radians gives

$$\theta = \frac{4\pi d}{\lambda} \cos\beta + \phi \tag{7.11}$$

where λ is the carrier wavelength. The composite signal at the antenna is the sum of the direct and reflected signal,

$$S = S_D + S_R = R\cos(\varphi + \psi) \tag{7.12}$$

It can be verified that resultant carrier phase voltage $R(A, \alpha, \theta)$ and the carrier phase multipath delay $\psi(\alpha, \theta)$ are

$$R(A, \alpha, \theta) = A\left(1 + 2\alpha\cos\theta + \alpha^2\right)^{1/2} \tag{7.13}$$

$$\psi(\alpha, \theta) = \tan^{-1}\left(\frac{\alpha \sin\theta}{1 + \alpha\cos\theta}\right) \tag{7.14}$$

Regarding notation, we used the symbols $d_{k,1}^p$ and $d_{k,2}^p$ in Chapter 5 to denote the total multipath, i.e., the multipath effect of all reflections on L1 and L2, respectively. If we consider the case of constant reflectivity, i.e., α is constant, the maximum path delay is found when $\partial\psi/\partial\theta = 0$. This occurs at $\theta(\psi_{\text{max}}) = \pm\cos^{-1}(-\alpha)$, the maximum value being $\psi_{\text{max}} = \pm\sin^{-1}\alpha$. The maximum multipath carrier phase error is only a function of the amplitude attenuation α in this particular case. The largest value is $\pm 90^\circ$ and occurs for $\alpha = 1$. This maximum corresponds to $\lambda/4$. If $\alpha \ll 1$ then ψ can be approximated by $\alpha \sin\theta$.

The multipath effect on pseudoranges depends among other things on the chipping rate T of the codes and the receiver's internal sampling interval S. A necessary step for each receiver is to correlate the received signal with an internally generated code replica. The offset in time that maximizes the correlation is a measure of the pseudorange. Avoiding the technical details, suffice it to say that time-shifting the internal code replica and determining the correlation for early, prompt, and late delays eventually determines the offset. The early and late delays differ from the prompt delay by $-S$ and S, respectively. When the early minus late correlation are zero, i.e., they have the same amplitude, the prompt delay is used as a measure of the pseudorange. Consult Kaplan (1996, p. 148) for additional details on the topic of code tracking loops and correlation. For a single multipath signal, the correlation function consists of the sum of two triangles, one for the direct signal and one for the multipath signal. This is conceptually demonstrated in Figure 7.6. The solid thin line and the dashed line represent the correlation functions of the direct and multipath signals, respectively. The thick solid line indicates the combined correlation function, i.e., the sum of the thin line and dashed line. The left figure refers to destructive reflection when the reflected signal arrives out of phase with respect to the direct signal. The right figure refers to constructive reflection when the reflected and direct signals are in phase. Let the combined signal be sampled at the early and late delays. The figure shows that the prompt delay would coincide with the maximum correlation for the direct signal and indicate the correct pseudorange but will be in error by the multipath-induced range error q for the combined signal. The resulting pseudorange measurement errors are negative for destructive reflection and positive for constructive reflection even though the reflected signal always arrives later than the direct one.

The pseudorange multipath error further depends on whether the sampling interval is greater or smaller than half the chipping period. Byun et al. (2002) provide the following expressions. If $S > T/2$ (wide sampling) then

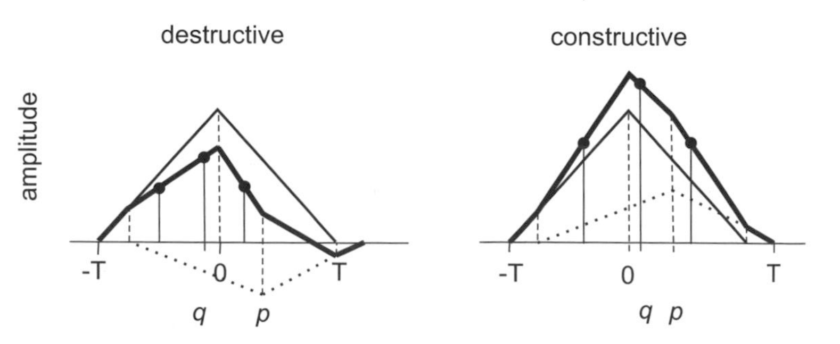

Figure 7.6 Correlation function in the presence of multipath. p denotes the time delay of the multipath signal and q is the multipath induced pseudorange error.

$$\Delta\tau_P = \begin{cases} \dfrac{\Delta\tau\,\alpha\cos(f\,\Delta\tau+\phi)}{1+\alpha\cos(f\,\Delta\tau+\phi)} & \text{if} \quad \Delta\tau < T - S + \Delta\tau_P \\[2ex] \dfrac{(T-S+\Delta\tau)\,\alpha\cos(f\,\Delta\tau+\phi)}{2+\alpha\cos(f\,\Delta\tau+\phi)} & \text{if} \quad T - S + \Delta\tau_P < \Delta\tau < S + \Delta\tau_P \\[2ex] \dfrac{(T+S-\Delta\tau)\,\alpha\cos(f\,\Delta\tau+\phi)}{2-\alpha\cos(f\,\Delta\tau+\phi)} & \text{if} \quad S + \Delta\tau_P < \Delta\tau < T + S + \Delta\tau_P \\[2ex] 0 & \text{if} \quad \Delta\tau > T + S + \Delta\tau_P \end{cases} \tag{7.15}$$

and if $S < T/2$ (narrow sampling) then

$$\Delta\tau_P = \begin{cases} \dfrac{\Delta\tau\,\alpha\cos(f\,\Delta\tau+\phi)}{1+\alpha\cos(f\,\Delta\tau+\phi)} & \text{if} \quad \Delta\tau < S + \Delta\tau_P \\[2ex] S\,\alpha\cos(f\,\Delta\tau+\phi) & \text{if} \quad S + \Delta\tau_P < \Delta\tau < T - S + \Delta\tau_P \\[2ex] \dfrac{(T+S-\Delta\tau)\,\alpha\cos(f\,\Delta\tau+\phi)}{2-\alpha\cos(f\,\Delta\tau+\phi)} & \text{if} \quad T - S + \Delta\tau_P < \Delta\tau < T + S \\[2ex] 0 & \text{if} \quad \Delta\tau > T + S \end{cases} \tag{7.16}$$

The pseudorange multipath error is $d_P = c\ \Delta\tau_P$, and $\Delta\tau$ denotes the time delay of the multipath signal. The expressions are valid for the P-codes and the C/A-code as long as the appropriate chipping period T is used.

Figure 7.7 shows an example of the envelope for the P1-code multipath range error $\Delta\tau_{P1}$ oscillations versus time delay $\Delta\tau$ for the wide-sampling case $S > T/2$. As the phase varies by π the multipath error changes from upper to lower bounds and vice versa. The distinct regions of Equation (7.15) are readily visible in the figure. Figure 7.8 shows an example of the C/A-code multipath range error for the narrow-sampling case $S < T/2$. The main difference between the wide and narrow sampling interval is that the latter has a constant peak at region 2. In fact, shortening the sampling interval S has long been recognized as a means to reduce the pseudorange multipath error.

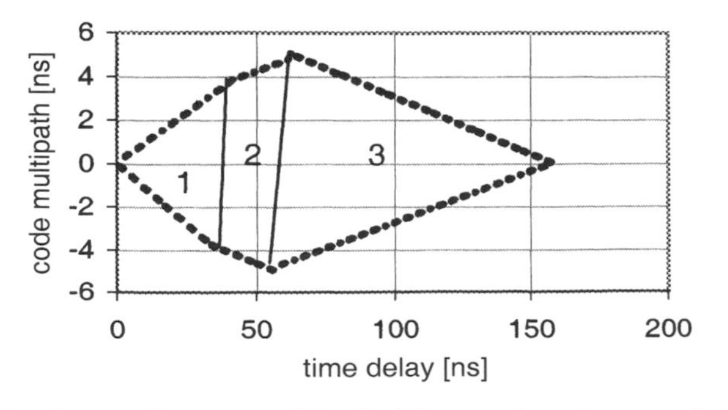

Figure 7.7 P1-code pseudorange multipath delay envelope in case of wide sampling. ($T = 98$ ns, $S = 60$ ns, $\alpha_1 = 0.1$, $\phi_1 = 0$)

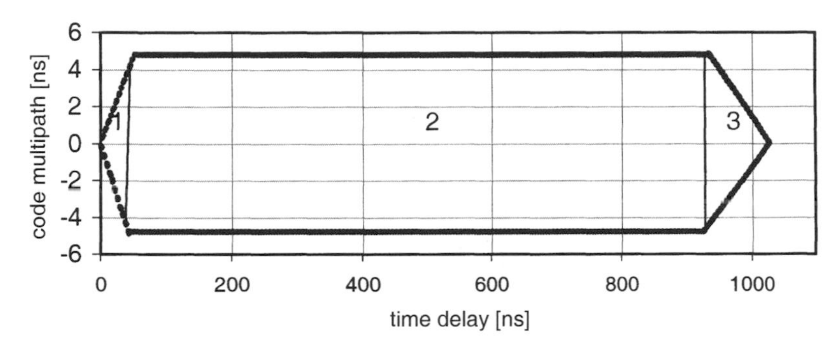

Figure 7.8 C/A-code pseudorange multipath delay envelope in case of narrow sampling. ($T = 980$ ns, $S = 48$ ns; $\alpha_1 = 0.1$, $\phi_1 = 0$)

See second component of Equation (7.16), where S appears as a factor. Comparing (7.15) and (7.16), we find that in region 1 the slopes of the envelopes are the same for wide and narrow correlating. Narrow correlation causes the bounds in region 2 to be smaller. Region 4, for which the multipath error is zero, is reached earlier the narrower the sampling (given the same chipping rate). The lower envelope in these figures corresponds to destructive reflection while the upper envelope refers to constructive reflection.

The multipath frequency f_ψ depends on the variation of the phase delay θ, as can be seen from (7.9), (7.14), (7.15), or (7.16). Differentiating (7.11) gives the expression for the multipath frequency,

$$f_\psi = \frac{1}{2\pi}\frac{d\theta}{dt} = \frac{2d}{\lambda}\sin\beta\,|\dot{\beta}| \tag{7.17}$$

The multipath frequency is a function of the elevation angle and is proportional to the distance d and the carrier frequency. For example, if we take $\dot{\beta} = 0.07$ mrad/sec (= one-half of the satellite's mean motion) and $\beta = 45°$, then the multipath period is about 5 minutes if $d = 10$ m and about 50 minutes if $d = 1$ m. The variation in satellite elevation angle causes the multipath frequency to become a function of time. According to (7.17), the ratio of the multipath frequencies for L1 and L2 equals that of the carrier frequencies, $f_{\psi,1}/f_{\psi,2} = f_1/f_2$.

As an example of a carrier phase multipath, consider a single multipath signal and the ionospheric phase observable (6.98). The effect of the multipath for this function is

$$\begin{aligned}\varphi_{MP} &\equiv \psi_1 - \frac{f_1}{f_2}\psi_2 \\ &= \tan^{-1}\left(\frac{\alpha\sin\theta_1}{1+\alpha\cos\theta_1}\right) - \frac{f_1}{f_2}\tan^{-1}\left(\frac{\alpha\sin\theta_2}{1+\alpha\cos\theta_2}\right)\end{aligned} \tag{7.18}$$

Figure 7.9 shows that the multipath φ_{MP} impacts the ionospheric observable in a complicated manner. The amplitude of the cyclic phase variations is nearly proportional to

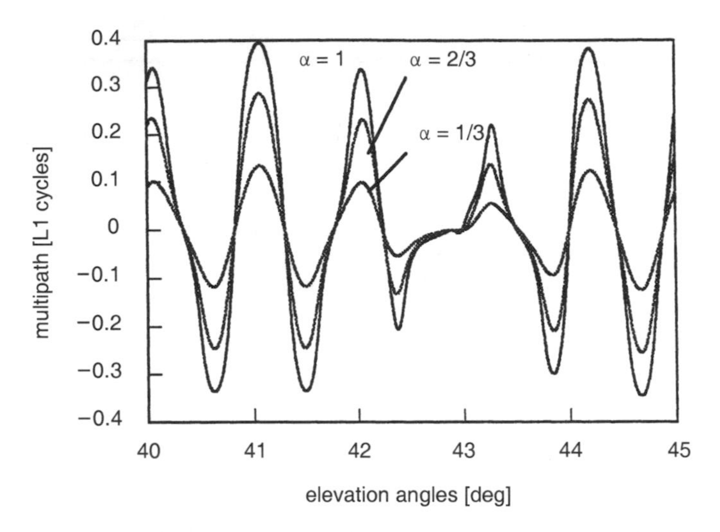

Figure 7.9 Example of multipath on the ionospheric carrier phase observable from a vertical planar surface. ($d = 10$ m, $\phi_1 = \phi_2 = 0$)

α. When analyzing the ionospheric observable in order to map the temporal variation of the ionospheric delay, the multipath signature (7.18) cannot be ignored. In fact, the multipath variation of (7.14) might occasionally impact our ability to fix the integer ambiguities, even for short baselines.

Figure 7.10 shows the effects of multipath on the pseudoranges P1 and P2, and the ionospheric free function (6.91). We are using the expression for region 1 in (7.15) or (7.16), since we consider the case of a nearby reflection. The time delay $\Delta\tau$ is a function of the satellite elevation angle and can be computed from (7.11). The figures show the multipath for a satellite that rises ($\beta = 0°$) until it passes overhead

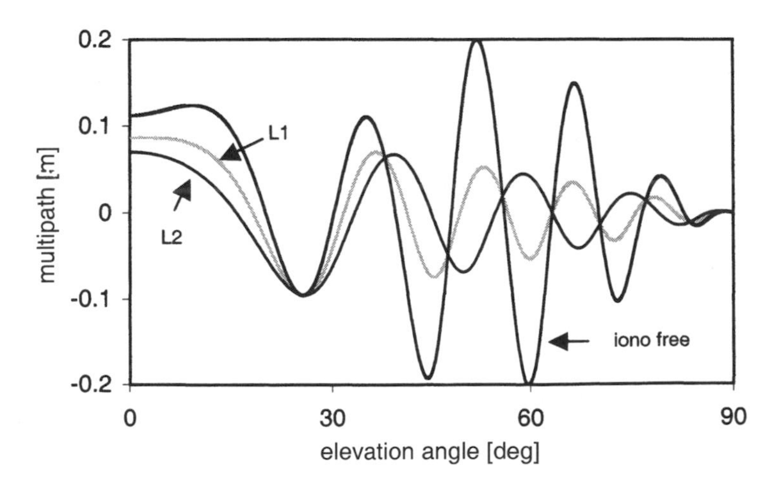

Figure 7.10 Pseudorange multipath from a single reflection on a vertical planar surface. ($\alpha = 0.1, d = 5\,\lambda_1, \phi_1 = \phi_2 = 0$)

($\beta = 90°$). The multipath is largest for a satellite in the horizon (reflection on vertical surface). In the case of reflection from a horizontal surface, the multipath has a reverse dependency, i.e., it is largest for satellites at the zenith, as can readily be verified.

Fenton et al. (1991) discuss one of the early implementations of narrow correlation in C/A-code receivers. Narrow correlator technology and on-receiver processing methods to reduce carrier phase and pseudorange multipath effects are extensively documented in the literature, e.g., van Dierendonck et al. (1992), Meehan and Young (1992), Veitsel et al. (1998), and Zhdanov et al. (2001). If the phase shift θ changes rapidly, one might even attempt to average the pseudorange measurements. In addition to sophisticated on-receiver signal processing, there are several external ways to mitigate multipath.

- Since multipath can also arrive from below the antenna (due to edge diffraction), a ground plate is helpful. The ground plate is usually a metallic surface of circular or rectangular form.
- Partial multipath rejection can be achieved by shaping the gain pattern of the antenna. Since a lot of multipath arrives from reflections near the horizon, multipath may be sharply reduced by using antennas having low gain in these directions.
- Improved multipath resistance is achieved with choke rings. These are metallic circular grooves with quarter-wavelength depth.
- Highly reflective surfaces change the polarization from right-hand circular (signal received directly from the GPS satellite) to left-hand circular. GPS antennas that are designed to receive right-hand polarized signals will attenuate signals of opposite polarization.
- Arrays of antennas can also be used to mitigate multipath. Due to a different multipath geometry, each antenna sees the multipath effect differently. Combined processing of signals from all antennas allows multipath mitigation (Fu et al., 2003). In a design proposed by Counselman the antenna elements are arranged along the vertical rather than the horizontal platter (Counselman, 1999).
- Since the geometry between a GPS satellite and a receiver-reflector repeats every sidereal day, multipath shows the same pattern between consecutive days. Such repetition is useful to verify the presence of multipath by analyzing the repeatability patterns and eventually model the multipath at the station. In relative positioning the double-difference observable is affected by multipath at both stations.

In practical applications, of course, the various satellite signals are reflected at different objects. The attenuation properties of these objects generally vary; in some cases attenuation might even depend on time. Since the angle of incident also affects attenuation, it can readily be appreciated that the multipath is a difficult error source to deal with. It is common practice not to observe satellites close to the horizon in order reduce multipath.

Equations (6.100) and (6.101) are useful to gage the multipath, in particular the multipath effect on the pseudoranges, if dual-frequency observations are available.

7.3 GEOMETRY-FREE SOLUTIONS

The undifferenced pseudorange (5.7) or (5.72) and carrier phase (5.10) equations make up the epoch solution. Using the short notation, which neglects the subscript for station k, the superscript for satellite p, and the epoch designator t, the geometry-free epoch solution using dual-frequency pseudoranges and carrier phases is written in the following form:

$$\begin{bmatrix} P_1 - cT_{GD} \\ P_2 - \alpha_f cT_{GD} \\ \Phi_1 \\ \Phi_2 \end{bmatrix} = \begin{bmatrix} 1 & 1 & 0 & 0 \\ 1 & \alpha_f & 0 & 0 \\ 1 & -1 & \lambda_1 & 0 \\ 1 & -\alpha_f & 0 & \lambda_2 \end{bmatrix} \begin{bmatrix} \rho + \Delta \\ I_{1,P} \\ N_1 \\ N_2 \end{bmatrix} + \begin{bmatrix} \delta_{1,P} \\ \delta_{2,P} \\ \delta_{1,\Phi} \\ \delta_{2,\Phi} \end{bmatrix} + \begin{bmatrix} \varepsilon_{1,P} \\ \varepsilon_{2,P} \\ \varepsilon_{1,\Phi} \\ \varepsilon_{2,\Phi} \end{bmatrix} \tag{7.19}$$

$$\Delta = -c\, d\underline{t} + c\, d\bar{t} + T \tag{7.20}$$

The carrier phases have been expressed in terms of distance values Φ_1 and Φ_2. In the usual notation, the interfrequency code offset at the satellite is T_{GD}, and ρ denotes the geometric topocentric distance from the receiver antenna to the satellite at the instant of signal transmission. The auxiliary parameter Δ combines the receiver clock correction $d\underline{t}$, satellite clock correction $d\bar{t}$, and the tropospheric delay T. Other parameters are the ionospheric delay $I_{1,P}$, and the ambiguities N_1 and N_2. The factor α_f is given in (5.13). The δ terms represent the hardware delays at the receiver and satellite and the signal multipath, and the epsilons are the noise. We can write (7.19) in matrix notation,

$$\boldsymbol{\ell}_b = \mathbf{Ax} + \boldsymbol{\delta} + \boldsymbol{\varepsilon} \tag{7.21}$$

Because the $\mathbf{A}$ matrix contains constants that do not depend on the receiver-satellite geometry, (7.19) is called the geometry-free model and is valid for static or moving receivers. Whereas the parameters $\rho + \Delta$ and $I_{1,P}$ change with time, the ambiguity parameters are constant unless there are cycle slips. The parameters can be estimated using least-squares or Kalman filtering. For example, Goad (1990) and Euler and Goad (1991) use the geometry-free model to study optimal filtering for the combined pseudorange and carrier phase observations for single and dual frequency. Neglecting the δ terms and the observational noise, the epoch solution for the parameters is

$$\mathbf{x} = \mathbf{A}^{-1}\boldsymbol{\ell}_b \tag{7.22}$$

$$\mathbf{A}^{-1} = \begin{bmatrix} \beta_f & -\gamma_f & 0 & 0 \\ -\gamma_f & \gamma_f & 0 & 0 \\ -\lambda_1^{-1}(\beta_f + \gamma_f) & 2\lambda_1^{-1}\gamma_f & \lambda_1^{-1}(\beta_f - \gamma_f) & 0 \\ -2\lambda_2^{-1}\beta_f & \lambda_2^{-1}(\beta_f + \gamma_f) & 0 & \lambda_2^{-1}(\beta_f - \gamma_f) \end{bmatrix} \tag{7.23}$$

The coefficients β_f and γ_f are given by (5.14) and (5.15). The geometry of the epoch solution is implicit in the covariance matrix. For computing the epoch covariance matrix, we assume that the observations $\boldsymbol{\ell}_b$ are not correlated. We further assume that the standard deviation of the carrier phases $\sigma_{1,\varphi}$ and $\sigma_{2,\varphi}$ are related as, $\sigma_{2,\Phi} = \sigma_{1,\Phi}\sqrt{\alpha_f}$, and that the standard deviations of the pseudorange and the linear carrier phases follow $k = \sigma_P/\sigma_\Phi$ for both frequencies. With these assumptions the covariance matrix $\boldsymbol{\Sigma}_{\ell_b}$ consists of diagonal elements k^2, $\alpha_f k^2$, 1, α_f, and a scalar σ_Φ^2. Applying the law of variance-covariance propagation (4.34), the covariance matrix of the parameters becomes

$$\boldsymbol{\Sigma}_x = \mathbf{A}^{-1}\,\boldsymbol{\Sigma}_{\ell_b}\,\left(\mathbf{A}^{-1}\right)^{\mathrm{T}} \tag{7.24}$$

If we set k equal to 154, which corresponds to the ratio of the L1 frequency and the P-code chipping rate, and use $\sigma_{1,\Phi} = 0.002$ m, then the standard deviations and the correlation matrix are

$$\left(\sigma_{\rho+\Delta}, \sigma_I, \sigma_{1,N}, \sigma_{2,N}\right) = (0.99\text{ m}, 0.77\text{ m}, 9.22\text{ cycL}_1, 9.22\text{ cycL}_2) \tag{7.25}$$

$$\mathbf{C}_x = \begin{bmatrix} 1 & -0.9697 & -0.9942 & -0.9904 \\ & 1 & 0.9904 & 0.9942 \\ & & 1 & 0.9995 \\ & & & 1 \end{bmatrix} \tag{7.26}$$

The standard deviations for the integer ambiguities have been converted to cycles in (7.25). Striking features of the epoch solution are the equality of the standard deviation for both ambiguities and the high correlation between all parameters. Of particular interest is the shape and orientation of the ellipse of standard deviation for the ambiguities. The general expressions (4.304) to (4.308) can be applied to the third and fourth parameters. The ellipse can be drawn with respect to the perpendicular N_1 and N_2 axes, which carry the units L1 cycles and L2 cycles. The computations show that the ellipse almost degenerates into a straight line with an azimuth of 45°, the semiminor and semimajor axes being 0.20 and 13.04, respectively.

A standard procedure for breaking high correlation is reparameterization by means of an appropriate transformation. For example, consider the transformation

$$\mathbf{z} = \mathbf{Z}\mathbf{x} \tag{7.27}$$

$$\mathbf{Z} = \begin{bmatrix} 1 & 0 & 0 & 0 \\ 0 & 1 & 0 & 0 \\ 0 & 0 & 1 & -1 \\ 0 & 0 & 1 & 0 \end{bmatrix} \tag{7.28}$$

which yields the following parameters and covariance matrix

$$\mathbf{z} = [\,\rho + \Delta \quad I_{1,P} \quad N_w \equiv N_1 - N_2 \quad N_1\,]^{\mathrm{T}} \tag{7.29}$$

$$\boldsymbol{\Sigma}_z = \mathbf{Z}\,\boldsymbol{\Sigma}_x\,\mathbf{Z}^{\mathrm{T}} \tag{7.30}$$

Using again the numerical values $k = 154$ and $\sigma_{1,\Phi} = 0.002$ m, the standard deviations and the correlation matrix become

$$\left(\sigma_{\rho+\Delta}, \sigma_I, \sigma_w, \sigma_{1,N}\right) = \left(0.99\,\mathrm{m}, 0.77\,\mathrm{m}, 0.28\,\mathrm{cycL_w}, 9.22\,\mathrm{cycL_1}\right) \tag{7.31}$$

$$\mathbf{C}_z = \begin{bmatrix} 1 & -0.9697 & -0.1230 & -0.9942 \\ & 1 & 0.1230 & 0.9904 \\ & & 1 & 0.0154 \\ \text{sym} & & & 1 \end{bmatrix} \tag{7.32}$$

The third parameter in (7.29) is the wide-lane ambiguity. We observed that there is little correlation between the wide-lane and L1 ambiguities. Furthermore, the correlations between the wide-lane ambiguity and both the topocentric distance and the ionospheric parameter have been reduced dramatically. Considering the small standard deviation for the wide-lane ambiguity in (7.31) and its low correlations with the other parameters, it should be possible to estimate the wide-lane ambiguity from epoch solutions. The semiaxes of the ellipse of standard deviation for the ambiguities are 9.22 and 0.28, respectively, and an orientation of 89.97° for the semimajor axis, i.e., the ellipse is elongated along the N_1 direction. The correlation matrix (7.32) still shows high correlations between N_1, the ionosphere, and the topocentric distance. If we consider the square root of the determinant of the covariance matrix to be a single number that measures correlation, then $(|\mathbf{C}_z|/|\mathbf{C}_x|)^{1/2} \approx 33$ implies a major decorrelation of the epoch parameters.

The solution (7.29) is also obtained if we express the carrier phases (7.19) in cycles and then multiply with $\mathbf{Z}$ from the left. In fact, the following popular expressions can be readily verified,

$$\begin{aligned} N_w &= \varphi_w - \frac{f_1 P_1 + f_2 P_2}{(f_1 + f_2)\,\lambda_w} + f(\delta_{1,P}, \delta_{2,P}, \delta_{1,\varphi}, \delta_{2,\varphi}) \\ &\approx \varphi_w - 0.65\,P_1 - 0.51\,P_2 + \cdots \end{aligned} \tag{7.33}$$

$$\varphi_w = \varphi_1 - \varphi_2 \tag{7.34}$$

$$\lambda_w = \frac{c}{f_w} = \frac{c}{f_1 - f_2} \approx 0.86\ \mathrm{m} \tag{7.35}$$

Equation (7.33) is frequently used to screen the observation for cycle slips. Alternatively, one might attempt to determine the wide-lane integer from a short set of observations, conceivably just one epoch, and then constrain that integer to resolve the integer for N_1 rapidly. As more frequencies are added to the satellites one will be able to carry out additional widelaning (Hatch et al., 2000).

Figure 7.11 shows the computed wide-lane ambiguity (7.33) for three consecutive days. The elevation angle of satellite PRN 07 changes from 50° to 5° during the 1.5 hours of observations. The plotted lines are shifted by integers for purpose of graphical separation. It can be readily seen that the wide-lane integer ambiguities can be estimated from these data. The origins of the plots are shifted 4 minutes each day in order to emphasize similarity in multipath disturbances. The same receiver-satellite geometry repeats about 4 minutes earlier each succeeding day.

The geometry-free epoch solution also applies to relative positioning. Using subscripts k and m, and the superscripts p and q to indicate differencing, we obtain from (7.19)

$$\begin{bmatrix} P^p_{km,1} \\ P^p_{km,2} \\ \Phi^p_{km,1} \\ \Phi^p_{km,2} \end{bmatrix} = \begin{bmatrix} 1 & 1 & 0 & 0 \\ 1 & \alpha_f & 0 & 0 \\ 1 & -1 & \lambda_1 & 0 \\ 1 & -\alpha_f & 0 & \lambda_2 \end{bmatrix} \begin{bmatrix} \rho^p_{km} - dt_{km} + T^p_{km} \\ I^p_{km,1,P} \\ N^p_{km,1} \\ N^p_{km,2} \end{bmatrix} + \begin{bmatrix} \delta^p_{km,1,P} \\ \delta^p_{km,2,P} \\ \delta^p_{km,1,\Phi} \\ \delta^p_{km,2,\Phi} \end{bmatrix} + \begin{bmatrix} \varepsilon^p_{km,1,P} \\ \varepsilon^p_{km,2,P} \\ \varepsilon^p_{km,1,\Phi} \\ \varepsilon^p_{km,2,\Phi} \end{bmatrix} \tag{7.36}$$

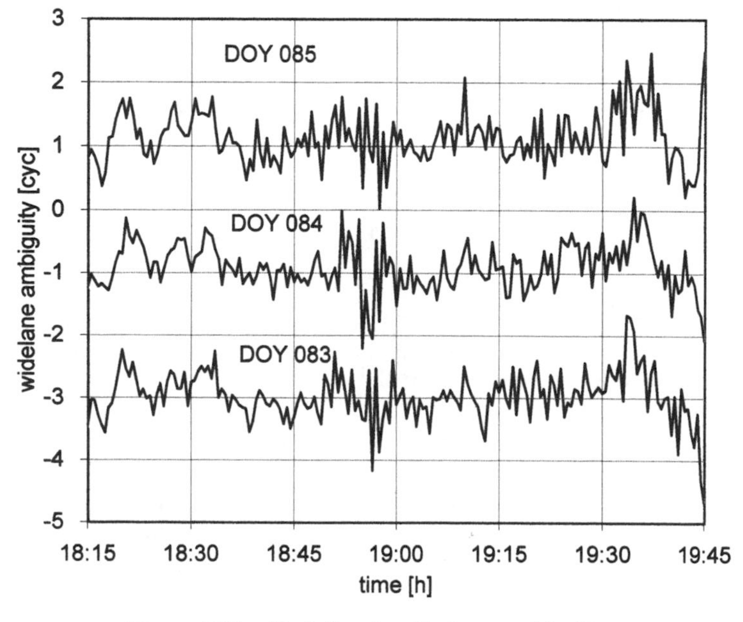

Figure 7.11 Variation in wide-lane ambiguity.

$$\begin{bmatrix} P^{pq}_{km,1} \\ P^{pq}_{km,2} \\ \Phi^{pq}_{km,1} \\ \Phi^{pq}_{km,2} \end{bmatrix} = \begin{bmatrix} 1 & 1 & 0 & 0 \\ 1 & \alpha_f & 0 & 0 \\ 1 & -1 & \lambda_1 & 0 \\ 1 & -\alpha_f & 0 & \lambda_2 \end{bmatrix} \begin{bmatrix} \rho^{pq}_{km} + T^{pq}_{km} \\ I^{pq}_{km,1,P} \\ N^{pq}_{km,1} \\ N^{pq}_{km,2} \end{bmatrix} + \begin{bmatrix} d^{pq}_{km,1,P} \\ d^{pq}_{km,2,P} \\ d^{pq}_{km,1,\Phi} \\ d^{pq}_{km,2,\Phi} \end{bmatrix} + \begin{bmatrix} \varepsilon^{pq}_{km,1,P} \\ \varepsilon^{pq}_{km,2,P} \\ \varepsilon^{pq}_{km,1,\Phi} \\ \varepsilon^{pq}_{km,2,\Phi} \end{bmatrix} \tag{7.37}$$

In single differencing the satellite clock error and the interfrequency bias T_{GD} cancel. In double differencing the receiver clock error cancels. The fast changing terms ρ^p_{km} and ρ^{pq}_{km} in the expressions above can be eliminated by subtracting the respective first equation. For example, for the case of single differencing we get

$$\begin{bmatrix} P^p_{km,2} - P^p_{km,1} \\ \Phi^p_{km,1} - P^p_{km,1} \\ \Phi^p_{km,2} - P^p_{km,1} \end{bmatrix} = \begin{bmatrix} \alpha_f - 1 & 0 & 0 \\ -2 & \lambda_1 & 0 \\ -\alpha_f - 1 & 0 & \lambda_2 \end{bmatrix} \begin{bmatrix} I^p_{km,1,P} \\ N^p_{km,1} \\ N^p_{km,2} \end{bmatrix} + \begin{bmatrix} \delta^p_{km,2,P} - \delta^p_{km,1,P} \\ \delta^p_{km,1,\Phi} - \delta^p_{km,1,P} \\ \delta^p_{km,2,\Phi} - \delta^p_{km,1,P} \end{bmatrix}$$

$$+ \begin{bmatrix} \varepsilon^p_{km,2,P} - \varepsilon^p_{km,1,P} \\ \varepsilon^p_{km,1,\Phi} - \varepsilon^p_{km,1,P} \\ \varepsilon^p_{km,2,\Phi} - \varepsilon^p_{km,1,P} \end{bmatrix} \tag{7.38}$$

This expression is especially useful for discovering and repairing cycle slips in single-difference ambiguities. The expression can be further simplified by recognizing that $I^p_{km,1,P} \approx 0$ for short baselines. For longer baselines, the ionospheric term can be modeled by a first-order polynomial in time (ionospheric bias and drift). The ambiguities can readily be transformed to $N^p_{km,w}$ and $N^p_{km,1}$ or $N^{pq}_{km,w}$ and $N^{pq}_{km,1}$, respectively, by using the **Z** matrix and thus providing the possibility of fixing the wide-lane ambiguities early. Expressing the carrier phases in cycles, the following expression can be readily verified from (7.37),

$$\begin{aligned} N^{pq}_{km,1} &= \varphi^{pq}_{km,1} + \frac{f_1}{f_1 - f_2}\left[N^{pq}_{km,w} - \varphi^{pq}_{km,w}\right] + \frac{f_1 + f_2}{f_2} I^{pq}_{km,1,\varphi} + f\left(d^{pq}_{km,1,\varphi}, d^{pq}_{km,2,\varphi}\right) \\ &\approx \varphi^{pq}_{km,1} + 4.5\left[N^{pq}_{km,w} - \varphi^{pq}_{km,w}\right] + \cdots \end{aligned} \tag{7.39}$$

This is the extra-wide-lane equation. If the wide-lane ambiguities are known from prior analysis, the double-differenced L1 ambiguity can be computed from the carrier phases only. Fortunately, Expression (7.39) does not depend on the large pseudorange multipath terms, but on the smaller carrier phase multipath terms. If the wide-lane ambiguity happens to be incorrectly identified by one, a situation that might occur for satellites at low elevation angles, the computed L1 ambiguity changes by 4.5 cycles. The first decimal of the computed L1 ambiguity would be close to 5. Because of prior knowledge that the L1 ambiguity is an integer, we can use that fact to decide between two candidate wide-lane ambiguities. This procedure is known as extra widelaning (Wübbena, 1990).

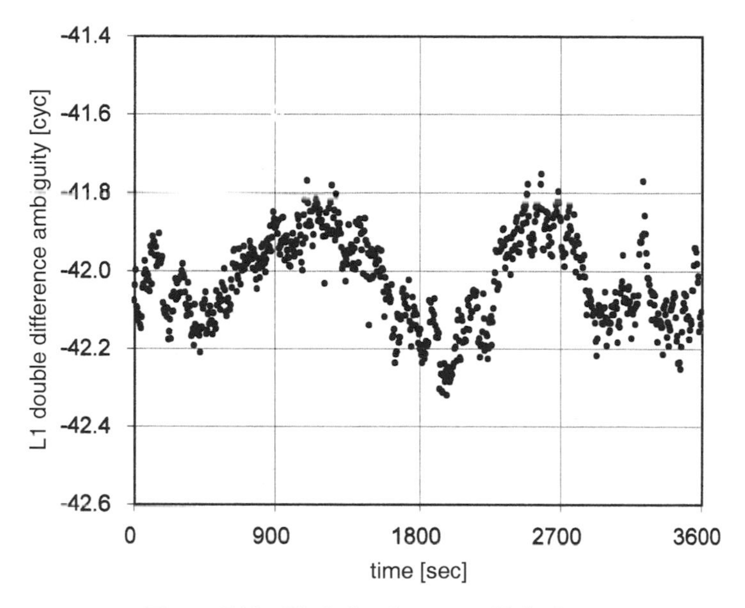

Figure 7.12 Variation in extra widelaning.

Figure 7.12 shows the variation of the computed L1 double-difference ambiguity using (7.39). For this 1-hour stretch of data the elevation angles of both satellites were above 45°. The integer double-difference wide-lane ambiguities were estimated first using (7.33) and then substituted in (7.39). The variation around the integer −42 is due to double differenced multipath. The ionospheric effects cancel because the baseline is so short. Studying figures like Figures 7.11 and 7.12 it can be readily appreciated that, under favorable circumstances, the L1 double-difference integer ambiguities can be determined from just one epoch of observations using the geometry-free solution.

We substitute the geometry-free epoch equation (7.19) into the ionospheric-free equations (6.91) and (6.94) and ionospheric equations (6.97) and (6.98). For single and double differences we obtain

$$\begin{bmatrix} P^p_{km,\mathrm{IF}} \\ P^p_{km,I} \\ \Phi^p_{km,\mathrm{IF}} \\ \Phi^p_{km,I} \end{bmatrix} \equiv \begin{bmatrix} \beta_f P^p_{km,1} - \gamma_f P^p_{km,2} \\ P^p_{km,1} - P^p_{km,2} \\ \beta_f \Phi^p_{km,1} - \gamma_f \Phi^p_{km,2} \\ \Phi^p_{km,1} - \Phi^p_{km,2} \end{bmatrix} = \tag{7.40}$$
$$\begin{bmatrix} 1 & 0 & 0 & 0 \\ 0 & 1-\alpha_f & 0 & 0 \\ 1 & 0 & \beta_f\lambda_1 & -\gamma_f\lambda_2 \\ 0 & 1-\alpha_f & \lambda_1 & -\lambda_2 \end{bmatrix} \begin{bmatrix} \rho^p_{km} - dt_{\_km} + T^p_{km} \\ I^p_{km,1,P} \\ N^p_{km,1} \\ N^p_{km,2} \end{bmatrix} + \begin{bmatrix} \delta^p_{km,\mathrm{IF},P} \\ \delta^p_{km,I,P} \\ \delta^p_{km,\mathrm{IF},\Phi} \\ \delta^p_{km,I,\Phi} \end{bmatrix}$$

$$
\begin{bmatrix} P^{pq}_{km,\mathrm{IF}} \\ P^{pq}_{km,I} \\ \Phi^{pq}_{km,\mathrm{IF}} \\ \Phi^{pq}_{km,I} \end{bmatrix} \equiv \begin{bmatrix} \beta_f P^{pq}_{km,1} - \gamma_f P^{pq}_{km,2} \\ P^{pq}_{km,1} - P^{pq}_{km,2} \\ \beta_f \Phi^{pq}_{km,1} - \gamma_f \Phi^{pq}_{km,2} \\ \Phi^{pq}_{km,1} - \Phi^{pq}_{km,2} \end{bmatrix} =
$$

$$
\begin{bmatrix} 1 & 0 & 0 & 0 \\ 0 & 1-\alpha_f & 0 & 0 \\ 1 & 0 & \beta_f \lambda_1 & -\gamma_f \lambda_2 \\ 0 & 1-\alpha_f & \lambda_1 & -\lambda_2 \end{bmatrix} \begin{bmatrix} \rho^{pq}_{km} + T^{pq}_{km} \\ I^{pq}_{km,1,P} \\ N^{pq}_{km,1} \\ N^{pq}_{km,2} \end{bmatrix} + \begin{bmatrix} d^{pq}_{km,\mathrm{IF},P} \\ d^{pq}_{km,I,P} \\ d^{pq}_{km,\mathrm{IF},\Phi} \\ d^{pq}_{km,I,\Phi} \end{bmatrix} \tag{7.41}
$$

The receiver and satellite hardware delay and signal multipath terms have been suitably transformed (note the subscripts of the δ terms).

7.4 POINT POSITIONING

According to the first equation in the geometry-free model (7.19), we estimate the sum $\rho + \Delta$ every epoch, using sequential least-squares or Kalman filtering. The topocentric distance ρ, of course, is a function of the receiver antenna position $\mathbf{x}_k$ and the satellite position $\mathbf{x}^p$. By using $\mathbf{x}_k$ and $\mathbf{x}^p$ explicitly in (7.19) we introduce the dependency on the receiver-satellite geometry. We further replace the auxiliary quantity Δ with the original definition (7.20), thus introducing the receiver and satellite clock errors explicitly.

Point positioning refers to the estimation of receiver antenna coordinates $\mathbf{x}_k$ and the receiver clock error $d\underline{t}_k$ using pseudorange observables. The role of carrier phases is limited to smoothing the pseudoranges, if used at all. More specifically, the term point positioning as used here implies several simplifying assumptions. The satellite positions $\mathbf{x}^p$ at transmission times are assumed known and available from the broadcast ephemeris. While we estimate the receiver clock error at every epoch, we neglect the residual satellite clock errors $d\bar{t}^p$. Of course, the satellite clock broadcast correction must be applied following (5.38). The satellite clocks are constantly monitored by the control center, which models the clock offsets by polynomials in time. The latter are part of the navigation message. The ionospheric and tropospheric delays are also computed from models, as explained in Chapter 6. Hardware delays and multipath are neglected.

The four unknowns $\mathbf{x}_k$ and $d\underline{t}_k$ can be computed using four pseudoranges measured simultaneously to four satellites. Using the simplifying assumptions made above we can write four equations of the type

$$
P^p_k - cT^p = \left\| \mathbf{x}^p - \mathbf{x}_k \right\| - c\, d\underline{t}_k = \rho^p_k - c\, d\underline{t}_k, \tag{7.42}
$$

one for each satellite (superscript p varies). The effect of the earth's rotation during the signal travel time must be incorporated in (7.42) following Section 5.3.2. Since the receiver clock error $d\underline{t}_k$ is solved together with the position of the receiver's antenna each epoch, a relatively inexpensive quartz crystal clock in the receiver is sufficient rather than an expensive atomic clock. The basic requirement, however, is that there are four satellites visible at a given epoch. This visibility requirement is a key factor in the design of the GPS-type of constellation that assures global coverage is available at any time.

Modifications of the basic point positioning solution can be readily envisioned. For example, for applications on the ocean it might be possible to determine the ellipsoidal height accurately from the height above water and the geoid undulation. Equations (7.42) can be expressed in terms of ellipsoidal latitude, longitude, and height using transformations (2.66) through (2.68) and the ellipsoidal height can be considered a known quantity. Therefore, at least in principle, pseudoranges of three satellites are sufficient to determine positions at sea. Other variations are possible.

Point positioning accuracy depends on the accuracy of the navigation message and the satellite constellation used. In practice, one prefers to observe not just four satellites but all satellites in view in order to achieve redundancy and better geometry. See the discussion on dilution of precision (DOP) below. The achievable accuracy is therefore subject to change as receiver technology keeps improving and the broadcast ephemeris gets more accurate. The modernization of GPS will, of course, have a major positive impact. Dual-frequency users can use the ionospheric-free function in Equation (7.42) and, therefore, eliminate the effect of the ionosphere. While single-frequency users can use the C/A-code pseudoranges, they unfortunately depend on the ionospheric model to reduce the impact of the ionosphere on the solutions.

7.4.1 Linearized Solution and DOPs

It has become common practice to use DOP factors to describe the effect of receiver-satellite geometry on the accuracy of point positioning. The DOP factors are simple functions of the diagonal elements of the covariance matrix of the adjusted parameters, derived from the linearized model. In general,

$$\sigma = \sigma_0 \, \mathrm{DOP} \tag{7.43}$$

where σ_0 denotes the standard deviation of the observed pseudoranges, and σ is a one-number representation of the standard deviation of position and/or time. When computing DOPs, the pseudorange observations are considered uncorrelated and of the same accuracy; i.e., the weight matrix is $\mathbf{P} = \mathbf{I}$. If the ordered set of parameters is

$$\mathbf{x}^{\mathrm{T}} = [\, dx_k \quad dy_k \quad dz_k \quad d\underline{t}_k \,] \tag{7.44}$$

then the design matrix follows from (7.42) after linearization around the nominal station location $\mathbf{x}_{k,0}$,

$$\mathbf{A} = \begin{bmatrix} \mathbf{e}_k^1 & c \\ \mathbf{e}_k^2 & c \\ \mathbf{e}_k^3 & c \\ \vdots & \vdots \end{bmatrix} \tag{7.45}$$

The $\mathbf{A}$ matrix has as many rows as there are satellites observed, which typically includes all satellites in view. The superscript i is a sequential identification for the satellite and not necessarily equal to the PRN number. The symbol $\mathbf{e}_k^i$ denotes the 1×3 row vector defined in (5.48). It contains the direction cosines for the vector from the nominal station location to satellite. The clock error parameter is often set to $\xi_k = c\,dt_{\underline{k}}$, making the elements in the second column of (7.45) unity. The cofactor matrix for the adjusted receiver position and receiver clock is

$$\mathbf{Q}_x = \left(\mathbf{A}^{\mathrm{T}}\mathbf{A}\right)^{-1} = \begin{bmatrix} q_x & q_{xy} & q_{xz} & q_{xt} \\ & q_y & q_{yz} & q_{yt} \\ & & q_z & q_{zt} \\ \text{sym} & & & q_t \end{bmatrix} \tag{7.46}$$

Since it is more convenient to interpret results in the local geodetic coordinate system (w) (consisting of the coordinates northing n, easting e, and up u), we transform the cofactor matrix (7.46) using (2.112) and (2.113) of Section 2.3.5.3. The result is

$$\mathbf{Q}_w = \begin{bmatrix} q_n & q_{ne} & q_{nu} \\ & q_e & q_{eu} \\ \text{sym} & & q_u \end{bmatrix} \tag{7.47}$$

The DOP factors are functions of the diagonal elements of (7.46) or (7.47). Table 7.2 shows the various dilution factors: vertical dilution of precision (VDOP) for the height, horizontal dilution of precision (HDOP) for horizontal positions, positional dilution of precision (PDOP), time dilution of precision (TDOP) and geometric dilution of precision (GDOP). GDOP is a composite measure reflecting the geometry of the position and time estimates.

The DOPs can be computed in advance, given the approximate receiver location and a predicted satellite ephemeris. The DOPs are useful for finding the best subset of satellites if a receiver has only four or five channels. Even though most receivers today observe all satellites in view, the DOPs are still useful to identify a temporal weakness in geometry in kinematic applications, in particular in the presence of signal obstruction.

TABLE 7.2 DOP Expressions

$\mathrm{VDOP} = \sqrt{q_h}$
$\mathrm{HDOP} = \sqrt{q_n + q_e}$
$\mathrm{PDOP} = \sqrt{q_n + q_e + q_u} = \sqrt{q_x + q_y + q_z}$
$\mathrm{TDOP} = \sqrt{q_t}$
$\mathrm{GDOP} = \sqrt{q_n + q_e + q_u + q_t\, c^2}$

7.4.2 Closed Solution

The closed-form point positioning solution has recently been treated in detail in Grafarend and Shan (2002) and Awange and Grafarend (2002a,b). The reader might consult these publications for in-depth study of closed expressions, for derivations, and as a good source of additional references. Bancroft's (1985) solution is a very early, if not the first, contribution on this topic. We merely summarize the solution using the notation of Goad (1998). In order to achieve compact expressions we define the following product of two arbitrary vectors **g** and **h** as

$$\langle \mathbf{g}, \mathbf{h} \rangle \equiv \mathbf{g}^{\mathrm{T}} \mathbf{M} \mathbf{h} \tag{7.48}$$

where **M** is the matrix

$$\mathbf{M} = \begin{bmatrix} {}_3\mathbf{I}_3 & \mathbf{O} \\ \mathbf{O} & -1 \end{bmatrix} \tag{7.49}$$

The relevant terms of the pseudorange (7.42) are, if the interfrequency bias T^i has been applied,

$$P_k^i + c\, dt_{\underline{k}} = \left\| \mathbf{x}^i - \mathbf{x}_k \right\|, \qquad 1 \le i \le 4 \tag{7.50}$$

Squaring both sides gives

$$\left(\mathbf{x}^i \cdot \mathbf{x}^i - P_k^{i2}\right) - 2\left(\mathbf{x}^i \cdot \mathbf{x}_k + P_k^i\, c\, dt_{\underline{k}}\right) = -\left(\mathbf{x}_k \cdot \mathbf{x}_k - c^2\, dt_{\underline{k}}^2\right) \tag{7.51}$$

As can be verified, the four pseudorange equations can be written in the compact form

$$\alpha - \mathbf{BM} \begin{bmatrix} \mathbf{x}_k \\ c\, dt_{\underline{k}} \end{bmatrix} + \Lambda \boldsymbol{\tau} = 0 \tag{7.52}$$

where

$$\Lambda = \frac{1}{2} \left\langle \begin{bmatrix} \mathbf{x}_k \\ c\, dt_{\underline{k}} \end{bmatrix}, \begin{bmatrix} \mathbf{x}_k \\ c\, dt_{\underline{k}} \end{bmatrix} \right\rangle \tag{7.53}$$

$$\alpha^i = \frac{1}{2} \left\langle \begin{bmatrix} \mathbf{x}^i \\ P_k^i \end{bmatrix}, \begin{bmatrix} \mathbf{x}^i \\ P_k^i \end{bmatrix} \right\rangle \tag{7.54}$$

$$\boldsymbol{\alpha}^{\mathrm{T}} = \begin{bmatrix} \alpha^1 & \alpha^2 & \alpha^3 & \alpha^4 \end{bmatrix} \tag{7.55}$$

$$\boldsymbol{\tau}^{\mathrm{T}} = [1 \quad 1 \quad 1 \quad 1] \tag{7.56}$$

$$\mathbf{B} = \begin{bmatrix} x^1 & y^1 & z^1 & -P_k^1 \\ x^2 & y^2 & z^2 & -P_k^2 \\ x^3 & y^3 & z^3 & -P_k^3 \\ x^4 & y^4 & z^4 & -P_k^4 \end{bmatrix} \tag{7.57}$$

The solution of (7.52) is

$$\begin{bmatrix} \mathbf{x}_k \\ c\,d\underline{t}_k \end{bmatrix} = \mathbf{M}\mathbf{B}^{-1}\left(\Lambda\boldsymbol{\tau} + \boldsymbol{\alpha}\right) \tag{7.58}$$

We note, however, that Λ is also a function of the unknowns $\mathbf{x}_k$ and $d\underline{t}_k$. We substitute (7.58) into (7.53), giving

$$\left\langle \mathbf{B}^{-1}\boldsymbol{\tau}, \mathbf{B}^{-1}\boldsymbol{\tau} \right\rangle \Lambda^2 + 2\left\{ \left\langle \mathbf{B}^{-1}\boldsymbol{\tau}, \mathbf{B}^{-1}\boldsymbol{\alpha} \right\rangle - 1 \right\} \Lambda + \left\langle \mathbf{B}^{-1}\boldsymbol{\alpha}, \mathbf{B}^{-1}\boldsymbol{\alpha} \right\rangle = 0 \tag{7.59}$$

This is a quadratic equation of Λ. Substituting its roots into (7.58) gives two solutions for the station coordinates $\mathbf{x}_k$. Converting the solution to geodetic coordinates and inspecting the respective ellipsoidal heights readily identifies the valid solution.

7.5 PRECISE POINT POSITIONING

Precise point positioning (PPP) refers to centimeter position accuracy of a single static receiver using a long observation series, and to subdecimeter accuracy of a roving receiver using the ionospheric-free pseudorange and carrier phase functions. The receiver clock error $d\underline{t}_k$ and the zenith tropospheric delay T_k (no superscript here) are estimated for each epoch, in addition to a constant R_k^p. When using PPP, one must avoid any simplifying assumptions, i.e., all known corrections must be applied to the observations and the corrections must be consistent. The satellite positions $\mathbf{x}^k$ at transmission times are typically computed from the postprocessed precise ephemeris available from the IGS or its associated processing centers. A crucial element in achieving centimeter position accuracy with PPP is accurate satellite clock corrections $d\bar{t}^p$, which are part of the precise ephemeris. The ionospheric effects are eliminated by using the ionospheric-free functions (6.91) and (6.94). The L1 and L2 integer ambiguities are combined into a rational constant R_k^p, which also serves to absorb unmodeled receiver and satellite hardware delays that might change with time, as well as the initial phase windup angles. The PPP model is

$$P_{k,\mathrm{IF}}^p = \rho_k^p - c\,d\underline{t}_k + T_{k,0}^p + dT_k\, m(\vartheta^p) \tag{7.60}$$

$$\varphi_{k,\mathrm{IF}}^p = \frac{f_1}{c}\rho_k^p - f_1\, d\underline{t}_k + R_k^p + \frac{f_1}{c}T_{k,0}^p + \frac{f_1}{c\sin\vartheta^p}\, dT_k \tag{7.61}$$

where the approximate tropospheric slant total delay (STD)

$$T_{k,0}^{p} = \text{ZHD}_k \, m_h(\vartheta^p) + \text{ZWD}_k \, m_{wv}(\vartheta^p) \tag{7.62}$$

can be computed from the zenith hydrostatic delay (ZHD) and the zenith wet delay (ZWD) models (6.17) and (6.18) and meteorological data. The mapping functions m_h and m_{wv} follow from (6.22). The estimated zenith total delay then becomes

$$T_k = \frac{T_{k,0}^{p}}{m(\vartheta^p)} + dT_k \tag{7.63}$$

Zumberge et al. (1998a) introduced PPP at centimeter level with GPS. One of their goals was to use postprocessed data from a permanently operating global network of stations to compute highly accurate positions for individual receivers that are not part of the permanent network. They also viewed PPP as a data compression strategy when they addressed the relationship between achievable position accuracy for individual stations as a function of the number of permanently observing network stations. Zumberge et al. (1998b) reported centimeter-level accuracy for static receivers and subdecimeter-level accuracy for kinematic receivers, even at a time when selective availability was still active. They modeled the receiver clock as white noise and the tropospheric delay as random walk in the Kalman filter.

JPL provides a free Internet processing service for PPP (Zumberge, 1998). Witchayangkoon and Segantine (1999) used this service to test the technique for various data sets varying from 1 hour to 24 hours. They reported generally 1 dm repeatability for 1-hour data sets and 1–2 cm repeatability for data sets greater than 4 hours. Of course the performance characteristics change for the better as the PPP model improves over time. The JPL service can be used to substitute baseline processing by submitting both data files of the baseline stations separately. Figure 7.13 further confirms the high accuracy of the PPP approach. In this particular test the solid earth tides corrections were not applied. Instead, the station coordinates were estimated every epoch together with the other parameters after the Kalman filter had initially stabilized. The top part of the figure shows the estimated variation in coordinates, estimated every 30 seconds using JPL's high clock rate ephemeris, whereas the bottom part shows the solid earth tides, computed from software downloaded from the IERS website. The outliers in the estimated up component are caused by the addition of a rising satellite.

Absolute positioning with only single-frequency observations is expected to be less accurate, especially in the height. An obvious reason for the degradation in accuracy is the effect of unmodeled ionospheric delays (Lachapelle et al., 1996). Øvstedal (2002) used the global ionospheric model provided by the IGS in connection with single-frequency observations and demonstrated a horizontal epoch-to-epoch accuracy of better than 1 m and a vertical accuracy of about 1 m.

As pointed out by Kouba (2001), the success of PPP depends on applying a consistent set of corrections. For example, whereas it is clear that the satellite antenna offsets must be carefully taken into consideration since there is no differencing of observations between stations that would effectively cancel the impact of these offsets, knowledge of the satellite receiver clock errors is extremely important for PPP positioning to be accurate. However, the estimates of satellite antenna offsets and

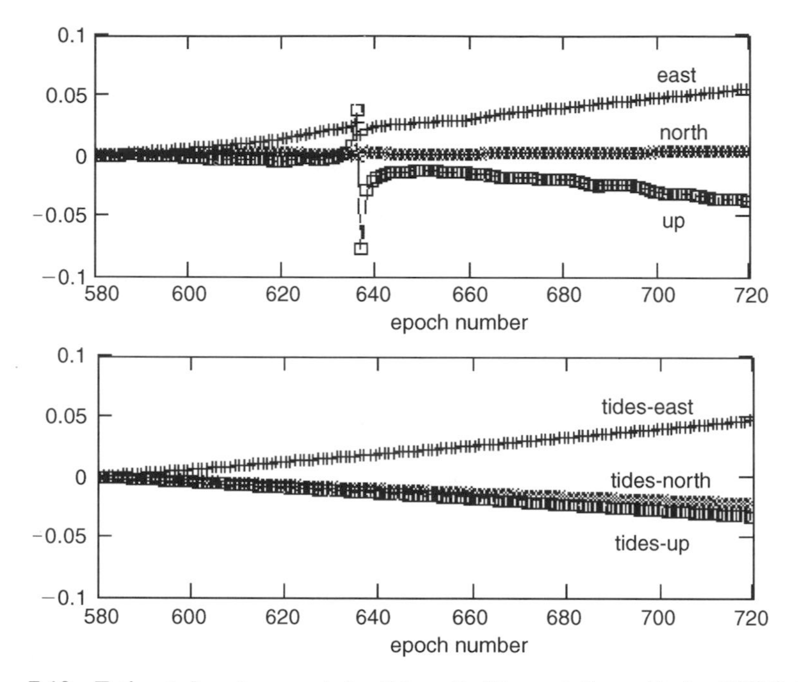

Figure 7.13 Estimated and computed solid earth tide variations. Station WES2, Massachusetts, DOY 2 (2000), epochs 17400–21600 seconds of day. (Witchayangkoon, 2000). Units are in meters.

satellite clock are highly correlated. One must, therefore, use the same antenna offsets that were used for generating the precise ephemeris (of which the satellite clock error is a part). The precise ephemeris must, of course, always refer to the same reference frame for positions derived with PPP to maintain their consistency. All errors that are typically expected to cancel in double differencing must be applied in PPP. For example, although the phase windup correction is little known to GPS users because its effect cancels in double differencing over short baselines, it directly affects PPP. The phase windup correction is especially important in kinematic applications where the antenna can readily rotate in azimuth. The absolute location of the receiver antenna phase center must be accurately known, in order to reduce the measurements to the height of the monument. The magnitude of the solid earth tides far exceeds the PPP accuracy and must therefore be incorporated. Even ocean loading is important for the most accurate applications of PPP.

7.6 REAL-TIME PRECISE POINT POSITIONING

Point positioning as described above refers to positioning of a single receiver using single- or dual-frequency pseudoranges and the broadcast ephemeris to compute satellite positions and satellite clock corrections. If we use the ionospheric-free pseudorange function, the postprocessed precise ephemeris, and precise satellite clock corrections, we call it precise point positioning. The latter term also applies when using the IGS rapid ephemeris that is currently available with a delay of several hours, or the predicted (ultrarapid) IGS ephemeris instead of the postprocessed ephemeris. Fortunately, the satellite clocks can now be more accurately predicted since selective availability has been turned off. If the satellite ephemeris and clocks are estimated from observations and are available in near real time, say with a latency of merely seconds, one speaks of real-time precise point positioning (real-time PPP). The latter approach requires (a) real-time orbit determination processing capability, a global network of tracking stations that forward dual-frequency observations to a processing center in real time and (b) communicating the results of the orbit determination, usually in the form of corrections, to the user in the field in real time.

Muellerschoen et al. (2000) describe the results of an Internet-based dual-frequency real-time precise point positioning system developed at JPL. The website IGDG (2002) contains several publications on the subject and offers a live demonstration of position determination at the JPL facility using Internet-based global differential GPS (IGDG), as the technique is called at JPL. Aspects of real-time orbital determination are addressed in Muellerschoen et al. (2001). JPL uses, at least originally when developing the technique, about fifty sites of NASA's global GPS network that is operated in batch mode over the Internet. Data forwarded by the stations include C/A-code and P1 and P2 code pseudoranges, the carrier phase observations, signal-to-noise ratios, and the receiver's point positioning solution. JPL's modified GIPSY program, called real-time GIPSY (RTG), is used to estimate orbits, polar motion, UT1-UTC, and tropospheric corrections at the known network reference stations once per minute. The satellite clock corrections are computed once per second. Figure 7.14

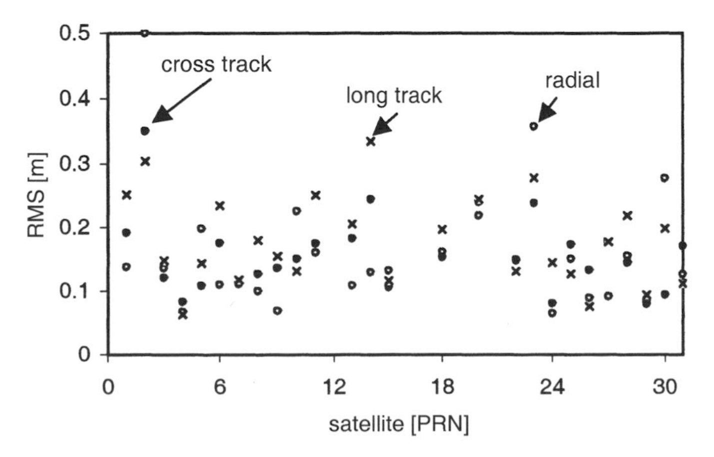

Figure 7.14 Radial, cross-track, and along-track RMS of difference between real-time orbits and IGS rapid solution for October 9, 2002. The computation interval is 15 minutes. (Data from Bar-Sever, JPL.)

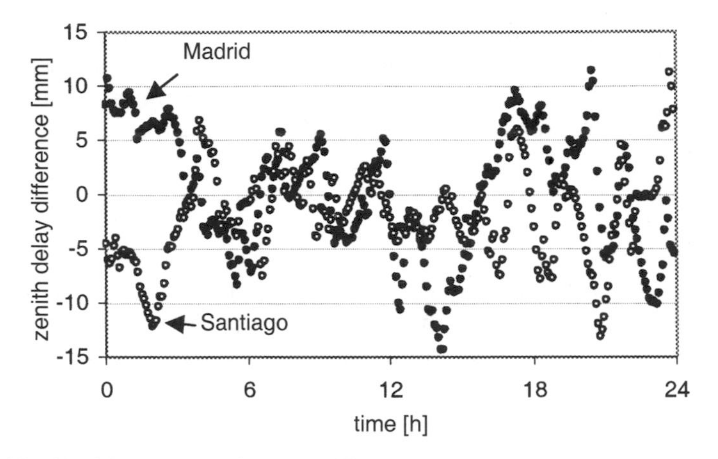

Figure 7.15 Zenith tropospheric delay difference for real-time estimates and computed values using the IGS rapid orbit and satellite clock solution. Data were recorded at the Madrid and Santiago reference stations on October 9, 2002. The estimation interval is 5 minutes. (Data from Bar-Sever, JPL.)

shows a comparison for one day between the real-time orbital solution and the IGS rapid ephemeris. Figure 7.15 is an example of real-time tropospheric delay estimation at two geographically widely separated stations; again, the comparison is with respect to the IGS rapid solution.

A global differential correction is computed for each satellite. The corrections represent the difference between the estimated precise real-time and broadcast ephemerides, and the estimated precise and broadcast satellite clock corrections. The corrections, parameterized in terms of Cartesian satellite coordinate corrections and satellite clock corrections, are available over the open Internet via a TCP server running at JPL. Hatch et al. (2002) describe a system, called StarFire, which transmits the correction via Inmarsat L-band communication frequency (1525–1565 MHz). It uses an L-band satellite communication receiver with a single, multifunction antenna designed to receive both GPS frequencies and the Inmarsat signals. Users apply the global differential corrections and use their dual-frequency pseudorange and ionospheric-free carrier phase observations to estimate the position and receiver clock per epoch and, optionally, estimate the troposphere (similar to PPP).

The global differential correction arrives at the user with some latency. It takes about 2 seconds to accumulate the global data set over the Internet and 0.5 second to process the clock solution. For Inmarsat one must add 1.5 seconds for uplink and downlink time. For real time, the correction must therefore be extrapolated over 4–5 seconds, which causes no substantial loss of accuracy, primarily because the satellite clocks are stable (when selective availability is off).

Both PPP and real-time PPP are uniformly accurate over the whole globe since both rely on the dual-frequency ionospheric-free function, which is not affected by the spatial and temporal variations of the ionosphere. The commercial version of real-

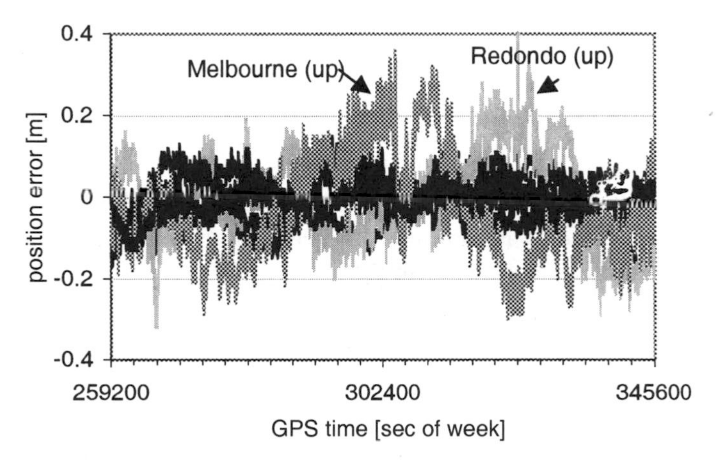

Figure 7.16 Example of real-time precise point positioning from Starfire at Redondo and Melbourne on September 18, 2002. The horizontal position error generally stays within 1 dm.

time PPP described in Hatch et al. (2002) applies the model tropospheric refraction, thus alleviating the need to estimate the troposphere by the user (which is adequate since StarFire does not offer centimeter-level accuracy but global decimeter accuracy in the horizontal, and perhaps slightly worse accuracy in the vertical). Figure 7.16 demonstrates the performance of StarFire.

7.7 RELATIVE POSITIONING

In relative positioning, the vector between two stations is determined when two receivers observe simultaneously. If more than two receivers observe at the same time, we speak of a session network consisting of all the co-observing stations. Session solutions result in a set of correlated vectors between the stations. In relative positioning, one tends to use double- or triple-difference observations, although single-difference observations could be used as well. Typically, one station is held fixed, i.e., coordinates $\mathbf{x}_k$ are known, and the coordinates $\mathbf{x}_m$ of the other station are estimated. We discussed the details as to the accuracy required for the location of the known station as a function of baseline length and satellite ephemeris accuracy in Section 5.35. The formulation given in that section can be readily generalized for the case when both receivers occupying the endpoints of a baseline are in motion.

7.7.1 Using Pseudoranges

In our customary notation the superscript p denotes the base satellite. If S satellites are observed at the same instant, then the superscript q takes on $S - 1$ values, i.e., there are $S - 1$ independent double-difference equations (7.38),

$$\begin{aligned} P_{km,1}^{pq} &= \left\| \mathbf{x}^p - \mathbf{x}_k \right\| - \left\| \mathbf{x}^p - \mathbf{x}_m \right\| - \left\{ \left\| \mathbf{x}^q - \mathbf{x}_k \right\| - \left\| \mathbf{x}^q - \mathbf{x}_m \right\| \right\} + d_{km,1,P}^{pq} + \varepsilon_{km,1,P}^{pq} \\ &= \rho_{km}^{pq} + d_{km,1,P}^{pq} + \varepsilon_{km,1,P}^{pq} \end{aligned} \tag{7.64}$$

available at each epoch. The hardware delay terms cancel in (7.64) and the unknown pseudorange multipath d_{km}^{pq}, while potentially large, is typically neglected. If the system (7.64) is solved by linearization and subsequent least-squares, then the row of the design matrix contains, respectively,

$$\frac{\partial P_{km}^{pq}}{\partial \mathbf{x}_m} = \mathbf{e}_m^p - \mathbf{e}_m^q \tag{7.65}$$

for each double difference.

In relative positioning with pseudoranges, the processing is usually carried out baseline by baseline and mathematical correlation between the double-differenced pseudorange observations are often neglected. These correlations are typically not neglected for carrier phase observations, which are the more accurate type of observations, and will be discussed below.

A closed solution is readily available for relative positioning with pseudoranges. Consider the three double differences that can be formed from the observations of four satellites,

$$P_{km}^{pi} = \left\| \mathbf{x}^p - \mathbf{x}_k \right\| - \left\| \mathbf{x}^p - \mathbf{x}_m \right\| - \left\{ \left\| \mathbf{x}^i - \mathbf{x}_k \right\| - \left\| \mathbf{x}^i - \mathbf{x}_m \right\| \right\}, \quad 1 \le i \le 3 \tag{7.66}$$

Let p denote the base satellite, in this case $p = 1$. Since the satellite coordinates and the station coordinates $\mathbf{x}_k$ are known, we can compute the auxiliary quantity Q,

$$Q_{km}^{pi} = P_{km}^{pi} - \left\| \mathbf{x}^p - \mathbf{x}_k \right\| + \left\| \mathbf{x}^i - \mathbf{x}_k \right\| \tag{7.67}$$

Comparing (7.66) and (7.67), we find that Q relates to the unknown $\mathbf{x}_m$ as

$$Q_{km}^{pi} = -\left\| \mathbf{x}^p - \mathbf{x}_m \right\| + \left\| \mathbf{x}^i - \mathbf{x}_m \right\| \tag{7.68}$$

Following Chaffee and Abel (1994), we translate the origin of the coordinate system to satellite p

$$\tilde{\mathbf{x}}^i = \mathbf{x}^i - \mathbf{x}^p \tag{7.69}$$

Noting that in the translated coordinate system $\tilde{\mathbf{x}}^p = 0$, we obtain from (7.68)

$$Q_{km}^{pi} + \|\tilde{\mathbf{x}}_m\| = \left\| \tilde{\mathbf{x}}^i - \tilde{\mathbf{x}}_m \right\| \tag{7.70}$$

Equations (7.70) and (7.50) are of the same form. Once $\tilde{\mathbf{x}}_m$ is computed the coordinates can be translated to $\mathbf{x}_m$ using (7.69). Squaring (7.70) gives

$$\left(\tilde{\mathbf{x}}^i \cdot \tilde{\mathbf{x}}^i - Q_{km}^{pi2}\right) - 2\left(\tilde{\mathbf{x}}^i \cdot \tilde{\mathbf{x}}_m + \|\tilde{\mathbf{x}}_m\| Q_{km}^{pi}\right) = 0 \tag{7.71}$$

This equation can be verified using

$$\Lambda^2 = \tilde{\mathbf{x}}_m \cdot \tilde{\mathbf{x}}_m \tag{7.72}$$

$$\alpha^i = \frac{1}{2}\left\langle \begin{bmatrix} \tilde{\mathbf{x}}^i \\ Q_{km}^{pi} \end{bmatrix}, \begin{bmatrix} \tilde{\mathbf{x}}^i \\ Q_{km}^{pi} \end{bmatrix} \right\rangle \tag{7.73}$$

$$\mathbf{B} = \begin{bmatrix} \tilde{x}^1 & \tilde{y}^1 & \tilde{z}^1 \\ \tilde{x}^2 & \tilde{y}^2 & \tilde{z}^2 \\ \tilde{x}^3 & \tilde{y}^3 & \tilde{z}^3 \end{bmatrix} \tag{7.74}$$

$$\boldsymbol{\tau}^{\mathrm{T}} = \begin{bmatrix} -Q_{km}^{p1} & -Q_{km}^{p2} & -Q_{km}^{p3} \end{bmatrix} \tag{7.75}$$

$$\tilde{\mathbf{x}}_k = \mathbf{B}^{-1}\left(\Lambda\boldsymbol{\tau} + \boldsymbol{\alpha}\right) \tag{7.76}$$

Substituting (7.76) in (7.72) gives the quadratic equation for Λ,

$$\left(\left\langle \mathbf{B}^{-1}\boldsymbol{\tau}, \mathbf{B}^{-1}\boldsymbol{\tau}\right\rangle - 1\right)\Lambda^2 + 2\left\langle \mathbf{B}^{-1}\boldsymbol{\tau}, \mathbf{B}^{-1}\boldsymbol{\alpha}\right\rangle \Lambda + \left\langle \mathbf{B}^{-1}\boldsymbol{\alpha}, \mathbf{B}^{-1}\boldsymbol{\alpha}\right\rangle = 0 \tag{7.77}$$

The two solutions for Λ are substituted in (7.76) to obtain two positions for station m. The ellipsoidal height can be used to decide which of the positions for m is the correct one.

The closed formulas can be generalized for more than four satellites. In this case the number of rows in $\mathbf{B}$ equals the number of satellites or the number of double differences. We multiply (7.52) from the left with $\mathbf{B}^{\mathrm{T}}$ and set $\bar{\boldsymbol{\alpha}} = \mathbf{B}^{\mathrm{T}}\boldsymbol{\alpha}$, $\bar{\mathbf{B}} = \mathbf{B}^{\mathrm{T}}\mathbf{B}$, and $\bar{\boldsymbol{\tau}} = \mathbf{B}^{\mathrm{T}}\boldsymbol{\tau}$. Equations (7.59) or (7.77) can then be rewrite in the bar-notation and solved for Λ.

7.7.2 Double-Difference Float and Triple-Difference Solutions

R receivers observing S satellites at T epochs generate at most RST carrier phase observations. In many cases, the data set might not be complete due to cycle slips and signal blockage. To see the symmetry of the expressions, we order the undifferenced phase observations $\boldsymbol{\psi}$ first by epoch, then by receiver, and then by satellite. For epoch i we have

$$\boldsymbol{\psi}_i = [\,\varphi_1^1(i) \quad \cdots \quad \varphi_1^S(i) \quad \cdots \quad \varphi_R^1(i) \quad \cdots \quad \varphi_R^S(i)\,]^{\mathrm{T}} \tag{7.78}$$

$$\boldsymbol{\psi} = \begin{bmatrix} \boldsymbol{\psi}_1 \\ \vdots \\ \boldsymbol{\psi}_T \end{bmatrix} \tag{7.79}$$

Regarding the stochastic model, we make the simplifying assumption that all carrier phase observations are uncorrelated and are of the same accuracy. Thus the complete $RST \times RST$ cofactor matrix of the undifferenced phase observations is

$$\mathbf{Q}_\varphi = \sigma_\varphi^2 \mathbf{I} \tag{7.80}$$

with σ_φ denoting the standard deviation of the phase measurement expressed in cycles.

The next task is to find the complete set of independent double-difference observations. We designate one station as the base station and one satellite as the base satellite. Without loss of generality, let station 1 be the base station, and satellite 1 be the base satellite. The session network of R stations is now thought of as consisting of $R - 1$ baselines emanating from the base station. There are $S - 1$ independent double differences for each baseline. Thus, a total of $(R - 1)(S - 1)$ independent double differences can be computed for the session network. On the basis of an ordered observation vector like (7.78), and the base station and base satellite ordering scheme, an independent set of double differences for epoch i is

$$\mathbf{\Delta}_i = \begin{bmatrix} \varphi_{12}^{12}(i) & \cdots & \varphi_{12}^{1S}(i) & \cdots & \varphi_{1R}^{12}(i) & \cdots & \varphi_{1R}^{1S}(i) \end{bmatrix}^{\mathrm{T}} \tag{7.81}$$

$$\mathbf{\Delta} = \begin{bmatrix} \mathbf{\Delta}_1 \\ \vdots \\ \mathbf{\Delta}_T \end{bmatrix} \tag{7.82}$$

The transformation from undifferenced to double-differenced observations is

$$\mathbf{\Delta} = \mathbf{D}\,\boldsymbol{\psi} \tag{7.83}$$

where $\mathbf{D}$ is the $(R - 1)(S - 1)T \times RST$ double-difference coefficient matrix. If we define the auxiliary matrix $\breve{\mathbf{I}}$ as

$$\breve{\mathbf{I}} = \begin{bmatrix} 1 & -1 & 0 & 0 \\ 1 & 0 & -1 & 0 \\ 1 & 0 & 0 & -1 \end{bmatrix} \tag{7.84}$$

then the pattern of $\mathbf{D}$ can be readily seen from Table 7.3. The boxes highlight the columns and rows that refer to a specific epoch. Each additional baseline adds another row and column to the highlighted area.

For the ordered vector of triple-difference observations, we have

$$\mathbf{\nabla}_i = \begin{bmatrix} \varphi_{12}^{12}(i+1,i) & \cdots & \varphi_{12}^{1S}(i+1,i) & \cdots & \varphi_{1R}^{12}(i+1,i) & \cdots & \varphi_{1R}^{1S}(i+1,i) \end{bmatrix}^{\mathrm{T}} \tag{7.85}$$

TABLE 7.3 Specification of the D Matrix

$\breve{\mathbf{I}}$	$-\breve{\mathbf{I}}$							
$\breve{\mathbf{I}}$		$-\breve{\mathbf{I}}$						
			$\breve{\mathbf{I}}$	$-\breve{\mathbf{I}}$				
			$\breve{\mathbf{I}}$		$-\breve{\mathbf{I}}$			
						$\breve{\mathbf{I}}$	$-\breve{\mathbf{I}}$	
						$\breve{\mathbf{I}}$		$-\breve{\mathbf{I}}$

Note: $R = 3$, $S = 4$, $T = 3$. The matrix $\mathbf{I}$ is of size 6.

$$\nabla = \begin{bmatrix} \nabla_1 \\ \vdots \\ \nabla_{T-1} \end{bmatrix} \tag{7.86}$$

$$\nabla = \mathbf{T}\,\boldsymbol{\Delta} = \mathbf{TD}\,\boldsymbol{\psi} \tag{7.87}$$

The matrix $\mathbf{T}$ might be called the epoch differencing coefficient matrix transforming double differences to triple differences. The product matrix $\mathbf{TD}$ might be called the triple-difference coefficient matrix that transforms single differences directly into double differences. The pattern of the $\mathbf{T}$ matrix is seen in Table 7.4. Each baseline adds one row and each epoch adds one column to this matrix.

The double- and the triple-difference observations are linear functions of the observed carrier phases. By applying covariance propagation and taking the cofactor matrix (7.80) into account, the respective cofactor matrices are

$$\mathbf{Q}_\Delta = \sigma_\varphi^2 \mathbf{D}\mathbf{D}^{\mathrm{T}} \tag{7.88}$$

$$\mathbf{Q}_\nabla = \mathbf{T}\mathbf{Q}_\Delta \mathbf{T}^{\mathrm{T}} \tag{7.89}$$

The double-difference cofactor matrix $\mathbf{Q}_\Delta$ is block-diagonal. The diagonal submatrix in the case of $R = 3$ and $S = 4$ is

TABLE 7.4 Specifications of the T Matrix

$-\mathbf{I}$	$\mathbf{I}$	
	$-\mathbf{I}$	$\mathbf{I}$

Note: $R = 3$, $S = 4$, $T = 3$. The matrix $\mathbf{I}$ is of size 6

$$\breve{\mathbf{Q}}_\Delta = \begin{bmatrix} 4 & 2 & 2 & 2 & 1 & 1 \\ 2 & 4 & 2 & 1 & 2 & 1 \\ 2 & 2 & 4 & 1 & 1 & 2 \\ 2 & 1 & 1 & 4 & 2 & 2 \\ 1 & 2 & 1 & 2 & 4 & 2 \\ 1 & 1 & 2 & 2 & 2 & 4 \end{bmatrix} \tag{7.90}$$

Each epoch adds a block to the diagonal of $\mathbf{Q}_\Delta$. The triple-difference cofactor matrix is in the case of $R = 3$, $S = 4$, and $T = 3$,

$$\breve{\mathbf{Q}}_\nabla = \begin{bmatrix} 2\breve{\mathbf{Q}}_\Delta & -\breve{\mathbf{Q}}_\Delta \\ -\breve{\mathbf{Q}}_\Delta & 2\breve{\mathbf{Q}}_\Delta \end{bmatrix} \tag{7.91}$$

The triple-difference cofactor matrix is band-diagonal for $T > 3$. The triple-difference observations between consecutive (adjacent) epochs are correlated. The inverse of the triple-difference cofactor matrix, which is required in the least-squares solution, is a full matrix. Eren (1987) gives an algorithm for computing the elements of the cofactor matrices (7.88) and (7.89), requiring no explicit matrix multiplication. The subscripts and superscripts of the undifferenced phase observations are used to compute the elements of the cofactor matrices directly.

The relevant terms of the double-difference carrier phase equation (5.25) are

$$\begin{aligned} \varphi_{km}^{pq} &= \frac{f}{c}\left\{\left\|\mathbf{x}^p - \mathbf{x}_k\right\| - \left\|\mathbf{x}^p - \mathbf{x}_m\right\| - \left\|\mathbf{x}^q - \mathbf{x}_k\right\| + \left\|\mathbf{x}^q - \mathbf{x}_m\right\|\right\} \\ &\quad + N_{km}^{pq} + d_{km,\varphi}^{pq} + \varepsilon_{km,\varphi}^{pq} \\ &= \frac{f}{c}\rho_{km}^{pq} + N_{km}^{pq} + d_{km,\varphi}^{pq} + \varepsilon_{km,\varphi}^{pq} \end{aligned} \tag{7.92}$$

The residual ionospheric and tropospheric terms are not explicitly listed in (7.92) since they are expected to cancel over short baselines. Notice the addition of the ambiguity term N_{km}^{pq} in (7.92) as compared to the expression (7.64) for pseudoranges. Assuming again that the station coordinates $\mathbf{x}_k$ are known, the parameters to be estimated are $\mathbf{x}_m$ and the double-difference ambiguities. Since the carrier phase multipath $d_{km,\varphi}^{pq}$ is not known it is typically treated as a model error and ignored. A row of the design matrix consists of

$$\frac{\partial \varphi_{km}^{pq}}{\partial \mathbf{x}_m} = \frac{f}{c}\left(\mathbf{e}_m^p - \mathbf{e}_m^q\right) \tag{7.93}$$

and contains a 1 in the column of the respective double-difference ambiguity parameter, and zero elsewhere. The least-squares solution or Kalman filter solution that estimates the parameters

$$\mathbf{x} = \begin{bmatrix} \mathbf{x}_m \\ \mathbf{b} \end{bmatrix} \tag{7.94}$$

$$\mathbf{b}^{\mathrm{T}} = \left[N_{12}^{12} \cdots N_{12}^{1S} \quad N_{13}^{12} \cdots N_{13}^{1S} \right] \tag{7.95}$$

is called the double-difference float solution.

Finally, the partial derivatives of triple differences follow from those of double differences by differencing

$$\frac{\partial \varphi_{km}^{pq}(j,i)}{\partial \mathbf{x}_m} = \frac{\partial \varphi_{km}^{pq}(j)}{\partial \mathbf{x}_m} - \frac{\partial \varphi_{km}^{pq}(i)}{\partial \mathbf{x}_m} \tag{7.96}$$

because the triple difference is the difference of two double differences. The design matrix of the triple difference contains no columns for the initial ambiguities, because these parameters cancel during the differencing.

7.7.3 Independent Baselines

The ordering scheme of base station and base satellite used for identifying the set of independent double-difference observations is not the only scheme available. It has been used here because of its simplicity. An example where the base station and base satellite scheme requires a slight modification occurs when the base station does not observe at a certain epoch due to temporary signal blockage or some other cause. If station 1 does not observe, then the double difference $\Delta\varphi_{23}^{pq}$ can be computed for this particular epoch. Because of the relationship

$$\varphi_{23}^{pq} = \varphi_{13}^{pq} - \varphi_{12}^{pq} \tag{7.97}$$

the ambiguity N_{23}^{pq} is related to the base station ambiguities as

$$N_{23}^{pq} = N_{13}^{pq} - N_{12}^{pq} \tag{7.98}$$

Introduction of N_{23}^{pq} as an additional parameter would create a singularity of the normal matrix because of the dependency expressed in (7.98). Instead of adding this new ambiguity, the base station ambiguities N_{12}^{pq} and N_{13}^{pq} are given the coefficients 1 and -1, respectively, in the design matrix. The partial derivatives with respect to the station coordinates can be computed as required by (7.97) and entered directly into the design matrix, because the respective columns are already there. A similar situation arises when the base satellite changes. The linear functions in this case are:

$$\varphi_{km}^{23} = \varphi_{km}^{13} - \varphi_{km}^{12} \tag{7.99}$$

$$N_{km}^{23} = N_{km}^{13} - N_{km}^{12} \tag{7.100}$$

The respective elements for the base satellite ambiguities in the design matrix are, again, 1 and -1.

One must identify $(R-1)(S-1)$ independent double-difference functions in network solutions. In session networks that contain a mixture of long and short baselines, it might be important to take advantage of short baselines because the respective unmodeled errors (troposphere, ionosphere, and possibly orbit) are expected to be small. Fixing the ambiguities to integers adds strength to the solution. This additional strength gained by fixing the ambiguities of a short baseline may also make it possible to fix the ambiguities for the next longer baseline, even though the ambiguity search algorithms might not have been successful without that constraint. The technique is sometimes referred to as "boot-strapping" from shorter to longer baselines. A suitable procedure would be to take baselines in all combinations and order them by increasing length and identify the set of independent baselines, starting with the shortest.

There are several schemes available to identify independent baselines and observations. Hilla and Jackson (2000) report using a tree structure and edges. Here we follow the suggestion of Goad and Mueller (1988) because it highlights yet another useful application of the Cholesky decomposition. Assume that matrix $\mathbf{D}$ of (7.83) reflects the ordering suggested here; i.e., the first rows of $\mathbf{D}$ refer to the double differences of the shortest baseline, the next set of rows refer to the second shortest baseline, and so on. We write the cofactor matrix (7.88) as

$$\mathbf{Q}_\Delta = \sigma_0^2\,\mathbf{D}\mathbf{D}^\mathrm{T} = \sigma_0^2\,\mathbf{L}\,\mathbf{L}^\mathrm{T} \tag{7.101}$$

where $\mathbf{L}$ denotes the Cholesky factor (A.94). The elements of the cofactor matrix $\mathbf{Q}_\Delta$ are

$$q_{ij} = \sum_k d_i(k)\,d_j(k) \tag{7.102}$$

where $d_i(k)$ denotes the ith row of the matrix $\mathbf{D}$. It is readily verified that the ith and jth columns of $\mathbf{Q}_\Delta$ are linearly dependent if the ith and jth rows of $\mathbf{D}$ are linearly dependent. In such a case $\mathbf{Q}_\Delta$ is singular. This situation exists when two double differences are linearly dependent. The diagonal element j of the Cholesky factor $\mathbf{L}$ will be zero. Thus, one procedure for eliminating the dependent observations is to carry out the computation of $\mathbf{L}$ and to discard those double differences that cause a zero on the diagonal. The matrix $\mathbf{Q}_\Delta$ can be computed row by row starting at the top; i.e., the double differences can be processed sequentially one at a time, from the top to the bottom. For each double difference, the respective row of $\mathbf{L}$ can be computed. In this way, the dependent observations can be immediately discovered and removed. Only the independent observations remain. The process ends as soon as $(R-1)\times(S-1)$ double differences have been found.

If all receivers observe all satellites for all epochs, this identification process needs to be carried out only once. The matrix $\mathbf{L}$, since it is now available, can be used to decorrelate the double differences. The corresponding residuals might be difficult to interpret, but could be transformed to the original observational space using $\mathbf{L}$ again.

7.7.4 Ambiguity Function

The least-squares techniques discussed above require partial derivatives and the minimization of $\mathbf{v}^T\mathbf{P}\mathbf{v}$, with $\mathbf{v}$ and $\mathbf{P}$ being the double-difference residuals and double-difference weight matrix. The derivatives and the discrepancy terms depend on the assumed approximate coordinates of the stations. The least-squares solution is iterated until the solution converges. In the case of the ambiguity function technique, we search for station coordinates that maximize the cosine of the residuals. Consider again the double-difference observation equation

$$v_{km}^{pq} = \varphi_{km,a}^{pq} - \varphi_{km,b}^{pq} = \frac{f}{c}\rho_{km,a}^{pq} + N_{km,a}^{pq} - \varphi_{km,b}^{pq} \tag{7.103}$$

In usual adjustment notation, the subscripts a and b denote the adjusted and the observed values, respectively. In (7.103) we have neglected again the residual double-difference ionospheric and tropospheric terms, as well as the signal multipath term. The residuals in units of radians are

$$\psi_{km}^{pq} = 2\pi\, v_{km}^{pq} \tag{7.104}$$

The key idea of the ambiguity function technique is to realize that a change in the integer N_{km}^{pq} changes the function ψ_{km}^{pq} by a multiple 2π and that the cosine of this function is not affected by such a change because

$$\cos\left(\psi_{km,L}^{pq}\right) = \cos\left(2\pi\, v_{km,L}^{pq}\right) = \cos\left[2\pi\left(v_{km,L}^{pq} + \Delta N_{km,L}^{pq}\right)\right] \tag{7.105}$$

where $\Delta N_{km,L}^{pq}$ denotes the arbitrary integer. The subscript L, denoting the frequency identifier, has been added for the purpose of generality.

There are $2\,(R-1)\,(S-1)$ double differences available for dual-frequency observations . If we further assume that all observations are equally weighted, then the sum of the squared residuals becomes, with the help of (7.104),

$$\mathbf{v}^T\mathbf{P}\mathbf{v}\left(\mathbf{x}_m, N_{km,L}^{pq}\right) = \sum_{L=1}^{2}\sum_{m=1}^{R-1}\sum_{q=1}^{S-1}\left(v_{km,L}^{pq}\right)^2 = \frac{1}{4\pi^2}\sum_{L=1}^{2}\sum_{m=1}^{R-1}\sum_{q=1}^{S-1}\left(\psi_{km,L}^{pq}\right)^2 \tag{7.106}$$

If the station coordinates $\mathbf{x}_k$ are known, the function could be minimized by varying the coordinates $\mathbf{x}_m$ and the ambiguities using least-squares estimation. The ambiguity function is defined as

$$\begin{aligned} AF(\mathbf{x}_m) &\equiv \sum_{L=1}^{2}\sum_{m=1}^{R-1}\sum_{q=1}^{S-1}\cos\left(\psi_{km,L}^{pq}\right) \\ &= \sum_{L=1}^{2}\sum_{m=1}^{R-1}\sum_{q=1}^{S-1}\cos\left\{2\pi\left[\frac{f_L}{c}\rho_{km,a}^{pq} + N_{km,L,a}^{pq} - \varphi_{km,L,b}^{pq}\right]\right\} \end{aligned}$$

$$= \sum_{L=1}^{2} \sum_{m=1}^{R-1} \sum_{q=1}^{S-1} \cos \left\{ 2\pi \left[\frac{f_L}{c} \rho_{km,a}^{pq} - \varphi_{km,L,b}^{pq} \right] \right\} \tag{7.107}$$

The small double-difference ionospheric, tropospheric, and multipath terms are not listed explicitly in this equation, although they are present and will affect the ambiguity function technique just as they do the other solution methods. Nevertheless, if we assume for a moment that these terms are negligible, and that the receiver positions are perfectly known, then Equation (7.107) shows that the maximum value of the ambiguity function is $2\,(R-1)\,(S-1)$ because the cosine of each term could be 1. Observational noise will cause the value of the ambiguity function to be slightly below the theoretical maximum. Since the ambiguity function does not depend on the ambiguities, it is also independent of cycle slips. This invariant property is the most outstanding feature of the ambiguity function and is unique among all the other solution methods.

Because the function $\psi_{km,L}^{pq}$ in (7.104) is small when good approximate coordinates are available (typically corresponding to several hundredths of a cycle), we can expand the cosine function in a series and neglect higher-order terms. Thus,

$$AF(\mathbf{x}_m) = \sum_{L=1}^{2} \sum_{m=1}^{R-1} \sum_{q=1}^{S-1} \cos \psi_{km,L}^{pq} = \sum_{L=1}^{2} \sum_{m=1}^{R-1} \sum_{q=1}^{S-1} \left[1 - \frac{\left(\psi_{km,L}^{pq}\right)^2}{2!} + \cdots \right]$$

$$= 2(R-1)\,(S-1) - \frac{1}{2} \sum_{L=1}^{2} \sum_{m=1}^{R-1} \sum_{q=1}^{S-1} \left(\psi_{km,L}^{pq}\right)^2 \tag{7.108}$$

$$= 2(R-1)\,(S-1) - 2\pi^2 \mathbf{v}^{\mathrm{T}}\mathbf{P}\mathbf{v}$$

The last part of this equation follows from (7.106). The ambiguity function and the least-squares solution are equivalent in the sense that the ambiguity function reaches maximum and $\mathbf{v}^{\mathrm{T}}\mathbf{P}\mathbf{v}$ minimum at the point of convergence, the correct $\mathbf{x}_m$ (Lachapelle et al., 1992).

There are several ways to initialize an ambiguity function solution. The simplest procedure is to use a search volume centered at some initial estimate of the station coordinates $\mathbf{x}_m$. Such an estimate could be computed from point positioning with pseudoranges; the size of the search volume would be a function of the accuracy of the estimate. This physical search volume is subdivided into a narrow grid of points with equal spacing. Each grid point is considered a candidate for the solution and used to compute the ambiguity function (7.107). The double-difference ranges $\rho_{km,a}^{pq}$, which are required in (7.107), are evaluated for the trial position. As the ambiguity function is computed by adding the individual cosine terms one double difference at a time, early exit strategies can be implemented to reduce the computational effort. For example, if the trial position differs significantly from the true position, the residuals are likely to be bigger than one would expect due to measurement noise, unmodeled ionospheric and tropospheric effects, and the multipath. An appropriate strategy could be to abandon the current trial position, i.e., to stop accumulating the ambiguity function and to begin with the next trial position. This would occur as soon as one term is below the cutoff criteria, e.g.,

$$\cos\left\{2\pi\left[\varphi_{km,L,a}^{pq}(t) - \varphi_{km,L,b}^{pq}(t)\right]\right\}_i < \varepsilon \tag{7.109}$$

The choice of the cutoff criteria ε is critical, not only for accelerating solutions, but also for assuring that the correct solution is not missed. This early exit strategy is unforgiving in the sense that once the correct (trial) position is rejected, the scanning of the remaining trial positions cannot yield the correct solution. The proper choice for ε is largely experimental.

A matter of concern is that the grid of trial positions is close enough to assure that the true solution is not missed. Of course, a very narrow spacing of the trial positions increases the computational load, despite the early exit strategy. The optimal spacing is somewhat related to the wavelength and to the number of satellites. On the other hand, the ambiguity function technique can be modified in several ways in order to increase its speed, such as using the double-difference widelanes. In this case, the trial positions can initially be widely spaced to reflect the wide-lane wavelength of 86 cm. These solutions could serve to identify a smaller physical search space, which can then be scanned using narrowly spaced trial positions.

The ambiguity function technique offers no opportunity to take the correlation between the double-difference observables into account. There is no direct accuracy measure for the final position that maximizes the ambiguity function, such as standard deviations of the coordinates. The quality of the solution is related to the spacing of the trial positions. If the trial positions, e.g., have a 1 cm spacing and a maximum of the ambiguity function is uniquely identified, then one could speak of centimeter accurate positioning. In order to arrive at a conventional accuracy measure, one can take the position that maximizes the ambiguity function and carry out a regular double-difference least-squares solution. Because the initial positions for this least-squares solution are already very accurate, a single iteration is sufficient and it should be possible to fix the integer. The fixed solution would give the desired statistical information.

The ambiguity function values of all trial positions are ordered by size and normalized (dividing by the number of observations). Often, peaks of lesser value surround the highest peak and it might be impossible to identify the maximum reliably. This situation typically happens when the observational strength is lacking. The solution can be improved by observing longer, selecting a better satellite configuration, using dual-frequency observations, etc.

The strength of the ambiguity function approach lies in the fact that the correct solution is obtained even if the data contain cycle slips. Remondi (1984) discusses the application of the ambiguity function technique to single differences. The geodetic use of the ambiguity function technique seems to be traceable to very long baseline interferometry (VLBI) observation processing. Counselman and Gourevitch (1981) present a very general ambiguity function technique and discuss in detail the patterns to be expected for various trial solutions.

7.7.5 Initialization on the Ground

A kinematic survey requires an initialization. This means the double-difference ambiguities are resolved first and then held fixed while other points are being surveyed, assuming of course that no cycle slips occurred while the rover moves or that cycle slips are repaired appropriately. A simple way for initial determination of ambiguities is to occupy two known stations. The procedure works best for short baselines where the ionospheric and tropospheric disturbances are negligible. The double-difference equation (7.92) can be readily solved for the ambiguity,

$$N_{km}^{pq} = \varphi_{km}^{pq} - \lambda^{-1} \rho_{km}^{pq} \tag{7.110}$$

when both receiver locations $\mathbf{x}_k$ and $\mathbf{x}_m$ are known. Usually simple rounding of the computed values is sufficient to obtain the integers. Once the initial ambiguities are known, the kinematic survey can begin. Let the subscripts k and m now denote the fixed and the moving receiver, then

$$\rho_m^{pq} = \rho_k^{pq} - \lambda \left\{ \varphi_{km}^{pq} - N_{km}^{pq} \right\} \tag{7.111}$$

If four satellites are observed simultaneously, there are three equations like (7.111) available to compute the coordinates of the moving receiver $\mathbf{x}_m$. If more than four satellites are available, the usual least-squares approach is applicable and cycle slips can be repaired from phase observations. In principle, if five satellites are observed we can repair one slip per epoch, if six satellites are observed, two slips can occur at the same time, etc.

Remondi (1985) introduced the antenna swap procedure in order to initialize the ambiguities for kinematic surveying. Assume that four or more satellites were observed at least for one epoch while receiver R_1 and its antenna were located at station k and receiver R_2 and its antenna were at station m. This is followed by the antenna swap, meaning that antenna R_1 moves to station m and antenna R_2 moves to station k, followed by at least one epoch of observations to the same satellites. The antennas remain connected to their respective receivers. During data processing, it is assumed that the antennas never moved. Using an expanded form of notation to identify the receiver and the respective observation, a double difference at epoch 1 when R_1 was at k and at epoch t when R_1 was at m can be written, respectively, as

$$\varphi_{km}^{pq}(R_2 - R_1, 1) = \lambda^{-1} \left[\rho_k^p(R_1, 1) - \rho_k^q(R_1, 1) - \rho_m^p(R_2, 1) + \rho_m^q(R_2, 1) \right] + N_{km}^{pq} \tag{7.112}$$

$$\varphi_{km}^{pq}(R_2 - R_1, t) = \lambda^{-1} \left[\rho_m^p(R_1, t) - \rho_m^q(R_1, t) - \rho_k^p(R_2, t) + \rho_k^q(R_2, t) \right] + N_{km}^{pq} \tag{7.113}$$

Notice the sequence of subscripts on the right-hand side of (7.113). Differencing both observations gives

$$\varphi_{km}^{pq}(R_2 - R_1, 1) - \varphi_{km}^{pq}(R_2 - R_1, t) = \lambda^{-1}\left[\rho_k^{pq}(t) - \rho_m^{pq}(t) + \rho_k^{pq}(1) - \rho_m^{pq}(1)\right]$$
$$\approx 2\lambda^{-1}\left[\rho_k^{pq}(t) - \rho_m^{pq}(t)\right] \qquad (7.114)$$

Equation (7.114) can be solved for $\mathbf{x}_m$, given $\mathbf{x}_k$ and observations to at least four satellites (three double differences). Once the position of m is known, the ambiguities can be computed from (7.110).

If the topocentric satellite distances did not change during the antenna swapping due to motion of the satellites, the antenna swap technique would yield a baseline vector of twice the actual length. The geometry of antenna swap can be readily visualized in a simplified one-dimensional situation. Consider a horizontal baseline and a satellite located somewhere along the extension of that baseline. As one antenna moves from one end of the baseline to the other, it will register, let's say, a positive accumulated carrier phase change equal to the length of the baseline. As the other antenna switches location, it will also register a carrier phase change equal to the negative of the length of the baseline. Both receivers together will register a motion of twice the length of the baseline.

Initialization by antenna swap on the ground is conveniently done for a very short baseline of a couple of meters. A typical point positioning solution for $\mathbf{x}_k$ is sufficient for such short baselines. See Equation (5.66) regarding the relationship between accuracy requirements for $\mathbf{x}_k$ as a function of the length of the baseline.

The initialization by antenna swap works for single- and dual-frequency receivers. At the time it was introduced it represented a major move forward in making kinematic surveying attractive in practice because at the time, surveyors operated mostly single-frequency receivers only. With dual-frequency receivers, all-in-view satellite observations, and optimized algorithms that include such optimal integer estimators as LAMBDA (to be discussed below), a baseline can be readily initialized without antenna swap. In fact, initialization, i.e., ambiguity fixing can even occur on-the-fly (OTF) while the roving receiver is moving. Under the right conditions, i.e., dual-frequency observation, many satellites visible, good antenna, etc. the ambiguities can be fixed every epoch, making not only antenna swap obsolete but also masking the difference between static and kinematic techniques. Including GLONASS satellites will improve the reliability of epoch-by-epoch ambiguity resolutions, as will future Galileo satellites and the addition of the L5 frequency to GPS, which is part of the GPS modernization.

7.7.6 GLONASS Carrier Phase Processing

The GLONASS satellites transmit at different carrier frequencies as specified by (3.91) and (3.92). Maintaining the notation introduced in Chapter 3 we use a superscript to identify the GLONASS satellite frequency, using the channel number. Since the GPS satellites use the same frequency, there is no need for this extra superscript to identify GPS satellite frequencies. In general, the superscript q and s are used to identify any of the S_{GPS} GPS or S_{GLO} GLONASS satellites, respectively. The superscripts p and r denote the respective base satellites. Following this notation, the single-difference observations for a GPS and a GLONASS satellite, respectively, can then be written as

$$\varphi^{q}_{km,1,\mathrm{GPS}} = \frac{f_1}{c}\rho^{q}_{km} + N^{q}_{km,1,\mathrm{GPS}} + d_{km,1,\mathrm{GPS}} - f_1\, dt_{\underline{k}m} \tag{7.115}$$

$$\varphi^{s}_{km,1,\mathrm{GLO}} = \frac{f^{s}_1}{c}\rho^{s}_{km} + N^{s}_{km,1,\mathrm{GLO}} + d_{km,1,\mathrm{GLO}} - f^{s}_1\, dt_{\underline{k}m} \tag{7.116}$$

These equations are based on the assumption that the receiver clock errors $dt_{\underline{k}m}$ are the same for both types of observations, GPS and GLONASS. The receiver hardware delays $d_{km,1,\mathrm{GPS}}$ and $d_{km,1,\mathrm{GLO}}$, on the contrary, are dealt with separately. We have neglected the signal multipath terms.

In case of GPS-only processing, one can combine the receiver delay $d_{km,1,\mathrm{GPS}}$ and clock $f_1 dt_{\underline{k}m}$ term into a new receiver-dependent term ξ_{km} that is estimated every epoch. The station coordinates, the single-difference ambiguities, and the epoch parameter ξ_{km} can then be estimated from observations to several satellites over a number of epochs using either least-squares or Kalman filtering. The usual ambiguity fixing techniques can be applied. In the case of combined processing of GPS and GLONASS single differences one uses the satellite-dependent parameterization,

$$\xi^{q}_{km,\mathrm{GPS}} = N^{q}_{km,1,\mathrm{GPS}} + d_{km,1,\mathrm{GPS}} \tag{7.117}$$

$$\xi^{s}_{km,\mathrm{GLO}} = N^{s}_{km,1,\mathrm{GLO}} + d_{km,1,\mathrm{GLO}} \tag{7.118}$$

We note that ξ parameters are constants in time but are not integers because of the receiver hardware delays. Using these auxiliary parameters the single-difference equations become

$$\varphi^{q}_{km,1,\mathrm{GPS}} = \frac{f_1}{c}\rho^{q}_{km} + \xi^{q}_{km,1,\mathrm{GPS}} - f_1\, dt_{\underline{k}m} \tag{7.119}$$

$$\varphi^{s}_{km,1,\mathrm{GLO}} = \frac{f^{s}_1}{c}\rho^{s}_{km} + \xi^{s}_{km,1,\mathrm{GLO}} - f^{s}_1\, dt_{\underline{k}m} \tag{7.120}$$

which allow us to estimate the station coordinates, the ξ constants, and the epoch clock parameters, again, using classical least-squares or Kalman filtering formulation. Unfortunately, the usual ambiguity fixing techniques cannot be directly applied to this single-difference formulation because the ξ parameters are not integers. However, one can still fix the double-difference ambiguities.

An immediate outcome of using (7.119) and (7.120) are the estimates $\hat{\xi}^{q}_{km,1,\mathrm{GPS}}$, $\hat{\xi}^{s}_{km,1,\mathrm{GLO}}$ and their respective variance-covariance matrix, denoted by the symbol $\mathbf{C}_{\xi}$. Using a matrix $\mathbf{D}$ containing elements 1, -1, and 0 at appropriate places, the estimated double-difference ambiguities with respect to the GPS reference satellite p and GLONASS reference satellite r are

$$\begin{bmatrix} \hat{N}^{pq}_{km,\text{GPS}} \\ \vdots \\ \hat{N}^{rs}_{km,1,\text{GLO}} \\ \vdots \end{bmatrix} = {}_{S_S+S_{\text{GLO}}-2}\mathbf{D}_{S_S+S_{\text{GLO}}} \begin{bmatrix} \hat{\xi}^{q}_{km,\text{GPS}} \\ \vdots \\ \hat{\xi}^{s}_{km,\text{GLO}} \\ \vdots \end{bmatrix} \tag{7.121}$$

Applying the covariance propagation (4.34), the covariance matrix of double-difference ambiguity is $\mathbf{\Sigma}_N = \mathbf{D}\mathbf{C}_{\xi}\mathbf{D}^{\mathrm{T}}$. Having $\mathbf{\Sigma}_N$ it is possible to determine the integer double-difference ambiguities using a technique such as LAMBDA. The subsequent constraint solution, in which the integer ambiguities are treated as known quantities, yields the final estimates for the station coordinates, the S_{GPS} parameters $\hat{\xi}^{q}_{km,1,\text{GPS}|N}$ and S_{GLO} parameters $\hat{\xi}^{s}_{km,1,\text{GLO}|N}$, as well as the final time estimates $d\hat{\underline{t}}_{km|N}$.

The variance-covariance propagation step can be avoided by using a parameterization in terms of $\xi^{p}_{km,1,\text{GPS}}$ for the base GPS satellite and the GPS double differences $\xi^{pq}_{km,1,\text{GPS}} \equiv N^{pq}_{km,1,\text{GPS}}$. The respective parameters for the GLONASS satellites are $\xi^{r}_{km,1,\text{GLO}}$ and $\xi^{rs}_{km,1,\text{GLO}} = N^{rs}_{km,1,\text{GLO}}$. A submatrix of the design matrix that reflects this parameterization is shown in Table 7.5. Ambiguity fixing can then be directly applied to the variance-covariance submatrix of the estimates $\hat{\xi}^{pq}_{km,1,\text{GPS}}$ and $\hat{\xi}^{rs}_{km,1,\text{GLO}}$.

Having the parameter estimation completed, using single-difference observations and fixing GPS-GPS and GLO-GLO double-difference ambiguities, we can inspect the double difference (GPS and GLO base satellites)

$$\begin{aligned} \Delta\hat{\xi}^{pr}_{km,1} &= \hat{\xi}^{p}_{km,1,\text{GPS}|N} - \hat{\xi}^{r}_{km,1,\text{GLO}|N} = N^{pr}_{km,1} + d_{km,1,\text{GLO}} - d_{km,1,\text{GPS}} \\ &= N^{pr}_{km,1} + \text{DDRB} \end{aligned} \tag{7.122}$$

TABLE 7.5 Submatrix for Alternate Parameterization of ξ

1								
1	1							
1		1						
1			1					
1				1				
					1			
					1	1		
					1		1	
					1			1

Note: The table shows the case for $S_{\text{GPS}} = 5$ and $S_{\text{GLO}} = 4$.

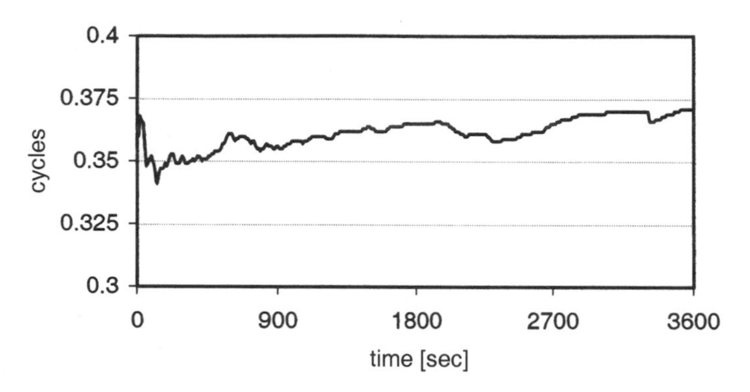

Figure 7.17 DDRB differences between GPS and GLONASS with fixed GPS-GPS and GLO-GLO double-difference ambiguities.

Since $N_{km,1}^{pr}$ is an integer the fractional part of $\Delta\hat{\xi}_{km,1}^{pr}$ is an estimate of the double-difference L1 receiver hardware bias, labeled DDRB, as applied to GPS and GLONASS. Figure 7.17 shows the estimated DDRB every 10 sec for a 1-hour series of observations of a 10 m baseline at Irwin, California, using 3S Navigation receivers that measured L1 pseudorange and carrier phases of GPS and GLONASS satellites on June 12, 1998. All GPS satellites were observed; there were four GLONASS satellites available during that hour. Figure 7.17 seems to suggest that one could further strengthen the combined GPS and GLONASS solution by modeling the DDRB as a constant.

The conventional double differencing of carrier phase observations has the well-known form

$$\varphi_{km,1,\text{GPS}}^{pq} = \frac{f_1}{c}\rho_{km}^{pq} + N_{km,1,\text{GPS}}^{pq} \tag{7.123}$$

$$\varphi_{km,1,\text{GLO}}^{rs} = \frac{f_1^r}{c}\rho_{km}^{r} - \frac{f_1^s}{c}\rho_{km}^{s} + N_{km,1,\text{GLO}}^{rs} - \left(f_1^r - f_1^s\right) d\underline{t}_{km} \tag{7.124}$$

The GLONASS double differences depend on the receiver clock error and the frequencies. This dependency is demonstrated in Figure 7.18, which shows the functions

$$\varphi_{km,1,b}^{ps} - \frac{f_1^s}{c}\rho_{km,0}^{s} + \frac{f_1}{c}\rho_{km,0}^{p} + \Delta_{km}^{ps} = -\left(f_1 - f_1^s\right) d\underline{t}_{km} \tag{7.125}$$

where the observations have been corrected for the topocentric satellite distances which have been evaluated for the known station coordinates and translated by Δ_{km}^{ps} for the lines to go through zero at the first epoch. The observational data are the same as those used in Figure 7.17. The order of the lines corresponds to that of the frequencies f_1^s. Equations (7.123) and (7.124) are suitable for estimating the double-difference integers, as long as the receiver clock differences are estimated at the same time.

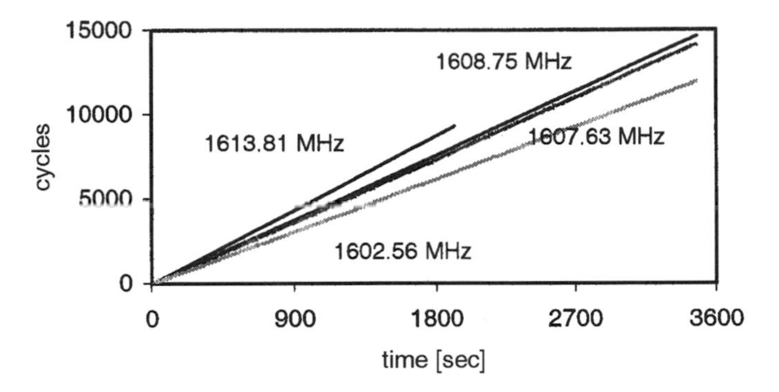

Figure 7.18 Impact of receiver clock errors on GLONASS double-differenced observations.

Scaling the carrier phases to distances, or to a mean GLONASS frequency, or to f_1^r or f_1^s eliminates the receiver clock term but introduces a linear combination of single-difference ambiguities whose coefficients are nonintegers. For example, we can write

$$\varphi_{km,1,\mathrm{GLO}}^{rs} = \varphi_{km,1,\mathrm{GLO}}^{r} - \frac{f_1^r}{f_1^s}\varphi_{km,1,\mathrm{GLO}}^{s} = \frac{f_1^r}{c}\rho_{km}^{rs} + \tilde{N}_{km,1,\mathrm{GLO}}^{rs} + \frac{f_1^r}{f_1^s}N_{km,1,\mathrm{GLO}}^{s} \tag{7.126}$$

The term $\tilde{N}_{km,1,\mathrm{GPS}}^{rs}$ is an integer. In practical applications, one can compute an approximate value $N_{km,1,\mathrm{GLO},0}^{r}$ for the single-difference ambiguity from (7.116) using station coordinates and receiver clock estimates computed from pseudoranges. Note that point positioning with GPS or GLONASS is conceptually the same, i.e., the GLONASS point positioning is not burdened with ambiguity issues or extra receiver clock complications. The double-difference GLONASS observation (7.126) can then be written as

$$\varphi_{km,1,\mathrm{GLO}}^{rs} - \frac{f_1^r}{f_1^s}N_{km,1,\mathrm{GLO},0}^{s} = \frac{f_1^r}{c}\rho_{km}^{rs} + \tilde{N}_{km,1,\mathrm{GLO}}^{rs} + \eta^{rs} \tag{7.127}$$

with

$$\eta^{rs} = \frac{f_1^r - f_1^s}{f_1^s}dN_{km,1,\mathrm{GLO}}^{s} \leq 0.01dN_{km,1,\mathrm{GLO}}^{s} \tag{7.128}$$

The size of the η term depends on the quality of the initial estimate $N_{km,1,\mathrm{GLO},0}^{s}$, since $dN_{km,1,\mathrm{GLO}}^{s} = N_{km,1,\mathrm{GLO}}^{s} - N_{km,1,\mathrm{GLO},0}^{s}$. If we neglect this term, (7.127) has the same form as a double-difference equation for GPS. However, neglecting the η term causes a model error that might make ambiguity fixing difficult, if not impossible, depending on the accuracy of the approximation $N_{km,1,\mathrm{GLO},0}^{s}$. The float solution does not require the η term.

The GLONASS broadcast ephemeris provides the satellite positions in the PZ90 coordinates system. Much effort has been made to compute accurate transformation parameters between PZ90 and WGS84. For example, see Bazlov et al. (1999a,b) and references therein. For small baselines any residual error due to the inaccurate transformation is likely to cancel when single differencing. When performing point positioning with pseudoranges it might be important to keep in mind that the GLONASS clocks are stirred according to UTC(SU) whereas GPS time follows UTC(USNO) within prescribed margins. In the case of the precise ephemeris computed by the IGS the reference frame is ITRF for both systems and time refers to a common standard.

Of course, the GLONASS observables can be used to form popular functions, such as the ionospheric, the ionospheric-free, the wide-lane ambiguity, and the multipath functions. Using the definition of the frequencies (3.91) and (3.92), we obtain the expressions that correspond to (5.13) to (5.16) for GLONASS

$$\alpha_{\text{GLO}} = \left(f_1^p / f_2^p\right)^2 = (9/7)^2 = 81/49 \tag{7.129}$$

$$\beta_{\text{GLO}} = \alpha_{\text{GLO}}/(\alpha_{\text{GLO}} - 1) = 81/32 \tag{7.130}$$

$$\gamma_{\text{GLO}} = 1/(\alpha_{\text{GLO}} - 1) = 49/32 \tag{7.131}$$

$$\delta_{\text{GLO}} = \sqrt{\alpha_{\text{GLO}}}/(\alpha_{\text{GLO}} - 1) = 63/32 \tag{7.132}$$

GLONASS has attracted a lot of attention, not only because of its potential to increase the number of usable satellites but also because of the fact that its satellites transmit at different carrier frequencies. The following is a sample of relevant literature: Raby and Daly (1993), Leick et al. (1995, 1998), Gourevitch et al. (1996), Povalyaev (1997), Pratt et al. (1997), Rapoport (1997), Kozlov and Tkachenko (1998), Roßbach (2001), and Wang et al. (2001).

7.7.7 Relative Positioning within CORS

The National Geodetic Survey (NGS) is the lead agency in establishing and operating the continuously operating reference station (CORS) system (NGS, 2002). The cooperation includes academic, commercial, and private organizations. While NGS does not guarantee that a particular site is operating at any given time, it tries to expand the system so that all points in the contiguous United States will be within a specified distance from an operational site. See Figure 7.19 for a map of CORS sites; at this writing there are more than 420 participating sites.

Tying to CORS stations is a practical way of connecting points determined by GPS to the geodetic frame. All CORS sites are known with centimeter accuracy in either the ITRF reference system or the NAD83 geodetic datum. The dual-frequency carrier phase and pseudorange observations of any CORS site, plus other ancillary station data and the precise ephemeris can be readily downloaded for postprocessing by the user. NGS also offers a processing service, online positioning user service (OPUS), via the Internet. The user simply uploads the observation files to NGS and receives

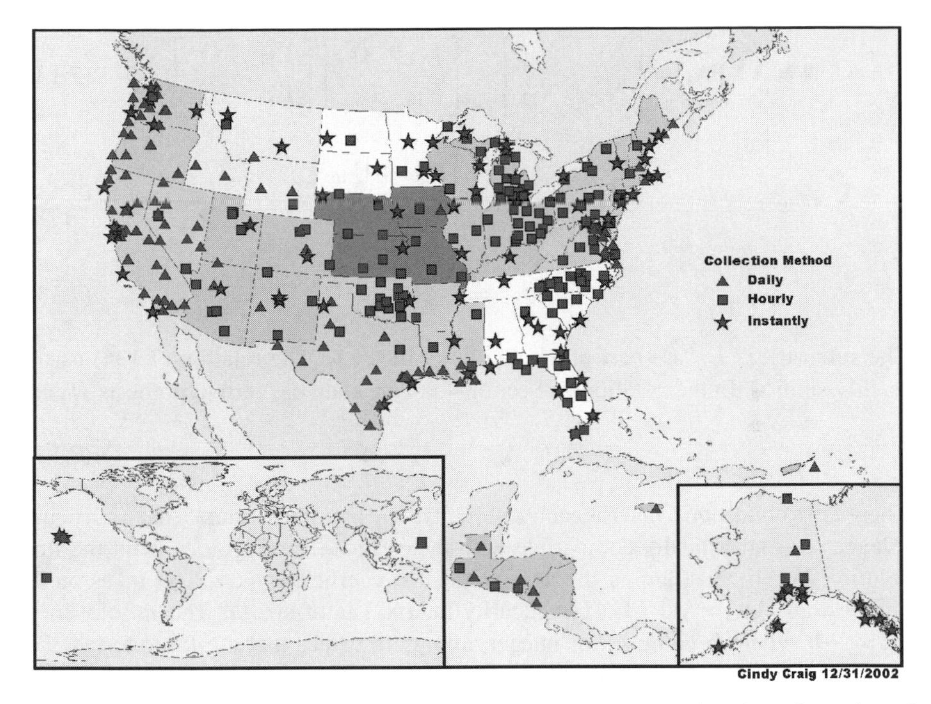

Figure 7.19 NGS CORS network. Data are received at NGS daily, hourly, and even in real time. In addition to serving the needs of postprocessing, the CORS system has the potential of becoming the backbone for real-time geodetic applications nationwide. (Courtesy of NGS.)

the results of the baseline processing via email. At this time, only dual-frequency processing is accepted.

7.8 AMBIGUITY FIXING

Fixing ambiguities implies converting real-valued ambiguity estimates to integers. The procedures follow the general linear hypothesis testing as described in Section 4.9.4. The objective is to constrain the estimated ambiguities of the float solution. Let's assume that the parameters are grouped as

$$\mathbf{x}^* = \begin{bmatrix} \hat{\mathbf{a}} \\ \hat{\mathbf{b}} \end{bmatrix} \tag{7.133}$$

The symbol $\hat{\mathbf{a}}$ then denotes the estimated station coordinates and possibly other parameters such as tropospheric refraction or receiver clock errors. $\hat{\mathbf{b}}$ denotes the estimated float ambiguities. Using the same partitioning, other relevant matrices from the float solution are

$$\mathbf{N} = \mathbf{A}_1^{\mathrm{T}}\,\mathbf{P}\mathbf{A}_1 = \begin{bmatrix} \mathbf{N}_{11} & \mathbf{N}_{21} \\ \mathbf{N}_{21} & \mathbf{N}_{22} \end{bmatrix} = \begin{bmatrix} \mathbf{L}_{11} & \mathbf{O} \\ \mathbf{L}_{12} & \mathbf{L}_{22} \end{bmatrix} \begin{bmatrix} \mathbf{L}_{11} & \mathbf{O} \\ \mathbf{L}_{12} & \mathbf{L}_{22} \end{bmatrix}^{\mathrm{T}} \tag{7.134}$$

$$\mathbf{Q}_{x^*} = \mathbf{N}^{-1} = \begin{bmatrix} \mathbf{Q}_a & \mathbf{Q}_{ab} \\ \mathbf{Q}_{ab}^{\mathrm{T}} & \mathbf{Q}_b \end{bmatrix} \tag{7.135}$$

$$\mathbf{Q}_b^{-1} = \mathbf{L}_{22}\mathbf{L}_{22}^{\mathrm{T}} \tag{7.136}$$

The submatrices $\mathbf{L}_{ij}$ are part of the Cholesky factor $\mathbf{L}$. The relation (7.136) can be readily verified. In the notation of Section 4.9.4 we state the zero hypothesis H_0 as

$$H_0\colon\ \mathbf{A}_2\mathbf{x}^* + \boldsymbol{\ell}_2 = \mathbf{o} \tag{7.137}$$

These are n conditions, one for each ambiguity. The hypothesis states that a particular integer set is statistically compatible with the estimated ambiguities from the float solution. When constraining the ambiguities the coefficient matrix $\mathbf{A}_2$ takes on the simple form $\mathbf{A}_2 = [\mathbf{O} \quad \mathbf{I}]$. The identity matrix $\mathbf{I}$ is of size n. The misclosure is $\boldsymbol{\ell}_2 = -\mathbf{b}$, where $\mathbf{b}$ is the set of integer ambiguity values that are to be tested. The change in $\mathbf{v}^{\mathrm{T}}\mathbf{P}\mathbf{v}$ due to the n constraints can be written according to (4.279)

$$\Delta\mathbf{v}^{\mathrm{T}}\mathbf{P}\mathbf{v} = \left[\hat{\mathbf{b}} - \mathbf{b}\right]^{\mathrm{T}} \mathbf{Q}_b^{-1} \left[\hat{\mathbf{b}} - \mathbf{b}\right] \tag{7.138}$$

which can be used in the F test (4.280)

$$\frac{\Delta\mathbf{v}^{\mathrm{T}}\mathbf{P}\mathbf{v}}{\mathbf{v}^{\mathrm{T}}\mathbf{P}\mathbf{v}^*}\,\frac{df}{n} \sim F_{n,df} \tag{7.139}$$

to test acceptance of H_0. $\mathbf{v}^{\mathrm{T}}\mathbf{P}\mathbf{v}^*$ comes from the float solution and df denotes the degree of freedom of the latter.

In the early days of GPS surveying, a test set $\mathbf{b}$ of integer values was obtained by simply rounding the estimated float ambiguities to the nearest integer. This approach works well for long observation times where many satellites can be observed and the change in satellite geometry over time significantly improves the float solution. In such cases, the estimated real-valued ambiguities are already close to integers and their estimated variances are small. The situation changes drastically when we attempt to shorten the time of observation, possibly down to the extreme of just one epoch. It is only the distribution of the satellites in the sky and the availability of observations at multiple frequencies that adds strength to the geometry in such a case. The estimated float ambiguities will not necessarily be close to integer, and the estimates will have large variances and be highly correlated in general. A possible solution is to find candidate sets $\mathbf{b}_i$ of integers and compute $\Delta\mathbf{v}^{\mathrm{T}}\mathbf{P}\mathbf{v}_i$ according to (7.138). Those with the smallest contribution are subjected to the test (7.139). There are two potential problems with this approach.

The first one is that we might have to test many sets $\mathbf{b}_i$ if the variances of the real-valued ambiguities are large. Let d_j denote the range for ambiguity j based on the estimated variance. If we form sets $\mathbf{b}_i$ for all possible combinations, then there are Πd_i such sets, $i = 1, \ldots, n$. An efficient algorithm is needed to shorten the computation time for ambiguity fixing. If ambiguities can be successfully fixed with just one epoch of observation, then the distinction between static and kinematic relative positioning becomes less relevant; the economic benefits of such a rapid survey technique are obvious. In addition, cycle slips would be rendered harmless because new ambiguities could be fixed every epoch. Much effort has gone into optimizing the computational approaches. The LAMBDA method has emerged as the favored method.

The second problem is that several candidate sets might pass the test (7.139). Naturally, one would like to identify the correct candidate as soon as possible and in doing so minimize the observation time. Discernibility of the candidate sets will be addressed in Section 7.8.3.

7.8.1 Early Efforts

Given the float solution and the respective covariance matrix, Frei and Beutler (1990) suggest a specific ordering scheme for the candidate ambiguity sets. The efficiency of their algorithms relies on the fact that if a certain ambiguity set is rejected, then a whole group of sets is identifiable that will also be rejected and consequently need not be computed explicitly. Euler and Landau (1992) and Blomenhofer et al. (1993) point out that the matrix $\mathbf{L}_{22}$ in (7.136) remains the same for all candidate sets. They further recommend computing (7.138) in two steps. If

$$\mathbf{g} = \mathbf{L}_{22}^{\mathrm{T}} \left[\hat{\mathbf{b}} - \mathbf{b} \right] \tag{7.140}$$

then $\Delta \mathbf{v}^{\mathrm{T}} \mathbf{P} \mathbf{v}$ can be written as

$$\Delta \mathbf{v}^{\mathrm{T}} \mathbf{P} \mathbf{v} = \mathbf{g}^{\mathrm{T}} \mathbf{g} = \sum_{i=1}^{n} g_i^2 \tag{7.141}$$

As soon as the first element g_1 has been computed, it can be squared and taken as the first estimate of the quadratic form. Note that $\Delta \mathbf{v}^{\mathrm{T}} \mathbf{P} \mathbf{v} \geq g_1^2$. The value $\Delta \mathbf{v}^{\mathrm{T}} \mathbf{P} \mathbf{v} = g_1^2$ is substituted in (7.139) to compute the test statistic, which is then compared with the critical F value. If that test fails, the respective trial ambiguity set can immediately be rejected. There is no need to compute the remaining g_i values. If the test passes, then the next value, g_2, is computed and the test statistic is computed based on $\Delta \mathbf{v}^{\mathrm{T}} \mathbf{P} \mathbf{v} = g_1^2 + g_2^2$. If this test fails, the ambiguity set is rejected; otherwise, g_3 is computed, etc. This procedure continues until either the zero hypothesis is rejected or all g_i are computed and the complete sum of the g square terms is known. This strategy can be combined with the ordering scheme mentioned above.

Chen and Lachapelle (1995) take advantage of the fact that integer ambiguity resolution accelerates if the number of candidates d_i for ambiguity i is small. The smaller these search ranges, the fewer ambiguity sets need to be tested. Their method leads to a sequential reduction of ambiguity range. The idea is readily demonstrated by sequential conditional adjustment in which constraints are imposed sequentially on the float solution. In general, the change in the cofactor matrix $\mathbf{Q}_{x^*}$ due to the constraints (7.137), according to Table 4.5, is

$$\Delta\mathbf{Q} = -\mathbf{N}^{-1}\mathbf{A}_2^{\mathrm{T}}\left(\mathbf{A}_2\mathbf{N}^{-1}\mathbf{A}_2^{\mathrm{T}}\right)^{-1}\mathbf{A}_2\mathbf{N}^{-1} \tag{7.142}$$

Because the diagonal elements of the matrix $\Delta\mathbf{Q}$ are negative, it follows that the diagonal elements of the updated cofactor matrix $\mathbf{Q}_{x|H_0} = \mathbf{Q}_{x^*} + \Delta\mathbf{Q}$ are smaller than those of $\mathbf{Q}_{x^*}$. Hence, any ranges derived from $\mathbf{Q}_{x|H_0}$ for the remaining ambiguities will be smaller than if one had continued to use $\mathbf{Q}_{x^*}$. The procedure starts with determining the range of the first ambiguity (which could be the one with smallest variance), using $\mathbf{Q}_{x^*}$ and identifying the j integer candidates $b_{1,j}$ that fall within a certain interval, say $\left[b_1^* \pm \sigma_1\right]$. The symbol σ_1 denotes the square root of the respective diagonal element of $\mathbf{Q}_{x^*}$. Next, one imposes the conditions $b_1^* = b_{1,j}$, using one of the j candidates within the range d_1. The impact of this condition on the other ambiguities is (Table 4.5)

$$\begin{bmatrix} \Delta\mathbf{x} \\ \Delta\mathbf{b} \end{bmatrix}_j = -\mathbf{N}^{-1}\mathbf{A}_2\left(\mathbf{A}_2\mathbf{N}^{-1}\mathbf{A}_2^{\mathrm{T}}\right)^{-1}\left(\mathbf{A}_2\mathbf{x}^* + \boldsymbol{\ell}_{2,j}\right) \tag{7.143}$$

Expression (7.137) represents just one equation in this case, and (7.142) does not change while the remaining candidates are used. The respective diagonal element of the updated cofactor matrix $\mathbf{Q}_{x|b_1}$ and all updates $b_2^* + \Delta b_{1,j}$ are used to determine the total range d_2 for the second ambiguity and the respective integers $b_{2,k}$ that fall within that range. Next, constrain all pairs $b_{1,j}$ and $b_{2,k}$ and determine the range for the third ambiguity. This process continues until the last ambiguity has been reached, each time using the updated covariance matrix and updated ambiguity to find the range. Chen and Lachapelle implement this strategy in a Kalman filter, called fast ambiguity search filter (FASF). They search all ambiguities every epoch. The attempt to fix the ambiguities is terminated if the number of possible ambiguity candidate sets exceeds a threshold value.

Melbourne (1985) discusses an approach in which station coordinates are eliminated from the observation equation prior to the search for the ambiguities. Let the 3×1 vector $\mathbf{x}$ contain only the station coordinates, then

$$\mathbf{v} = \mathbf{A}\mathbf{x} + \mathbf{b} + \boldsymbol{\ell} \tag{7.144}$$

should represent n double-difference observation equations at one epoch. Let $\mathbf{G}$ denote an $n \times (n-3)$ matrix that fulfills $\mathbf{G}^{\mathrm{T}}\mathbf{A} = \mathbf{O}$, then (7.144) becomes

$$\mathbf{G}^{\mathrm{T}}\left(\mathbf{b} - \mathbf{v} + \boldsymbol{\ell}\right) = \mathbf{o} \tag{7.145}$$

The columns of the matrix $\mathbf{G}$ span the null space of $\mathbf{A}$ or $\mathbf{A}\mathbf{A}^{\mathrm{T}}$; such a matrix always exists.

If we consider $\mathbf{v} = \mathbf{o}$, then (7.145) represents $n - 3$ conditions for the n double-difference ambiguities. Through trial and error one could attempt to identify the correct set of ambiguities. Each epoch adds another set of $n-3$ equations to (7.145). The elements of $\mathbf{G}$ change with time as the coefficients in $\mathbf{A}$ change with the motion of the satellites. Eventually, enough epochs will be available with different $\mathbf{G}$ matrices to allow a unique identification of the ambiguity. Only the correct set of ambiguities will always fulfill (7.145). In actual application where the residuals are not zero but are small, one would be looking for ambiguity values that are close to integers. Alternatively, applying Equation (7.145) to several epochs can be readily used to build a mixed-model least-squares solution to estimate $\hat{\mathbf{b}}$. The receivers could even be in motion as long as the locations are known well enough to compute the coefficients in $\mathbf{A}$.

Hatch (1990) suggests a scheme that divides satellites into primary and secondary ones. Consider four satellites, called the primary satellites. The respective three double-difference equations contain the station coordinates and three double-difference ambiguities. When the satellite geometry changes over time, it is possible to estimate all of these parameters. Any satellites in addition to these four satellites, called the secondary satellites, are strictly speaking redundant, although we know that these extra satellites improve the overall solution geometry. The extra satellites are used to develop yet another procedure for rapidly identifying integer ambiguities.

The primary and secondary satellites are identified below by subscripts p and s, respectively. We group the observation equations accordingly, i.e.,

$$\mathbf{v}_p = \mathbf{A}_p\mathbf{x} + \left(\mathbf{b}_p + \boldsymbol{\ell}_p\right) = \mathbf{A}_p\mathbf{x} + \tilde{\boldsymbol{\ell}}_p \tag{7.146}$$

$$\mathbf{v}_s = \mathbf{A}_s\mathbf{x} + (\mathbf{b}_s + \boldsymbol{\ell}_s) = \mathbf{A}_s\mathbf{x} + \tilde{\boldsymbol{\ell}}_s \tag{7.147}$$

Note that the 3×1 vector $\mathbf{x}$ contains only coordinate parameters. Each of the two groups may contain observations from one or several epochs. The method assumes that the ambiguities $\mathbf{b}_p$ are known and evaluates the effect of that assumption. The procedure starts by computing trial sets $\mathbf{b}_{p,i}$ for the three primary ambiguities using an initial position estimate $\mathbf{x}_0$, obtained from the point positioning solution or from the float solution if several epochs of observations are available and the receivers do not move (see below). We can compute the change in position with respect to $\mathbf{x}_0$ for a given set of primary trial ambiguities using the usual least-squares formulation

$$\mathbf{x}_{p,i} = -\mathbf{N}_p^{-1}\mathbf{A}_p^{\mathrm{T}}\mathbf{P}_p\tilde{\boldsymbol{\ell}}_{p,i} \tag{7.148}$$

with $\mathbf{N}_p = \alpha_p^{\mathrm{T}}\mathbf{P}_p\mathbf{A}_p$. This is a nonredundant solution because only three observation equations are available. Each ambiguity trial set gives a different position $\mathbf{x}_{p,i}$ while the matrices $\mathbf{A}_p$ and $\mathbf{P}_p$ do not change. The coefficients of $\mathbf{A}_p$ are evaluated for $\mathbf{x}_0$. Using $\mathbf{x}_{p,i}$ the ambiguities for the secondary satellites, $\mathbf{b}_{s,i}$, can be derived from

$$N_{km,s}^{1q} = \varphi_{km,s}^{1q} - \rho_{km,p}^{1q}(\mathbf{x}_p) \tag{7.149}$$

where we have denoted the base satellite with a superscript 1; the superscript q varies over the secondary satellites. These estimates are rounded to the nearest integer. Next we compute the correction to the positions $\mathbf{x}_{p,i}$ using sequential least-squares (Table 4.2):

$$\Delta\mathbf{x}_{p,i} = \mathbf{x}_{p,i} - \mathbf{N}_p^{-1}\mathbf{A}_s^{\mathrm{T}}\left(\mathbf{P}_s + \mathbf{A}_s\mathbf{N}_p^{-1}\mathbf{A}_s^{\mathrm{T}}\right)^{-1}\left(\mathbf{A}_s\mathbf{x}_{p,i} + \tilde{\boldsymbol{\ell}}_{s,i}\right) \tag{7.150}$$

The dimension of $\tilde{\boldsymbol{\ell}}_{s,i}$ equals the number of additional satellites. If the set of primary ambiguities used to generate $\mathbf{x}_{p,i}$ is the correct one, then the respective secondary ambiguities $\mathbf{b}_{s,i}$ are correct and, consequently, $\Delta\mathbf{x}_{p,i}$ should be zero. If $\Delta\mathbf{x}_{p,i}$ falls outside a tolerance region, whose size is a function of the accuracy of $\mathbf{x}_0$, then $\mathbf{b}_{s,i}$ and consequently $\mathbf{b}_{p,i}$ are rejected and the search continues with (7.148) using a different trial set $\mathbf{b}_{p,j}$. If the set is acceptable, the residuals for the combined solution of primary and secondary observations are

$$\mathbf{v}_p = \mathbf{A}_p\Delta\mathbf{x}_P \tag{7.151}$$

$$\mathbf{v}_s = \mathbf{A}_s\left(\mathbf{x}_p + \Delta\mathbf{x}_P\right) + \tilde{\boldsymbol{\ell}}_s \tag{7.152}$$

The quadratic function

$$\mathbf{v}^{\mathrm{T}}\mathbf{P}\mathbf{v} = \mathbf{v}_p^{\mathrm{T}}\mathbf{P}_p\,\mathbf{v}_p + \mathbf{v}_s^{\mathrm{T}}\mathbf{P}_s\,\mathbf{v}_s \tag{7.153}$$

can be used to discern several qualifying solutions.

7.8.2 LAMBDA

Teunissen (1993) introduced the least-squares ambiguity decorrelation adjustment (LAMBDA) method. The LAMBDA technique, which has been referred to as the integer least-squares estimator, is the estimator that has the highest probability of correct integer estimation among all possible admissible integer estimators (Teunissen, 1999). This probabilistic justification of LAMBDA in addition to its speed has resulted in a high popularity and general acceptance of the technique. This section merely highlights some features of the LAMBDA algorithm. The reader is refered to Jonge and Tiberius (1996) for details of implementation. At the core of LAMBDA is the Z transformation

$$\mathbf{z} = \mathbf{Z}^{\mathrm{T}}\,\mathbf{b} \tag{7.154}$$

$$\hat{\mathbf{z}} = \mathbf{Z}^{\mathrm{T}}\,\hat{\mathbf{b}} \tag{7.155}$$

$$\mathbf{Q}_z = \mathbf{Z}^{\mathrm{T}}\mathbf{Q}_b\mathbf{Z} \tag{7.156}$$

where $\mathbf{Z}$ is a regular and square matrix. In order for integers to be preserved, i.e., the integers $\mathbf{b}$ should be mapped into integers $\mathbf{z}$ and vice versa, it is necessary that the elements of both matrices $\mathbf{Z}$ and $\mathbf{Z}^{-1}$ are integers. The condition $|\mathbf{Z}| = \pm 1$ assures that the inverse contains only integer elements if $\mathbf{Z}$ contains integers. Simply consider this: if all elements of $\mathbf{Z}$ are integers, then this is also true for the cofactor matrix $\mathbf{C}$. Therefore, the inverse

$$\mathbf{Z}^{-1} = \frac{\mathbf{C}^{\mathrm{T}}}{|\mathbf{Z}|} \tag{7.157}$$

has integer elements because of the condition $|\mathbf{Z}| = \pm 1$. The latter condition also implies that

$$|\mathbf{Q}_z| = \left|\mathbf{Z}^{\mathrm{T}}\mathbf{Q}_b\mathbf{Z}\right| = \left|\mathbf{Z}^{\mathrm{T}}\right| |\mathbf{Q}_b| \, |\mathbf{Z}| = |\mathbf{Q}_b| \tag{7.158}$$

The quadratic form also remains invariant with respect to the Z transformation. Substituting (7.154) and (7.155) in (7.138) and using the inverse of (7.156) gives

$$\begin{aligned}\Delta\mathbf{v}^{\mathrm{T}}\mathbf{P}\mathbf{v} &= \left[\hat{\mathbf{b}} - \mathbf{b}\right]^{\mathrm{T}} \mathbf{Q}_b^{-1} \left[\hat{\mathbf{b}} - \mathbf{b}\right] \\ &= \left[\hat{\mathbf{z}} - \mathbf{z}\right]^{\mathrm{T}} \left(\mathbf{Z}^{-1}\right)^{\mathrm{T}} \mathbf{Q}_b^{-1}\mathbf{Z}^{-1} \left[\hat{\mathbf{z}} - \mathbf{z}\right] \\ &= \left[\hat{\mathbf{z}} - \mathbf{z}\right]^{\mathrm{T}} \mathbf{Q}_z^{-1} \left[\hat{\mathbf{z}} - \mathbf{z}\right]\end{aligned} \tag{7.159}$$

Consider the following example with two random integer variables $\hat{\mathbf{b}} = [\hat{b}_1 \quad \hat{b}_2]^{\mathrm{T}}$. Let the respective covariance matrix be

$$\boldsymbol{\Sigma}_b = \begin{bmatrix} \sigma_{b_1}^2 & \sigma_{b_1 b_2} \\ \sigma_{b_2 b_1} & \sigma_{b_2}^2 \end{bmatrix} \tag{7.160}$$

The transformation $\mathbf{z} = \mathbf{Z}^{\mathrm{T}}\mathbf{b}$ utilizes a transformation matrix of the special form

$$\mathbf{Z}^{\mathrm{T}} = \begin{bmatrix} 1 & \beta \\ 0 & 1 \end{bmatrix} \tag{7.161}$$

where $\hat{\mathbf{z}} = [\hat{z}_1 \quad \hat{z}_2]^{\mathrm{T}}$. Note that $|\mathbf{Z}| = 1$. The element β is obtained by rounding $-\sigma_{b_1 b_2}/\sigma_{b_2}^2$ to the nearest integer $\beta = \operatorname{int}\left(-\sigma_{b_1 b_2}/\sigma_{b_2}^2\right)$. Because β is an integer, the transformed z variables will also be integers. Variance-covariance propagation gives

$$\boldsymbol{\Sigma}_z = \mathbf{Z}^{\mathrm{T}} \, \boldsymbol{\Sigma}_b \, \mathbf{Z} = \begin{bmatrix} \beta^2\sigma_{b_2}^2 + 2\beta\sigma_{b_1 b_2} + \sigma_{b_1}^2 & \beta\sigma_{b_2}^2 + \sigma_{b_1 b_2} \\ \beta\sigma_{b_2}^2 + \sigma_{b_1 b_2} & \sigma_{b_2}^2 \end{bmatrix} \tag{7.162}$$

Let ε denote the change due to the rounding, i.e., $\varepsilon = \sigma_{b_1 b_2}/\sigma_{b_2}^2 + \beta$. Using (7.162), the variance σ_z^2 of the transformed variable z_1 can be written as

$$\sigma_{z_1}^2 = \sigma_{b_1}^2 - \left(\frac{\sigma_{b_1 b_2}^2}{\sigma_{b_2}^4} - \varepsilon^2 \right) \sigma_{b_2}^2 \tag{7.163}$$

This expression shows that the variance of the transformed variable decreases compared to the original one, i.e., $\sigma_{z_1}^2 < \sigma_{b_1}^2$ whenever

$$\left| \sigma_{b_1 b_2} / \sigma_{b_2}^2 \right| > 0.5 \tag{7.164}$$

and that both are equal when $\left|\sigma_{b_1 b_2}/\sigma_{b_2}^2\right| = |\,\varepsilon\,| = 0.5$. The property of decreasing the variance while preserving the integer makes the transformation (7.161) a favorite to resolve ambiguities because it minimizes the search. It is interesting to note that z_1 and z_2 would be completely decorrelated if one were to choose $\beta = -\sigma_{b_1 b_2}/\sigma_{b_2}^2$. However, such a selection is not permissible because it would not preserve the integer property of the transformed variables.

When implementing LAMBDA, the $\mathbf{Z}$ matrix is constructed from the $n \times n$ submatrix $\mathbf{Q}_b$ (7.135). There are n variables $\hat{\mathbf{b}}$ that must be transformed. Using the Cholesky decomposition we find

$$\mathbf{Q}_b = \mathbf{H}^{\mathrm{T}} \mathbf{K} \mathbf{H} \tag{7.165}$$

The matrix $\mathbf{H}$ is the modified Cholesky factor that contains 1 at the diagonal and follows (7.136). $\mathbf{K}$ is a diagonal matrix containing the diagonal squared terms of the Cholesky factor. Assume that we are dealing with ambiguities i and $i+1$ and partition these two matrices accordingly,

$$\mathbf{H} = \left[\begin{array}{cc|cc|cc} 1 & & & & & \\ \vdots & \ddots & & & & \\ \hline h_{i,1} & \cdots & 1 & & & \\ h_{i+1,1} & \cdots & h_{i+1,i} & 1 & & \\ \hline \vdots & \cdots & \vdots & \vdots & \ddots & \\ h_{n,1} & \cdots & h_{n,i} & h_{n,i+1} & \cdots & 1 \end{array} \right] = \begin{bmatrix} \mathbf{H}_{11} & \mathbf{O} & \mathbf{O} \\ \mathbf{H}_{21} & \mathbf{H}_{22} & \mathbf{O} \\ \mathbf{H}_{31} & \mathbf{H}_{32} & \mathbf{H}_{33} \end{bmatrix} \tag{7.166}$$

$$\mathbf{K} = \left[\begin{array}{cc|cc|cc} k_{1,1} & & & & & \\ & \ddots & & & & \\ \hline & & k_{i,i} & & & \\ & & & k_{i+1,i+1} & & \\ \hline & & & & \ddots & \\ & & & & & k_{n,n} \end{array} \right] = \begin{bmatrix} \mathbf{K}_{11} & \mathbf{O} & \mathbf{O} \\ \mathbf{O} & \mathbf{K}_{22} & \mathbf{O} \\ \mathbf{O} & \mathbf{O} & \mathbf{K}_{33} \end{bmatrix} \tag{7.167}$$

The transformation matrix $\mathbf{Z}$ is partitioned similarly,

$$\mathbf{Z}_1 = \left[\begin{array}{c|cc|c} \mathbf{I} & & & \\ \hline & 1 & 0 & \\ & \beta & 1 & \\ \hline & & & \mathbf{I} \end{array}\right] = \begin{bmatrix} \mathbf{I}_{11} & \mathbf{O} & \mathbf{O} \\ \mathbf{O} & \mathbf{Z}_{22} & \mathbf{O} \\ \mathbf{O} & \mathbf{O} & \mathbf{I}_{33} \end{bmatrix} \tag{7.168}$$

where $\beta = -\text{int}\left(h_{i+1,i}\right)$ represents the negative of the rounded value of $h_{i+1,i}$.

$$\hat{\mathbf{z}}_1 = \mathbf{Z}_1^{\mathrm{T}} \hat{\mathbf{b}} \tag{7.169}$$

$$\mathbf{Q}_{z,1} = \mathbf{Z}_1^{\mathrm{T}} \mathbf{Q}_b \mathbf{Z}_1 = \mathbf{Z}_1^{\mathrm{T}} \mathbf{H}^{\mathrm{T}} \mathbf{K} \mathbf{H} \mathbf{Z}_1 = \mathbf{H}_1^{\mathrm{T}} \mathbf{K}_1 \mathbf{H}_1 \tag{7.170}$$

It can be shown that the specific form of $\mathbf{Z}_1$ and the choice of $\mathbf{Z}_{22}$ imply the following updates

$$\mathbf{Q}_{z,1} = \begin{bmatrix} \mathbf{Q}_{11} & & \text{sym} \\ \mathbf{Z}_{22}^{\mathrm{T}} \mathbf{Q}_{21} & \mathbf{Z}_{22}^{\mathrm{T}} \mathbf{Q}_{22} \mathbf{Z}_{22} & \\ \mathbf{Q}_{31} & \mathbf{Q}_{32} \mathbf{Z}_{22} & \mathbf{Q}_{33} \end{bmatrix} \tag{7.171}$$

$$\mathbf{H}_1 = \mathbf{H}\mathbf{Z}_1 = \begin{bmatrix} \mathbf{H}_{11} & \mathbf{O} & \mathbf{O} \\ \mathbf{H}_{21} & \bar{\mathbf{H}}_{22} & \mathbf{O} \\ \mathbf{H}_{31} & \bar{\mathbf{H}}_{32} & \mathbf{H}_{33} \end{bmatrix} \tag{7.172}$$

$$\bar{\mathbf{H}}_{22} = \begin{bmatrix} 1 & 0 \\ h_{i+1,i} + \beta & 1 \end{bmatrix} \tag{7.173}$$

$$\bar{\mathbf{H}}_{32} = \begin{bmatrix} h_{i+2,i} + \beta h_{i+2,i+1} & h_{i+2,i+1} \\ h_{i+3,i} + \beta h_{i+3,i+1} & h_{i+3,i+1} \\ \vdots & \vdots \\ h_{n,i} + \beta h_{n,i+1} & h_{n,i+1} \end{bmatrix} \tag{7.174}$$

$$\mathbf{K}_1 = \mathbf{K} \tag{7.175}$$

The matrix $\mathbf{K}$ does not change due to this decorrelation transformations.

If $\beta = 0$ the transformation (7.169) is not necessary. However, it is necessary to check whether or not the ambiguities i and $i+1$ should be permuted to achieve further decorrelation. Consider the permutation transformation

$$\mathbf{Z}_2 = \begin{bmatrix} \mathbf{I} & & \\ & \begin{matrix} 0 & 1 \\ 1 & 0 \end{matrix} & \\ & & \mathbf{I} \end{bmatrix} = \begin{bmatrix} \mathbf{I}_{11} & \mathbf{O} & \mathbf{O} \\ \mathbf{O} & \mathbf{P} & \mathbf{O} \\ \mathbf{O} & \mathbf{O} & \mathbf{I}_{33} \end{bmatrix} \tag{7.176}$$

This specific choice for $\mathbf{Z}_2$ leads to

$$\bar{\mathbf{H}}_{22} = \begin{bmatrix} 1 & 0 \\ h'_{i+1,i} & 1 \end{bmatrix} = \begin{bmatrix} 1 & 0 \\ \dfrac{h_{i+1,i} k_{i+1,i+1}}{k_{i,i} + h^2_{i+1,i} k_{i+1,i+1}} & 1 \end{bmatrix} \tag{7.177}$$

$$\bar{\mathbf{H}}_{21} = \begin{bmatrix} -h_{i+1,i} & 1 \\ \dfrac{k_{i,i}}{k_{i,i} + h^2_{i+1,i} k_{i+1,i+1}} & h'_{i+1,j} \end{bmatrix} \mathbf{H}_{21} \tag{7.178}$$

$$\bar{\mathbf{H}}_{32} = \begin{bmatrix} h_{i+2,i+1} & h_{i+2,i} \\ h_{i+3,i+1} & h_{i+3,i} \\ \vdots & \vdots \\ h_{n,i+1} & h_{n,i} \end{bmatrix} \tag{7.179}$$

$$\bar{\mathbf{K}}_{22} = \begin{bmatrix} k'_{i,i} & 0 \\ 0 & k'_{i+1,i+1} \end{bmatrix} = \begin{bmatrix} k_{i+1,i+1} - \dfrac{h^2_{i+1,i} k^2_{i+1,i+1}}{k_{i,i} + h^2_{i+1,i} k_{i+1,i+1}} & 0 \\ 0 & k_{i,i} + h^2_{i+1,i} k_{i+1,i+1} \end{bmatrix} \tag{7.180}$$

Permutation changes the matrix $\mathbf{K}$ at $\bar{\mathbf{K}}_{22}$. To achieve full decorrelation, the terms $k'_{i+1,i+1}$ and $k_{i+1,i+1}$ must be inspected while the ith and $(i + 1)$th ambiguity are considered. Permutation is required if $k'_{i+1,i+1} < k_{i+1,i+1}$. If permutation occurs, the procedure again starts with the last pair of the $(n-1)$th and nth ambiguities and tries to reach the first and second ambiguities. A new $\mathbf{Z}$ transformation matrix is constructed whenever decorrelation takes place or the order of two ambiguities is permuted. This procedure is completed when no diagonal elements are interchanged.

The result of the Z transformations can be written as

$$\hat{\mathbf{z}} = \mathbf{Z}_q^{\mathrm{T}} \cdots \mathbf{Z}_2^{\mathrm{T}} \mathbf{Z}_1^{\mathrm{T}} \hat{\mathbf{b}} \tag{7.181}$$

$$\mathbf{Q}_{z,q} = \mathbf{Z}_q^{\mathrm{T}} \cdots \mathbf{Z}_2^{\mathrm{T}} \mathbf{Z}_1^{\mathrm{T}} \mathbf{Q}_b \mathbf{Z}_1 \mathbf{Z}_2 \cdots \mathbf{Z}_q = \mathbf{H}_q^{\mathrm{T}} \mathbf{K}_q \mathbf{H}_q \tag{7.182}$$

The matrices $\mathbf{H}_q$ and $\mathbf{K}_q$ are obtained as part of the consecutive transformations. The permuting steps assure that $\mathbf{K}_q$ contains decreasing diagonal elements, the smallest element being located at the lower right corner. As a measure of decorrelation between the ambiguities, we might consider the scalar (Teunissen, 1994)

$$r = |\mathbf{R}|^{1/2} \qquad 0 \le r \le 1 \tag{7.183}$$

where $\mathbf{R}$ represents a correlation matrix. Applying (7.183) to $\mathbf{Q}_b$ and $\mathbf{Q}_{z,q}$ will give a relative sense of the decorrelation achieved. A value of r close to 1 implies a high decorrelation. Therefore, we expect $r_b < r_{z,q}$. The scalar r is called the ambiguity decorrelation number.

The search step entails finding candidate sets of $\hat{\mathbf{z}}_i$ given ($\hat{\mathbf{z}}$, $\mathbf{Q}_{z,q}$), which minimize

$$\Delta\mathbf{v}^{\mathrm{T}}\mathbf{P}\mathbf{v} = \left[\hat{\mathbf{z}} - \mathbf{z}\right]^{\mathrm{T}} \mathbf{Q}_{z,q}^{-1} \left[\hat{\mathbf{z}} - \mathbf{z}\right] \tag{7.184}$$

A possible procedure would be to use the diagonal elements of $\mathbf{Q}_{z,q}$, construct a range for each ambiguity centered around $\hat{\mathbf{z}}_{\mathbf{i}}$, form all possible sets $\mathbf{z}_i$, evaluate the quadratic form for each set, and keep track of those sets that produce the smallest $\Delta\mathbf{v}^{\mathrm{T}}\mathbf{P}\mathbf{v}$. A more organized and efficient approach is achieved by transforming the $\hat{\mathbf{z}}$ variables into variables $\hat{\mathbf{w}}$ that are stochastically independent. First, we decompose the inverse of $\mathbf{Q}_{z,q}$ as

$$\mathbf{Q}_{\hat{z},q}^{-1} = \mathbf{M}\,\mathbf{S}\,\mathbf{M}^{\mathrm{T}} \tag{7.185}$$

where $\mathbf{M}$ denotes the lower triangular matrix with 1's along the diagonal, and $\mathbf{S}$ is a diagonal matrix containing positive values that increase toward the lower right corner. The latter property follows from the fact that $\mathbf{S}$ is the inverse of $\mathbf{K}_q$. The transformed variables $\hat{\mathbf{w}}$,

$$\hat{\mathbf{w}} = \mathbf{M}^{\mathrm{T}} \left[\hat{\mathbf{z}} - \mathbf{z}\right] \tag{7.186}$$

are distributed as $\hat{\mathbf{w}} \sim N(\mathbf{o}, \mathbf{S}^{-1})$. Because $\mathbf{S}$ is a diagonal matrix the variables $\hat{\mathbf{w}}$ are stochastically independent. Using (7.186) and (7.185) the quadratic form (7.184) can be written as

$$\Delta\mathbf{v}^{\mathrm{T}}\mathbf{P}\mathbf{v} \equiv \hat{\mathbf{w}}^{\mathrm{T}}\,\mathbf{S}\,\hat{\mathbf{w}} = \sum_{i=1}^{n} \hat{w}_i^2 s_{i,i} \le \chi^2 \tag{7.187}$$

The symbol χ^2 acts as a scalar; additional explanations will be given below. Finally, we introduce the auxiliary quantity, also called the conditional estimate,

$$\hat{w}_{i|I} = \sum_{j=i+1}^{n} m_{j,i} \left(\hat{z}_j - z_j\right) \tag{7.188}$$

The symbol $| I$ indicates the values for z_j have already been selected, i.e., are known. Note that the subscript j goes from $i+1$ to n. Since $m_{i,i} = 1$ and using (7.188) and (7.186), we can write the ith component as

$$\hat{w}_i = \hat{z}_i - z_i + \hat{w}_{i|I} \qquad i = 1, n-1 \tag{7.189}$$

The bounds of the $\hat{w}$ parameters follow from (7.187). We begin with the nth level to determine the bounds for the nth ambiguity and then proceed to level 1, establishing the bound for the other ambiguities. Using the term with $\hat{w}_n s_{n,n}$ in (7.187), and knowing that the matrix element $m_{n,n} = 1$ in (7.186), we find

$$\hat{w}_n^2 s_{n,n} = \left(z_n - \hat{z}_n\right)^2 s_{n,n} \leq \chi^2 \tag{7.190}$$

The bounds are

$$\hat{z}_n - \left(\chi^2/s_{n,n}\right)^{1/2} \leq z_n \leq \hat{z}_n + \left(\chi^2/s_{n,n}\right)^{1/2} \tag{7.191}$$

Using the terms from i to n in (7.187) and (7.189), we obtain for level i,

$$\hat{w}_i^2 s_{i,i} = \left(\hat{z}_i - z_i + \hat{w}_{i|I}\right)^2 s_{i,i} \leq \left[\chi^2 - \sum_{j=i+1}^{n} \hat{w}_j^2 s_{j,j}\right] \tag{7.192}$$

$$\begin{aligned} \hat{z}_i + \hat{w}_{i|I} - \frac{1}{\sqrt{s_{i,i}}}\left[\chi^2 - \sum_{j=i+1}^{n} \hat{w}_j^2 s_{j,j}\right]^{1/2} &\leq z_i \\ \leq \hat{z}_i + \hat{w}_{i|I} + \frac{1}{\sqrt{s_{i,i}}}\left[\chi^2 - \sum_{j=i+1}^{n} \hat{w}_j^2 s_{j,j}\right]^{1/2} & \end{aligned} \tag{7.193}$$

The bounds (7.191) and (7.193) can contain one or several integer values z_n or z_i. All values must be used when locating the bounds and integer values at the next lower level. The process stops when level 1 is reached. For certain combinations, the process stops earlier if the square root in (7.193) becomes negative.

Figure 7.20 demonstrates how one can proceed systematically, trying to reach the first level. At a given level, one proceeds from the left to the right while reaching a lower level. This example deals with $n = 4$ ambiguities z_1, z_2, z_3, and z_4. The fourth level produced the qualifying values $z_4 = \{-1, 0, 1\}$. Using $z_4 = -1$ does not produce a solution at level 3 and the branch terminates. Using $z_4 = 0$ gives $z_3 = \{-1, 0\}$ at level 3. Using $z_3 = -1$ and $z_4 = 0$, or in short notation $z = (-1, 0)$, one gets $z_2 = 0$ at level 2. The combination $z = (0, -1, 0)$ does not produce a solution at level 1; the branch terminates. Returning to level 3, we try the combination $z = (0, 0)$, $z_3 = 0$, giving $z_2 = \{-1, 0, 1\}$ at level 2. Trying the left branch with $z = (-1, 0, 0)$ gives no solution and the branch terminates. Using $z = (0, 0, 0)$ gives

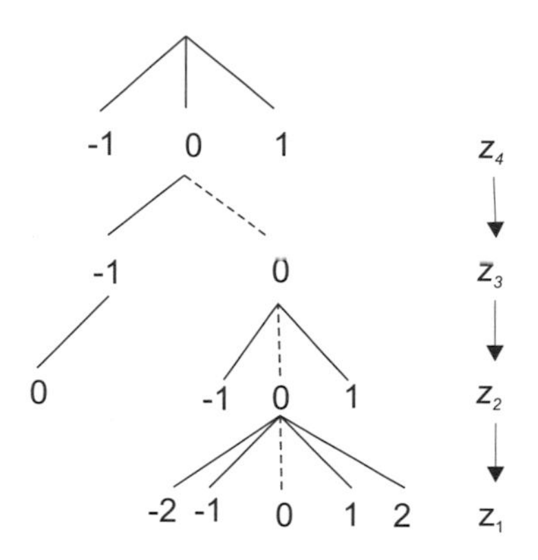

Figure 7.20 Candidate ambiguities encountered during the search procedure with decorrelation.

$z_1 = \{-2, -1, 0, 1, 2\}$ at the first level. The last possibility, using $z = (1, 0, 0)$, gives no solution. We conclude that five ambiguity sets $\mathbf{z}_i = (z_1, 0, 0, 0)$ satisfy the condition (7.187). In general, several branches can reach the first level. Because $s_{n,n}$ is the largest value in $\mathbf{S}$, the number of z_n candidates is correspondingly small, thus lowering the number of branches that originate from level n and assuring that not many branches reach level 1.

The change $\Delta\mathbf{v}^{\mathrm{T}}\mathbf{P}\mathbf{v}_i$ can be computed efficiently from (7.187) because all $\hat{\mathbf{w}}_i$ sets become available as part of computing the candidate ambiguity sets. The matrix $\mathbf{S}$ does not change. The qualifying candidates $\mathbf{z}_i$ are converted back to $\mathbf{b}_i$ using the inverse of (7.181).

If the constant χ^2 for ambiguity search is chosen improperly, it is possible that the search procedure may not find any candidate vector or that too many candidate vectors are obtained. The latter results in time-consuming searches. This dilemma can be avoided if the constant is set close to the $\Delta\mathbf{v}^{\mathrm{T}}\mathbf{P}\mathbf{v}$ value of the best candidate ambiguity vector. To do so, the real-valued ambiguities of the float solution are rounded to the nearest integer, and then substituted into (7.184). The constant is then taken to be equal to $\Delta\mathbf{v}^{\mathrm{T}}\mathbf{P}\mathbf{v}$. This approach guarantees obtaining at least one candidate vector, which consequently is probably the best candidate vector because the decorrelated ambiguities have such a high precision. One can compute a new constant χ^2 by adding or subtracting an increment to one of the nearest integer entries. Using this procedure results in only a few candidate integer ambiguity vectors and guarantees that at least two vectors are obtained.

LAMBDA is a general procedure that requires only the covariance submatrix and the float estimates of the ambiguities. Therefore, the LAMBDA procedure applies even if other parameters are estimated at the same time, such as station coordinates, tropospheric parameters, and clock errors. LAMBDA readily applies to dual-frequency observations, or even future situations, when observations from more than two frequencies become available. Since only the covariance submatrix matters, the observations can come from any available satellite system such as GPS, GLONASS, or even Galileo. Even more generally, LAMBDA applies to any least-squares integer estimation, regardless of what the physical meaning of the integer parameters is.

LAMBDA is also applicable to estimating a subset of ambiguities. For example, in the case of dual-frequency ambiguities one might parameterize in terms of the widelane and the L1 ambiguities. LAMBDA could operate initially on the wide-lane covariance submatrix and fix the wide-lane ambiguities immediately, and then attempt to fix the L1 ambiguities as sufficient geometry becomes available. Teunissen (1997) shows that the Z transformation always includes the widelane but goes far beyond that to achieve an even better decorrelation.

In order to judge the expected performance of the ambiguity resolution, one can compute the success rate, i.e., the probability of correct integer estimation. The success rate depends on the covariance matrix and as such on the geometry embedded in the functional and stochastic model (Teunissen, 1998).

7.8.3 Discernibility

The ambiguity testing outlined above is a repeated application of null hypotheses testing for each ambiguity set. The procedure tests the changes $\Delta \mathbf{v}^T \mathbf{P} \mathbf{v}$ due to the constraints. The decision to accept or to reject the null hypothesis is based on the probability of the type-I error, which is usually taken to be $\alpha = 0.05$. In many cases, several of the null hypotheses will pass, thus identifying several qualifying ambiguity sets. This happens if there is not enough information in the observations to determine the integers uniquely and reliably. Additional observations might help resolve the situation. The ambiguity set that generates the smallest $\Delta \mathbf{v}^T \mathbf{P} \mathbf{v}$ fits the float solution best and, consequently, is considered the most favored fixed solution. The goal of additional statistical considerations is to provide conditions that make it possible to discard all but one of the ambiguity sets that passed the null hypotheses test.

The alternative hypothesis H_a is always relative to the null hypothesis H_0. The formalism for the null hypothesis is given in Section 4.9.4. In general, the null and alternative hypotheses are

$$H_0\colon \quad \mathbf{A}_2 \mathbf{x}^* + \boldsymbol{\ell}_2 = \mathbf{o} \tag{7.194}$$

$$H_a\colon \quad \mathbf{A}_2 \mathbf{x}^* + \boldsymbol{\ell}_2 + \mathbf{w}_2 = \mathbf{o} \tag{7.195}$$

Under the null hypothesis the expected value of the constraint is zero. See also Equation (4.270). Thus,

$$E\left(\mathbf{z}_{H_0}\right) \equiv E\left(\mathbf{A}_2 \mathbf{x}^* + \boldsymbol{\ell}_2\right) = \mathbf{o} \tag{7.196}$$

Because $\mathbf{w}_2$ is a constant, it follows that

$$E\left(\mathbf{z}_{H_a}\right) \equiv E\left(\mathbf{A}_2 \mathbf{x}^* + \boldsymbol{\ell}_2 + \mathbf{w}_2\right) = \mathbf{w}_2 \tag{7.197}$$

The random variable $\mathbf{z}_{H_a}$ is multivariate normal distributed with mean $\mathbf{w}_2$, i.e.,

$$\mathbf{z}_{H_a} \sim N_{n-r}\left(\mathbf{w}_2,\ \sigma_0^2\,\mathbf{T}^{-1}\right) \tag{7.198}$$

See Equation (4.272) for the corresponding expression for the zero hypothesis. The matrix $\mathbf{T}$ has the same meaning as in Section 4.9.4, i.e.,

$$\mathbf{T} = \left(\mathbf{A}_2\mathbf{N}_1^{-1}\mathbf{A}_2^{\mathrm{T}}\right)^{-1} \tag{7.199}$$

The next step is to diagonalize the covariance matrix of $\mathbf{Z}_{H_a}$ and to compute the sum of the squares of the transformed random variables. These newly formed random variables have a unit variate normal distribution with a nonzero mean. According to Section A.5.2, the sum of the squares has a noncentral chi-square distribution. Thus,

$$\frac{\Delta\mathbf{v}^{\mathrm{T}}\mathbf{P}\mathbf{v}}{\sigma_0^2} = \frac{\mathbf{z}_{H_a}^{\mathrm{T}}\,\mathbf{T}\,\mathbf{z}_{H_a}}{\sigma_0^2} \sim \chi^2_{n_2,\lambda} \tag{7.200}$$

where the noncentrality parameter is

$$\lambda = \frac{\mathbf{w}_2^{\mathrm{T}}\,\mathbf{T}\,\mathbf{w}_2}{\sigma_0^2} \tag{7.201}$$

The reader is referred to the statistical literature, such as Koch (1988), for additional details on noncentral distributions and their respective derivations. Finally, the ratio

$$\frac{\Delta\mathbf{v}^{\mathrm{T}}\mathbf{P}\mathbf{v}}{\mathbf{v}^{\mathrm{T}}\mathbf{P}\mathbf{v}^*}\ \frac{n_1 - r}{n_2} \sim F_{n_2,\,n_1-r,\lambda} \tag{7.202}$$

has a noncentral F distribution with noncentrality λ. If the test statistic computed under the specifications of H_0 fulfills $F \leq F_{n_2,\,n_1-r,\alpha}$, then H_0 is accepted with a type-I error of α. The alternative hypothesis H_a can be separated from H_0 with the power $1 - \beta(\alpha, \lambda)$. The type-II error is

$$\beta\,(\alpha, \lambda) = \int_0^{F_{n_2,\,n_1-r,\,1-\alpha}} F_{n_2,n_1-r,\lambda}\ dx \tag{7.203}$$

The integration is taken over the noncentral F-distribution function from zero to the value $F_{n_2,\,n_1-r,\alpha}$, which is specified by the significance level α.

Because the noncentrality is different for each alternative hypothesis according to (7.201), the type-II error $\beta\,(\alpha, \lambda)$ also varies with H_a. Rather than using the individual type-II errors to make decisions, Euler and Schaffrin (1990) propose using the ratio of noncentrality parameters. They designate the float solution as the common alternative hypothesis H_a, for all null hypotheses. In this case, the value $\mathbf{w}_2$ in (7.195) is

$$\mathbf{w}_2 = -\left(\mathbf{A}_2\mathbf{x}^* + \boldsymbol{\ell}_2\right) \tag{7.204}$$

and the noncentrality parameter becomes

$$\lambda \equiv \frac{\mathbf{w}_2^{\mathrm{T}} \mathbf{T} \mathbf{w}_2}{\sigma_0^2} = \frac{\Delta \mathbf{v}^{\mathrm{T}} \mathbf{P} \mathbf{v}}{\sigma_0^2} \tag{7.205}$$

where $\Delta \mathbf{v}^{\mathrm{T}} \mathbf{P} \mathbf{v}$ is the change of the sum of squares due to the constraint of the null hypothesis.

Let the null hypothesis that causes the smallest change $\Delta \mathbf{v}^{\mathrm{T}} \mathbf{P} \mathbf{v}$ be denoted by H_{sm}. The change in the sum of the squares and the noncentrality are $\Delta \mathbf{v}^{\mathrm{T}} \mathbf{P} \mathbf{v}_{\mathrm{sm}}$ and λ_{sm}, respectively. For any other null hypothesis we have $\lambda_j > \lambda_{\mathrm{sm}}$. If

$$\frac{\Delta \mathbf{v}^{\mathrm{T}} \mathbf{P} \mathbf{v}_j}{\Delta \mathbf{v}^{\mathrm{T}} \mathbf{P} \mathbf{v}_{\mathrm{sm}}} = \frac{\lambda_j}{\lambda_{\mathrm{sm}}} \geq \lambda_0 \left(\alpha, \beta_{\mathrm{sm}}, \beta_j\right) \tag{7.206}$$

then the two ambiguity sets comprising the null hypotheses H_{sm} and H_j are sufficiently discernible. Both hypotheses are sufficiently different to be distinguishable by means of their type-II errors. Because of its better compatibility with the float solution, the ambiguity set of the H_{sm} hypothesis is kept, and the set comprising H_j is discarded.

Figure 7.21 shows the ratio $\lambda_0(\alpha, \beta_{\mathrm{sm}}, \beta_j)$ as a function of the degree of freedom and the number of conditions. Euler and Schaffrin (1990) recommend a ratio between 5 and 10, which reflects a relatively large β_{sm} and a smaller β_j. Since H_{sm} is the hypothesis with the least impact on the adjustment, i.e., the most compatible with the float solution, it is desirable to have $\beta_{\mathrm{sm}} > \beta_j$ (recall that the type-II error equals the probability of accepting the wrong null hypothesis). Observing more satellites reduces the ratio for given type-II errors.

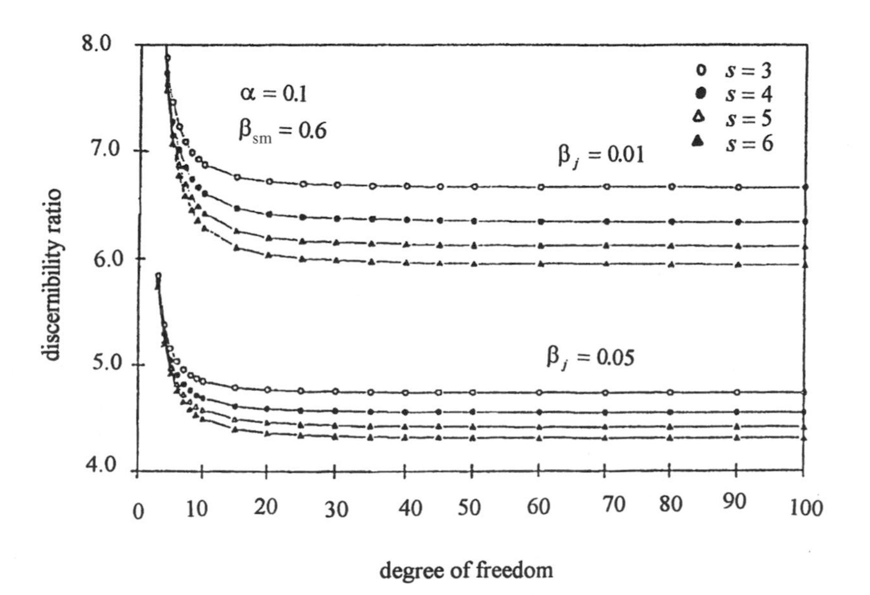

Figure 7.21 Discernibility ratio. (Permission by Springer Verlag.)

Many software packages implement a fixed value for the ratio of the best and the second-best solutions, e.g.,

$$\frac{\Delta \mathbf{v}^{\mathrm{T}}\mathbf{P}\mathbf{v}_{\text{2nd smallest}}}{\Delta \mathbf{v}^{\mathrm{T}}\mathbf{P}\mathbf{v}_{\text{sm}}} > 3 \tag{7.207}$$

to decide on discernibility. The explanations given above lend some theoretical justification to this commonly used practice, at least for a high degree of freedom. Other discrimination tests are proposed in Wang et al. (1998).

7.9 REAL-TIME RELATIVE POSITIONING

Transmitting the pseudorange and/or the carrier phase observations from a reference station to a moving receiver allows the latter to compute its location in real time. On-site computations allow for real-time quality assurance of kinematic applications and precise navigation to a known location. Various approaches have become available that apply to local areas, regions, or even the globe. Local area approaches generally aim to transmit sufficiently accurate information to allow a mobile user to fix ambiguities and therefore determine its position at the centimeter level with respect to the reference station. There are various options available for transmitting the data, i.e., cell phones, dedicated ground transmitters, geostationary satellites, and the Internet.

7.9.1 Carrier Phase and Pseudorange Corrections

Transmitting corrections is less of a telemetry load than transmitting the raw observations, because the dynamic range of the corrections is small. For every satellite p observed at station k, we determine an integer number K_k^p,

$$\begin{aligned} K_k^p = \operatorname{int}\left(\frac{P_{k,b}^p(1) - \Phi_{k,b}^p(1)}{\lambda}\right) &= \operatorname{int}\big(2I_{k,P}^p(1) - \lambda N_k^p(1) + cT_{\text{GD}}^p \\ &\quad + \delta_{k,P}^p(1) - \delta_{k,\Phi}^p(1)\big) \end{aligned} \tag{7.208}$$

using the observed pseudoranges and carrier phases at some initial epoch, and then compute the carrier phase range $\Theta_k^p(t)$ at subsequent epochs as

$$\Theta_{k,b}^p(t) = \Phi_{k,b}^p(t) + \lambda K_k^p \tag{7.209}$$

The numerical value of the carrier phase range is close to that of the pseudorange, differing primarily because of the ionosphere, as can be seen from the right side of (7.208). The discrepancy for the carrier phase range at epoch t is

$$
\begin{aligned}
\ell_k^p &= \Theta_{k,0}^p - \Theta_{k,b}^p = \rho_{k,0}^p - \Theta_{k,b}^p = \rho_{k,0}^p - \left(\Phi_{k,b}^p + \lambda K_k^p\right) \\
&= -\lambda\left(N_k^p + K_k^p\right) + c\,d\underline{t}_k - c\,d\bar{t}^p - I_{k,\Phi}^p - T_k^p - \delta_{k,\Phi}^p \\
&= \lambda\,\Delta N_k^p + c\,d\underline{t}_k - c\,d\bar{t}^p - I_{k,\Phi}^p - T_k^p - \delta_{k,\Phi}^p
\end{aligned}
\tag{7.210}
$$

The term ΔN_k^p is present because K_k^p only approximates N_k^p. The mean discrepancy μ_k of all satellites observed at the site and epoch t is

$$
\mu_k(t) = \frac{1}{S}\sum_{p=1}^{S} \ell_k^p(t) \tag{7.211}
$$

where S denotes the number of satellites. This mean discrepancy is driven primarily by the receiver clock error. The carrier phase correction at epoch t is

$$
\Delta\Phi_k^p = \rho_{k,0}^p - \Theta_{k,b}^p - \mu_k = \rho_{k,0}^p - \left(\Phi_{k,b}^p + \lambda K_k^p\right) - \mu_k \tag{7.212}
$$

The second part of this equation follows by substituting (7.209) for the carrier phase range. The phase correction (7.212) is transmitted to the moving receiver m. The rover's carrier phase Φ_m^p is corrected by adding the carrier phase correction, which was computed at receiver k,

$$
\bar{\Phi}_m^p = \Phi_m^p + \Delta\Phi_k^p \tag{7.213}
$$

Let us consider the single-difference observable (5.12) in the form

$$
\Phi_k^p - \Phi_m^p = \rho_k^p - \rho_m^p + \lambda N_{km}^p - c(d\underline{t}_k - d\underline{t}_m) + I_{km,\Phi}^p + T_{km}^p + \delta_{km,\Phi}^p \tag{7.214}
$$

Equation (7.212) can be solved for Φ_k^p and substituted into (7.214). After rearrangement, one obtains

$$
-\bar{\Phi}_m^p = -\rho_m^p + \lambda\left(N_{km}^p + K_k^p\right) - c(d\underline{t}_k - d\underline{t}_m) + \mu_k + I_{km,\Phi}^p + T_{km}^p + \delta_{km,\Phi}^p \tag{7.215}
$$

The left side of this equation is equal to the negative of the corrected carrier phase $\bar{\Phi}_m^p$. The differencing equation (7.215) between two satellites gives an expression that corresponds to the double-difference observable

$$
\bar{\Phi}_m^{qp} \equiv \bar{\Phi}_m^p - \bar{\Phi}_m^q = \rho_m^q - \rho_m^p + \lambda\left(N_{km}^{pq} + K_k^p - K_k^q\right) + I_{km,\Phi}^{pq} + T_{km}^{pq} + d_{km,\Phi}^{pq} \tag{7.216}
$$

The position of station m can now be computed at site m using the corrected observation $\bar{\Phi}_m^p$ to at least four satellites and forming three equations like (7.216). These equations differ from their conventional double-difference form by the fact that the modified ambiguity

$$\bar{N}_{km}^{pq} = N_{km}^{pq} + K_k^p - K_k^q \tag{7.217}$$

is estimated instead of N_{km}^{pq}.

The telemetry load can be further reduced if it is possible to increase the time between transmissions of the carrier phase corrections. For example, if the change in the discrepancy from one epoch to the next is smaller than the measurement accuracy at the moving receiver, or if the variations in the discrepancy are too small to affect adversely the required minimal accuracy for the moving receiver's position, it is possible to average carrier phase corrections over time and to transmit the averages. It might be desirable to transmit the rate of correction $\partial\Delta\Phi/\partial t$. If t_0 denotes the reference epoch, the user can interpolate the correctors over time as

$$\Delta\Phi_k^p(t) = \Delta\Phi_k^p(t_0) + \frac{\partial\Delta\Phi_k^p}{\partial t}(t - t_0) \tag{7.218}$$

One way to reduce the size and the slope of the discrepancy is to use the best available coordinates for the fixed receiver and a good satellite ephemeris. Clock errors affect the discrepancies directly, as is seen in Equation (7.210). Connecting a rubidium clock to the fixed receiver can effectively control the variations of the receiver clock error dt_k. Prior to its termination, selective availability was the primary cause of satellite clock error $d\bar{t}^p$ and was a determining factor that limited modeling like (7.218).

In the case of pseudorange corrections, we obtain similarly

$$\ell_k^p = \rho_k^p - P_k^p \tag{7.219}$$

$$\Delta P_k^p = \rho_{k,0}^p - P_k^p - \mu_k \tag{7.220}$$

$$\bar{P}_m^p(t) = P_m^p(t) + \Delta P_k^p(t) \tag{7.221}$$

$$\bar{P}_m^{qp}(t) \equiv \bar{P}_m^p - \bar{P}_m^q = \rho_m^q(t) - \rho_m^p(t) + I_{km,P}^{pq} + T_{km}^{pq} + d_{km,P}^{pq} \tag{7.222}$$

The approach described here is applicable to the L1 and L2 carrier phases and to all three codes.

7.9.2 Local Network Corrections

The impact of the troposphere, the ionosphere, and orbital errors on the single- and double-difference observables are at the same level as the carrier phase measurement resolution for short baselines or less. In fact, the definition of short baselines is directly linked to this cancellation of tropospheric and ionospheric effects and orbital errors on the single-difference observables, i.e., $T_{km}^p \approx 0$, $I_{km}^p \approx 0$, and $d\rho_{km}^p \approx 0$. It is common practice for short baselines to fix the double-difference ambiguities to integers using the LAMBDA procedure, yielding centimeter-accurate baselines. Traditionally, RTK techniques are applied to short baselines involving one base station and one roving receiver, using double differencing and employing some ambiguity fixing technique. Since RTK positioning is very economical, it is desirable to extend the reach of RTK over longer baselines. Because of the high spatial correlation of troposphere, ionosphere, and orbital errors, one expects that over a sufficiently small region the error terms T_{km}^p, I_{km}^p, and $d\rho_{km}^p$ depend on the distance between receivers. Wübbena et al. (1996) took advantage of this dependency and suggested the use of reference station networks to extend the reach of RTK.

There are two requirements at the heart of multiple reference station RTK. First, the positions of the reference stations must be accurately known at the centimeter level. This can be readily accomplished using postprocessing and long observation times. The second requirement is that the single- or double-difference integer ambiguities for baselines between reference stations are also known. It is then possible to compute tropospheric and ionospheric corrections and transmit these to an RTK user, to be applied to the rover's observations. For the discussion below we assume that the tropospheric term T_{km}^p includes the orbital error $d\rho_{km}^p$, which is permissible according to (7.19) and (7.20).

There are several variations available regarding the practical implementation of multiple reference station RTK. Because of its prevailing use with short baselines and because of the design of existing software, the initial implementation of multiple reference station networks was derived from double-difference observations. Below we give a general description using single differences. One could begin with the dual-frequency pseudorange and carrier phase observations and use (7.33) to compute the wide-lane ambiguity integers $N_{km,w}^p = N_{km,1}^p - N_{km,2}^p$ between network reference stations and for every satellite. One could then compute the tropospheric delay T_{km}^p using measured meteorological data, the tropospheric models such as (6.17) and (6.18) for the vertical dry and wet delays, and the tropospheric model mapping function (6.22). Alternatively, a continuously running Kalman filter can be used on the ionospheric-free function (7.41)

$$\Phi_{km,\mathrm{IF}}^p - \rho_{km,0}^p = T_k m\left(\vartheta_k^p\right) - T_m m\left(\vartheta_m^p\right) + c\, dt_{km} + R_{km}^p \tag{7.223}$$

to estimate the vertical tropospheric delays T_k and T_m, the receiver clock difference dt_{km}, and the ambiguity constant R_{km}^p using observations from all satellites. A simpler parameterization $T_{km} m(\vartheta_k^p)$ or $T_{km} m(\vartheta_k^q)$ may be permissible for the relative tropospheric correction. The subscript zero in $\rho_{km,0}^p$ indicates that the known reference station coordinates are used to compute the ranges. The individual ambiguities $N_{km,1}^p$ and $N_{km,2}^p$ can then be estimated from

$$R_{km}^p = \beta_f \lambda_1 N_{km,1}^p - \gamma_f \lambda_2 N_{km,2}^p \tag{7.224}$$

$$N_{km,w}^p = N_{km,1}^p - N_{km,2}^p \tag{7.225}$$

The ionospheric term $I_{km,1,P}^p$ can then be computed from the ionospheric function (7.40)

$$I_{km,1,P}^{p} = \left(1 - \alpha_f\right)^{-1} \left(\Phi_{km,I}^{p} - \lambda_1 N_{km,1}^{p} + \lambda_2 N_{km,2}^{p}\right) \tag{7.226}$$

Let k now denote the master reference station and m the other reference stations. The master reference station generates its own observations and receives observations from the other reference stations in real time. The Kalman filter, which runs at the master reference station, generates the corrections $T_{km}^{p} = T_k m(\vartheta_k^p) - T_m m(\vartheta_m^p)$ and $I_{km,1,P}^{p}$ at every epoch, for all reference stations and all satellites. These corrections are used to predict the respective corrections at a roving receiver's location. Various models are in use for computing these corrections. For example, the parameterization could be in terms of latitude, longitude, and height and using different models for T_{km}^{p} and $I_{km,1,P}^{p}$ to consider their characteristic spatial and temporal behavior. One of the simplest location-dependent models is a plane

$$T_{km}^{p}(t) = a_1^p(t) + a_2^p(t)\, n_m + a_3^p(t)\, e_m + a_4^p(t)\, u_m \tag{7.227}$$

$$I_{km1,P}^{p}(t) = b_1^p(t) + b_2^p(t)\, n_m + b_3^p(t)\, e_m + b_4^p(t)\, u_m \tag{7.228}$$

The symbols n_m, e_m, and u_m denote northing, easting, and up coordinates in the geodetic horizon at the master reference station k. The symbol m varies to include all other reference stations in the network. A set of coefficients $a_i^p(t)$ and $b_i^p(t)$, also called the network coefficients, are estimated by least-squares for every satellite p and, in principle, every epoch. Because of the high temporal correlation of the troposphere and ionosphere, one might model these coefficients over time, thus reducing the amount of data to be transmitted. The master reference station k transmits its own carrier phase observations, or alternatively, the carrier phase corrections as described by (7.212) and the network coefficients $\{a_i, b_i\}$ over the network. A rover n applies the tropospheric and ionospheric corrections (7.227) and (7.228) for its approximate position, and determines its precise location by least-squares from the series of double-difference observations

$$\Phi_{kn,\mathrm{IF}}^{pq}(t) - T_{kn}^{pq}(t) = \rho_{kn}^{pq}(t) + \beta_f \lambda_1 N_{kn,1}^{pq} - \gamma_f \lambda_2 N_{kn,2}^{pq} + d_{kn,I,\Phi}^{pq} \tag{7.229}$$

$$\Phi_{kn,I}^{pq}(t) - \left(1 - \alpha_f\right) I_{kn,1,P}^{pq}(t) = \lambda_1 N_{kn,1}^{pq} - \lambda_2 N_{kn,2}^{pq} + d_{kn,\mathrm{IF},\Phi}^{pq} \tag{7.230}$$

using the standard ambiguity fixing techniques.

Rather than transmitting network coefficients $\{a_i^p, b_i^p\}$, one might consider transmitting corrections $\{T_{km}^{p}, I_{km,1,P}^{p}\}$ for a grid of points at known locations within the network. The mobile user would interpolate the corrections for the rover's approximate location and apply them to the observations. Vollath et al. (2000) suggest the use of virtual reference stations (VRSs) to avoid changing existing software that double-differences the original observations directly. The VRS concept requires that the rover transmit its approximate location to the master reference station, which computes the corrections $\{T_{km}^{p}, I_{km,1,P}^{p}\}$ for the rover's approximate location. In addition, the master reference station computes virtual observations for the approximate rover location using its own observations and then corrects them for troposphere and ionosphere, i.e.,

$$\begin{bmatrix} P^p_{v,1} \\ P^p_{v,2} \\ \Phi^p_{v,1} \\ \Phi^p_{v,2} \end{bmatrix} = \begin{bmatrix} P^p_{k,1} \\ P^p_{k,2} \\ \Phi^p_{k,1} \\ \Phi^p_{k,2} \end{bmatrix} + \begin{bmatrix} 1 & 1 \\ 1 & \alpha_f \\ 1 & -1 \\ 1 & \alpha_f \end{bmatrix} \begin{bmatrix} \rho^p_{vk} + T^p_{vk} \\ I^p_{vk,1,P} \end{bmatrix} \tag{7.231}$$

and transmits the corrected, virtual observations to the rover. The rover merely has to double-difference its own observations with those received from the master reference station. No additional tropospheric or ionospheric corrections/interpolations are required at the rover because the effective, virtual baseline is very short, typically in the range of meters corresponding to the rover's initial determination of its approximate location from pseudoranges. In Equation (7.231) the subscript v in ρ^p_{vk} indicates that the distances are evaluated for the location of the virtual reference station v. The need for the rover to transmit data can be eliminated if the master reference station transmits corrected virtual observation to an evenly spaced grid of predetermined points within the network. The rover can determine its position with respect to the nearest virtual grid point. The grid approach supports many mobile users, since they all use the same data sent from the master reference station.

The multiple reference station techniques described above depend on the master reference station operator's skill in modeling the spatial and temporal corrections (7.227) and (7.228). The success of fixing the ambiguities correctly at the rover directly depends on the validity of the tropospheric and ionospheric corrections. Raquet (1998), Lachapelle et al. (2000), and Fortes (2002) compute a covariance function from the double-difference carrier phase discrepancies of the known network baselines. They then use least-squares collocation to compute undifferenced corrections for each satellite at all reference stations and predict undifferenced corrections for a grid of known locations. The conversion of corrections from the double-difference domain to the undifferenced domain is carried out based on covariance functions associated with spatial differential errors (for troposphere/orbits and ionosphere) and assigning the absolute errors equal to zero at a reference point normally located close to the center of the region covered by the network (considering that the user software normally implements the double-difference model, what matters is how the residual errors change from one location to the other and not their actual absolute values). These covariance functions are then used to compute covariance matrices to be applied in the prediction of the errors at the user location using least-squares collocation. The master reference station transmits these corrections to the other reference stations, where they are applied to the undifferenced observations. These corrected undifferenced observations are broadcast over the network (in addition to the predicted gridded corrections). Zebhauser et al. (2002) suggest transmitting the observation of the master reference station and the observation differences between pairs of reference stations. The latter would be corrected for location, receiver clock, and ambiguities, i.e.,

$$\Phi^{p}_{km,\mathrm{IF}} - \rho^{p}_{km} - dt_{\_km} - R^{p}_{km} = T^{p}_{km} + d^{p}_{km,\mathrm{IF},\Phi} \tag{7.232}$$

$$\Phi^{p}_{km,I}(t) - \lambda_1 N^{p}_{km,1} + \lambda_2 N^{p}_{km,2} = \left(1 - \alpha_f\right) I^{p}_{km,1,P}(t) + d^{p}_{km,I,\Phi} \tag{7.233}$$

The user at the roving station is free to use any modeling and interpolation model to compute the respective tropospheric and ionospheric corrections.

The message formats for data exchange between a single base station and a single rover generally follow the standards set by the Radio Technical Commission for Maritime Services (RTCM). RTCM is a nonprofit scientific and educational organization consisting of international member organizations that include manufacturers, marketing, service providers, maritime user entities representing interests from small recreational craft to deep-sea shipping, educational institutions, labor unions, and government agencies (RTCM, 2002). RTCM special committees address in-depth radiocommunication and radionavigation areas of concern to the RTCM members. The reports prepared by these committees are usually published as RTCM recommendations. The RTCM special committee 104 deals with global navigation satellite systems. It has issued Standards for Differential GNSS (currently version 2.3) and Standards for Differential Navstar GNSS Reference Stations (currently Version 1.1). It is expected that RTCM standards will be available in the near future, including all message types needed for real-time RTK within multiple reference station networks.

7.9.3 WADGPS

The modeling in (7.227) and (7.228) and the achievable accuracy for the corrections T^{p}_{kn} and $I^{p}_{kn,1,P}$ usually determine the size of the area over which real-time RTK is possible, unless accurate corrections are available from other sources. As the area increases, the ambiguities cannot be fixed and the carrier phases are used to smooth the pseudoranges. The tropospheric and ionospheric corrections are typically parameterized by latitude and longitude and transmitted to the user via geostationary satellites for such wide area differential GPS (WADGPS) networks. Also, the tropospheric corrections and the orbital satellite errors are typically dealt with separately. Early work on WADGPS is found in Brown (1989), Kee et al. (1991), and Ashkenazi et al. (1992).

Several WADGPS systems have been implemented around the world; e.g., Whitehead et al. (1998) describe a system that is privately operated to support precision agriculture. The Federal Aviation Administration (FAA) is developing a WADGPS called WAAS (wide area augmentation system). WAAS is a satellite-based augmentation system (SBAS), meaning that the differential corrections and other relevant data important for enhancing reliability and integrity of the system are transmitted via satellites. WAAS will provide guidance to aircraft at thousands of airports and airstrips where there has previously been no precision landing capability (Loh et al., 1995). Other SBASs have been developed in Europe and Japan under the names European Geostationary Navigation Overlay Service (EGNOS) and MTSAT Satellite-based Augmentation System (MSAS). Several U.S. agencies, such as the Federal Railroad Administration, the U.S. Coast Guard, the Federal Highway Administration, and the Office of the Secretary of Transportation are developing the nationwide differential global positioning system (NDGPS). The system began as an expansion of the

U.S. Coast Guard's maritime differential GPS service and incorporated the ground wave emergency network (GWEN) sites, which became available at the end of the cold war (Allen, 1999; Cook, 2000). NDGPS utilizes powerful ground transmitters to broadcast the corrections.

As the coverage area of a WADGPS further increases, it eventually will become a global system, being conceptually similar to the one discussed in Section 7.6.

CHAPTER 8

NETWORK ADJUSTMENTS

This chapter deals with minimal or inner constraint solutions for a polyhedron of stations, i.e., the GPS vector network. The relative locations in such networks are usually more accurate than the geocentric location of the polyhedron. Typically, the relative position accuracy is derived from fixed carrier phase solutions, whereas the geocentric location is obtained from point positioning with pseudoranges. Despite the versatility of GPS, there are still many situations in which terrestrial observations such as angles and distances measured with theodolite and electronic distance measurement (EDM) are useful for supplementing GPS vectors. We discuss combination solutions that use additional parameters, such as three rotations and a scale factor. The rotation parameters can absorb rotational misalignment between the coordinate system of the GPS ephemeris and the terrestrial coordinate system. If applied in local networks, these rotation parameters may also be useful for absorbing a geoid slope if geoid undulations are not available in the terrestrial system.

The 3D geodetic model discussed in Chapter 2 is the most natural one to be used for these network adjustments. Since GPS gives accurate geodetic height differences, the clear distinction between orthometric and geodetic (ellipsoidal) heights is always important. This is particularly true when traditional leveling is replaced with GPS height determination.

The chapter contains three examples of vector adjustments. While observing GPS vector networks has become a routine occurrence, these examples have some "historic value" as they helped establish GPS as a tool for accurate surveying. The Montgomery County geodetic network densification demonstrated the utility of GPS to densify classical first-order horizontal geodetic networks in terms of the achievable accuracy and the high degree of flexibility in network design. The Stanford Linear Collider (SLC) engineering survey pioneered in the sense that millimeter accuracy was achieved using satellite techniques and that an independent method for verifying this accuracy was available. The Orange County densification demonstrated the use of least-squares to quality control large vector data sets.

8.1 GPS VECTOR NETWORKS

In the case of two receivers observing, carrier phase processing gives the vector between the stations, expressed in the reference frame of the ephemeris, and the 3×3 covariance matrix of the coordinate differences. The covariance matrix of all vector observations is block-diagonal, with 3×3 submatrices along the diagonal. In a session solutions, in which case R receivers observe the same satellites simultaneously, the results are $(R-1)$ independent vectors, and a $3(R-1) \times 3(R-1)$ covariance matrix. The covariance matrix is still block-diagonal, but the size of the nonzero diagonal matrices is a function of R.

Like any other survey, a GPS survey that has determined the relative locations of a cluster of stations should be subjected to a minimal or inner constraint adjustment for purposes of quality control. For example, the network should not contain unconnected vectors whose endpoints are not tied to other parts of the network. At the network level, the quality of the derived vector observations can be assessed, the geometric strength of the overall network can be analyzed, internal and external reliability can be computed, and blunders may be discoverable and removable. For example, a blunder in an antenna height will not be discovered when processing a single baseline, but it will be noticeable in the network solution if stations are reoccupied independently. Covariance propagation for computing distances, angles, or other functions of the coordinates should be done, as usual, with the minimal or inner constraint solution.

The mathematical model is the standard observation equation model, i.e.,

$$\boldsymbol{\ell}_a = \mathbf{f}(\mathbf{x}_a) \tag{8.1}$$

where $\boldsymbol{\ell}_a$ contains the adjusted observations and $\mathbf{x}_a$ denotes the adjusted station coordinates. The mathematical model is linear if the parameterization of receiver positions is in terms of Cartesian coordinates. In this case the vector observation between stations k and m is modeled simply as

$$\begin{bmatrix} \Delta x_{km} \\ \Delta y_{km} \\ \Delta z_{km} \end{bmatrix} = \begin{bmatrix} x_k - x_m \\ y_k - y_m \\ z_k - z_m \end{bmatrix} \tag{8.2}$$

The relevant portion of the design matrix $\mathbf{A}$ for the model (8.2) is

$$\mathbf{A}_{km} = \begin{array}{c} \begin{matrix} x_k & y_k & z_k & x_m & y_m & z_m \end{matrix} \\ \begin{bmatrix} 1 & 0 & 0 & -1 & 0 & 0 \\ 0 & 1 & 0 & 0 & -1 & 0 \\ 0 & 0 & 1 & 0 & 0 & -1 \end{bmatrix} \end{array} \tag{8.3}$$

The design matrix looks like one for a leveling network. The coefficients are either 1, -1, or 0. Each vector contributes three rows. Because vector observations contain information about the orientation and scale, one only needs to fix the translational location of the polyhedron. Minimal constraints for fixing the origin can be imposed by simply deleting the three coordinate parameters of one station, holding that particular station effectively fixed.

Inner constraints must fulfill the condition

$$\mathbf{Ex} = \mathbf{o} \tag{8.4}$$

according to (4.201), or, what amounts to the same condition,

$$\mathbf{E}^{\mathrm{T}}\mathbf{A} = \mathbf{O} \tag{8.5}$$

It can be readily verified that

$$\mathbf{E} = [{}_3\mathbf{I}_3 \quad {}_3\mathbf{I}_3 \quad {}_3\mathbf{I}_3 \quad \cdots] \tag{8.6}$$

fulfills these conditions. The matrix $\mathbf{E}$ consists of a row of 3×3 identity matrices. There are as many identity matrices as there are stations in the network. The inner constraint solution uses the pseudoinverse (4.203)

$$\mathbf{N}^{+} = \left(\mathbf{A}^{\mathrm{T}}\mathbf{PA} + \mathbf{E}^{\mathrm{T}}\mathbf{E}\right)^{-1} - \mathbf{E}^{\mathrm{T}}\left(\mathbf{EE}^{\mathrm{T}}\mathbf{EE}^{\mathrm{T}}\right)^{-1}\mathbf{E} \tag{8.7}$$

of the normal matrix. If one sets the approximate coordinates to zero, which can be done since the mathematical model is linear, then the origin of the coordinate system is at the centroid of the cluster of stations. For nonzero approximate coordinates, the coordinates of the centroid remain invariant; i.e., the values are the same whether computed from the approximate coordinates or the adjusted coordinates. The standard ellipsoid reflects the true geometry of the network and the satellite constellation. See Chapter 4 for a discussion on which quantities are variant or invariant with respect to different choices of minimal constraints.

The GPS-determined coordinates refer to the coordinate system of the satellite positions (ephemeris). The broadcast ephemeris coordinate system is given in WGS84, and the precise ephemeris is in ITRF. Both coordinate systems agree at the couple-of-centimeters level.

The primary result of a typical GPS survey is best viewed as a polyhedron of stations whose relative positions have been accurately determined (to the centimeter or even the millimeter level), but the translational position of the polyhedron is typically known only at the meter level (point positioning with pseudoranges). The orientation of the polyhedron is implied by the vector observations. The Cartesian coordinates (or coordinate differences) of the GPS survey can, of course, be converted to geodetic latitude, longitude, and height. If geoid undulations are available, the orthometric heights (height differences) can be readily computed. The variance-covariance components of the adjusted parameters can be transformed to the local geodetic system for ease of interpretation using (2.208).

8.2 TRANSFORMING NEARLY ALIGNED COORDINATE SYSTEMS

The transformation of three-dimensional coordinate systems has been given much attention ever since geodetic satellite techniques made it possible to relate local and geocentric geodetic datums. Some of the pertinent works are Veis (1960), Molodenskii et al. (1962), Badekas (1969), Vaniček and and Wells (1974), Leick and van Gelder (1975), and Soler and van Gelder (1987). We assume that the Cartesian coordinates of points on the earth's surface are available in two systems. Often it might be difficult to obtain the Cartesian coordinates in the local geodetic datum because the geoid undulations with respect to the local datum might not be accurately known.

Figure 8.1 shows the coordinate system $(x) = (x, y, z)$, which is related to the coordinate system $(u) = (u, v, w)$ by the translation vector $\mathbf{t} = [\Delta x \quad \Delta y \quad \Delta z]^{\mathrm{T}}$ between the origins of the two coordinate systems and the small rotations $(\varepsilon, \psi, \omega)$ around the (u, v, w) axes, respectively. The transformation equation expressed in the (x) coordinate system can be seen from Figure 8.1:

$$\mathbf{t} + (1 + s)\,\mathbf{Ru} - \mathbf{x} = \mathbf{o} \tag{8.8}$$

where $1 + s$ denotes the scale factor between the systems and $\mathbf{R}$ is the product of three consecutive orthogonal rotations around the axes of (u):

$$\mathbf{R} = \mathbf{R}_3(\omega)\mathbf{R}_2(\psi)\mathbf{R}_1(\varepsilon) \tag{8.9}$$

The symbol $\mathbf{R}_i$ denotes the rotation matrix for a rotation around axis i (see Section A.2). The angles $(\varepsilon, \psi, \omega)$ are positive for counterclockwise rotations about the respective (u, v, w) axes, as viewed from the end of the positive axis. For nearly aligned coordinate systems these rotation angles are differentially small, allowing the following simplification

$$\mathbf{R} = \mathbf{I} + \mathbf{Q} = \mathbf{I} + \begin{bmatrix} 0 & \omega & -\psi \\ -\omega & 0 & \varepsilon \\ \psi & -\varepsilon & 0 \end{bmatrix} \tag{8.10}$$

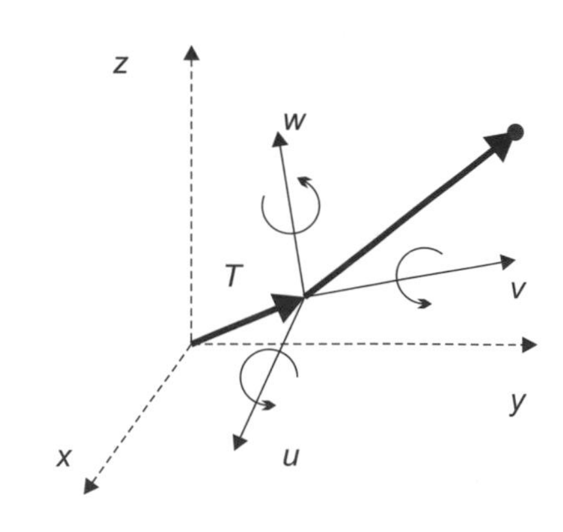

Figure 8.1 Differential transformation between Cartesian coordinate systems.

Combining (8.8) and (8.10) gives the linearized form

$$\mathbf{t} + \mathbf{u} + s\mathbf{u} + \mathbf{Q}\mathbf{u} - \mathbf{x} = \mathbf{o} \tag{8.11}$$

For the purpose of distinguishing various approaches, we call the transformation (8.8) model 1. The seven transformation parameters $(\Delta x, \Delta y, \Delta z, s, \varepsilon, \psi, \omega)$ can be estimated by a least-squares. The Cartesian coordinates $\mathbf{u}$ and $\mathbf{x}$ are the observations. Equation (8.11) represents a mixed model $\mathbf{f}(\boldsymbol{\ell}_a, \mathbf{x}_a) = \mathbf{o}$. See Section 4.4 for additional explanations of the mixed model adjustment. Each station contributes three equations to (8.8).

A variation of (8.8), called model 2, is

$$\mathbf{t} + \mathbf{u}_0 + (1 + s)\,\mathbf{R}\,(\mathbf{u} - \mathbf{u}_0) - \mathbf{x} = \mathbf{o} \tag{8.12}$$

where $\mathbf{u}_0$ is the vector in the system (u) to a point located somewhere within the network that is to be transformed. A likely choice for $\mathbf{u}_0$ is the centroid. All other notation is the same as in Equation (8.8). If one follows the same procedure as described for the previous model, i.e., omitting second-order terms in scale and rotation and their products, then (8.12) becomes

$$\mathbf{t} + \mathbf{u} + s\,(\mathbf{u} - \mathbf{u}_0) + \mathbf{Q}\,(\mathbf{u} - \mathbf{u}_0) - \mathbf{x} = \mathbf{o} \tag{8.13}$$

Model 3 uses the same rotation point $\mathbf{u}_0$ as model 2, but the rotations are about the axes (n, e, u) of the local geodetic coordinate system at $\mathbf{u}_0$. The n axis is tangent to the geodetic meridian, but the positive direction is toward the south; the e axis is perpendicular to the meridian plane and is positive eastward. The u axis is along the ellipsoidal normal with its positive direction upward, forming a right-handed system with n and e. Similar to Equation (8.12), one obtains

$$\mathbf{t} + \mathbf{u}_0 + (1 + s)\,\mathbf{M}\,(\mathbf{u} - \mathbf{u}_0) - \mathbf{x} = \mathbf{o} \tag{8.14}$$

If (η, ξ, α) denote positive rotations about the (n, e, u) axes and if $(\varphi_0, \lambda_0, h_0)$ are the geodetic coordinates for the point of rotation $\mathbf{u}_0$, it can be verified that the $\mathbf{M}$ matrix is

$$\mathbf{M} = \mathbf{R}_3^T(\lambda_0)\mathbf{R}_2^T(90 - \varphi_0)\mathbf{R}_3(\alpha)\mathbf{R}_2(\xi)\mathbf{R}_1(\eta)\mathbf{R}_2(90 - \varphi_0)\mathbf{R}_3(\lambda_0) \tag{8.15}$$

Since the rotation angles (η, ξ, α) are differentially small, the matrix $\mathbf{M}$ simplifies to

$$\mathbf{M}(\lambda_0, \varphi_0, \eta, \xi, \alpha) = \alpha\mathbf{M}_\alpha + \xi\mathbf{M}_\xi + \eta\mathbf{M}_\eta + \mathbf{I} \tag{8.16}$$

where

$$\mathbf{M}_\alpha = \begin{bmatrix} 0 & \sin\varphi_0 & -\cos\varphi_0\sin\lambda_0 \\ -\sin\varphi_0 & 0 & \cos\varphi_0\cos\lambda_0 \\ \cos\varphi_0\sin\lambda_0 & -\cos\varphi_0\cos\lambda_0 & 0 \end{bmatrix} \tag{8.17}$$

$$\mathbf{M}_\xi = \begin{bmatrix} 0 & 0 & -\cos\lambda_0 \\ 0 & 0 & -\sin\lambda_0 \\ \cos\lambda_0 & \sin\lambda_0 & 0 \end{bmatrix} \tag{8.18}$$

$$\mathbf{M}_\eta = \begin{bmatrix} 0 & -\cos\varphi_0 & -\sin\varphi_0\sin\lambda_0 \\ \cos\varphi_0 & 0 & \sin\varphi_0\cos\lambda_0 \\ \sin\varphi_0\sin\lambda_0 & -\sin\varphi_0\cos\lambda_0 & 0 \end{bmatrix} \tag{8.19}$$

If, again, second-order terms in scale and rotations and their products are neglected, the model (8.14) becomes

$$\mathbf{t} + \mathbf{u} + s\,(\mathbf{u} - \mathbf{u}_0) + (1 + s)\,(\mathbf{M} - \mathbf{I})\,(\mathbf{u} - \mathbf{u}_0) - \mathbf{x} = \mathbf{o} \tag{8.20}$$

Models 2 and 3 differ in that the rotations in model 3 are around the local geodetic coordinate axes at $\mathbf{u}_0$. The rotations (η, ξ, α) are (ε, ψ, ω) as related as follows:

$$\begin{bmatrix} \eta \\ \xi \\ \alpha \end{bmatrix} = \mathbf{R}_2(90 - \varphi_0)\mathbf{R}_3(\lambda_0) \begin{bmatrix} \varepsilon \\ \psi \\ \omega \end{bmatrix} \tag{8.21}$$

Models 1 and 2 use the same rotation angles. The translations for models 1 and 2 are related as

$$\mathbf{t}_2 = \mathbf{t}_1 - \mathbf{u}_0 + (1 + s)\mathbf{R}\,\mathbf{u}_0 \tag{8.22}$$

according to (8.8) and (8.12). Only $\mathbf{t}_1$, i.e., the translation vector of the origin as estimated from model 1, corresponds to the geometric vector between the origins of the coordinate systems (x) and (u). The translational component of model 2, $\mathbf{t}_2$, is a function of $\mathbf{u}_0$, as shown in (8.22). Because models 2 and 3 use the same $\mathbf{u}_0$, both yield identical translational components. It is not necessary that all seven parameters always be estimated. In small areas it might be sufficient to estimate only the translation components.

8.3 COMBINATION THROUGH ROTATION AND SCALING

Assume a situation in which a network of terrestrial observations, such as horizontal angles, slant distances, zenith angles, leveled height differences, and geoid undulations, are available. Assume further that the relative positions of some of these network stations have been determined with GPS. As a first step one could carry out separate minimal or inner constraint solutions for the terrestrial observations and the GPS vectors, as a matter of quality control. When combining both sets of observations in one adjustment, the definition of the coordinate systems might become important.

For example, consider the case that coordinates of some of the stations are known in the "local datum" (u) and that (u) does not coincide with (x), i.e., the coordinate system of the GPS vectors. Let it be further required that if the adjusted coordinates should be expressed in (u), then the following model

$$\boldsymbol{\ell}_{1a} = \mathbf{f}_1(\mathbf{x}_u) \tag{8.23}$$

$$\boldsymbol{\ell}_{2a} = \mathbf{f}_2(s, \eta, \xi, \alpha, \mathbf{x}_a) \tag{8.24}$$

might be applicable. The model (8.23) pertains to the terrestrial observations, denoted here as the $\boldsymbol{\ell}_1$ set. This model is discussed in Chapter 2. In adjustment notation the parameters $\mathbf{x}_a$ denote station coordinates in the geodetic system (u). The observations for Model (8.24) are the Cartesian coordinate differences between stations as obtained from GPS carrier phase processing. The additional parameters in (8.24) are the scale correction s and three rotation angles. The rotation angles are small as they relate the nearly aligned geodetic coordinate systems (u) and (x). Because GPS yields the coordinate differences, there is no need to include the translation parameter $\mathbf{t}$. If the coordinate systems (u) and (x) coincide, then the estimate of the rotation angles should statistically be zero. Even if $\boldsymbol{\ell}_1$ in (8.23) does not contain observations at all, some of the station coordinates in the (u) system can still be treated as observed parameters and thus allow the estimation of the scale and rotation parameters. This is a simple way to implement the GPS vector observations into the existing network.

The mathematical model (8.24) follows directly from the transformation expression (8.14). Applying this expression to the coordinate differences for stations k and m yields

$$(1 + s)\,\mathbf{M}(\lambda_0, \varphi_0, \eta, \xi, \alpha)\,(\mathbf{u}_k - \mathbf{u}_m) - (\mathbf{x}_k - \mathbf{x}_m) = \mathbf{o} \tag{8.25}$$

The coordinate differences

$$\mathbf{x}_{km} = \mathbf{x}_k - \mathbf{x}_m \tag{8.26}$$

represent the observed GPS vector between stations k and m. Thus the mathematical model (8.24) can be written as

$$\mathbf{x}_{km} = (1 + s)\,\mathbf{M}(\lambda_0, \varphi_0, \eta, \xi, \alpha)\,(\mathbf{u}_k - \mathbf{u}_m) \tag{8.27}$$

After substituting (8.16) into (8.27), we readily obtain the partial derivatives of the design matrix. Table 8.1 lists the partial derivatives with respect to the station

TABLE 8.1 Design Submatrix for Stations Occupied with Receivers

Parameterization	Station m	Station k
(u)	$(1+s)\,\mathbf{M}$	$-(1+s)\,\mathbf{M}$
(φ, λ, h)	$(1+s)\,\mathbf{MJ}(\varphi_m, \lambda_m)$	$-(1+s)\,\mathbf{MJ}(\varphi_k, \lambda_k)$
(n, e, u)	$(1+s)\,\mathbf{MJ}(\varphi_m, \lambda_m)\mathbf{H}^{-1}(\varphi_m)$	$-(1+s)\,\mathbf{MJ}(\varphi_k, \lambda_k)\mathbf{H}^{-1}(\varphi_k)$

TABLE 8.2 Design Submatrix for the Transformation Parameters

s	η	ξ	α
$\mathbf{u}_m - \mathbf{u}_k$	$\mathbf{M}_\eta\,(\mathbf{u}_m - \mathbf{u}_k)$	$\mathbf{M}_\xi\,(\mathbf{u}_m - \mathbf{u}_k)$	$\mathbf{M}_\alpha\,(\mathbf{u}_m - \mathbf{u}_k)$

coordinates for (a) Cartesian parameterization, (b) parameterization in terms of geodetic latitude, longitude, and height, and (c) parameterization in terms of the local geodetic coordinate systems. The transformation matrices **J** and **H** referred to in the table are those of (2.106) and (2.108). Table 8.2 contains the partial derivatives of the transformation parameters.

8.4 GPS NETWORK EXAMPLES

In these examples only independent vectors between stations are included; i.e., if three receivers observe simultaneously, only two vectors are used. The stochastic model does not include the mathematical correlation between simultaneously observed vectors. The variance-covariance matrix of the observed vectors is 3×3 block-diagonal. Craymer and Beck (1992) discuss various aspects of session versus single-baseline processing. They also point out that inclusion of the trivial (dependent) baselines distorts the formal accuracy by increasing the redundancy in the model artificially, resulting in overly optimistic covariance matrices. The covariance information used was obtained directly from baseline processing and does not accommodate small uncertainties in eccentricity, i.e., setting up the antenna over the mark. Only single-frequency carrier phases were available at the time the observations were made.

8.4.1 Montgomery County Geodetic Network

At the time of the Montgomery County (Pennsylvania) geodetic network densification, the window of satellite visibility was about 5 hours for GPS, just long enough to allow two sessions with the then state-of-the-art static approach (Collins and Leick, 1985). Much liberty was taken in the network design (Figure 8.2) by taking advantage of GPS's insensitivity to the shape of the network (as compared to the many rules of classical triangulation and trilateration). The longest baseline observed was about 42 km. Six horizontal stations with known geodetic latitude and longitude and seven vertical stations with known orthometric height were available for tying the GPS survey to the existing geodetic networks. Accurate geoid information was not available at the time.

Figure 8.3 shows two intersections of the ellipsoid of standard deviation for the inner constraint least-squares solution. The top set of ellipses shows the horizontal intersection (i.e., the ellipses of standard deviation in the geodetic horizon), and the bottom set of ellipses shows the vertical intersection in the east-west direction. The figure also shows the daily satellite visibility plot for the time and area of the project.

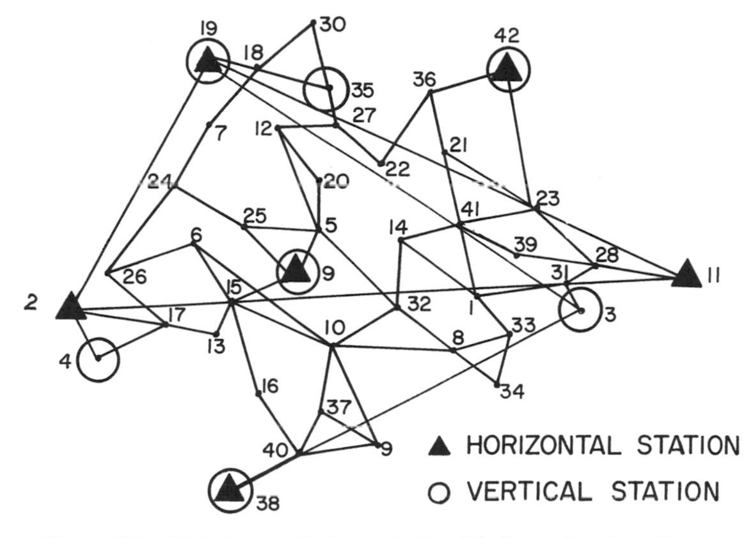

Figure 8.2 Existing geodetic control and independent baselines.

The dots in that figure represent the directions of the semimajor axis of the ellipsoids of standard deviation for each station. These directions tend to be located around the center of the satellite constellation. The standard ellipses show a systematic orientation in both the horizontal and the vertical planes. This dependency of the shape of the ellipses with the satellite constellation enters into the adjustment through the 3×3 correlation matrices. With a better distribution of the satellites over the hemisphere, the alignments seen in Figure 8.3 for the horizontal ellipses do not occur. Because satellites are observed above the horizon, the ellipses will still be stretched along the vertical.

The coordinates of the polyhedron of stations are given in the coordinate system of the broadcast ephemeris; at the time of the Montgomery County survey this was WGS72 (today this would be WGS84 or the latest ITRF). The positions of the polyhedron stations can be expressed in terms of geodetic coordinates relative to any ellipsoid, as long as the location of the ellipsoid is specified. For example, the minimal constraints could be specified by equating the geodetic and astronomic latitude and longitude of station 29, and equating the ellipsoidal height and the orthometric height. The ellipsoid defined in that manner is tangent to the geoid at station 29. By comparing the resulting geodetic heights with known orthometric heights at the vertical stations, we can construct a geoid undulation map (with respect to the thus defined ellipsoid). The geoid undulations at other stations can be interpolated to give orthometric height from the basic relation $H = h - N$.

The method described above can be generalized by not using the astronomic position for station 29. The geodetic latitude and longitude, such as the NAD83 positions, can be used as minimal constraints. The thus defined local ellipsoid is not tangent to the geoid at station 29. The undulations with respect to such an ellipsoid are shown in Figure 8.4.

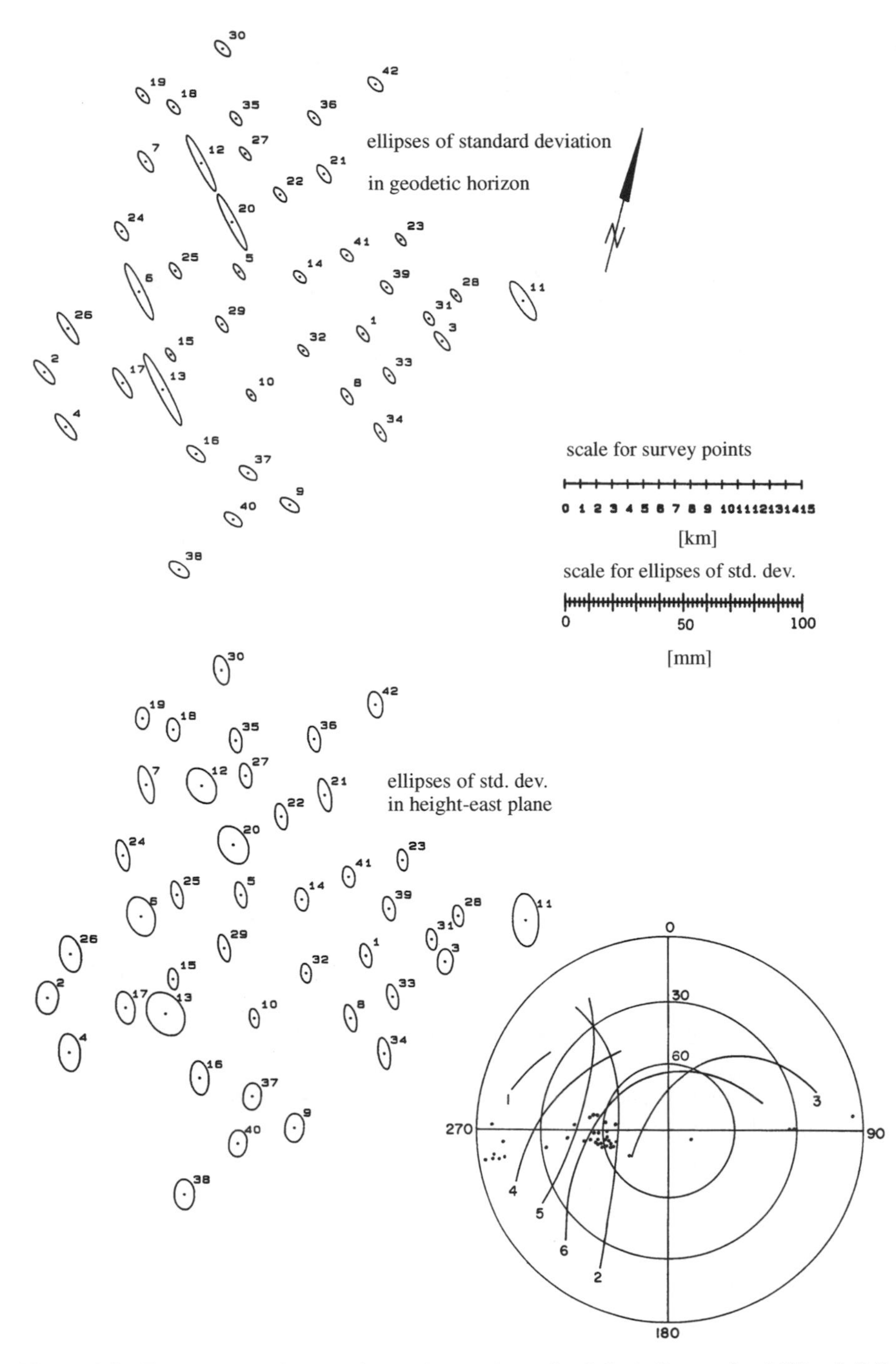

Figure 8.3 Inner constraint solution, ellipses of standard deviation, and satellite visibility plot.

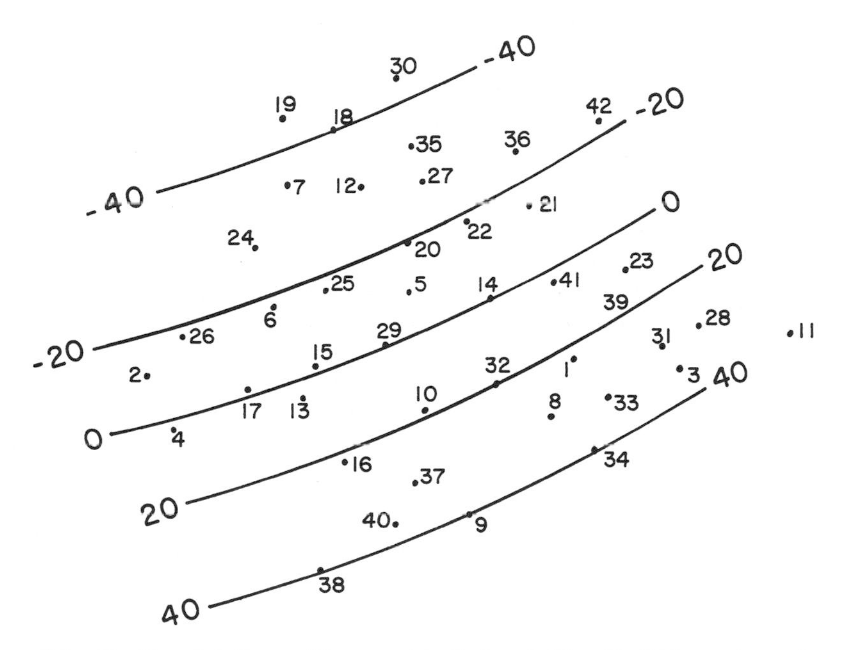

Figure 8.4 Geoid undulations with respect to the local ellipsoid. Units are in centimeters.

Alternatively, one can estimate the topocentric rotations (η, ξ, α) and a scale factor implied by model (8.27). There are seven minimal constraints required in this case, e.g., the geodetic latitude and longitude for two stations and the geodetic heights for three stations distributed well over the network. If one uses orthometric heights for these three stations instead, the angles (ξ, η) reflect the average deflection of the vertical angles. Using orthometric heights instead of geodetic heights forces the ellipsoid to coincide locally with the geoid (as defined or implied by the orthometric heights at the vertical stations). The rotation in azimuth α is determined by the azimuthal difference between the two stations held fixed and the GPS vector between the same stations. The scale factor is also determined by the two stations held fixed; it contains the possible scale error of the existing geodetic network and the effect of a constant but unknown undulation (i.e., geoid undulations with respect to the ellipsoid of the existing geodetic network).

Simple geometric interpolation of geoid undulations has its limits, of course. For example, any error in a given orthometric height will result inevitably in an erroneous geoid feature. As a result, the orthometric heights computed from the interpolated geoid undulations will be in error. Depending on the size of the survey area and the "smoothness" of the geoid in that region, such erroneous geoid features might or might not be discovered from data analysis. These difficulties can be avoided if an accurate geoid model is available.

8.4.2 SLC Engineering Survey

A GPS survey was carried out in 1984 to support construction of the Stanford Linear Collider (SLC) with the objective of achieving millimeter relative positional accuracy with GPS and combining GPS vector with terrestrial observations (Ruland and Leick, 1985). Because the network was only 4 km long, the broadcast ephemeris errors as well as the impact of the troposphere and ionosphere cancel. The position accuracy in such small networks is limited by the carrier phase measurement accuracy, the phase center variation of the receiver antenna, and the multipath. We used the Macrometer antenna, which is known for its good multipath rejection property and accurate definition of the phase center.

The network is shown in Figure 8.5. Stations 1, 10, 19, and 42 are along the 2-mile-long linear accelerator (linac); the remaining stations of the "loop" were to be determined with respect to these linac stations. The disadvantageous configuration of this network, in regard to terrestrial observations such as angles and distances, is obvious. To improve this configuration, one would have to add stations adjacent to the linac; this would have been costly because of the local topography and construction. For GPS positioning such a network configuration is acceptable because the accuracy of positioning depends primarily on the satellite configuration and not on the shape of the network. Figure 8.6 shows the horizontal ellipses of standard deviation and the satellite visibility plot for the inner constraint vector solution. The dark spot on the visibility plot represents the directions of the semimajor axes of the standard ellipsoids.

This survey offered an interesting comparison. For the frequent realignment of the linear accelerator, the linac laser alignment system was installed. This system is capable of determining positions perpendicular to the axis of the linac to better than ±0.1 mm over the total length of 3050 m. A comparison of the linac stations 1, 10, 19, and 42, as determined from the GPS vector solution with respect to the linac alignment system, was done by means of a transformation. The discrepancies did not exceed ±1 mm for any of the four linac stations.

8.4.3 Orange County Densification

The Orange County GPS survey was comprised of more than 7000 vectors linking 2000 plus stations at about 0.5 mile spacing. With that many vector observations, it is beneficial to use graphics to analyze observations, the adjustments, and other

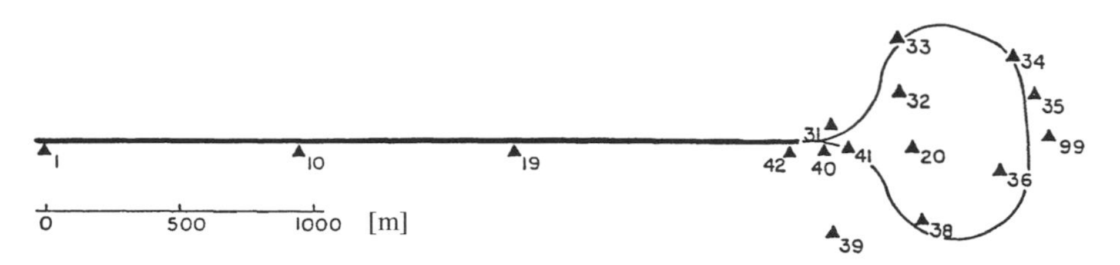

Figure 8.5 The SLC network configuration.

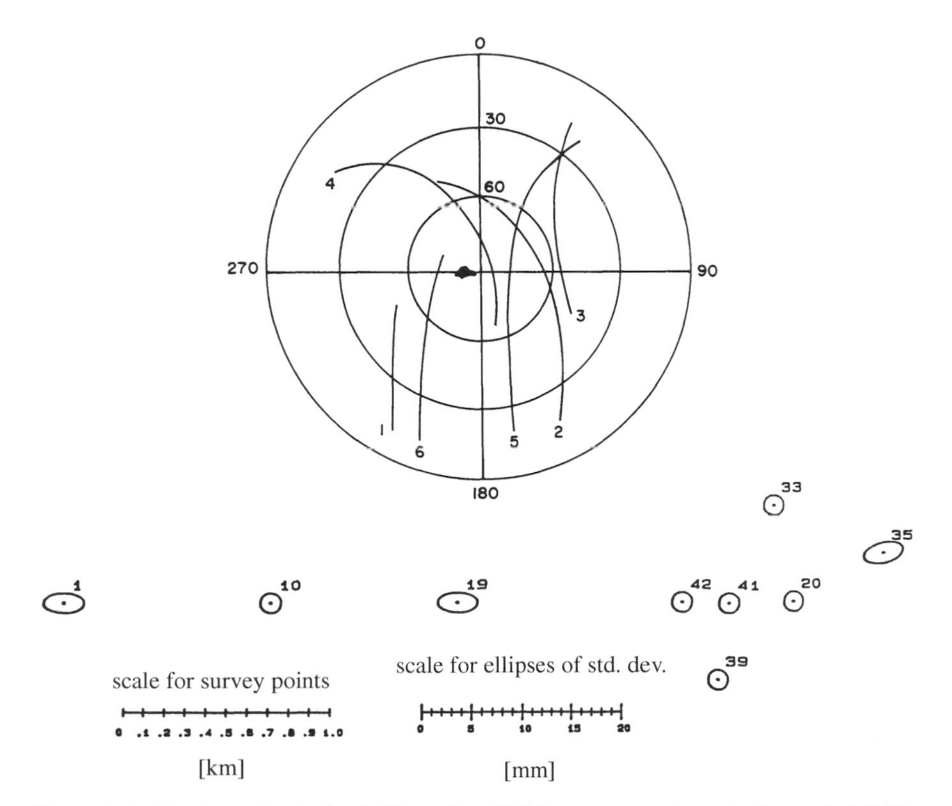

Figure 8.6 Horizontal standard ellipses for GPS inner constraint solution and visibility plot.

relevant quantities. Some of these plots indicate outliers (i.e., a deviation from an otherwise systematic variation). These outliers are the prime candidates for in-depth studies and analysis. Redundancy number and internal reliability plots appear useful for identifying weak portions of the network (which may result from a deweighting of observations during automated blunder detection). The variance-covariance matrix of vector observations is the determining factor that shapes most of the functions. The graphs below refer to the minimal constraint solutions only. Other aspects of the solution are given in Leick and Emmons (1994).

A Priori Stochastic Information The study begins with the variance-covariance matrices of the estimated vectors from the phase processing step. A simple function of the a priori statistics such as

$$\sigma_k = \sqrt{\sigma_{k1}^2 + \sigma_{k2}^2 + \sigma_{k3}^2} \tag{8.28}$$

is sufficient, where k identifies the vector. Other simple functions, such as the trace of the variance-covariance matrix, can be used as well. The symbols on the right-hand side of (8.28) denote the diagonal elements of the 3×3 variance-covariance

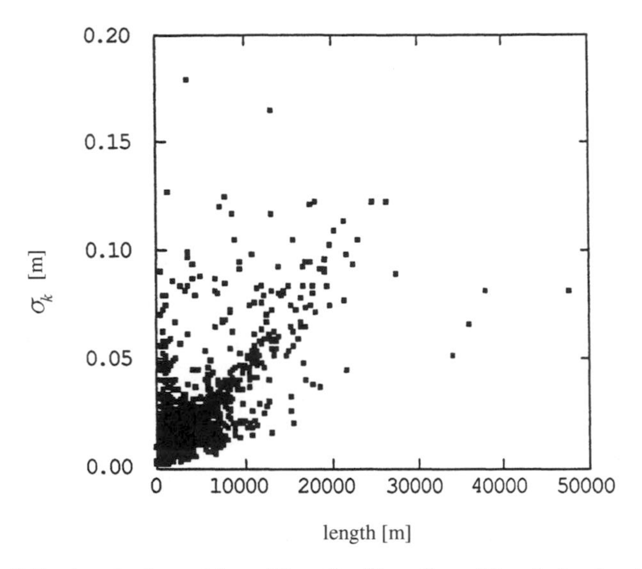

Figure 8.7 A priori precision of length of baseline. (Permission by ASCE.)

matrix. Figure 8.7 displays σ_k as a function of the length of the vectors. For longer lines, there appears to be a weak length dependency of about 1:200,000. Several of the shorter baselines show larger-than-expected values. While is not necessarily detrimental to include vectors with large variances in an adjustment, they are unlikely to contribute to the strength of the network solution. Analyzing the averages of σ_k for all vectors of a particular station is useful in discovering stations that might be connected exclusively to low-precision vector observations.

Variance Factor Figures 8.8 and 8.9 show the square root of the estimated variance factor f_k for each vector k. The factor is computed from

$$f_k = \sqrt{\frac{\bar{\mathbf{v}}^{\mathrm{T}}\bar{\mathbf{v}}_k}{R_k}} \tag{8.29}$$

with

$$R_k = \bar{r}_{k1} + \bar{r}_{k2} + \bar{r}_{k3} \qquad 0 \leq R_k \leq 3 \tag{8.30}$$

where $\bar{\mathbf{v}}_k$ denote the decorrelated residuals and $\bar{r}_{k1}$, $\bar{r}_{k2}$, and $\bar{r}_{k3}$ are the redundancy numbers of the decorrelated vector components. See Equation (4.375) regarding the decorrelation of vector observations. The estimates of f_k are shown in Figures 8.8 and 8.9 as a function of the baseline length and the a priori statistics σ_k. The scale factor f_k in Figure 8.10 is computed following the procedure of automatic deweighting observations discussed in Section 4.11.3 (i.e., if the ratio of residual and standard deviation is beyond a threshold value, the scaling factor is computed from

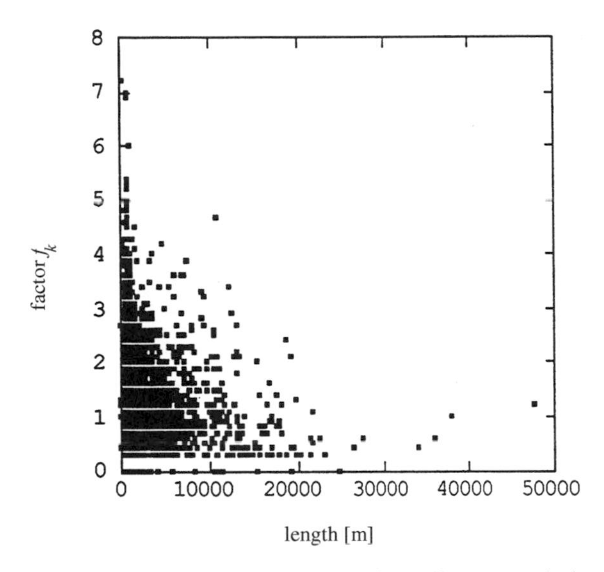

Figure 8.8 Variance factor versus length of baseline. (Permission ASCE.)

an empirical rule and the residuals). All components of the vector are multiplied with the same factor (the largest of three). Figures 8.8 to 8.10 show that the largest factors are associated with the shortest baselines or lines with small σ_k (which tend to be the shortest baselines). For short baselines the centering errors of the antenna and the separation of the electronic and geometric center of the antenna are important; neither is reflected by the stochastic model used here. The variance-covariance submatrices

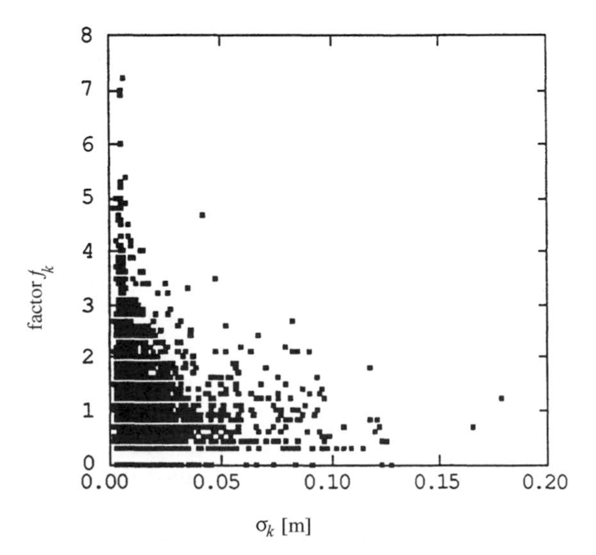

Figure 8.9 Variance factor versus precision of baseline. (Permission ASCE.)

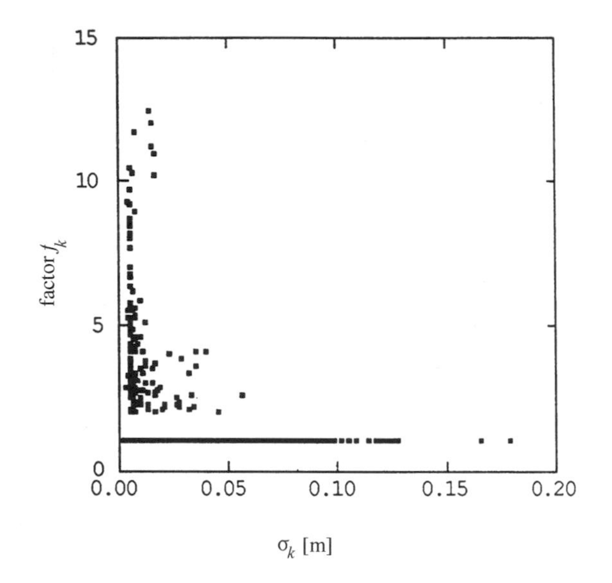

Figure 8.10 Applied variance scale factor. (Permisson ASCE.)

should be scaled only after the observations with the largest computed factors have been analyzed and possibly been corrected. If proper justification can be found, the factors can be applied. The factors f_k of Figure 8.10 were used in these solutions.

Redundancy Numbers The vector redundancy number R_k in (8.30) varies between 3 and zero. Values close to 3 indicate maximum contribution to the redundancy and minimum contribution to the solution, i.e., the observation is literally redundant. Such observations contribute little, if anything at all, to the adjustment because of the presence of other, usually much more accurate, observations. A redundancy of zero indicates an uncontrolled observation, which occurs, e.g., if a station is determined by one observation only. A small redundancy number implies little contribution to the redundancy but a big contribution to the solution. Such observations "overpower" other observations and usually have small residuals. As a consequence of their "strength," blunders in these observations might not be discovered.

The ordered redundancy numbers in Figure 8.11 exhibit a distinctly sharp decrease as the smallest values are reached. Inspection of the data indicates that these very small redundancies occur whenever there is only one good vector observation left to a particular station, while the other vectors to that station have been deweighted by scaling the variance-covariance matrices as part of the automatic blunder detection procedure. Typically, the scaled vectors have a high redundancy number, indicating their diminished contribution. The only remaining unscaled observation contributes the most; therefore, the respective residuals are very small, usually in the millimeter range. Consequently, a danger of automated blunder detection and deweighting is that parts of the network might become uncontrolled.

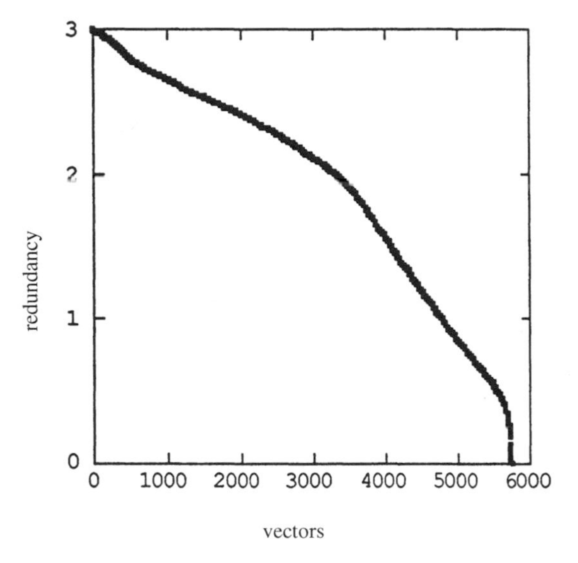

Figure 8.11 Ordered vector redundancy. (Permission ASCE.)

Figure 8.12 indicates that long vectors have large redundancy numbers. The shapes in this figure suggest that it might be possible to identify vectors that can be deleted from the adjustment without affecting the strength of the solution. The steep slope suggests that the assembly of short baselines determines the shape of the network. Mixing short and long baselines is useful only if long baselines have been determined with an accuracy comparable to that of shorter lines. This can possibly be

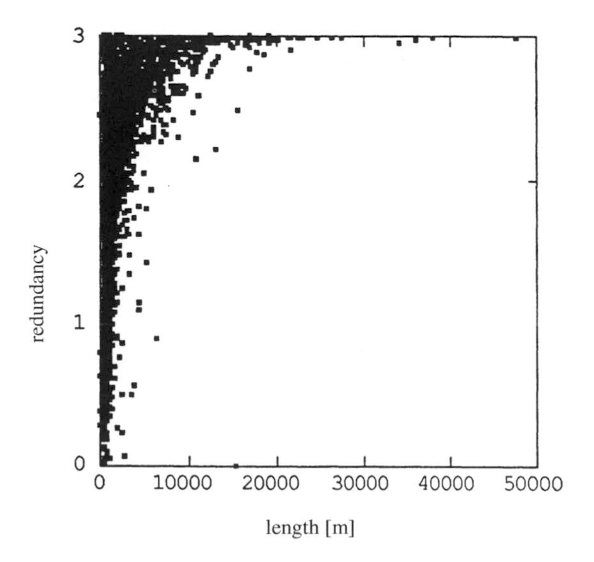

Figure 8.12 Vector redundancy versus length of baseline. (Permission ASCE.)

accomplished through longer observation times, using dual-frequency receivers, and processing with a precise ephemeris.

Internal Reliability Internal reliability values are shown in Figure 8.13. These values are a function of the internal reliability vector components as follows:

$$I_k = \sqrt{I_{k1}^2 + I_{k2}^2 + I_{k3}^2} \tag{8.31}$$

The internal reliability components are computed according to Equation (4.363) for the decorrelated vector observations, and are then transformed back to the physical observation space. The values plotted are based on the factor $\delta_0 = 4.12$. There is essentially a linear relationship between internal reliability and the quality of the observations as expressed by σ_k. The slope essentially equals δ_0. The outliers in Figure 8.13 are associated with small σ_k and pertain to a group of "single vectors" that result when the other vectors to the same station have been deweighted. The linear relationship makes it possible to identify the outliers for further inspection and analysis. Furthermore, this linear relationship nicely confirms that internal reliability is not a function of the shape of the GPS network.

Blunders and Absorption Figure 8.14 shows blunders as predicted by the respective residuals. As detailed in (4.366), a relationship exists between computed blunders, residuals, and redundancies. The figure shows the blunder function

$$B_k = \sqrt{B_{k1}^2 + B_{k2}^2 + B_{k3}^2} \tag{8.32}$$

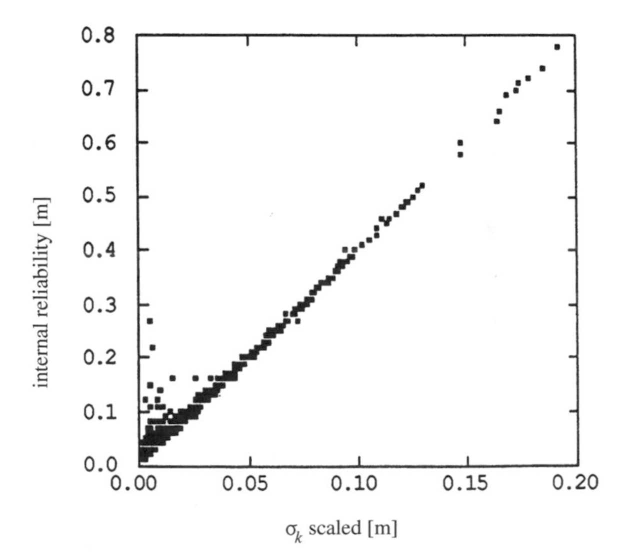

Figure 8.13 Internal reliability versus precision of baseline. (Permission ASCE.)

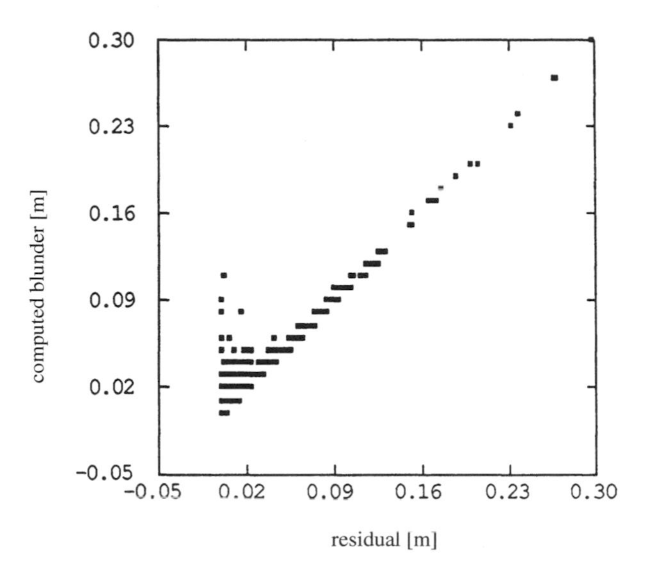

Figure 8.14 Computed blunders versus residuals. (Permission ASCE.)

versus the residual function

$$v_k = \sqrt{v_{k1}^2 + v_{k2}^2 + v_{k3}^2} \tag{8.33}$$

The computed blunder and the residuals refer to the physical observation space. This relationship appears to be primarily linear with slope 1:1 (at least for the larger

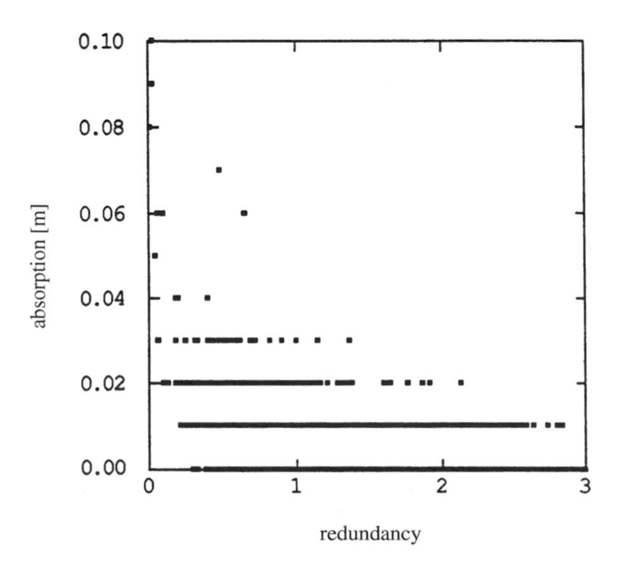

Figure 8.15 Absorption versus redundancy. (Permission ASCE.)

residuals). The outliers seen for small residuals refer to the group of observations with smallest redundancy numbers.

Figure 8.15 shows absorption versus redundancy. Absorption specifies that part of a blunder which is absorbed in the solution, i.e., absorption indicates falsification of the solution. The values

$$A_k = v_k + B_k \tag{8.34}$$

are plotted. As expected, the observations with lowest redundancy tend to absorb the most. In the extreme case, the absorption is infinite for zero redundancy, and zero for a redundancy of 3 (vector observations). Clearly, very small redundancies reflect an insensitivity to blunders, which is not desirable.

As is the case for terrestrial observation, it is not sufficient to limit quality control to residuals and normalized residuals. It is equally important that the quality of the network be presented in terms of redundancy and reliability measures. These functions are, among other things, useful in judging the implications of deweighting. The consequences of deweighting are not always readily apparent in large networks.

CHAPTER 9

TWO-DIMENSIONAL GEODETIC MODELS

Computations on the ellipsoid and the conformal mapping plane became popular when K. F. Gauss significantly advanced the field of differential geometry and least-squares. Gauss used his many talents to develop geodetic computations on the ellipsoidal surface and on the conformal map. The problem presented itself to Gauss when he was asked to observe and compute a geodetic network in northern Germany. Since the curvature of the ellipsoidal changes with latitude, the mathematics of computing on the ellipsoidal surface becomes mathematically complex. With the conformal mapping approach, additional mathematical developments are needed. Both approaches require a new element not discussed thus far, the geodesic (the shortest distance between two points on a surface). Developing expressions for the geodesic on the ellipsoidal surface and its image on the map requires advanced mathematical skills.

Computations on either the ellipsoidal surface or the conformal map are inherently two-dimensional. The stations are parameterized in terms of geodetic latitude and longitude or conformal mapping coordinates. The third dimension, the height, does not appear explicitly as a parameter but has been "used up" during the reduction of the spatial observations to the computation surface. Networks on the ellipsoidal surface or the conformal map have historically been labeled "horizontal networks" and treated separately from a one-dimensional "vertical network." Such a separation was justified at a time when the measurement tools could be readily separated into those that measured primarily "horizontal information" and those that yielded primarily "vertical information." GPS breaks this separation because it provides accurate three-dimensional positions.

Because the two-dimensional geodetic models have such a long tradition and were the backbone of geodetic computations prior to the introduction of geodetic space techniques, the respective solutions belong to the most classical of all geodetic theories and are appropriately documented in the literature. Unfortunately, many of the references on this subject are out of print. We therefore summarize the Gauss midlatitude solution, the transverse Mercator mapping, and Lambert conformal mapping in Appendixes B and C. Supporting material from differential geometry is also provided in order to appreciate the "roots and flavor" of the mathematics involved. Additional derivations are available in Leick (2002) that support lectures on the subject. The following literature has been found helpful: Dozier (1980), Heck (1987), Kneissl (1959), Grossman (1976), Hristow (1955), Lambert (1772), Lee (1976), Snyder (1982), and Thomas (1952). Publication of many of these "classical" references has been discontinued.

The ellipsoidal and conformal mapping expressions are generally given in the form of mathematical series that are a result of multiple truncations at various steps during the development. These truncations affect the computational accuracy of the expressions and their applicability to the size of the area. The expressions given here are sufficiently accurate for typical applications in surveying and geodesy. Some terms may even be negligible when applied over small areas. For unusual applications covering large areas, one might have to use more accurate expressions found in the specialized literature. In all cases, however, given today's powerful computers, one should not be overly concerned about a few unnecessary algebraic operations.

Two types of observations apply to computations on a surface: azimuth (angle) and distance. The reductions, partial derivatives, and other quantities that apply to angles can be conveniently obtained through differencing the respective expressions for azimuths.

9.1 THE ELLIPSOIDAL MODEL

This section contains the mathematical formulations needed to carry out computations on the ellipsoidal surface. We introduce the geodesic line and reduce the 3D geodetic observations to geodesic azimuth and distance. The direct and inverse solutions are based on the Gauss midlatitude expressions. Finally, the partial derivatives are given that allow network adjustment on the ellipsoid.

9.1.1 Reduction of Observations

The geodetic azimuth α of Section 2.3.5 is the angle between two normal planes that have the ellipsoidal normal in common; the geodetic horizontal angle δ is defined similarly. These 3D model observations follow from the original observation upon corrections for the deflection of the vertical. Spatial distances can be used directly in the 3D model presented in Section 2.3.5. However, angles and distances must be reduced further in order to obtain model observables on the ellipsoidal surface with respect to the geodesic.

9.1.1.1 Angular Reduction to Geodesic Figure 9.1 shows the reduction of azimuth. The geodetic azimuth, α, is shown in the figure as the azimuth of the normal

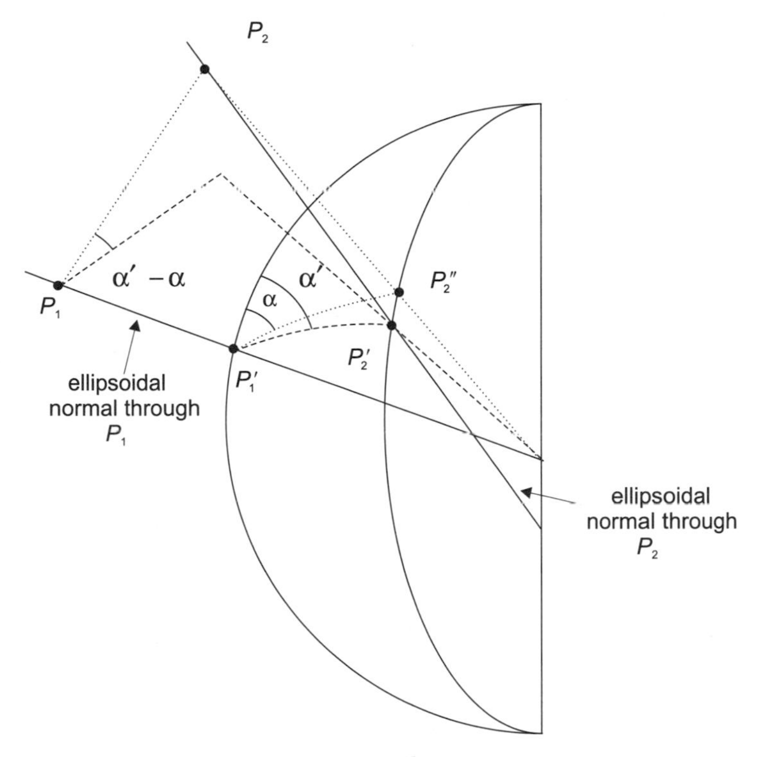

Figure 9.1 Normal section azimuth versus height of target.

plane defined by the ellipsoidal normal of P_1 and the space point P_2. The representatives of these space points are located along the respective ellipsoidal normals on the surface of the ellipsoid and are denoted by P_1' and P_2'. The dotted line P_1' to P_2'' denotes the intersection of this normal plane with the ellipsoid. The azimuth of the normal section defined by the ellipsoidal normal at P_1 and the surface point P_2' is α'. The angular difference $(\alpha' - \alpha)$ is the reduction in azimuth due to height of P_2; the expression is given in Table 9.1. The height of the observing station P_1 does not affect the reduction because α is the angle between planes.

The need for another angular reduction follows from Figure 9.2. Assume that two ellipsoidal surface points P_1 and P_2 (labeled P_1' and P_2' in Figure 9.1) are located at

TABLE 9.1 Reducing Geodetic Azimuth to Geodesic Azimuth

$$\left(\alpha_1' - \alpha_1\right)_{[\text{arcs}]} = 0.108 \cos^2 \varphi_1 \sin 2\alpha_1 \, h_{1[\text{km}]} \tag{a}$$

$$\left(\widehat{\alpha}_1 - \alpha_1\right)_{[\text{arcs}]} = -0.028 \cos^2 \varphi_1 \sin 2\alpha_1 \left(\frac{\widehat{s}_{[\text{km}]}}{100}\right)^2 \tag{b}$$

$$\Delta\alpha_{[\text{arcs}]} = 0.108 \cos^2 \varphi_1 \sin 2\alpha_1 h_{1[\text{km}]} - 0.028 \cos^2 \varphi_1 \sin 2\alpha_1 \left(\frac{\widehat{s}_{[\text{km}]}}{100}\right)^2 \tag{c}$$

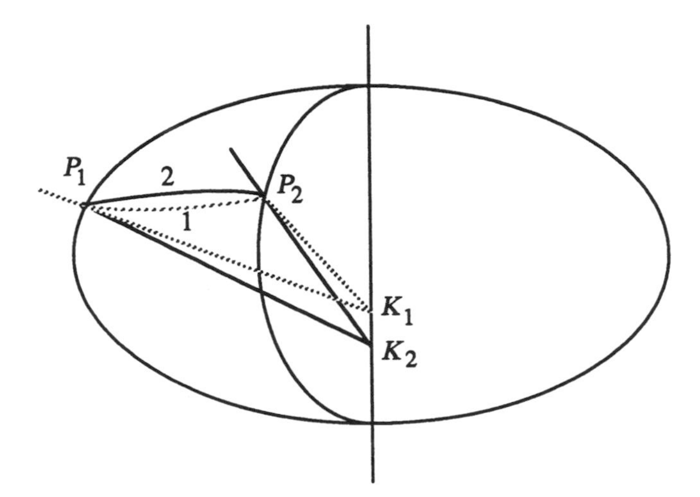

Figure 9.2 Normal sections on the ellipsoid.

different latitudes. Line 1 is the normal section from P_1 to P_2 and line 2 indicates the normal section from P_2 to P_1. It can be readily seen that these two normal sections do not coincide, because the curvature of the ellipsoidal meridian changes with latitude. The question is, which of these two normal sections should be adopted for the computations? Introducing the geodesic, which connects these two points in a unique way, solves this dilemma. There is only one geodesic from P_1 to P_2. Figure 9.3 shows the approximate geometric relationship between the normal sections and the geodesic. The angular reduction $(\widehat{\alpha} - \alpha')$ is required to get the azimuth $\widehat{\alpha}$ of the geodesic. The expression is listed in Table 9.1 (note that approximate values for the azimuth α and length $\widehat{s}$ of the geodesic are sufficient for expressions on the right-hand side of Table 9.1).

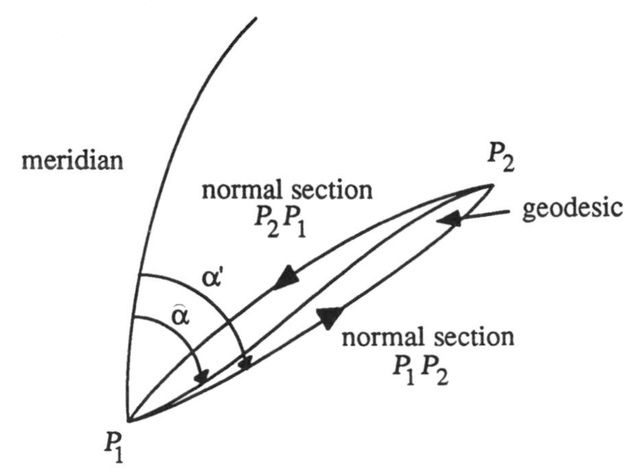

Figure 9.3 Normal section azimuth versus geodesic azimuth.

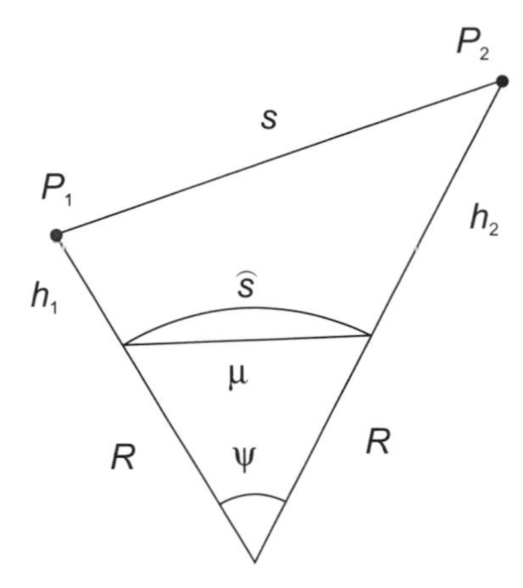

Figure 9.4 Slant distance versus geodesic.

9.1.1.2 Distance Reduction to Geodesic The slant distance s (not to be confused with the scale correction of Section 8.2 which uses the same symbol) must be reduced to the length of a geodesic $\widehat{s}$. Figure 9.4 shows an ellipsoidal section along the line of sight. The expression for the length $\widehat{s}$ of the geodesic is typically based on a spherical approximation of the ellipsoidal arc. At this level of approximation, there is no need to distinguish between the lengths of the geodesic and the normal section. The radius R, which is evaluated according to Euler's equation (B.8) for the center of the line, serves as radius of curvature of the spherical arc. The expressions in Table 9.2 relate the slant distance s to the lengths of the geodesic $\widehat{s}$.

One should note that computing the length of the geodesic requires knowledge of the ellipsoidal heights. Using orthometric heights might introduce errors in the distance reduction. The height difference $\Delta h = h_2 - h_1$ in Expression (e) of Table 9.2 must be accurately known for lines with a large slope. Differentiating this expression gives the approximate relation

$$d\mu \approx -\frac{\Delta h}{\mu} d\Delta h \tag{9.1}$$

where $d\Delta h$ represents the error in the height difference. Surveyors often reduce the slant distance in the field to the local geodetic horizon using the elevation angle

TABLE 9.2 Reducing Slant Distance to Geodesic

$$\frac{1}{R} = \frac{\cos^2 \alpha}{M} + \frac{\sin^2 \alpha}{N} \tag{d}$$

$$\mu = \sqrt{\frac{s^2 - \Delta h^2}{\left(1 + \frac{h_1}{R}\right)\left(1 + \frac{h_2}{R}\right)}} \tag{e}$$

$$\widehat{s} = R\psi = 2R \sin^{-1}\left(\frac{\mu}{2R}\right) \tag{f}$$

TABLE 9.3 Relative Distance Error versus Height

$h_{m[m]}$	h_m/R
6.37	1:1000000
63.7	1:100000
100	1:64000
500	1:13000
637	1:10000
1000	1:6300

that is measured together with the slant distance. For observations reduced in such a manner, Δh is small (although not zero), but there is now a corresponding accuracy requirement regarding the measured elevation angle.

If both stations are located at about the same height $h_1 \approx h_2 \approx h_m$, one obtains from (e)

$$\frac{\mu(h_m) - s}{\mu(h_m)} = \frac{h_m}{R} \tag{9.2}$$

This equation relates the relative error in distance reduction to the mean height of the line. Table 9.3 shows that just 6 m in height error causes a 1 ppm error in the reduction. Networks are routinely achieved that accurately with GPS.

Since modern EDM instruments are very accurate, it is desirable to apply the height corrections consistently. It is good to remember the rule of thumb that *a 6 m error in height of the line causes a relative change in distance of 1 ppm.* We recognize that geodetic heights are required, not orthometric heights. Since geoid undulations can be as large as 100 m, it is clear that they must be taken into account for high-precision surveying.

9.1.2 Direct and Inverse Solutions on the Ellipsoid

The reductions discussed above produce the geodesic observables, i.e., the geodesic azimuths $\widehat{\alpha}$, the geodesic distance $\widehat{s}$, and the angle between geodesics $\widehat{\delta}$. At the heart of computations on the ellipsoidal surface are the so-called direct and inverse problems, which are summarized in Table 9.4. For the direct problem, the geodetic latitude and longitude of one station, say, $P_1(\varphi_1, \lambda_1)$, and the geodesic azimuth $\widehat{\alpha}_{12}$ and geodesic distance $\widehat{s}_{12}$ to another point P_2 are given; the geodetic latitude and longitude of station $P_2(\varphi_2, \lambda_2)$, and the back azimuth $\widehat{\alpha}_{21}$ must be computed. For the inverse problem, the geodetic latitudes and longitudes of $P_1(\varphi_1, \lambda_1)$ and $P_2(\varphi_2, \lambda_2)$ are given, and the forward and back azimuth and the length of the geodesic are required. Note that $\widehat{s}_{12} = \widehat{s}_{21}$ but $\widehat{\alpha}_{12} \neq \widehat{\alpha}_{21} \pm 180°$. There are many solutions available in the literature for the direct and inverse problems. Some of these solutions are valid for geodesics that go all around the ellipsoid. We use the Gauss midlatitude (GML) functions given in Table B.2.

Because the GML functions are a result of series developments and, as such, subject to truncation errors, Figure 9.5 has been prepared to provide some insight into

TABLE 9.4 Direct and Inverse Solutions on the Ellipsoid

Direct Solution	Inverse Solution
$P_1(\varphi_1, \lambda_1), \widehat{\alpha}_{12}, \widehat{s}_{12}$	$P_1(\varphi_1, \lambda_1), P_2(\varphi_2, \lambda_2)$
$\downarrow$	$\downarrow$
$\left(\varphi_2, \lambda_2, \widehat{\alpha}_{21}\right)$	$\left(\widehat{\alpha}_{12}, \widehat{s}_{12}, \widehat{\alpha}_{21}\right)$

the accuracy of these expressions. The center of the graph is at ($\varphi = 45°, \lambda = 0$) and covers the 4° × 4° test area $-43° < \varphi < 47°$ and $-2° < \lambda < 2°$. The first step is to compute the inverse solution between the center $P(\varphi = 45°, \lambda = 0)$ and various points $P_i(\varphi_i, \lambda_i)$ within the area. The resulting geodesic distances and azimuths $(\widehat{s}_i, \widehat{\alpha}_i)$ are then put in the direct solution to compute the positions $P_{i,c}(\varphi_{i,c}, \lambda_{i,c})$. The figure shows contour lines for the difference

$$\delta = R\sqrt{\left(\varphi_i - \varphi_{i,c}\right)^2 + \left(\lambda_i - \lambda_{i,c}\right)^2 \cos^2 \varphi_i} \tag{9.3}$$

The contour lines increase from zero at the center to 0.15 mm at the edge. These values represent the accumulated effect of truncations made when developing the GML expressions. This truncation error is essentially zero for points close to the meridian of the center point. There is a large area of about 1° × 1° around the center where the truncation error is less than 0.1 mm. As one departs from this central area,

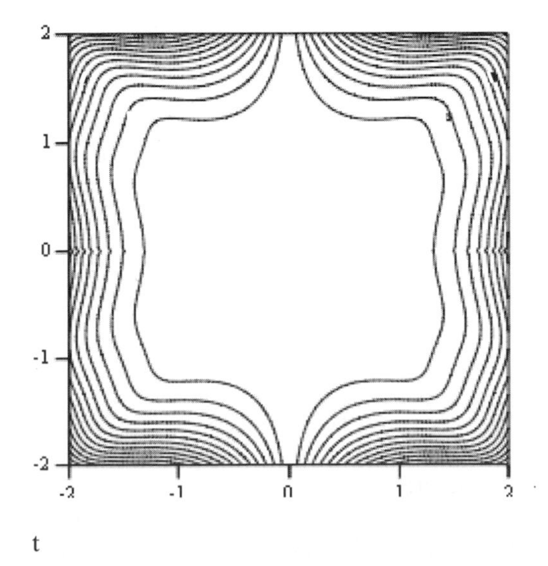

Figure 9.5 Accuracy of GML functions within a 4 x 4 degree area. Contour increment is 1/10 mm.

the truncation error increases rapidly. The basic shape of this figure and the values do not change significantly for different latitudes of the center. Because the ellipsoid is a figure of rotation, the longitude of the center does not matter.

The GML solution satisfies typical geodetic applications. In the unlikely case that they are not sufficient because long lines are involved, one can always replace them with other solutions that are valid for long geodesics.

9.1.3 Network Adjustment on the Ellipsoid

The geodesic azimuths, geodesic distances, and the angles between geodesics form a network of stations on the ellipsoidal surface that can be adjusted using standard least-squares techniques. The ellipsoidal network contains no explicit height information. The height information was used during the transition of the 3D geodetic observables to the geodesic observables on the ellipsoid. Conceptually, this is expressed by $\{\varphi, \lambda, h\} \rightarrow \{\varphi, \lambda\}$ and $\{\alpha, \delta, \beta, s, \Delta h, \Delta N\} \rightarrow \{\widehat{\alpha}, \widehat{\delta}, \widehat{s}\}$. The geodetic height h is no longer a parameter, and geodesic observables do not include quantities that directly correspond to the geodetic vertical angle, the geodetic height difference Δh, or the geoid undulation difference ΔN.

Least-squares techniques are discussed in detail in Chapter 4. For discussion in this section, we use the observation equation model

$$\mathbf{v} = \mathbf{A}\mathbf{x} + (\boldsymbol{\ell}_0 - \boldsymbol{\ell}_b) \tag{9.4}$$

In the familiar adjustment notation the symbol $\mathbf{v}$ denotes the residuals, $\mathbf{A}$ is the design matrix, and $\mathbf{x}$ represents the corrections to the approximate parameters $\mathbf{x}_0$. The symbol $\boldsymbol{\ell}_b$ denotes the observations, in this case the geodesic observables, and $\boldsymbol{\ell}_0$ represents the observables as computed from the approximate parameters

$$\mathbf{x}_0 = \begin{bmatrix} \cdots & \varphi_{i,0} & \lambda_{i,0} & \cdots \end{bmatrix}^T \tag{9.5}$$

using the GML functions. If we further use the (2-1-3) subscript notation to denote the angle measured at station 1 from station 2 to station 3 in a clockwise sense, then the geodesic observables can be expressed as

$$\widehat{\alpha}_{12,b} = \alpha_{12,b} + \Delta\alpha_{12} \tag{9.6}$$

$$\widehat{\delta}_{213,b} = \delta_{213,b} + \Delta\alpha_{13} - \Delta\alpha_{12} \tag{9.7}$$

$$\widehat{s}_{12} = s\,(s_{12}, R, h_1, h_2) \tag{9.8}$$

In order to make the interpretation of the coordinate (parameter) shifts easier, it is advantageous to reparameterize the parameters to northing ($dn_i = M_i\, d\varphi_i$) and easting ($de_i = N_i \cos\varphi_i\, d\lambda_i$). Using the partial derivatives in Table B.3, the observation equations for the geodesic observables become

$$v_{\widehat{\alpha}} = \frac{\sin\widehat{\alpha}_{12,0}}{\widehat{s}_{12,0}} dn_1 + \frac{\cos\widehat{\alpha}_{21,0}}{\widehat{s}_{12,0}} de_1 + \frac{\sin\widehat{\alpha}_{21,0}}{\widehat{s}_{12,0}} dn_2 - \frac{\cos\widehat{\alpha}_{21,0}}{\widehat{s}_{12,0}} de_2 + \left(\widehat{\alpha}_{12,0} - \widehat{\alpha}_{12,b}\right) \tag{9.9}$$

$$\begin{aligned} v_{\widehat{\delta}} = {} & \left(\frac{\sin\widehat{\alpha}_{13,0}}{\widehat{s}_{13,0}} - \frac{\sin\widehat{\alpha}_{12,0}}{\widehat{s}_{12,0}}\right) dn_1 + \left(\frac{\cos\widehat{\alpha}_{31,0}}{\widehat{s}_{13,0}} - \frac{\cos\widehat{\alpha}_{21,0}}{\widehat{s}_{12,0}}\right) de_1 \\ & - \frac{\sin\widehat{\alpha}_{21,0}}{\widehat{s}_{12,0}} dn_2 + \frac{\cos\widehat{\alpha}_{21,0}}{\widehat{s}_{12,0}} de_2 \\ & + \frac{\sin\widehat{\alpha}_{31,0}}{\widehat{s}_{13,0}} dn_3 - \frac{\cos\widehat{\alpha}_{31,0}}{\widehat{s}_{13,0}} de_3 + \left(\widehat{\alpha}_{13,0} - \widehat{\alpha}_{13,b}\right) \end{aligned} \tag{9.10}$$

$$v_{\widehat{s}} = -\cos\widehat{\alpha}_{12,0}\, dn_1 + \sin\widehat{\alpha}_{21,0}\, de_1 - \cos\widehat{\alpha}_{21,0}\, dn_2 - \sin\widehat{\alpha}_{21,0}\, de_2 + \left(\widehat{s}_{12,0} - \widehat{s}_{12,b}\right) \tag{9.11}$$

The quantities $(\widehat{\alpha}_0, \widehat{\beta}_0, \widehat{s}_0)$ are computed from the inverse solution. The GLM functions are particularly suitable for this purpose because the inverse solution is noniterative. The results of the adjustment of the ellipsoidal network are the adjusted observations $(\widehat{\alpha}_a, \widehat{\beta}_a, \widehat{s}_a)$ and the adjusted coordinates

$$\mathbf{x}_a = [\cdots \quad \varphi_{i,a} \quad \lambda_{i,a} \quad \cdots]^T \tag{9.12}$$

The partial derivatives (9.9) to (9.11) are a result of series expansion and are therefore approximations and subject to truncation errors. It is of course necessary that the partial derivatives and the GML functions have the same level of accuracy. Figure 9.6 shows a test computation for the $4^\circ \times 4^\circ$ test area. We first use the GML inverse solution to compute $(\widehat{s}_{ci}, \widehat{\alpha}_{ci})$ between center $P_c(\varphi_c, \lambda_c)$ and $P_i(\varphi_i, \lambda_i)$, and then $(\widehat{s}_{ci,d}, \widehat{\alpha}_{ci,d})$ between $P_c(\varphi_c + d\varphi_c, \lambda_c + d\lambda_c)$ and $P_i(\varphi_i + d\varphi_i, \lambda_i + d\lambda_i)$. The differentials cause a shift on the ellipsoidal surface of

$$s_{i,\text{GML}} = \sqrt{\left(\widehat{s}_{ci} - \widehat{s}_{ci,d}\right)^2 + \left(\widehat{\alpha}_{ci} - \widehat{\alpha}_{ci,d}\right)^2 \widehat{s}_{ci}^2} \tag{9.13}$$

Using the linear forms (9.9) and (9.11) (see also Table B.3), we compute

$$s_{i,lin} = \sqrt{d\widehat{s}_{ci}^2 + d\widehat{\alpha}_{ci}^2\, \widehat{s}_{ci}^2} \tag{9.14}$$

The differences $s_{i,\text{GML}} - s_{i,lin}$ are contoured in Figure 9.6 for the values $d\varphi_c = d\lambda_c = d\varphi_i = d\lambda_i = 1$ m. The straight lines at the center in north-south and east-west directions are the zero contour lines. The other contour lines increase in steps of 0.1 mm starting at -0.4 mm in the southeast and southwest corners and ending at

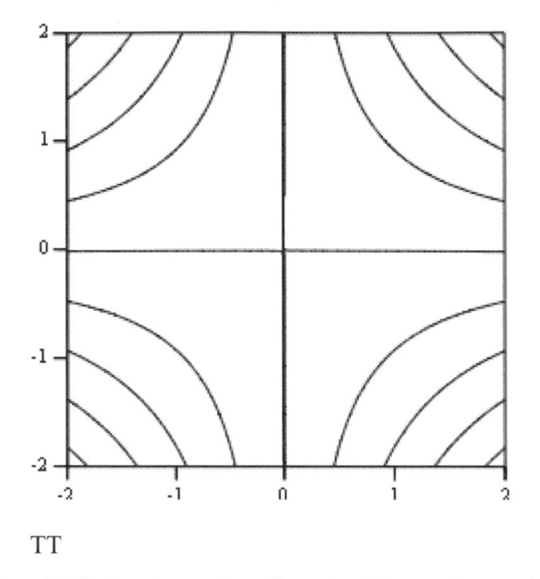

TT

Figure 9.6 Comparing GML functions minus linearized form. Contour increment is 1/10 mm.

0.4 mm in the northwest and northeast corners. There is an agreement to better than 1 mm within the test area. The shape of the contour lines changes even with the sign of the differentials $(d\varphi_i, d\lambda_i)$. This figure, therefore, serves only as an example of what discrepancies to expect for differentials of 1 m. There is a proportional relationship between the magnitude of the discrepancies and the magnitude of the differentials, at least for reasonable values of the differentials. For example, differentials of 10 m cause discrepancies in the range of 4 mm. Since it is easy to establish approximate coordinates of about 1 m with GPS, we can readily state that GML functions and the differentials are consistent within 1 mm in the $4° \times 4°$ test area.

9.2 THE CONFORMAL MAPPING MODEL

If the goal is to map the ellipsoid onto a plane in order to display it on the computer screen or to assemble overlays of spatial data, any unique mapping from the ellipsoid to the plane may be used. In conformal mapping, we map the ellipsoidal surface conformally onto a plane. The conformal property preserves angles. An angle between two curves, say, two geodesics on the ellipsoid, is defined as the angle between the tangents on these curves. Therefore, conformal mapping preserves the angle between the tangents of curves on the ellipsoid and the respective mapped images. The conformal property makes conformal maps useful for computations because the directional elements between the ellipsoid and the map have a known relationship.

Users who prefer to work with plane mapping coordinates rather than geodetic latitude and longitude can still use the 3D adjustment procedures developed in Chapter 2. The given mapping coordinates can be transformed to the ellipsoidal and then used, together with heights, in the 3D geodetic adjustment. The adjusted geodetic positions can subsequently be mapped to the conformal plane. A user might not even be aware that geodetic coordinates had been used in the adjustment.

9.2.1 Reduction of Observations

Figure 9.7 shows the mapping elements (γ, Δt, Δs) that link the geodesic observations ($\widehat{s}$, $\widehat{\alpha}$) to the corresponding observables on the mapping plane ($\bar{d}$, $\bar{t}$). Note that $\widehat{s}$ is the length of the geodesic on the ellipsoid and not the length of the mapped geodesic that does not enter any of the equations below. The mapping plane must not be confused with the local astronomic or geodetic horizon. It is simply the outcome of mapping the ellipsoidal surface conformally into a plane. One can generate many such mapping planes for the same ellipsoidal surface area.

In Figure 9.7, the Cartesian coordinate system in the mapping plane is denoted by (x, y). The points $P_1(x_1, y_1)$ and $P_2(x_2, y_2)$ are the images of corresponding points on the ellipsoid. Consider for a moment the geodesic that connects the points P_1 and P_2 on the ellipsoid. This geodesic can be mapped point by point; the result is the mapped geodesic as shown in the figure. This image is a smooth but mathematically complicated curve. The straight line between the images P_1 and P_2 is the rectilinear chord. The image of ellipsoidal meridian may or may not be a straight line on the map. In order to be general, Figure 9.7 shows the tangent on the mapped meridian. The angle between the y axis and the mapped meridian is the meridian convergence γ; it is generally counted positive in the counterclockwise sense. Because of conformal property, the geodetic azimuth of the geodesic is preserved during the mapping and it must be equal to the angle between the tangents on the mapped meridian and mapped geodesic as shown. The symbols $\bar{T}$ and $\bar{t}$ denote the grid azimuth of the mapped geodesic and the rectilinear chord, respectively. The small angle $\Delta t = \bar{T} - \bar{t}$ is called the arc-to-chord correction. It is related to the grid azimuth $\bar{t}$, the meridian convergence γ, and the azimuth of the geodesic $\widehat{\alpha}$ by

$$\Delta t = \widehat{\alpha} - \gamma - \bar{t} \tag{9.15}$$

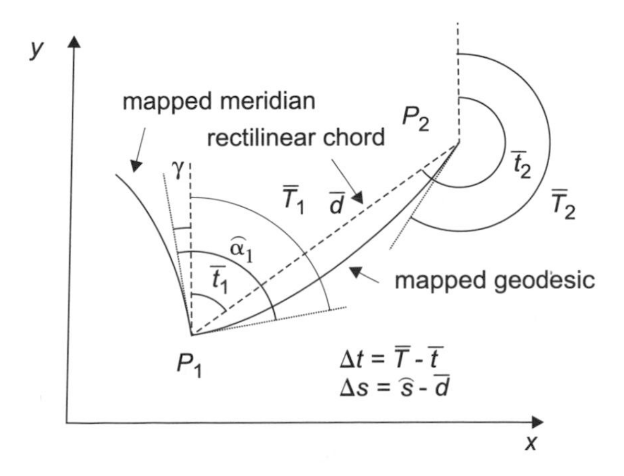

Figure 9.7 Mapping elements.

There is, of course, no specification in conformal mapping as to the preservation of the length of the geodesic. Typically, one is not explicitly interested in the length of the projected geodesic $\bar{s}$, but needs the length of the rectilinear chord $\bar{d}$. The map distance reduction, $\Delta s = \widehat{s} - \bar{d}$, follows readily from the lengths of the geodesic on the ellipsoid and the rectilinear chord. The factor

$$k_L = \frac{\bar{d}}{\widehat{s}} \tag{9.16}$$

is called the line scale factor. It must not be confused with the point scale factor k. See Equation (C.42).

The angle between rectilinear chords on the map at station i is

$$\bar{\delta}_i = \bar{t}_{i,i+1} - \bar{t}_{i,i-1} + 2\pi = \bar{T}_{i,i+1} - \Delta t_{i,i+1} - \left(\bar{T}_{i,i-1} - \Delta t_{i,i-1}\right) + 2\pi \tag{9.17}$$

The angle between the geodesics on either the ellipsoid or their mapped images is

$$\widehat{\delta}_i = \widehat{\alpha}_{i,i+1} - \widehat{\alpha}_{i,i-1} + 2\pi = \bar{T}_{i,i+1} + \gamma_i - \left(\bar{T}_{i,i-1} + \gamma_i\right) + 2\pi = \bar{T}_{i,i+1} - \bar{T}_{i,i-1} + 2\pi \tag{9.18}$$

The difference

$$\Delta\delta_i \equiv \widehat{\delta}_i - \bar{\delta}_i = \Delta t_{i,i+1} - \Delta t_{i,i-1} \tag{9.19}$$

is the angular arc-to-chord reduction. Equations (9.17) to (9.19) do not depend on the meridian convergence.

Even though the term map distortion has many definitions, one associates a small Δt and Δs with small distortions, meaning that the respective reductions in angle and distance are small and perhaps even negligible. It is important to note that the mapping elements change in size and sign with the location of the line and with its orientation. In order to keep Δt and Δs small, we limit the area represented in a single mapping plane in size, thus the need for several mappings to cover large regions of the globe. In addition, the mapping elements are also functions of elements specified by the designer of the map, e.g., the factor k_0, the location of the central meridian, or the standard parallel.

9.2.2 The Angular Excess

The angular reduction can be readily related to the ellipsoidal angular excess. The sum of the interior angles of a polygon of rectilinear chords on the map (see Figure 9.8) is

$$\sum_i \bar{\delta}_i = (n-2) \times 180^\circ \tag{9.20}$$

The sum of the interior angles of the corresponding polygon on the ellipsoid, consisting of geodesics, is

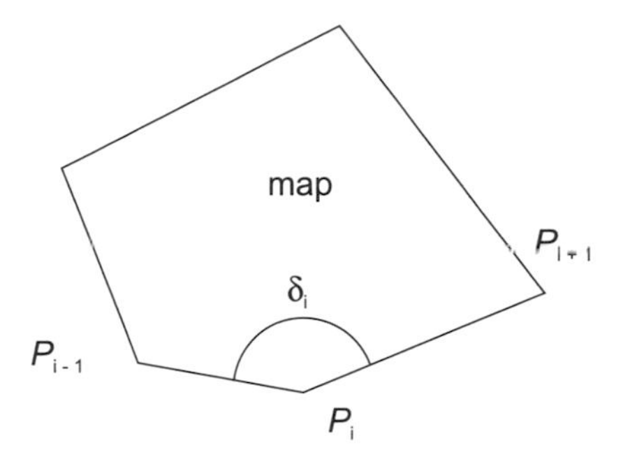

Figure 9.8 Angle on the map.

$$\sum_i \widehat{\delta}_i = (n-2) \times 180^\circ + \varepsilon \tag{9.21}$$

where ε denotes the ellipsoidal angular excess. It follows from (9.19) to (9.21) that

$$\varepsilon = \sum_i \Delta t_{i,i+1} - \sum_i \Delta t_{i,i-1} \tag{9.22}$$

The angular excess can therefore be computed from either the sum of interior geodesic angles, Equation (9.22), or Expression (B.65), which uses the Gauss curvature.

9.2.3 Direct and Inverse Solutions on the Map

Having the grid azimuth $\bar{t}$ and the length of the rectilinear chord $\bar{d}$, or the angle $\bar{\delta}$ between rectilinear chords, the rules of plane trigonometry apply in a straightforward manner. In case the geodetic latitude and longitude are given, one can use the mapping equations to compute the map coordinates first. The direct and inverse solutions are shown in Figures 9.9 and 9.10, respectively.

The accuracy of the expressions for the transverse Mercator mapping is sampled in Figure 9.11. First, the geodetic latitudes and longitudes (φ_i, λ_i) were mapped to (x_i, y_i) and then computed back to the ellipsoid giving $(\varphi_{ic}, \lambda_{ic})$, using the inverse mapping functions. The distance (9.3) is plotted. The figure shows the case $(\lambda = 2^\circ, 0^\circ < \varphi_i < 90^\circ)$. The discrepancies are symmetric with respect to longitude. In the region $-2^\circ < \lambda_i < 2^\circ$ the discrepancies are even smaller. The maximum values for $\lambda = 3^\circ$ or $\lambda = 4^\circ$ are about 0.4 mm and 3 mm, respectively.

The discrepancies seen in Figure 9.11 are the result of the combined contribution of truncation errors in the transverse Mercator mapping functions of Tables C.1 and C.2, as well as Expression (B.42) for the elliptic arc. The Lambert conformal mapping

$$\left\{ P_1(x_1, y_1) \Leftrightarrow P_1(\varphi_1, \lambda_1), \bar{d}_{12}, \bar{t}_{12} \right\}$$

$$\downarrow$$

$$\left\{ \begin{array}{l} x_2 = x_1 + \bar{d}_{12} \sin \bar{t}_{12} \\ y_2 = y_1 + \bar{d}_{12} \cos \bar{t}_{12} \end{array} \right\}$$

Figure 9.9 Direct solution on the map.

$$\{P_1(x_1, y_1), P_2(x_2, y_2)\} \Leftrightarrow \{P_1(\varphi_1, \lambda_1), P_2(\varphi_2, \lambda_2)\}$$

$$\downarrow$$

$$\left\{ \begin{array}{l} \bar{d}_{12} = d(x_1, y_1, x_2, y_2) \\ \bar{t}_{12} = t(x_1, y_1, x_2, y_2) \end{array} \right\}$$

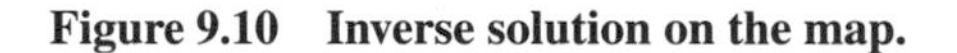

Figure 9.10 Inverse solution on the map.

is not affected by truncation errors because the conversion from the isometric latitude to the geodetic latitude is done iteratively and is only limited by number of significant digits carried by the computer.

9.2.4 Network Adjustment on the Map

The fact that plane trigonometry can be used makes network adjustments on the conformal plane especially attractive. The observed geodesic azimuth, angle, and distance $(\widehat{\alpha}, \widehat{\delta}, \widehat{s})$ are further corrected by $(\Delta t, \Delta\delta, \Delta s)$ to obtain the respective observables on the map. During the adjustment, the current point of expansion, denoted by the subscript 0 in the expressions below, should be used for computing the reductions. At any time during the adjustment, one may choose to deal with the geodetic latitude and longitude or the mapping coordinates, because both sets are related by the mapping equations. This scheme of reduction is shown in Figure 9.12. It requires that the GML functions be used to compute the azimuth $\widehat{\alpha}_{12,0}$.

Just to be sure that there are no misunderstandings about the term *plane,* let us review what created the situation that allows us to use plane trigonometry. The conformal mapping model builds upon the 3D geodetic and 2D ellipsoidal models as visualized by the transition of parameters $\{\varphi, \lambda, h\} \rightarrow \{\varphi, \lambda\} \rightarrow \{x, y\}$ and observables $\{\alpha, \delta, \beta, s, \Delta h, \Delta N\} \rightarrow \{\widehat{\alpha}, \widehat{\delta}, \widehat{s}\} \rightarrow \{\bar{t}, \bar{\delta}, \bar{d}\}$. The height parameter and the vertical observations are not present in the conformal mapping model.

Using again the (2-1-3) subscript notation for angles and standard adjustment notation otherwise, the mapping observables are

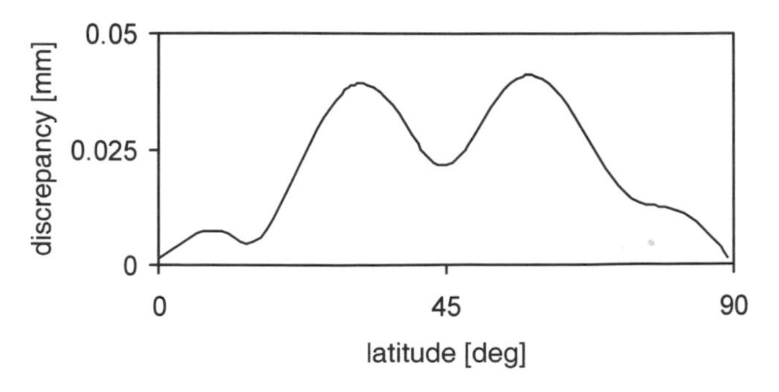

Figure 9.11 Accuracy of direct and inverse mapping for TM.

$$\left\{P_1\left(\varphi_{1,0}, \lambda_{1,0}, x_{1,0}, y_{1,0}\right), P_2\left(\varphi_{2,0}, \lambda_{2,0}, x_{2,0}, y_{2,0}\right)\right\}$$

$$\downarrow$$

$$\left\{\widehat{s}_{12,0}, \widehat{\alpha}_{12,0}, \gamma_{1,0}, \bar{d}_{12,0}, \bar{t}_{12,0}\right\}$$

$$\downarrow$$

$$\left\{\begin{array}{l} \Delta t_{12,0} = \widehat{\alpha}_{12,0} - \gamma_{1,0} - t_{12,0} \\ \Delta s_{12,0} = \widehat{s}_{12,0} - d_{12,0} \end{array}\right\}$$

Figure 9.12 Reducing observations for plane network adjustments.

$$\bar{t}_{12,b} = \widehat{\alpha}_{12,b} - \gamma_{1,0} - \Delta t_{12,0} \tag{9.23}$$

$$\bar{\delta}_{213,b} = \widehat{\delta}_{213,b} - \Delta t_{13,0} + \Delta t_{12,0} \tag{9.24}$$

$$\bar{d}_{12,b} = \widehat{s}_{12,b} - \Delta s_{12,0} \tag{9.25}$$

The observation equations are

$$v_{\bar{t}} = \frac{\sin \bar{t}_{12,0}}{\bar{d}_{12,0}} dy_1 - \frac{\cos \bar{t}_{12,0}}{\bar{d}_{12,0}} dx_1 - \frac{\sin \bar{t}_{12,0}}{\bar{d}_{12,0}} dy_2 + \frac{\cos \bar{t}_{12,0}}{\bar{d}_{12,0}} dx_2 + \left(\bar{t}_{12,0} - \bar{t}_{12,b}\right) \tag{9.26}$$

$$\begin{aligned} v_{\bar{\delta}} = {} & \left(\frac{\sin \bar{t}_{13,0}}{\bar{d}_{13,0}} - \frac{\sin \bar{t}_{12,0}}{\bar{d}_{12,0}}\right) dy_1 - \left(\frac{\cos \bar{t}_{13,0}}{\bar{d}_{13,0}} - \frac{\cos \bar{t}_{12,0}}{\bar{d}_{12,0}}\right) dx_1 + \frac{\sin \bar{t}_{12,0}}{\bar{d}_{12,0}} dy_2 \\ & - \frac{\cos \bar{t}_{12,0}}{\bar{d}_{12,0}} dx_2 - \frac{\sin \bar{t}_{13,0}}{\bar{d}_{13,0}} dy_3 + \frac{\cos \bar{t}_{13,0}}{\bar{d}_{13,0}} dx_3 + \left(\bar{t}_{13,0} - \bar{t}_{13,b}\right) \end{aligned} \tag{9.27}$$

$$v_{\bar{d}} = -\cos \bar{t}_{12,0}\, dy_1 - \sin \bar{t}_{12,0}\, dx_1 + \cos \bar{t}_{12,0}\, dy_2 + \sin \bar{t}_{12,0}\, dx_2 + \left(\bar{t}_{12,0} - \bar{t}_{12,b}\right) \tag{9.28}$$

The scheme in Figure 9.12 suggests that the reduction elements Δt and Δs be computed from the approximate coordinates and that the GML functions be used to compute the geodesic azimuth. The use of the GML type of functions can be avoided if explicit functions for Δt and Δs are available.

9.2.5 The Δt and Δs Functions

The literature contains functions that relate Δt and Δs explicitly to the geodetic latitude and longitude or mapping coordinates, i.e.,

$$\Delta t = t_{\text{geod}}(\varphi_1, \lambda_2, \varphi_1, \lambda_2) = t_{\text{map}}(x_1, y_2, x_1, y_2) \tag{9.29}$$

$$\Delta s = s_{\text{geod}}(\varphi_1, \lambda_2, \varphi_1, \lambda_2) = s_{\text{map}}(x_1, y_2, x_1, y_2) \tag{9.30}$$

$$\left\{P_1\left(\varphi_{1,0}, \lambda_{1,0}, x_{1,0}, y_{1,0}\right), P_2\left(\varphi_{2,0}, \lambda_{2,0}, x_{2,0}, y_{2,0}\right)\right\}$$
$$\downarrow$$
$$\left\{\bar{d}_{12,0}, \bar{t}_{12,0}, \gamma_{1,0}, k_{1,0}, k_{2,0}, k_{m,0}, \Delta t_{12,0}, \Delta s_{12,0}\right\}$$

Figure 9.13 Mapping reductions using functions for Δt and Δs.

The available functions $(t_{\text{geod}}, s_{\text{geod}})$ or $(t_{\text{map}}, s_{\text{map}})$ are typically again a result of series expansions and truncations. The derivation of (9.29) and (9.30) requires at least as much algebraic work as the derivation of the GML functions and the mapping equations because the mapped geodesic is involved. These functions are, strictly speaking, not needed if we use the GML functions as demonstrated above. The reduction scheme using the explicit functions of Δt and Δs is shown in Figure 9.13. It does not contain the geodesic azimuth $\widehat{\alpha}_{12,0}$ explicitly. The point scale factor k serves merely as an auxiliary quantity to express Δs in a compact form. The subscripts of k indicate the point of evaluation. In the case of m, k is evaluated at the midpoint $[(\varphi_1+\varphi_2)/2, (\lambda_1+\lambda_2)/2]$. See Table 9.5 for a listing of the expressions.

Computing the angular access of a polygon is a convenient way to verify the accuracy of the Δt expressions. First, we compute the angular excess of a geodesic polygon on the ellipsoid using the GML functions. For reasons of convenience, we choose equally spaced points from a geodesic circle as the vertices of the geodesic polygon (points on the geodesic circle and the center of the circle are connected by a geodesic with the same length). The sides of the polygon must, of course, be geodesics. The latter is automatically achieved since the geodesic angle between the sides of the polygon is computed from the GML functions. The conformal mappings are specified by the central meridian and the standard parallel that go through the center of the circle, as well as taking $k_0 = 1$. The radius of the geodesic circle, i.e., the size of the polygon, was varied from zero to the equivalent of about 2°. Figure 9.14 shows the differences in the angular excess for these polygons as computed from the GML functions and the explicit functions of Table 9.5. The figure shows that the TM and LC expressions in Table 9.5 are of the same accuracy within the region of computation and that they agree to about 0.1 arcsec with the GML computation. Since the angular excess is independent of the specific conformal mapping the lines in Figure 9.14 coincide, in theory.

TABLE 9.5 Explicit Functions for Δt and Δs

TM: $\Delta t_1 = \dfrac{(x_2+2x_1)(y_2-y_1)}{6k_0^2R_1^2}$	LC: $\Delta t_1 = \dfrac{(2y_1+y_2)(x_1-x_2)}{6k_0^2R_0^2}$

$$\frac{1}{k_L} \equiv \frac{\widehat{s}}{\bar{d}} = \frac{1}{6}\left(\frac{1}{k_1} + \frac{4}{k_m} + \frac{1}{k_2}\right)$$

$$\Delta s = \widehat{s}\,(1 - k_L)$$

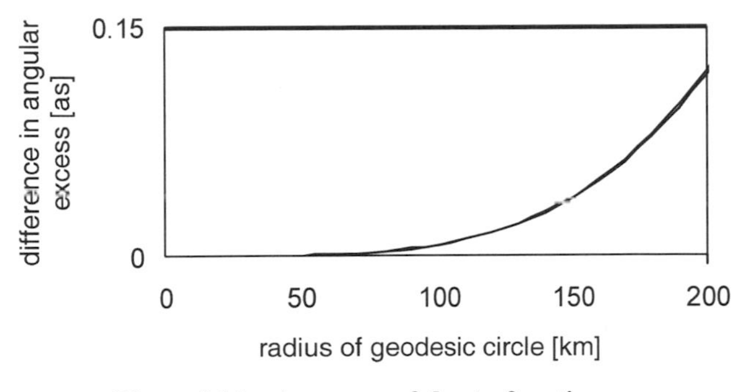

Figure 9.14 Accuracy of the Δt functions.

The accuracy of the expressions for Δs can be verified similarly. First, we map the center of the geodesic circle and the polygon vertices using the direct mapping equations. Next, we compute the map distances $\bar{d}_i$ between the mapped center and mapped polygon points, and form the difference $\Delta s_{i,dm} = \widehat{s}_i - \bar{d}_i$. The subscript dm indicates that these values were obtained by using the direct mapping equations. Next, the values Δs_i are computed from the explicit expressions in Table 9.5. Figure 9.15 shows the differences $\Delta s_i - \Delta s_{i,dm}$ for both TM and LC. The same conformal mapping specifications have been used as given above, and, again, the radius of the geodesic circle covers up to 2°. The figure demonstrates millimeter-level agreement in the range of the test area.

Expressions for Δt and Δs that are even more accurate are available in the literature.

9.2.6 Similarity Revisited

In Appendix C, the conformal property is identified as similarity between infinitesimally small figures. It is, of course, difficult to interpret such a statement because

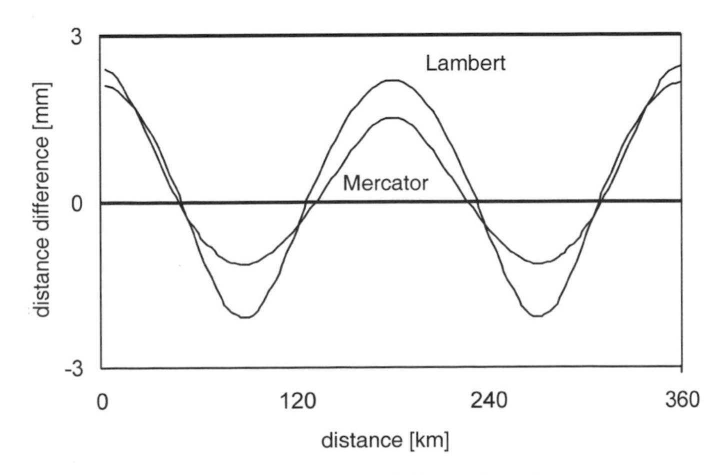

Figure 9.15 Accuracy of the Δs functions.

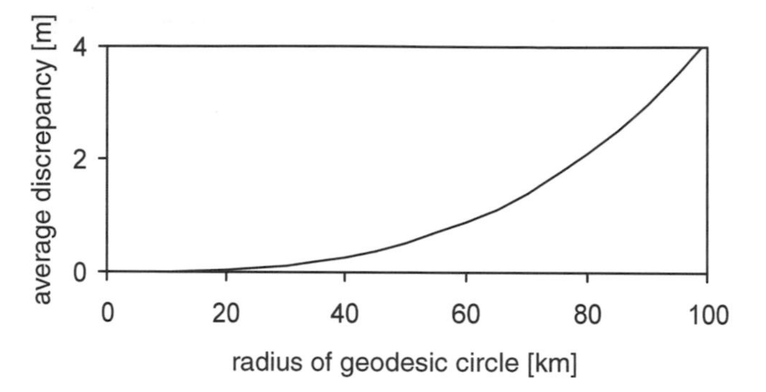

Figure 9.16 Similarity transformation of two mapped geodesic circles as a function of radius.

one typically does not think in terms of infinitesimally small figures. Transforming two clusters of points that were generated with two different maps can readily expose the degree of similarity. If the discrepancies exceed a specified limit, then similarity transformation cannot be used in practice to transform clusters of points between different maps.

We use n equally spaced points on a geodesic circle on the ellipsoid with its center at $\varphi_0 = 45^\circ$ and central meridian and then map these points with the transverse Mercator and Lambert conformal mapping functions. We use $k_0 = 1$. These two sets of map coordinates are input to a least-squares solution that estimates the parameters of a similarity transformation, i.e., two translations, one scale factor, and one rotation angle. The least-squares solution also generates residuals v_{x_i} and v_{y_i} for station i, which we use to compute the station discrepancy $d_i = (v_{x_i}^2 + v_{x_i}^2)^{1/2}$. We use the average of the d_i as a measure of fit. The radius of the geodesic circle is incremented from 10 to 100 km for the solutions shown in Figure 9.16. The 1 m average is reached just beyond the 50 km radius. The shape of this curve and the magnitude of the

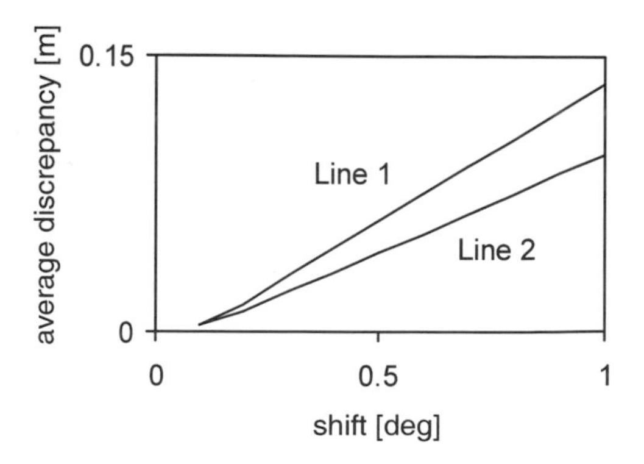

Figure 9.17 Similarity transformation of two mapped geodesic circles as a function of location.

discrepancies depend on the specifications of the mappings. The figure shows an optimal situation because the circle is centered at the origin of the Lambert conformal mapping and at the central meridian of the Mercator mapping. With $k_0 = 1$ the area around the center of the circle has the least distortion and the similarity model fits relatively well.

Figure 9.17 shows discrepancies for different locations of the geodesic circle within the mapping area while the radius remains constant at 10 km. For line 1 (LC), the standard parallel of the Lambert conformal mapping shifts from 45° to 46° while the center of the geodesic circle remains at latitude 45°. In the case of line 2 (TM), the center of the geodesic circle moves from 0° to 1° in longitude while maintaining a latitude of 45°.

APPENDIX A

GENERAL BACKGROUND

This appendix provides mathematical material that is handy to have available in a classroom situation to support key derivations or conclusions of the main chapters. It begins with a listing of expressions from spherical trigonometry. The rotation matrices are given along with brief definitions of positive and negative rotations. The sections on linear algebra, linearization, and statistics contain primary reference material for the least-squares adjustment given in Chapter 4. The subsection on the distribution of sums of variables is particularly useful when deriving the distribution of $\mathbf{v}^T\mathbf{P}\mathbf{v}$.

A.1 SPHERICAL TRIGONOMETRY

The sides of a spherical triangle are defined by great circles. A great circle is an intersection of the sphere with a plane that passes through the center of the sphere. It follows from geometric consideration of the special properties of the sphere that great circles are normal sections and geodesic lines. Figure A.1 shows a spherical triangle with corners (A, B, C), sides (a, b, c), and angles (α, β, γ). Notice that the sequence of the elements in the respective triplets is consistent, counterclockwise in this case. The sides of the spherical triangle are given in angular units. In many applications, one of the corners of the spherical triangle represents the North or South Pole. Documentation of the expressions listed below is readily available from the mathematical literature. Complete derivations can be found in Sigl (1977).

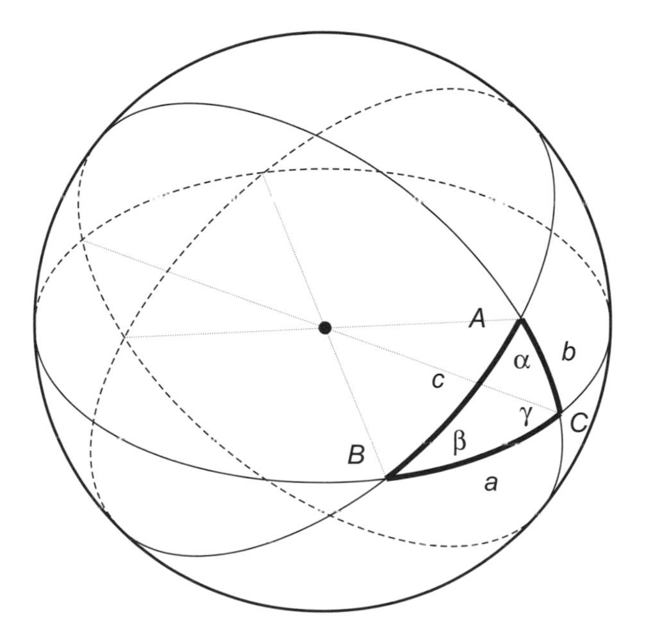

Figure A.1 The spherical triangle.

Law of Sine

$$\frac{\sin a}{\sin \alpha} = \frac{\sin b}{\sin \beta} \tag{A.1}$$

$$\frac{\sin a}{\sin \alpha} = \frac{\sin c}{\sin \gamma} \tag{A.2}$$

Law of Cosine for Sides

$$\left.\begin{aligned} \cos a &= \cos b \cos c + \sin b \sin c \cos \alpha \\ \cos b &= \cos c \cos a + \sin c \sin a \cos \beta \\ \cos c &= \cos a \cos b + \sin a \sin b \cos \gamma \end{aligned}\right\} \tag{A.3}$$

Law of Cosine for Angles

$$\left.\begin{aligned} \cos \alpha &= -\cos \beta \cos \gamma + \sin \beta \sin \gamma \cos a \\ \cos \beta &= -\cos \gamma \cos \alpha + \sin \gamma \sin \alpha \cos b \\ \cos \gamma &= -\cos \alpha \cos \beta + \sin \alpha \sin \beta \cos c \end{aligned}\right\} \tag{A.4}$$

Five Argument Formulas

$$\left.\begin{aligned}\sin a\cos\beta &= \cos b\sin c - \sin b\cos c\cos\alpha\\ \sin b\cos\gamma &= \cos c\sin a - \sin c\cos a\cos\beta\\ \sin c\cos\alpha &= \cos a\sin b - \sin a\cos b\cos\gamma\end{aligned}\right\}\tag{A.5}$$

$$\left.\begin{aligned}\sin a\cos\gamma &= \cos c\sin b - \sin c\cos b\cos\alpha\\ \sin b\cos\alpha &= \cos a\sin c - \sin a\cos c\cos\beta\\ \sin c\cos\beta &= \cos b\sin a - \sin b\cos a\cos\gamma\end{aligned}\right\}\tag{A.6}$$

$$\left.\begin{aligned}\sin\alpha\cos b &= \cos\beta\sin\gamma + \sin\beta\cos\gamma\cos a\\ \sin\beta\cos c &= \cos\gamma\sin\alpha + \sin\gamma\cos\alpha\cos b\\ \sin\gamma\cos a &= \cos\alpha\sin\beta + \sin\alpha\cos\beta\cos c\end{aligned}\right\}\tag{A.7}$$

$$\left.\begin{aligned}\sin\alpha\cos c &= \cos\gamma\sin\beta + \sin\gamma\cos\beta\cos a\\ \sin\beta\cos a &= \cos\alpha\sin\gamma + \sin\alpha\cos\gamma\cos b\\ \sin\gamma\cos b &= \cos\beta\sin\alpha + \sin\beta\cos\alpha\cos c\end{aligned}\right\}\tag{A.8}$$

Four Argument Formulas

$$\left.\begin{aligned}\sin\alpha\cot\beta &= \cot b\sin c - \cos c\cos\alpha\\ \sin\alpha\cot\gamma &= \cot c\sin b - \cos b\cos\alpha\\ \sin\beta\cot\gamma &= \cot c\sin a - \cos a\cos\beta\\ \sin\beta\cot\alpha &= \cot a\sin c - \cos c\cos\beta\\ \sin\gamma\cot\alpha &= \cot a\sin b - \cos b\cos\gamma\\ \sin\gamma\cot\beta &= \cot b\sin a - \cos a\cos\gamma\end{aligned}\right\}\tag{A.9}$$

Gauss (Delambre, Mollweide) Formulas—not all permutations listed

$$\sin\frac{\alpha}{2}\sin\frac{b+c}{2} = \sin\frac{a}{2}\cos\frac{\beta-\gamma}{2}\tag{A.10}$$

$$\sin\frac{\alpha}{2}\cos\frac{b+c}{2} = \cos\frac{a}{2}\cos\frac{\beta+\gamma}{2}\tag{A.11}$$

$$\cos\frac{\alpha}{2}\sin\frac{b-c}{2} = \sin\frac{a}{2}\sin\frac{\beta-\gamma}{2}\tag{A.12}$$

$$\cos\frac{\alpha}{2}\cos\frac{b-c}{2} = \cos\frac{a}{2}\sin\frac{\beta+\gamma}{2}\tag{A.13}$$

Napier Analogies—not all permutations listed

$$\tan \frac{a+b}{2} = \tan \frac{c}{2} \frac{\cos \dfrac{\alpha - \beta}{2}}{\cos \dfrac{\alpha + \beta}{2}} \tag{A.14}$$

$$\tan \frac{a-b}{2} = \tan \frac{c}{2} \frac{\sin \dfrac{\alpha - \beta}{2}}{\sin \dfrac{\alpha + \beta}{2}} \tag{A.15}$$

$$\tan \frac{\alpha+\beta}{2} = \cot \frac{\gamma}{2} \frac{\cos \dfrac{a-b}{2}}{\cos \dfrac{a+b}{2}}$$

$$\tan \frac{\alpha-\beta}{2} = \cot \frac{\gamma}{2} \frac{\sin \dfrac{a-b}{2}}{\sin \dfrac{a+b}{2}}$$

Half Angle Formulas

$$s = (a+b+c)/2 \tag{A.16}$$

$$k = \sqrt{\frac{\sin(s-a)\sin(s-b)\sin(s-c)}{\sin s}} \tag{A.17}$$

$$\tan \frac{\alpha}{2} = \frac{k}{\sin(s-a)} \tag{A.18}$$

$$\tan \frac{\beta}{2} = \frac{k}{\sin(s-b)} \tag{A.19}$$

$$\tan \frac{\gamma}{2} = \frac{k}{\sin(s-c)} \tag{A.20}$$

Half Side Formulas

$$\sigma = (\alpha+\beta+\gamma)/2 \tag{A.21}$$

$$k' = \sqrt{\frac{\cos(\sigma-\alpha)\cos(\sigma-\beta)\cos(\sigma-\gamma)}{-\cos\sigma}} \tag{A.22}$$

$$\tan\frac{a}{2} = \frac{\cos(\sigma - \alpha)}{k'} \tag{A.23}$$

$$\tan\frac{b}{2} = \frac{\cos(\sigma - \beta)}{k'} \tag{A.24}$$

$$\tan\frac{c}{2} = \frac{\cos(\sigma - \gamma)}{k'} \tag{A.25}$$

L'Huilier-Serret Formulas

$$M = \sqrt{\frac{\tan\dfrac{s-a}{2}\cdot\tan\dfrac{s-b}{2}\cdot\tan\dfrac{s-c}{2}}{\tan\dfrac{s}{2}}} \tag{A.26}$$

$$\tan\frac{\varepsilon}{4} = M\cdot\tan\frac{s}{2} \tag{A.27}$$

$$\tan\left(\frac{\alpha}{2} - \frac{\varepsilon}{4}\right) = M\cdot\cot\frac{s-a}{2} \tag{A.28}$$

$$\tan\left(\frac{\beta}{2} - \frac{\varepsilon}{4}\right) = M\cdot\cot\frac{s-b}{2} \tag{A.29}$$

$$\tan\left(\frac{\gamma}{2} - \frac{\varepsilon}{4}\right) = M\cdot\cot\frac{s-c}{2} \tag{A.30}$$

The symbol ε denotes the spherical angular excess. The area of spherical triangle can be expressed as

$$\Delta = \varepsilon\, r^2 \tag{A.31}$$

where r denotes the radius of the sphere.

A.2 ROTATION MATRICES

Rotations between coordinate systems are very conveniently expressed in terms of rotation matrices. The rotation matrices

$$R_1(\theta) = \begin{bmatrix} 1 & 0 & 0 \\ 0 & \cos\theta & \sin\theta \\ 0 & -\sin\theta & \cos\theta \end{bmatrix} \tag{A.32}$$

$$R_2(\theta) = \begin{bmatrix} \cos\theta & 0 & -\sin\theta \\ 0 & 1 & 0 \\ \sin\theta & 0 & \cos\theta \end{bmatrix} \tag{A.33}$$

$$R_3(\theta) = \begin{bmatrix} \cos\theta & \sin\theta & 0 \\ -\sin\theta & \cos\theta & 0 \\ 0 & 0 & 1 \end{bmatrix} \tag{A.34}$$

describe rotations by the angle θ of a right-handed coordinate system around the first, second, and third axes, respectively. The rotation angle is positive for a counterclockwise rotation, as viewed from the positive end of the axis about which the rotation takes place. The result of successive rotations depends on the specific sequence of the individual rotations. An exception to this rule is differentially small rotations for which the sequence of rotations does not matter.

A.3 LINEAR ALGEBRA

Surveying computations rest heavily upon concepts from linear algebra. In general, there is a nonlinear mathematical relationship between the observations and other quantities, such as coordinates, height, area, and volume. Seldom is there a natural linear relation between observations as there is in spirit leveling. Least-squares adjustment and statistical treatment require that nonlinear mathematical relations be linearized; i.e., the nonlinear relationship is replaced by a linear relationship. Possible errors caused by neglecting the nonlinear portion are eliminated through appropriate iteration. The result of linearization is a set of linear equations that is subject to further analysis, thus the need to know the elements of linear algebra. The use of linear algebra in the derivation and analysis of surveying measurements fortunately does not require the memorization of all possible proofs and theorems. Strang and Borre (1997) is a very useful reference that emphasizes linear algebra as it applies to geodesy and GPS. This appendix merely summarizes some elements from linear algebra for the sake of completeness.

A.3.1 Determinants and Matrix Inverse

Let the elements of a matrix $\mathbf{A}$ be denoted by a_{ij}, where the subscript i denotes the row and j the column. A $u \times u$ square matrix $\mathbf{A}$ has a uniquely defined determinant, denoted by $|\mathbf{A}|$, and said to be of order u. The determinant of a 1×1 matrix equals the matrix element. The determinant of $\mathbf{A}$ is expressed as a function of determinants of submatrices of size $(u-1) \times (u-1)$, $(u-2) \times (u-2)$, etc. until the size 2 or 1 is reached. The determinant is conveniently expressed in terms of minors and cofactors.

The minor can be computed for each element of the matrix. It is equal to the determinant after the respective row and column have been deleted. For example, the minor for $i = 1$ and $j = 2$ is

$$m_{12} = \begin{vmatrix} a_{21} & a_{23} & \cdots & a_{2u} \\ a_{31} & a_{33} & \cdots & a_{3u} \\ \vdots & \vdots & \cdots & \vdots \\ a_{u1} & a_{u3} & \cdots & a_{uu} \end{vmatrix} \tag{A.35}$$

The cofactor c_{ij} is equal to plus or minus the minor, depending on the subscripts i and j,

$$c_{ij} = (-1)^{i+j}\, m_{ij} \tag{A.36}$$

The determinant of $\mathbf{A}$ can now be expressed as

$$|\mathbf{A}| = \sum_{j=1}^{u} a_{kj}\, c_{kj} \tag{A.37}$$

The subscript k is fixed in (A.37) but can be any value between 1 and u; i.e., the determinant can be computed based on the minors for any one of the u rows or columns. Of course, the determinant (A.35) can be expressed as a function of determinants of matrixes of size $(u-2) \times (u-2)$, etc.

Determinants have many useful properties. For example, the rank of a matrix equals the order of the largest nonsingular square submatrix, i.e., the largest order for a nonzero determinant that can be found. The determinant is zero and the matrix is singular if the columns or rows of $\mathbf{A}$ are linearly dependent. The inverse of the square matrix can be expressed as

$$\mathbf{A}^{-1} = \frac{1}{|\mathbf{A}|}\mathbf{C}^{\mathrm{T}} \tag{A.38}$$

where $\mathbf{C}$ is the cofactor matrix consisting of the elements c_{ij} given in (A.36). The product of the matrix and its inverse equals the identity matrix, i.e., $\mathbf{A}\mathbf{A}^{-1} = \mathbf{I}$ and $\mathbf{A}^{-1}\mathbf{A} = \mathbf{I}$. These simple relations do not hold for generalized matrix inverses that can be computed for singular or even rectangular matrices. Information on generalized inverses is available in the standard mathematical literature. The inverse of a nonsingular square matrix $\mathbf{A}$, $\mathbf{B}$, $\mathbf{C}$, follows the simple rules

$$(\mathbf{ABC})^{-1} = \mathbf{C}^{-1}\mathbf{B}^{-1}\mathbf{A}^{-1} \tag{A.39}$$

Computation techniques for inverting nonsingular square matrices abound in linear algebra textbooks. In many cases the matrices to be inverted show a definite pattern and are often sparsely populated. When solving large systems of equations, it might be necessary to take advantage of these patterns in order to reduce the computation load (George and Liu, 1981). Very useful subroutines are available in the public domain, e.g., Milbert (1984). Some applications might produce ill-conditioned (numerically near-singular) matrices that require special attention.

A.3.2 Eigenvalues and Eigenvectors

Let $\mathbf{A}$ denote a $u \times u$ matrix and $\mathbf{x}$ be a $u \times 1$ vector. If $\mathbf{x}$ fulfills the equation

$$\mathbf{A}\,\mathbf{x} = \lambda\,\mathbf{x} \tag{A.40}$$

it is called an eigenvector, and the scalar λ is the corresponding eigenvalue. Equation (A.40) can be rewritten as

$$(\mathbf{A} - \lambda\mathbf{I})\,\mathbf{x} = \mathbf{o} \tag{A.41}$$

If $\mathbf{x}_0$ denotes a solution of (A.41) and α is a scalar, then $\alpha\mathbf{x}_0$ is also a solution. It follows that (A.41) provides only the direction of the eigenvector. There exists a nontrivial solution for $\mathbf{x}$ if the determinant is zero; i.e.,

$$|\mathbf{A} - \lambda\mathbf{I}| = 0 \tag{A.42}$$

This is the characteristic equation. It is a polynomial of the uth order in λ, providing u solutions λ_i, with $i = 1, \cdots, u$. Some of the eigenvalues can be zero, equal (multiple solution), or even complex. Equation (A.41) provides an eigenvector $\mathbf{x}_i$ for each eigenvalue λ_i.

For a symmetric matrix, all eigenvalues are real. Although the characteristic equation might have multiple solutions, the number of zero eigenvalues is equal to the rank defect of the matrix. The eigenvectors are mutually orthogonal,

$$\mathbf{x}_i^{\mathrm{T}}\mathbf{x}_j = 0 \tag{A.43}$$

For positive-definite matrices all eigenvalues are positive. Let the normalized eigenvectors $\mathbf{x}_i / \|\mathbf{x}_i\|$ be denoted $\mathbf{e}_i$; we can combine the normalized eigenvectors into a matrix,

$$\mathbf{E} = [\mathbf{e}_1 \quad \mathbf{e}_2 \quad \cdots \quad \mathbf{e}_u] \tag{A.44}$$

The matrix $\mathbf{E}$ is an orthonormal matrix for which

$$\mathbf{E}^{\mathrm{T}} = \mathbf{E}^{-1} \tag{A.45}$$

holds.

A.3.3 Diagonalization

Consider again a $u \times u$ matrix $\mathbf{A}$ and the respective matrix $\mathbf{E}$ that consists of the normalized eigenvectors. The product of these two matrices is

$$\begin{aligned} \mathbf{AE} &= [\mathbf{Ae}_1 \quad \mathbf{Ae}_2 \quad \cdots \quad \mathbf{Ae}_u] \\ &= [\lambda_1 \mathbf{e}_1 \quad \lambda_2 \mathbf{e}_2 \quad \cdots \quad \lambda_u \mathbf{e}_u] \\ &= \mathbf{E}\,\mathbf{\Lambda} \end{aligned} \tag{A.46}$$

where $\mathbf{\Lambda}$ is a diagonal matrix with λ_i as elements at the diagonal. Multiplying this equation by $\mathbf{E}^{\mathrm{T}}$ from the left and making use of Equation (A.45), one gets

$$\mathbf{E}^{\mathrm{T}}\mathbf{AE} = \mathbf{\Lambda} \tag{A.47}$$

Taking the inverse of both sides by applying the rule (A.39) and using (A.45) gives

$$\mathbf{E}^{\mathrm{T}}\mathbf{A}^{-1}\mathbf{E} = \mathbf{\Lambda}^{-1} \tag{A.48}$$

Equation (A.47) simply states that if a matrix $\mathbf{A}$ is premultiplied by $\mathbf{E}^{\mathrm{T}}$ and postmultiplied by $\mathbf{E}$, where the columns of $\mathbf{E}$ are the normalized eigenvectors, then the product is a diagonal matrix whose diagonal elements are the eigenvalues of $\mathbf{A}$. Equation (A.47) is further modified by

$$\mathbf{\Lambda}^{-1/2}\mathbf{E}^{\mathrm{T}}\mathbf{AE}\mathbf{\Lambda}^{-1/2} = \mathbf{I} \tag{A.49}$$

Defining the matrix $\mathbf{D}$ as

$$\mathbf{D} \equiv \mathbf{E}\mathbf{\Lambda}^{-1/2} \tag{A.50}$$

then

$$\mathbf{D}^{\mathrm{T}}\mathbf{AD} = \mathbf{I} \tag{A.51}$$

If the $u \times u$ matrix $\mathbf{A}$ is positive-semidefinite with rank $R(\mathbf{A}) = r < u$, an equation similar to (A.47) can be found. Consider the matrix

$$\mathbf{E} = \left[{}_u\mathbf{F}_r \; {}_u\mathbf{G}_{u-r}\right] \tag{A.52}$$

where the column of $\mathbf{F}$ consists of the normalized eigenvectors that pertain to the r nonzero eigenvalues. The submatrix $\mathbf{G}$ consists of $u - r$ eigenvectors that pertain to the $u - r$ zero eigenvalues. The columns of $\mathbf{F}$ and $\mathbf{G}$ span the column and null space, respectively, of the matrix $\mathbf{A}$. Because of Equation (A.40) it follows that

$$\mathbf{A}\,\mathbf{G} = \mathbf{O} \tag{A.53}$$

Applying Equations (A.52) and (A.53) gives

$$\mathbf{E}^{\mathrm{T}}\mathbf{A}\mathbf{E} = \begin{bmatrix} \mathbf{F}^{\mathrm{T}} \\ \mathbf{G}^{\mathrm{T}} \end{bmatrix} \mathbf{A}\,[\mathbf{F} \quad \mathbf{G}] = \begin{bmatrix} \mathbf{F}^{\mathrm{T}}\mathbf{A}\mathbf{F} & \mathbf{O} \\ \mathbf{O} & \mathbf{O} \end{bmatrix} = \begin{bmatrix} \mathbf{\Lambda} & \mathbf{O} \\ \mathbf{O} & \mathbf{O} \end{bmatrix} \tag{A.54}$$

The submatrix contains the r nonzero eigenvalues. If

$$\mathbf{D} = (\mathbf{E}\mathbf{\Lambda}^{-1/2} \,\vdots\, \mathbf{G}) \tag{A.55}$$

it follows that

$$\mathbf{D}^{\mathrm{T}}\mathbf{A}\mathbf{D} = \begin{bmatrix} \mathbf{I} & \mathbf{O} \\ \mathbf{O} & \mathbf{O} \end{bmatrix} \tag{A.56}$$

where the symbol $\mathbf{I}$ denotes an $r \times r$ identity matrix.

A.3.4 Quadratic Forms

Let $\mathbf{A}$ denote a $u \times u$ matrix and $\mathbf{x}$ be a $u \times 1$ vector. Then

$$v = \mathbf{x}^{\mathrm{T}}\mathbf{A}\mathbf{x} \tag{A.57}$$

is a quadratic form. The matrix $\mathbf{A}$ is called positive semidefinite if for all $\mathbf{x}$

$$\mathbf{x}^{\mathrm{T}}\mathbf{A}\mathbf{x} \geq 0 \tag{A.58}$$

and positive definite if for all $\mathbf{x}$

$$\mathbf{x}^{\mathrm{T}}\mathbf{A}\mathbf{x} > 0 \tag{A.59}$$

The following are some of the properties that are valid for a positive definite matrix $\mathbf{A}$:

1. $R(\mathbf{A}) = u$ (full rank).
2. $a_{ii} > 0$ for all i.
3. The inverse $\mathbf{A}^{-1}$ is positive definite.
4. Let $\mathbf{B}$ be an $n \times u$ matrix with rank $u < n$. Then the matrix $\mathbf{B}^{\mathrm{T}}\mathbf{A}\mathbf{B}$ is positive definite. If $R(\mathbf{B}) = r < u$, then $\mathbf{B}^{\mathrm{T}}\mathbf{A}\mathbf{B}$ is positive semidefinite.
5. Let $\mathbf{D}$ be a $q \times q$ matrix formed by deleting $u - p$ rows and the corresponding $u - p$ columns of $\mathbf{A}$. Then $\mathbf{D}$ is positive definite.

A necessary and sufficient condition for a symmetric matrix to be positive definite is that the principal minor determinants be positive; i.e.,

$$a_{11} > 0, \qquad \begin{vmatrix} a_{11} & a_{12} \\ a_{21} & a_{22} \end{vmatrix} > 0, \qquad \ldots, \qquad |\mathbf{A}| > 0 \tag{A.60}$$

or that all eigenvalues are real and positive.

If $\mathbf{A}$ is positive definite, then (A.57) is the equation of a u-dimensional ellipsoid expressed in a Cartesian coordinate system (x). The center of the ellipsoid is at $\mathbf{x} = \mathbf{o}$. Transforming (rotating) the coordinate system (x)

$$\mathbf{x} = \mathbf{E}\,\mathbf{y} \tag{A.61}$$

expresses the quadratic form in the (y) coordinate system,

$$\mathbf{y}^{\mathrm{T}}\mathbf{E}^{\mathrm{T}}\mathbf{A}\mathbf{E}\mathbf{y} = v \tag{A.62}$$

Since the matrix $\mathbf{E}$ consists of normalized eigenvectors we can use (A.47) to obtain the simple expressions

$$\begin{aligned} v &= \mathbf{y}^{\mathrm{T}}\boldsymbol{\Lambda}\,\mathbf{y} \\ &= y_1^2\,\lambda_1 + y_2^2\,\lambda_2 + \cdots + y_u^2\,\lambda_u \end{aligned} \tag{A.63}$$

This expression can be written as

$$\frac{y_1^2}{v/\lambda_1} + \frac{y_2^2}{v/\lambda_2} + \cdots + \frac{y_u^2}{v/\lambda_u} = 1 \tag{A.64}$$

This is the equation for the u-dimensional ellipsoid in the principal axes form; i.e., the coordinate system (y) coincides with the principal axes of the hyperellipsoid, and the lengths of the principal axes are proportional to the reciprocal of the square root of the eigenvalues. All eigenvalues are positive because the matrix $\mathbf{A}$ is positive definite. Equation (A.61) determines the orientation between the (x) and (y) coordinate systems. If $\mathbf{A}$ has a rank defect, the dimension of the hyperellipsoid is $R(\mathbf{A}) = r < u$.

Let the vectors $\mathbf{x}$ and $\mathbf{y}$ be of dimension $u \times 1$ and let the $u \times u$ matrix $\mathbf{A}$ contain constants. Consider the quadratic form

$$w = \mathbf{x}^{\mathrm{T}}\mathbf{A}\,\mathbf{y} = \mathbf{y}^{\mathrm{T}}\mathbf{A}^{\mathrm{T}}\,\mathbf{x} \tag{A.65}$$

Because (A.65) is a 1×1 matrix, the expression can be transposed. This fact is used frequently to simplify expressions when deriving least-squares solutions. The total differential dw is

$$dw = \frac{\partial w}{\partial \mathbf{x}}d\mathbf{x} + \frac{\partial w}{\partial \mathbf{y}}d\mathbf{y} \tag{A.66}$$

The vectors $d\mathbf{x}$ and $d\mathbf{y}$ contain the differentials of the components of $\mathbf{x}$ and $\mathbf{y}$, respectively. From (A.66) and (A.65) it follows that

$$dw = \mathbf{y}^{\mathrm{T}}\mathbf{A}^{\mathrm{T}}d\mathbf{x} + \mathbf{x}^{\mathrm{T}}\mathbf{A}\,d\mathbf{y} \tag{A.67}$$

If the matrix $\mathbf{A}$ is symmetric, then the total differential of

$$\phi = \mathbf{x}^{\mathrm{T}}\mathbf{A}\,\mathbf{x} \tag{A.68}$$

is, according to (A.67),

$$d\phi = 2\mathbf{x}^{\mathrm{T}}\mathbf{A}\,d\mathbf{x} \tag{A.69}$$

The gradient of ϕ with respect to $\mathbf{x}$ is

$$\frac{\partial \phi}{\partial \mathbf{x}} \equiv \begin{bmatrix} \dfrac{\partial \phi}{\partial x_1} & \cdots & \dfrac{\partial \phi}{\partial x_u} \end{bmatrix}^{\mathrm{T}} = 2\,\mathbf{A}\,\mathbf{x} \tag{A.70}$$

Equation (A.70) can be readily verified by computing the partial derivatives $\partial\phi/\partial x_t$ at the tth component,

$$\begin{aligned} \frac{\partial \mathbf{x}^{\mathrm{T}}\mathbf{A}\,\mathbf{x}}{\partial x_t} &= \frac{\partial}{\partial x_i}\left(\sum_{j=1}^{k}\sum_{i=1}^{k} x_i x_j a_{ij}\right) \\ &= \sum_{j=1}^{k} x_j a_{tj} + \sum_{i=1}^{k} x_i a_{it} = 2\sum_{j=1}^{k} x_j a_{tj} \\ &= [2\mathbf{A}\mathbf{x}]_t \end{aligned} \tag{A.71}$$

because $\mathbf{A}$ is symmetric.

Equation (A.70) is the foundation for deriving least-squares solutions, which requires locating the stationary point (minimum) for a quadratic function. The procedure is to take the partial derivatives with respect to all variables and equate them to zero. While the details of the least-squares derivations are given in Chapter 4, the following example serves to demonstrate the principle of minimization using matrix notation.

Let $\mathbf{B}$ denote an $n \times u$ rectangular matrix with $n > u$, $\boldsymbol{\ell}$ is an $n \times 1$ vector, and $\mathbf{P}$ an $n \times n$ symmetric weight matrix that can include the special case $\mathbf{P} = \mathbf{I}$. The elements of $\mathbf{B}$, $\boldsymbol{\ell}$, and $\mathbf{P}$ are constants. The least-squares solution of

$$\mathbf{v} = \mathbf{B}\,\mathbf{x} + \boldsymbol{\ell} \tag{A.72}$$

requires $\phi(\mathbf{x}) \equiv \mathbf{v}^{\mathrm{T}}\mathbf{P}\mathbf{v} = \min$. First, we compute the gradient (column vector)

$$\begin{aligned} \frac{\partial \mathbf{v}^{\mathrm{T}}\mathbf{P}\mathbf{v}}{\partial \mathbf{x}} &= \frac{\partial}{\partial \mathbf{x}}\left[(\mathbf{B}\mathbf{x} + \boldsymbol{\ell})^{\mathrm{T}}\,\mathbf{P}\,(\mathbf{B}\mathbf{x} + \boldsymbol{\ell})\right] \\ &= \frac{\partial}{\partial \mathbf{x}}\left(2\boldsymbol{\ell}^{\mathrm{T}}\,\mathbf{P}\mathbf{B}\mathbf{x} + \mathbf{x}^{\mathrm{T}}\,\mathbf{B}^{\mathrm{T}}\,\mathbf{P}\mathbf{B}\mathbf{x} + \boldsymbol{\ell}^{\mathrm{T}}\,\mathbf{P}\boldsymbol{\ell}\right) \\ &= 2\mathbf{B}^{\mathrm{T}}\,\mathbf{P}\mathbf{B}\mathbf{x} + 2\mathbf{B}^{\mathrm{T}}\,\mathbf{P}\boldsymbol{\ell} \end{aligned} \tag{A.73}$$

and equate it to zero,

$$\frac{\partial \mathbf{v}^{\mathrm{T}}\mathbf{P}\mathbf{v}}{\partial \mathbf{x}} = \mathbf{o} \tag{A.74}$$

to assure that the least-squares solution for $\mathbf{x}$, denoted by $\hat{\mathbf{x}}$,

$$\hat{\mathbf{x}} = -\left(\mathbf{B}^{\mathrm{T}}\,\mathbf{P}\mathbf{B}\right)^{-1}\mathbf{B}^{\mathrm{T}}\,\mathbf{P}\boldsymbol{\ell} \tag{A.75}$$

at least represents a stationary point of $\phi(\mathbf{x})$. In Chapter 4 we verify that indeed a minimum has also been achieved.

A.3.5 Matrix Partitioning

Consider the following partitioning of the nonsingular square matrix $\mathbf{N}$,

$$\mathbf{N} = \begin{bmatrix} \mathbf{N}_{11} & \mathbf{N}_{12} \\ \mathbf{N}_{21} & \mathbf{N}_{22} \end{bmatrix} \tag{A.76}$$

where $\mathbf{N}_{11}$ and $\mathbf{N}_{22}$ are square matrices, although not necessarily of the same size. Let's denote the inverse matrix by $\mathbf{Q}$ and partition it accordingly; i.e.,

$$\mathbf{Q} = \mathbf{N}^{-1} = \begin{bmatrix} \mathbf{Q}_{11} & \mathbf{Q}_{12} \\ \mathbf{Q}_{21} & \mathbf{Q}_{22} \end{bmatrix} \tag{A.77}$$

so that the sizes of $\mathbf{N}_{11}$ and $\mathbf{Q}_{11}$, $\mathbf{N}_{12}$ and $\mathbf{Q}_{12}$, etc., are respectively the same. Equations (A.76) and (A.77) imply the following four relations:

$$\mathbf{N}_{11}\mathbf{Q}_{11} + \mathbf{N}_{12}\mathbf{Q}_{21} = \mathbf{I} \tag{A.78}$$

$$\mathbf{N}_{11}\mathbf{Q}_{12} + \mathbf{N}_{12}\mathbf{Q}_{22} = \mathbf{O} \tag{A.79}$$

$$\mathbf{N}_{21}\mathbf{Q}_{11} + \mathbf{N}_{22}\mathbf{Q}_{21} = \mathbf{O} \tag{A.80}$$

$$\mathbf{N}_{21}\mathbf{Q}_{12} + \mathbf{N}_{22}\mathbf{Q}_{22} = \mathbf{I} \tag{A.81}$$

The solutions for the submatrices $\mathbf{Q}_{ij}$ are carried out according to the standard rules for solving a system of linear equations, with the restriction that the inverse is defined only for square submatrices. Multiplying (A.78) from the left by $\mathbf{N}_{21}\mathbf{N}_{11}^{-1}$ and subtracting the product from (A.80) gives

$$\mathbf{Q}_{21} = -\left(\mathbf{N}_{22} - \mathbf{N}_{21}\mathbf{N}_{11}^{-1}\mathbf{N}_{12}\right)^{-1}\mathbf{N}_{21}\mathbf{N}_{11}^{-1} \tag{A.82}$$

Multiplying (A.79) from the left by $\mathbf{N}_{21}\mathbf{N}_{11}^{-1}$ and subtracting the product from (A.81) gives

$$\mathbf{Q}_{22} = \left(\mathbf{N}_{22} - \mathbf{N}_{21}\mathbf{N}_{11}^{-1}\mathbf{N}_{12}\right)^{-1} \tag{A.83}$$

Substituting (A.83) in (A.79) gives

$$\mathbf{Q}_{12} = -\mathbf{N}_{11}^{-1}\mathbf{N}_{12}\left(\mathbf{N}_{22} - \mathbf{N}_{21}\mathbf{N}_{11}^{-1}\mathbf{N}_{12}\right)^{-1} \tag{A.84}$$

Substituting (A.82) in (A.78) gives

$$\mathbf{Q}_{11} = \mathbf{N}_{11}^{-1} + \mathbf{N}_{11}^{-1}\mathbf{N}_{12}\left(\mathbf{N}_{22} - \mathbf{N}_{21}\mathbf{N}_{11}^{-1}\mathbf{N}_{12}\right)^{-1}\mathbf{N}_{21}\mathbf{N}_{11}^{-1} \tag{A.85}$$

An alternative solution for $(\mathbf{Q}_{11}, \mathbf{Q}_{12}, \mathbf{Q}_{21}, \mathbf{Q}_{22})$ is readily obtained. Multiplying (A.80) from the left by $\mathbf{N}_{12}\mathbf{N}_{22}^{-1}$ and subtracting the product from (A.78) gives

$$\mathbf{Q}_{11} = \left(\mathbf{N}_{11} - \mathbf{N}_{12}\mathbf{N}_{22}^{-1}\mathbf{N}_{21}\right)^{-1} \tag{A.86}$$

Substituting (A.86) in (A.80) gives

$$\mathbf{Q}_{21} = -\mathbf{N}_{22}^{-1}\mathbf{N}_{21}\left(\mathbf{N}_{11} - \mathbf{N}_{12}\mathbf{N}_{22}^{-1}\mathbf{N}_{21}\right)^{1} \tag{A.87}$$

Premultiplying (A.81) by $\mathbf{N}_{12}\mathbf{N}_{22}^{-1}$ and subtracting (A.79) gives

$$\mathbf{Q}_{12} = -\left(\mathbf{N}_{11} - \mathbf{N}_{12}\mathbf{N}_{22}^{-1}\mathbf{N}_{21}\right)^{-1}\mathbf{N}_{12}\mathbf{N}_{22}^{-1} \tag{A.88}$$

Substituting (A.88) in (A.81) gives

$$\mathbf{Q}_{22} = \mathbf{N}_{22}^{-1} + \mathbf{N}_{22}^{-1}\mathbf{N}_{21}\left(\mathbf{N}_{11} - \mathbf{N}_{12}\mathbf{N}_{22}^{-1}\mathbf{N}_{21}\right)^{-1}\mathbf{N}_{12}\mathbf{N}_{22}^{-1} \tag{A.89}$$

Usually the above partitioning technique is used to reduce the size of large matrices that must be inverted or to derive alternative expressions. Because these matrix identities are frequently used, and because they look somewhat puzzling unless one is aware of the simple solutions given above, they are summarized here again to be able to view them at a glance;

$$\left(\mathbf{N}_{11} - \mathbf{N}_{12}\mathbf{N}_{22}^{-1}\mathbf{N}_{21}\right)^{-1} = \mathbf{N}_{11}^{-1} + \mathbf{N}_{11}^{-1}\mathbf{N}_{12}\left(\mathbf{N}_{22} - \mathbf{N}_{21}\mathbf{N}_{11}^{-1}\mathbf{N}_{12}\right)^{-1}\mathbf{N}_{21}\mathbf{N}_{11}^{-1} \tag{A.90}$$

$$\mathbf{N}_{11}^{-1}\mathbf{N}_{12}\left(\mathbf{N}_{22} - \mathbf{N}_{21}\mathbf{N}_{11}^{-1}\mathbf{N}_{12}\right)^{-1} = \left(\mathbf{N}_{11} - \mathbf{N}_{12}\mathbf{N}_{22}^{-1}\mathbf{N}_{21}\right)^{-1}\mathbf{N}_{12}\mathbf{N}_{22}^{-1} \tag{A.91}$$

$$\left(\mathbf{N}_{22} - \mathbf{N}_{21}\mathbf{N}_{11}^{-1}\mathbf{N}_{12}\right)^{-1}\mathbf{N}_{21}\mathbf{N}_{11}^{-1} = \mathbf{N}_{22}^{-1}\mathbf{N}_{21}\left(\mathbf{N}_{11} - \mathbf{N}_{12}\mathbf{N}_{22}^{-1}\mathbf{N}_{21}\right)^{-1} \tag{A.92}$$

$$\left(\mathbf{N}_{22} - \mathbf{N}_{21}\mathbf{N}_{11}^{-1}\mathbf{N}_{12}\right)^{-1} = \mathbf{N}_{22}^{-1} + \mathbf{N}_{22}^{-1}\mathbf{N}_{21}\left(\mathbf{N}_{11} - \mathbf{N}_{12}\mathbf{N}_{22}^{-1}\mathbf{N}_{21}\right)^{-1}\mathbf{N}_{12}\mathbf{N}_{22}^{-1} \tag{A.93}$$

A.3.6 Cholesky Factor

For positive definite matrices the square root method, also known as the Cholesky method, is an efficient way to solve systems of equations and to invert the matrix.

Because $\mathbf{N}$ is positive definite, it is written as the product of a lower triangular matrix $\mathbf{L}$ and an upper triangular matrix $\mathbf{L}^T$:

$$\mathbf{N} = \mathbf{L}\mathbf{L}^T \tag{A.94}$$

It is readily seen that if $\mathbf{E}$ is an orthonormal matrix having the property (A.45) then the new matrix $\mathbf{B} = \mathbf{LE}$ is also a Cholesky factor because $\mathbf{B}\mathbf{B}^T = \mathbf{L}\mathbf{E}\mathbf{E}^T\mathbf{L}^T = \mathbf{L}\mathbf{L}^T$.

The lower and upper triangular matrices have several useful properties. For example, the eigenvalues of a triangular matrix equal the diagonal elements, and the determinant of the triangular matrix equals the product of the diagonal elements. Because the determinant of a matrix product is equal to the product of the determinants of the factors, it follows that $\mathbf{N}$ is singular if one of the diagonal elements of $\mathbf{L}$ is zero. This fact can be used advantageously during the computation of $\mathbf{L}$ to eliminate parameters that cause a singularity.

The Cholesky algorithm provides the instruction for computing the lower triangular matrix $\mathbf{L}$. The elements of $\mathbf{L}$ are

$$l_{jk} = \begin{cases} \sqrt{n_{jj} - \sum_{m=1}^{j-1} l_{jm}^2} & \text{for} \quad k = j \\ \frac{1}{l_{kk}}\left(n_{jk} - \sum_{m=1}^{k-1} l_{jm} l_{km}\right) & \text{for} \quad k < j \\ 0 & \text{for} \quad k > j \end{cases} \tag{A.95}$$

where $1 \le j \le u$ and $1 \le k \le u$. The Cholesky algorithm preserves the pattern of leading zeros in the rows and columns of $\mathbf{N}$, as can be readily verified. For example, if the first x elements in row y of $\mathbf{N}$ are zero, then the first x elements in row y of $\mathbf{L}$ are also zero. Taking advantage of this fact speeds up the computation of $\mathbf{L}$ for a large system that exhibits significant patterns of leading zeros. The algorithm (A.95) begins with the element l_{11}. Subsequently, the columns (or rows) can be computed sequentially from 1 to u, whereby previously computed columns (or rows) remain unchanged while the next one is computed.

The inverse of the submatrix $\mathbf{Q}_{22}$—see (A.77)—can be conveniently expressed as a function of the submatrix $\mathbf{L}_{22}$. Using $\mathbf{L}$,

$$\mathbf{L} = \begin{bmatrix} \mathbf{L}_{11} & \mathbf{O} \\ \mathbf{L}_{21} & \mathbf{L}_{22} \end{bmatrix} \tag{A.96}$$

to express $\mathbf{N}$ in terms of submatrices,

$$\mathbf{N} = \begin{bmatrix} \mathbf{L}_{11}\mathbf{L}_{11}^T & \mathbf{L}_{11}\mathbf{L}_{21}^T \\ \mathbf{L}_{21}\mathbf{L}_{11}^T & \mathbf{L}_{21}\mathbf{L}_{21}^T + \mathbf{L}_{22}\mathbf{L}_{22}^T \end{bmatrix} \tag{A.97}$$

and applying (A.83) readily gives

$$\mathbf{Q}_{22}^{-1} = \mathbf{L}_{22}\mathbf{L}_{22}^{\mathrm{T}} \tag{A.98}$$

Depending on the application one might group the parameters such that (A.98) can be used directly, i.e., the needed inverse is a simple function of the Cholesky factors that had been computed previously.

The diagonal elements of **L** are not necessarily unity. Consider a new matrix **G** with elements taken from **L** such that $g_{jk} = l_{jk}/l_{jj}$ and a new diagonal matrix **D** such that $d_{jj} = l_{jj}^2$; then

$$\mathbf{L} = \mathbf{G}\sqrt{\mathbf{D}} \tag{A.99}$$

$$\mathbf{N} = \mathbf{L}\mathbf{L}^{\mathrm{T}} = \mathbf{G}\mathbf{D}\mathbf{G}^{\mathrm{T}} \tag{A.100}$$

Because **N** is a positive definite matrix, the diagonal elements **G** are +1.

The unknown **x** can be solved without explicitly inverting the matrix. Assume that we must solve the system of equations

$$\mathbf{N}\mathbf{x} = \mathbf{u} \tag{A.101}$$

The first step in the solution of (A.101) is to substitute (A.94) for **N** and premultiply with $\mathbf{L}^{-1}$ to obtain the triangular equations

$$\mathbf{L}^{\mathrm{T}}\mathbf{x} = \mathbf{L}^{-1}\mathbf{u} \tag{A.102}$$

Denoting the right-hand side of (A.102) by $\mathbf{c}_u$ and multiplying by **L**, we can write

$$\mathbf{L}^{\mathrm{T}}\mathbf{x} = \mathbf{c}_u \tag{A.103}$$

$$\mathbf{L}\,\mathbf{c}_u = \mathbf{u} \tag{A.104}$$

We solve $\mathbf{c}_u$ from (A.104) starting with the first element. Using **L** and $\mathbf{c}_u$, the solution of (A.103) yields the parameters **x**, starting with the last element.

In least-squares, the auxiliary quantity

$$l = -\mathbf{c}_u^{\mathrm{T}}\mathbf{c}_u = -\left(\mathbf{L}^{-1}\mathbf{u}\right)^{\mathrm{T}}\left(\mathbf{L}^{-1}\mathbf{u}\right) = -\mathbf{u}^{\mathrm{T}}\mathbf{N}^{-1}\mathbf{u} \tag{A.105}$$

is useful for computing $\mathbf{v}^{\mathrm{T}}\mathbf{P}\mathbf{v}$ (see Table 4.1) to assess the quality of the adjustment. The Cholesky algorithm provides l from $\mathbf{c}_u$ without explicitly using the inverses of **N** and **L**.

Computing the inverse requires a much bigger computational effort than merely solving the system of equations. The first step is to make u solutions of the type (A.104) to obtain the columns of **C**,

$$\mathbf{L}\mathbf{C} = \mathbf{I} \tag{A.106}$$

where $\mathbf{I}$ is the $u \times u$ identity matrix. This is followed by u solutions of the type (A.103), using the columns of $\mathbf{C}$ for $\mathbf{C}_u$, to obtain the respective u columns of the inverse of $\mathbf{N}$.

The Cholesky factor $\mathbf{L}$ can be used directly to compute uncorrelated observations. From (A.94) it follows that premultiplying $\mathbf{N}$ with $\mathbf{L}^{-1}$ and postmultiplying it with the transpose gives the identity matrix. Therefore, the Cholesky factor $\mathbf{L}$ can be used in ways similar to the matrix $\mathbf{D}$ in (A.51). A frequent application is the decorrelation of observations. In this case the inverse $\mathbf{L}^{-1}$ is not required explicitly. For example, if we let $\mathbf{L}$ now denote the Cholesky factor of the covariance matrix of the observations $\boldsymbol{\Sigma}_{\ell_b}$, then the transformation (4.235) can be written as

$$\mathbf{L}^{-1}\mathbf{v} = \mathbf{L}^{-1}\mathbf{A}\mathbf{x} + \mathbf{L}^{-1}\boldsymbol{\ell} \tag{A.107}$$

Denoting the transformed observations by a bar, we get

$$\bar{\mathbf{v}} = \bar{\mathbf{A}}\mathbf{x} + \bar{\boldsymbol{\ell}} \tag{A.108}$$

$$\mathbf{L}\,\bar{\boldsymbol{\ell}} = \boldsymbol{\ell} \tag{A.109}$$

$$\mathbf{L}\,\bar{\mathbf{A}}_\alpha = \mathbf{A}_\alpha \tag{A.110}$$

The matrix $\bar{\mathbf{A}}$ and the vector $\bar{\boldsymbol{\ell}}$ can be computed directly with $\mathbf{L}$ using (A.110) and (A.109), respectively. The subscript α in (A.110) indicates the column. Upon completion of the adjustment the residuals follow from

$$\mathbf{L}\,\bar{\mathbf{v}} = \mathbf{v} \tag{A.111}$$

It is at times advantageous to work with decorrelated observations. Examples are horizontal angle observations or even GPS vectors. Decorrelated observations can be added one at a time to the adjustment, whereas correlated observations should be added by sets. See also Section 4.10.6 for a discussion of decorrelated redundancy numbers.

A.4 LINEARIZATION

Observations are often related by nonlinear functions of unknown parameters. The adjustment algorithm uses a linear functional relationship between the observations and the parameters and uses iterations to account for the nonlinearity. To perform an adjustment, one must therefore linearize these relationships. Expanding the functions in a Taylor series and retaining only the linear terms accomplishes this. Consider the nonlinear function

$$y = f(x) \tag{A.112}$$

which has one variable x. The Taylor series expansion of this function is

$$y = f(x_o) + \left.\frac{\partial y}{\partial x}\right|_{x_0} dx + \frac{1}{2!}\left.\frac{\partial^2 y}{\partial x^2}\right|_{x_0} dx^2 + \cdots \tag{A.113}$$

The linear portion is given by the first two terms

$$\bar{y} = f(x_o) + \left.\frac{\partial y}{\partial x}\right|_{x_0} dx \tag{A.114}$$

The derivative is evaluated at the point of expansion x_0. At the point of expansion, the linearized and the nonlinear functions are tangent. They separate by

$$\varepsilon = y - \bar{y} \tag{A.115}$$

as x departs from the expansion point x_0. The linear form (A.114) is a sufficiently accurate approximation of the nonlinear relation (A.112) only in the vicinity of the point of expansion.

The expansion of a two-variable function

$$z = f(x, y) \tag{A.116}$$

is

$$z = f(x_0, y_0) + \left.\frac{\partial z}{\partial x}\right|_{x_0, y_0} dx + \left.\frac{\partial z}{\partial y}\right|_{x_0, y_0} dy + \cdots \tag{A.117}$$

The point of expansion is $P(x = x_0, y = y_0)$. The linearized form

$$\bar{z} = f(x_0, y_0) + \left.\frac{\partial z}{\partial x}\right|_{x_0, y_0} dx + \left.\frac{\partial z}{\partial y}\right|_{x_0, y_0} dy \tag{A.118}$$

represents the tangent plane on the surface (A.116) at the expansion point. A generalization for the expansion of multivariable functions is readily seen. If n functions are related to u variables as in

$$\mathbf{y} = \mathbf{f}(\mathbf{x}) = \begin{bmatrix} f_1(\mathbf{x}) \\ f_2(\mathbf{x}) \\ \vdots \\ f_n(\mathbf{x}) \end{bmatrix} = \begin{bmatrix} f_1(x_1, x_2, \cdots, x_u) \\ f_2(x_1, x_2, \cdots, x_u) \\ \vdots \\ f_n(x_1, x_2, \cdots, x_u) \end{bmatrix} \tag{A.119}$$

the linearized form is

$$\bar{\mathbf{y}} = \mathbf{f}(\mathbf{x}_0) + \left.\frac{\partial \mathbf{f}}{\partial \mathbf{x}}\right|_{x_0} d\mathbf{x} \tag{A.120}$$

where

$$\frac{\partial \mathbf{f}}{\partial \mathbf{x}} = {}_n\mathbf{G}_u = \begin{bmatrix} \dfrac{\partial f_1}{\partial x_1} & \dfrac{\partial f_1}{\partial x_2} & \cdots & \dfrac{\partial f_1}{\partial x_u} \\ \dfrac{\partial f_2}{\partial x_1} & \dfrac{\partial f_2}{\partial x_2} & \cdots & \dfrac{\partial f_2}{\partial x_u} \\ \vdots & \vdots & \cdots & \vdots \\ \dfrac{\partial f_n}{\partial x_1} & \dfrac{\partial f_n}{\partial x_2} & \cdots & \dfrac{\partial f_n}{\partial x_u} \end{bmatrix} \tag{A.121}$$

The point of expansion is $P(\mathbf{x} = \mathbf{x}_0)$. Every component of $\mathbf{y}$ is differentiated with respect to every variable. Thus, the matrix $\mathbf{G}$ has as many columns as there are parameters, and as many rows as there are components in $\mathbf{y}$. The components of $\mathbf{f}(\mathbf{x}_0)$ are equal to the respective functions evaluated at $\mathbf{x}_0$.

A.5 STATISTICS

Brief explanations are given on one-dimensional distributions and hypothesis testing. The material of this appendix can be found in the standard literature on statistics. The expressions for the noncentral distribution are given, e.g., in Koch (1988).

A.5.1 One-Dimensional Distributions

The chi-square density function is given by

$$f(x) = \begin{cases} \dfrac{1}{2^{r/2}\,\Gamma(r/2)}\, x^{(r/2)-1} e^{-x/2} & x > 0 \\ 0 & \text{elsewhere} \end{cases} \tag{A.122}$$

The symbol r denotes a positive integer and is called the degree of freedom. The mean, i.e., the expected value, equals r, and the variance equals $2r$. The degree of freedom is sufficient to describe completely the chi-square distribution. The symbol Γ denotes the well-known gamma function, which is dealt with in books on advanced calculus and can be written as

$$\Gamma(g) = (g-1)! \tag{A.123}$$

$$\Gamma\left(g + \frac{1}{2}\right) = \frac{\sqrt{\pi}}{2^{2g-1}} \frac{\Gamma(2g)}{\Gamma(g)} \tag{A.124}$$

for positive integer g. Examples of the chi-square distribution for small degrees of freedom are given in Figure A.2. The probability that the random variable $\tilde{x}$ is less than w_α is

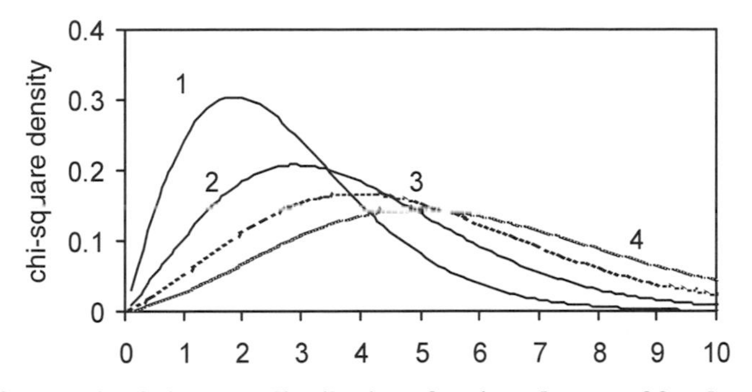

Figure A.2 Chi-square distribution of various degrees of freedom.

$$P(\tilde{x} < w_\alpha) = \int_0^{w_\alpha} f(x)\, dx = 1 - \alpha \tag{A.125}$$

Equation (A.125) implies that to the right of w_α there is the probability α; integrating from w_α to infinity gives α. If $\tilde{x}$ has a chi-square distribution, then this is expressed by the notation $\tilde{x} \sim \chi_r^2$.

The distribution (A.122) is more precisely called the central chi-square distribution. The noncentral chi-square is a generalization of this distribution. The density function does not have a simple closed form; it consists of an infinite sum of terms. If $\tilde{x}$ has a noncentral chi-square distribution, this is expressed by $\tilde{x} \sim \chi_{r,\lambda}^2$ where λ denotes the noncentrality parameter. The mean is

$$E(\tilde{x}) = r + \lambda \tag{A.126}$$

as opposed to just r for the central chi-square distribution.

The density function of the normal distribution is

$$f(x) = \frac{1}{\sigma\sqrt{2\pi}} e^{(x-\mu)^2/2\sigma^2} \qquad -\infty < x < \infty \tag{A.127}$$

where μ and σ^2 denote the mean and the variance. The notation $\tilde{x} \sim n(\mu, \sigma^2)$ is usually used. The two parameters μ and σ completely describe the normal distribution. See Figure A.3. The normal distribution has the following characteristics:

1. The distribution is symmetric about the mean.
2. The maximum density is at the mean.
3. For small variances, the maximum density is larger and the slopes are steeper than in the case of large variances.
4. The inflection points are at $x = \mu \pm \sigma$.

If $\tilde{x} \sim n(\mu, \sigma^2)$, then the transformed variable

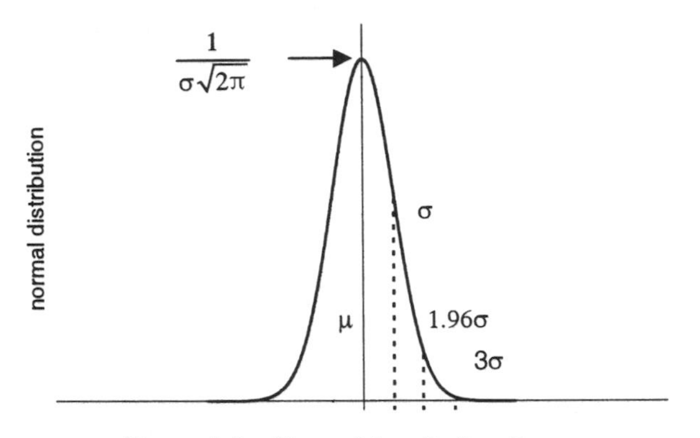

Figure A.3 Normal density function.

$$\tilde{w} = \frac{\tilde{x} - \mu}{\sigma} \sim n(0, 1) \tag{A.128}$$

has a normal distribution with zero mean and unit variance. The random variable $\tilde{w}$ is said to have a standardized normal distribution. The density function for $\tilde{w}$ is

$$f(w) = \frac{1}{\sqrt{2\pi}} e^{-w^2/2} \qquad -\infty < w < \infty \tag{A.129}$$

The probability that the random variable $\tilde{x}$ is less than w_α is

$$P(\tilde{x} < w_\alpha) = \int_0^{w_\alpha} f(w)\, dw \tag{A.130}$$

Table A.1 lists selected values that are frequently quoted. For a normal distribution, in about 68% of all cases the observations fall within one standard deviation from the mean, and only every 370th observation deviates from the mean by more than 3σ. Therefore, the 3σ value is sometimes taken as the limit to what is regarded as random error. Any larger deviation from the mean is usually considered a blunder. Statistically, large errors cannot be avoided, but their occurrence is unlikely. The 3σ criteria is not necessarily applicable in least-squares adjustments because the pertinent random variables have multivariate distributions and are correlated, thus reflecting the geometry of the adjustment. Further details are given in Sections 4.10 and 4.11.

Assume that $\tilde{w} \sim n(0, 1)$ and $\tilde{v} \sim \chi_r^2$ are two stochastically independent random variables with unit normal and chi-square distribution, respectively; then the random variable

TABLE A.1 Selected Values from the Normal Distribution

x	σ	2σ	3σ	0.674σ	1.645σ	1.960σ
$N(x) - N(-x)$	0.6827	0.9544	0.9973	0.5	0.90	0.95

$$\tilde{t} = \frac{\tilde{w}}{\sqrt{\tilde{v}/r}} \tag{A.131}$$

has a t distribution with r degrees of freedom. The distribution function is

$$f(t_r) = \frac{\Gamma\left[(r+1)/2\right]}{\sqrt{\pi r}\ \Gamma(r/2)} \left[1 + \frac{t^2}{r}\right]^{-(r+1)/2} \qquad -\infty < t < \infty \tag{A.132}$$

The density function (A.132) is symmetric with respect to $t = 0$. See Figure A.4. Furthermore, if $r = \infty$ then the t distribution is identical to the standardized normal distribution; i.e.,

$$t_\infty = n(0, 1) \tag{A.133}$$

The density in the vicinity of the mean (zero) is smaller than for the unit normal distribution, whereas the reverse is true at the extremities of the distribution. The t distribution converges rapidly toward the normal distribution. If the random variable $\tilde{w} \sim n(\delta, 1)$ is normal distributed with unit variance but with a nonzero mean, then the function (A.131) has a noncentral t distribution with r degrees of freedom and a noncentrality parameter δ.

Consider two stochastically independent random variables, $\tilde{u} \sim \chi^2_{r_1}$ and $\tilde{v} \sim \chi^2_{r_2}$, distributed with r_1 and r_2 degrees of freedom, respectively; then the random variable

$$\tilde{F} = \frac{\tilde{u}/r_1}{\tilde{v}/r_2} \tag{A.134}$$

has the density function

$$f(F_{r_1,r_2}) = \frac{\Gamma\left[(r_1+r_2)/2\right](r_1/r_2)^{r_1/2}}{\Gamma(r_1/2)\ \Gamma(r_2/2)} \frac{F^{(r_1/2)-1}}{(1 + r_1 F/r_2)^{(r_1+r_2)/2}} \qquad 0 < F < \infty \tag{A.135}$$

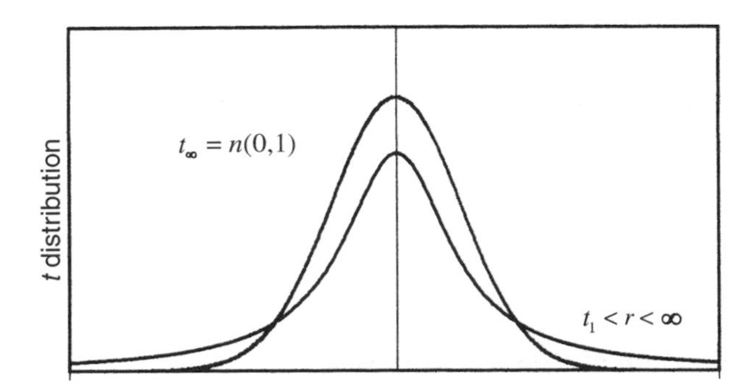

Figure A.4 The probability density function of the t distribution.

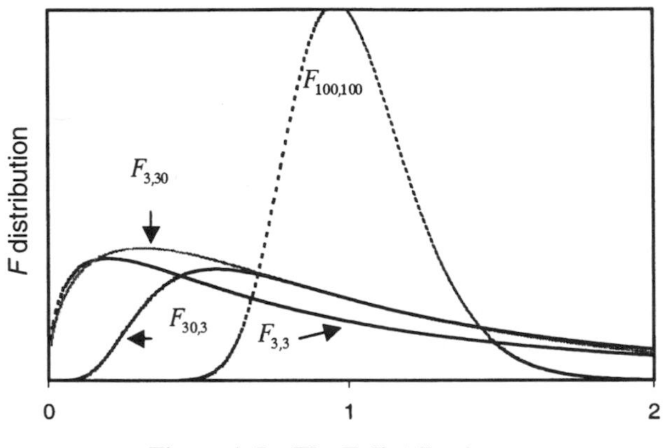

Figure A.5 The F distribution.

This is the F distribution with r_1 and r_2 degrees of freedom. The mean, or the expected value, is

$$E(F_{r_1,r_2}) = \frac{r_2}{r_2 - 2} \tag{A.136}$$

Care should always be taken to identify the degrees of freedom properly since the density function is not symmetric in these variables. See Figure A.5. The following relationship holds:

$$F_{r_1,r_2,\alpha} = \frac{1}{F_{r_2,r_1,1-\alpha}} \tag{A.137}$$

The F distribution is related to the chi-square and the t distributions as follows:

$$\frac{\chi_r^2}{r} \sim F_{r,\infty} \tag{A.138}$$

$$t_r^2 \sim F_{1,r} \tag{A.139}$$

If $\tilde{u} \sim \chi^2_{r_1,\lambda}$ has a noncentral chi-square distribution with r_1 degrees of freedom and a noncentrality parameter λ, then the function F in (A.134) has a noncentral F distribution with r_1 and r_2 degrees of freedom and noncentrality parameter λ. The mean for the noncentral distribution is

$$E(F_{r_1,r_2,\lambda}) = \frac{r_2}{r_2 - 2}\left(1 + \frac{\lambda}{r_1}\right) \tag{A.140}$$

A.5.2 Distribution of Sums of Variables

The following functions of random variables are required in the derivation of distributions of key random variables in least-squares estimation.

Assume that $(\tilde{x}_1, \tilde{x}_2, \cdots, \tilde{x}_n)$ are n stochastically independent variables, each having a normal distribution, with different means μ_i and variances σ^2. Then the linear function

$$\tilde{y} = k_1\tilde{x}_1 + k_2\tilde{x}_2 + \cdots + k_n\tilde{x}_n \tag{A.141}$$

is distributed as

$$\tilde{y} \sim n\left(\sum_i^n k_i\mu_i, \sum_i^n k_i^2\sigma_i^2\right) \tag{A.142}$$

If the random variable $\tilde{w}$ has a standardized normal distribution, i.e., $\tilde{w} \sim (0, 1)$, then the square of the standardized normal distribution

$$\tilde{v} = \tilde{w}^2 \sim \chi_1^2 \tag{A.143}$$

has a chi-square distribution with one degree of freedom.

Assume that $(\tilde{x}_1, \tilde{x}_2, \cdots, \tilde{x}_n)$ are n stochastically independent random variables, each having a chi-square distribution. The degrees of freedom r_i can differ. Then the random variable

$$\tilde{y} = \tilde{x}_1 + \tilde{x}_2 + \cdots + \tilde{x}_n \tag{A.144}$$

is distributed

$$\tilde{y} \sim \chi_{\Sigma r_i}^2 \tag{A.145}$$

The degree of freedom equals the sum of the individual degrees of freedom.

Assume $(\tilde{x}_1, \tilde{x}_2, \cdots, \tilde{x}_n)$ are n stochastically independent random variables, each having a normal distribution. The means are nonzero. Then

$$\tilde{y} \sim \sum \tilde{w}^2 = \sum^n \left(\frac{\tilde{x}_i - \mu_i}{\sigma_i}\right)^2 \sim \chi_n^2 \tag{A.146}$$

Assume that $(\tilde{x}_1, \tilde{x}_2, \cdots, \tilde{x}_n)$ are n stochastically independent normal random variables with different means μ_i and variances σ_i^2. Then the sum of squares

$$\tilde{y} = \sum \tilde{x}_i^2 \sim \chi_{n,\lambda}^2 \tag{A.147}$$

has a noncentral chi-square distribution. The degree of freedom is n and the noncentrality parameter is

$$\lambda = \sum \frac{\mu_i^2}{\sigma_i^2} \tag{A.148}$$

A.5.3 Hypothesis Tests

A hypothesis is a statement about the parameters of a distribution. A test of a hypothesis is a rule that, based on the sample values, leads to a decision to accept or reject the null hypothesis. A test statistic is computed from the sample values (the observations) and from the specifications of the null hypothesis. If the test statistic falls within a critical region, the null hypothesis is rejected. For example, $\tilde{\mathbf{v}}^T \mathbf{P}\tilde{\mathbf{v}}$ is a test statistic having a chi-square distribution. The computed test statistic is $\mathbf{v}^T\mathbf{P}\mathbf{v}$. The specification of the zero hypothesis could be that the a posteriori variance of unit weight has a certain numerical value that, in turn, specifies the variance-covariance matrix of the observations (which is a parameter of the multivariate normal distribution of the observations).

Because the sample statistic is computed from sample values (observations), the computed value may fall inside the critical region even though the null hypothesis H_0 is true. There is a probability α that this can happen. One speaks of a type-I error if the hypothesis H_0 is rejected although it is true; the probability of a type-I error is α, which, incidentally, is also the significance level of the test. However, there is a probability that the sample statistics fall in the critical region when H_0 is false (and hence H_1 is true). That probability is denoted by $1 - \beta$ in Figure A.6. If the sample statistic does not fall in the critical region, but the alternative hypothesis H_1 is true, one would mistakenly accept H_0 and commit a type-II error. The probability of committing a type-II error is β.

Figure A.6 displays the probability density functions of the test statistics under the specifications of the null hypothesis H_0 and the alternative hypothesis H_1. The figure also shows the critical region for which the null hypothesis is rejected, and

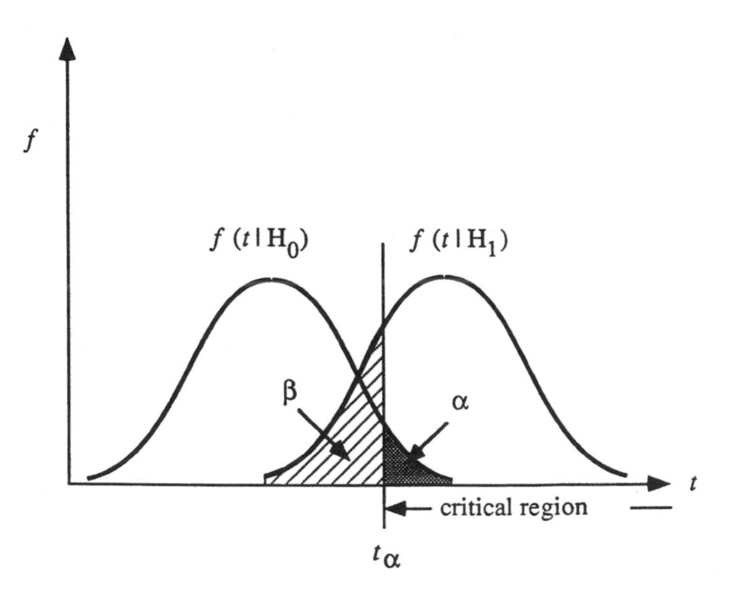

Figure A.6 Example of probability distributions of test statistics and critical region.

the alternative hypothesis is accepted if the computed sample statistics t falls in that region. Thus, reject H_0 if

$$t > t_\alpha \tag{A.149}$$

The shape and location of the density function of the test statistics under the alternative hypothesis depend on the specifications of the alternative hypothesis. Thus, the probability of a type-II error, β, depends on the specifications of H_1. A desirable approach in statistical testing would be to minimize the probability of both types of errors. However, this is not practical, because all distributions of the alternative hypotheses, which, in general, are of the noncentral type, would have to be computed. Figure A.6 shows that the probability β increases as α decreases. A common procedure is to fix the probability of a type-I error to, say, $\alpha = 0.05$, and not compute β.

The rule (A.149) is a one-tail test in the upper end of the distribution. Depending on the situation, it might be desirable to employ a two-tail test. In that case the null hypothesis is rejected if

$$|\, t \,| > t_{\alpha/2} \tag{A.150}$$

and the distribution H_0 is symmetric. It is rejected if

$$t > t_{\alpha/2} \tag{A.151}$$

$$t < t_{1-\alpha/2} \tag{A.152}$$

and the distribution is not symmetric. The critical regions are at both tails of the distribution, with each tail covering a probability area of $\alpha/2$.

However, much effort has gone into research as to how the magnitude of β can be controlled (Baarda, 1968). After all, committing a type-II error implies accepting the null hypothesis even though the alternative hypothesis is true. For example, it could mean that it has been concluded that no deformation took place even though actual deformations occurred. Such an error could be costly in many respects. In Section 4.10.2 some consideration is given to the type-II error in regards to blunder detection and internal and external reliability, again based on Baarda's work. Section 7.8.3 considers type-II errors in regard to ambiguity fixing.

The goodness-of-fit test is a simple and useful example of statistical testing. Assume we wish to test a series of observations to determine whether they come from a certain population with a specified distribution. We subdivide the observation series into n bins. Let n_i denote the number of observations in bin i. The subdivision should be such that $n_i \geq 5$. Compute for each bin the expected number d_i of observations based on the hypothetical distribution. It can be shown that

$$\chi^2 = \sum_{i=1}^{n} \frac{(n_i - d_i)^2}{d_i} \tag{A.153}$$

is distributed approximately as χ^2_{n-1}. The zero hypothesis states that the sample is from the specified distribution. Reject H_0 at a 100α% significance level if

$$\chi^2 > \chi^2_{n-1,\alpha} \tag{A.154}$$

This test could be used to verify that normalized residuals belong to $n(0, 1)$.

APPENDIX B

THE ELLIPSOID

The ellipsoid of rotation serves as the geometric structure for mathematical formulations and computations. For example, the observables of the 3D geodetic model refer to the ellipsoidal normal and the geodetic horizon, whereas the observables of the ellipsoidal model are the angle between geodesics and the length of the geodesic on the ellipsoidal surface. In the case of the conformal mapping model, the ellipsoidal surface is mapped conformally. Details of these mathematical models are given in Chapter 9. Because the ellipsoid is so important as a computational reference and as a means to express position coordinates, the ellipsoid and the related geometry are summarized here. Because only ellipsoids of rotation have been adopted in practical geodesy and surveying and triaxial ellipsoids have been limited to theoretical studies, we will use the term *ellipsoid* for reasons of brevity to mean ellipsoid of rotation. Such an ellipsoid is generated when rotating an ellipse around the semiminor axis.

The expressions for computing on the ellipsoidal surface and on the conformal mapping plane are deeply rooted in differential geometry. Working expressions typically utilize series expansions that are simplified by truncating insignificant terms (having specific applications in terms of accuracy and area in mind). The algebraic work necessary to arrive at working expressions is considerable and not at all obvious to the novice. Prior to the introduction of electronic computers, there was a strong interest in producing computationally efficient expressions. The expressions have been extensively documented in the geodetic literature, although some of this literature is now old and is even out of print. Many of the derivations are documented in Leick (2002).

The mathematical literature offers plenty of excellent texts on differential geometry. Differential geometry of course deals in general terms with surfaces. While this section focuses on the ellipsoid, the universality of expressions will occasionally be emphasized. The reader is advised to consult the mathematical literature if a more precise and comprehensive exposition of differential geometry is desired than is offered in the "tailored" approach of this appendix.

B.1 GEODETIC LATITUDE, LONGITUDE, AND HEIGHT

A popular way to give positions in 3D space is by means of geodetic latitude, geodetic longitude, and geodetic height. To be sure, these quantities are often referred to as ellipsoidal latitude, ellipsoidal longitude, and ellipsoidal height. Regardless of what one calls them, it is important to realize that they refer to an ellipsoid and not to a sphere, and thus are conceptually and numerically different from spherical latitude, longitude, and height. Another popular way of giving positions in space is Cartesian coordinates. It follows that the geodetic and Cartesian coordinates triplets are mathematically related.

Figure B.1 shows an ellipse with semimajor axis a and semiminor axis b. In the (ξ, η) coordinate system, the equation of the ellipse has the familiar form

$$\frac{\xi^2}{a^2} + \frac{\eta^2}{b^2} = 1 \tag{B.1}$$

Two parameters are sufficient to define the ellipse. Often the semimajor axis a and the flattening f, or a and the eccentricity e, are used to define an ellipse. These auxiliary quantities are related by

$$f = \frac{a - b}{a} \tag{B.2}$$

$$e^2 = 2f - f^2 \tag{B.3}$$

The figure also shows the tangent to the ellipse at some point A. The normal to this tangent intersects the semiminor axis at point C. The symbol N is used to denote the segment $\overline{AC}$. The angle φ is the geodetic latitude and equals the angle between the normal and the semimajor axis. It follows readily that

$$\xi = N \cos \varphi \tag{B.4}$$

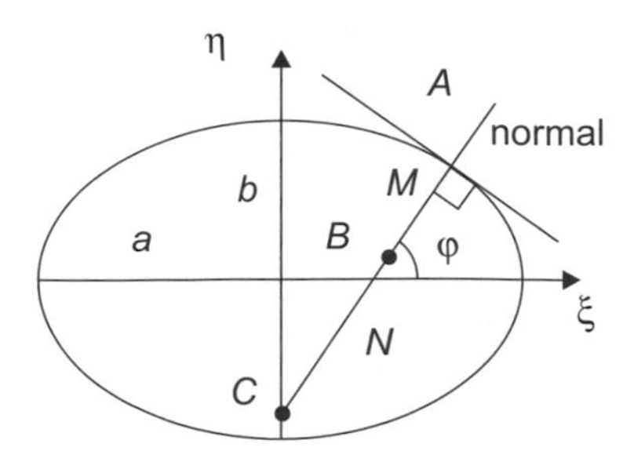

Figure B.1 Elements of the ellipse.

Upon stepping deeper into the geometry of the ellipse, it is found that

$$\eta = N\left(1 - e^2\right)\sin\varphi \tag{B.5}$$

and

$$N = \frac{a}{\left(1 - e^2 \sin^2 \varphi\right)^{1/2}} \tag{B.6}$$

Additional interpretation of the N will be given below. The symbol M in Figure B.1 denotes the segment $\overline{AB}$ taken along the normal, i.e., the perpendicular of the tangent. M equals the radius of curvature of the ellipse at point A. Stepping again deeper into the geometry of the ellipse, we find that the radius of curvature can be expressed as

$$M = \frac{a\left(1 - e^2\right)}{\left(1 - e^2 \sin^2 \varphi\right)^{3/2}} \tag{B.7}$$

Note that the variable in Expressions (B.4) to (B.7) is the geodetic latitude.

Rotating the ellipse of Figure B.1 around the η axis generates the ellipsoid of rotation, or simply the ellipsoid. Figure B.2 shows such an ellipsoid and the associated Cartesian and geodetic coordinates. The Cartesian coordinate system $(\mathrm{x}) = (x, y, z)$ has its origin at the center of the ellipsoid, the z axis coincides with the semiminor axis, and the x and y axes are located in the equatorial plane of the ellipsoid. The directions of the x and z axes and the center of the ellipsoid are typically fixed by conventions. The ellipsoidal normal through a space point P, i.e., a point on the physical earth surface, intersects with the z axis because of the rotational symmetry of the ellipsoid; however, it does not pass through the origin of the Cartesian coordinate system because of the flattening of the ellipsoid. The length of the ellipsoidal normal from P to the ellipsoid is the geodetic height h. The angle between the ellipsoidal

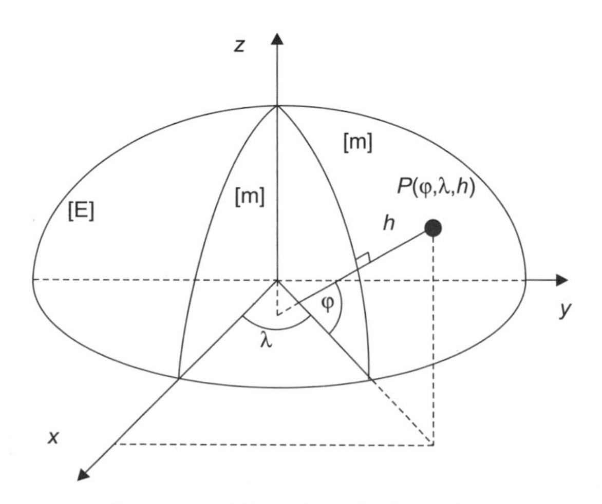

Figure B.2 The ellipsoid of rotation.

normal and the equatorial plane is the geodetic latitude φ in accordance with the definition given earlier.

According to the construction of the ellipsoid, any intersection of the ellipsoid $[E]$ with a plane that contains the z axis generates an ellipse that is called the *geodetic* meridian $[m]$. The geodetic longitude λ is then defined as the angle between two geodetic meridian planes and counted positive eastward starting at the x axis. Therefore, the triplet of geodetic coordinates (φ, λ, h) completely describes the position of a point in space.

The plane at the earth's surface point P, which is perpendicular to the ellipsoidal normal, defines the local geodetic horizon. This is the primary horizontal reference plane in the 3D geodetic model. Notice the distinction between the local geodetic horizon and the local astronomic horizon introduced elsewhere in this book (the latter is perpendicular to the plumb line at P).

Constant geodetic latitude and longitude trace the familiar lattice of meridians and parallels on the surface $[E]$. In differential geometry, such lines are called curvilinear lines $[\varphi]$ and $[\lambda]$, and (φ, λ) are called curvilinear coordinates. It is to be understood that the term *curvilinear* refers to general surface and not just to the ellipsoid. Differential geometry also offers the key for interpreting N in Equation (B.6). A plane that contains the surface normal, in this case the ellipsoidal surface normal, is called a normal plane. The intersection of a normal plane with the surface (the ellipsoid) is a normal section.

With this terminology, the geodetic meridians $[\lambda]$ are simply normal sections generated by a normal plane that contains the z axis. Consider the special case of a normal plane at P that is rotated with respect to the plane of the meridian by 90°. This is called the prime vertical normal plane. It also intersects the ellipsoid along a normal section, denoted by $[pv]$. N is the radius of curvature of that normal section $[pv]$. In fact, the student of differential geometry will recognize Equation (B.4) as an application of the famous Meusnier theorem, which relates the radius of curvature of a general surface curve to the radius of curvature of the normal section when both curves have a common tangent. In this case, the general surface curve is the parallel $[\varphi]$.

Given the geometric interpretation of the radius of curvatures of the meridian and the prime vertical normal sections, the curious student probably suspects another important relationship. Consider that the radius of curvature R of a normal section in direction α is given by the Euler's equation as

$$\frac{1}{R} = \frac{\cos^2 \alpha}{M} + \frac{\sin^2 \alpha}{N} \tag{B.8}$$

The symbol α denotes the geodetic azimuth, i.e., the angle between two normal planes having the ellipsoidal normal at P in common. This is precisely the azimuth used in the 3D geodetic model. Equations (B.6), (B.7), and (B.8) imply $M \leq R \leq N$. Deeper study of differential geometry would reveal that the directions of the meridian and the prime vertical belong to the special group of directions that are perpendicular to each other and for which the curvatures (reciprocal of radius of curvatures) take on maximum and minimum values. These are the principal directions.

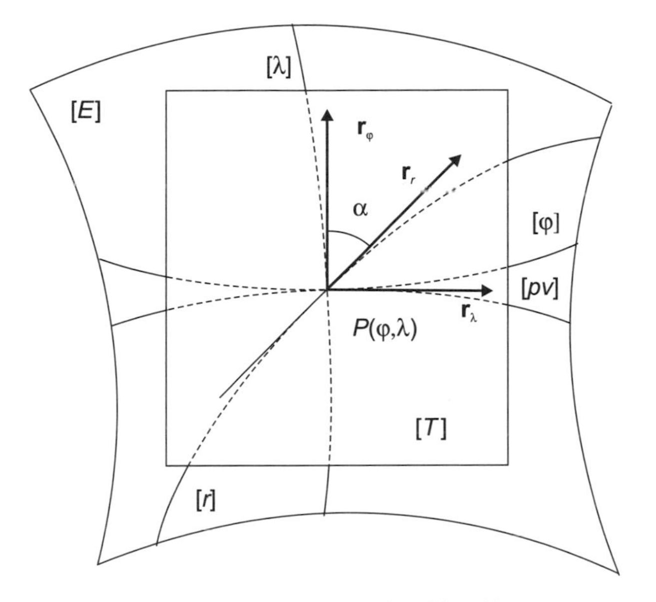

Figure B.3 Sections on the ellipsoid.

Figure B.3 shows various intersections. The tangent plane $[T]$ of the ellipsoidal surface $[E]$ at $P(\varphi, \lambda, h = 0)$ is spanned by the tangent vectors $\mathbf{r}_\varphi$ and $\mathbf{r}_\lambda$ of the meridian $[\lambda]$ and parallel $[\varphi]$. The non-normal section $[\varphi]$ and the normal section $[pv]$ have the tangent $\mathbf{r}_\lambda$ in common. The azimuth α of the general normal section $[r]$ is the angle between the respective normal planes or, equivalently, the angle in the tangent plane between $\mathbf{r}_\varphi$ and $\mathbf{r}_r$. The angle between $\mathbf{r}_\varphi$ and $\mathbf{r}_\lambda$ is 90° because they represent the principal directions.

The Cartesian coordinates $(\mathrm{x}) = (x, y, z)$ can be expressed as a function of the geodetic coordinates (φ, λ, h) using Equations (B.4) and (B.5), and the geometry shown in Figure B.2, as follows:

$$x = (N + h) \cos\varphi \cos\lambda \tag{B.9}$$

$$y = (N + h) \cos\varphi \sin\lambda \tag{B.10}$$

$$z = \left[N\left(1 - e^2\right) + h\right] \sin\varphi \tag{B.11}$$

The inverse solution, i.e., expressing the triplet (φ, λ, h) as a function of (x, y, z) involves a nonlinear mathematical relationship. The longitude follows straightforwardly from (B.9) and (B.10) as

$$\tan\lambda = \frac{y}{x} \tag{B.12}$$

One only needs to pay attention to the quadrant of the longitude λ. In geodesy the longitude is typically positive eastward starting from the x axis and counting from

0° to 360°, i.e., $0^\circ \le \lambda < 360^\circ$. Others give east $0^\circ \le \lambda(\text{E}) \le 180^\circ$ or west $0^\circ < \lambda(\text{W}) < 180^\circ$ longitudes counting from 0° to 180°, respectively, or give negative values in the region $-180^\circ < \lambda < 0^\circ$. The geodetic latitude follows from the nonlinear equation (B.11) using some iterative technique. For this purpose it is convenient to rewrite (B.11) as

$$\tan\varphi = \frac{z}{\sqrt{x^2+y^2}}\left(1 + \frac{e^2 N \sin\varphi}{z}\right) \tag{B.13}$$

and use

$$\varphi_{\text{initial}} = \tan^{-1}\left[\frac{z}{\left(1-e^2\right)\sqrt{x^2+y^2}}\right] \tag{B.14}$$

on the right-hand side of (B.13) to start the iteration. The iteration stops after successive solutions yield negligible changes in the geodetic latitude. After convergence, the geodetic height follows from

$$h = \frac{\sqrt{x^2+y^2}}{\cos\varphi} - N \tag{B.15}$$

as can be readily verified.

The differential relations between the Cartesian and geodetic coordinates are

$$\begin{bmatrix} dx \\ dy \\ dz \end{bmatrix} = \mathbf{J}(\varphi, \lambda, h)\begin{bmatrix} d\varphi \\ d\lambda \\ dh \end{bmatrix} \tag{B.16}$$

with transformation matrix $\mathbf{J}(\varphi, \lambda, h)$ being

$$\mathbf{J}(\varphi, \lambda, h) = \begin{bmatrix} -(M+h)\cos\lambda\sin\varphi & -(N+h)\cos\varphi\sin\lambda & \cos\varphi\cos\lambda \\ -(M+h)\sin\lambda\sin\varphi & (N+h)\cos\varphi\cos\lambda & \cos\varphi\sin\lambda \\ (M+h)\cos\varphi & 0 & \sin\varphi \end{bmatrix} \tag{B.17}$$

Obtaining the partial derivatives in such compact forms requires some algebraic work. For these and other compact forms to be developed later, it is helpful to take note of the following partial derivatives:

$$\frac{\partial(N\cos\varphi)}{\partial\varphi} = -M\sin\varphi \tag{B.18}$$

$$\frac{\partial(N\sin\varphi)}{\partial\varphi} = \frac{M\cos\varphi}{1-e^2} \tag{B.19}$$

$$\frac{\partial(M\sin\varphi)}{\partial\varphi} = \frac{M}{N\cos\varphi}\left[(2N-3M)\sin^2\varphi + N\right] \tag{B.20}$$

TABLE B.1 Dimensions of Important Ellipsoids

Datum	Ellipsoid	a [m]	$1/f$
NAD27	Clarke 1866	6378206.4	294.9786982
WGS72	WGS72	6378135.0	298.26
NAD83	GRS80	6378137.0	298.257222101
WGS84	WGS84	6378137.0	298.257223563

$$\frac{\partial(M\cos\varphi)}{\partial\varphi} = \frac{M}{N}(2N - 3M)\sin\varphi \tag{B.21}$$

It might be comforting to know that the formulations given above are all that is needed to deal with the 3D geodetic model. Curvature is the only element that has thus far been taken from the realm of differential geometry. The elements of the geodesic or even conformal mapping have not yet been needed. These facts account for the relative mathematical simplicity of the 3D geodetic model.

Table B.1 lists the defining values of a sample of ellipsoids that are in use today or have some historical relevancy. The size of the ellipsoid is usually identified with a name. One speaks of a *datum* if the size of the ellipsoid and its location with respect to the earth is defined. The semiaxes of a typical earth ellipsoid differ by about $a - b \approx 21$ km. If the ellipsoid is scaled to 1 m, this difference is just 3 mm.

B.2 COMPUTATIONS ON THE ELLIPSOIDAL SURFACE

The two-dimensional ellipsoidal and conformal mapping models require the geodesic line and the solution of geodesic triangles (triangles whose sides are geodesic lines) on the ellipsoidal surface. Because the respective expressions are based on differential geometry, this section offers a brief summary of the relevant material. Several expressions are given in general form and are valid to any smooth surface whose second derivatives exist and are continuous. While (φ, λ) continue to represent the geodetic latitude and longitude of the ellipsoid, they could easily be more generally interpreted as curvilinear coordinates on other surfaces.

B.2.1 Fundamental Coefficients

Equations (B.9) to (B.11) for the ellipsoid can be written in a compact and general form as

$$\mathbf{r}(\varphi, \lambda) = \begin{bmatrix} x(\varphi, \lambda) \\ y(\varphi, \lambda) \\ z(\varphi, \lambda) \end{bmatrix} \tag{B.22}$$

In fact, we can view (B.22) as the equation of general surface whose second derivatives exist and are continuous. The tangent vector to the λ curvilinear line is given by

$$\mathbf{r}_\varphi = \frac{\partial \mathbf{r}}{\partial \varphi} = \begin{bmatrix} \dfrac{\partial x(\varphi, \lambda)}{\partial \varphi} \\ \dfrac{\partial y(\varphi, \lambda)}{\partial \varphi} \\ \dfrac{\partial z(\varphi, \lambda)}{\partial \varphi} \end{bmatrix} \tag{B.23}$$

Similarly, the tangent vector to the φ curvilinear line is

$$\mathbf{r}_\lambda = \frac{\partial \mathbf{r}(\varphi, \lambda)}{\partial \lambda} \tag{B.24}$$

The surface expression (B.22) can formally be expanded in a Taylor series. Let the point of expansion be at $\mathbf{r}\,(\varphi, \lambda)$ and let the differential increments be denoted as $d\varphi$ and $d\lambda$. Limiting the expansion to second-order terms gives

$$\begin{aligned} \mathbf{r}(\varphi + d\varphi, \lambda + d\lambda) &= \mathbf{r}(\varphi, \lambda) + \mathbf{r}_\varphi d\varphi + \mathbf{r}_\lambda\, d\lambda \\ &\quad + \frac{1}{2}\left\{\mathbf{r}_{\varphi\varphi}\, d\varphi^2 + 2\mathbf{r}_{\varphi\lambda}\, d\varphi\, d\lambda + \mathbf{r}_{\lambda\lambda}\, d\lambda^2\right\} + \ldots \end{aligned} \tag{B.25}$$

It can be readily visualized that the first part of this expression,

$$\mathbf{t}(\varphi, \lambda) = \mathbf{r}(\varphi, \lambda) + \mathbf{r}_\varphi\, d\varphi + \mathbf{r}_\lambda\, d\lambda \tag{B.26}$$

represents the tangent plane $[T]$ which is located at $\mathbf{r}(\varphi, \lambda)$ and spanned by the vectors $\mathbf{r}_\varphi$ and $\mathbf{r}_\lambda$. The total differential

$$d\mathbf{r} = \mathbf{r}_\varphi\, d\varphi + \mathbf{r}_\lambda\, d\lambda \tag{B.27}$$

is a vector in the tangent plane and represents the linearized surface distance on $[E]$ from $P(\varphi, \lambda)$ to $P(\varphi + d\varphi, \lambda + d\lambda)$. See Figure B.4. The square of the length of the total differential is

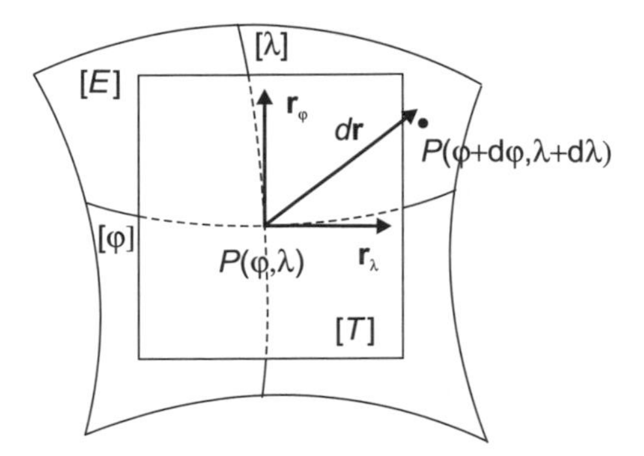

Figure B.4 The total differential.

$$
\begin{aligned}
ds^2 &= d\mathbf{r} \cdot d\mathbf{r} \\
&= \mathbf{r}_\varphi \cdot \mathbf{r}_\varphi \, d\varphi^2 + 2\mathbf{r}_\varphi \cdot \mathbf{r}_\lambda \, d\varphi \, d\lambda + \mathbf{r}_\lambda \cdot \mathbf{r}_\lambda \, d\lambda^2 \\
&= E \, d\varphi^2 + 2F \, d\varphi \, d\lambda + G \, d\lambda^2
\end{aligned} \tag{B.28}
$$

This is the first fundamental form. The quantities E, F, G are called, since Gauss, the first fundamental coefficients. Properties of the surface that can be expressed as a function of the first fundamental coefficients are called intrinsic properties. The totality of intrinsic properties of the surface is called the intrinsic geometry of the surface. Using vector identities, one can verify that

$$
\begin{aligned}
EG - F^2 &= (\mathbf{r}_\varphi \cdot \mathbf{r}_\varphi)(\mathbf{r}_\lambda \cdot \mathbf{r}_\lambda) - (\mathbf{r}_\varphi \cdot \mathbf{r}_\lambda)^2 = (\mathbf{r}_\varphi \times \mathbf{r}_\lambda) \cdot (\mathbf{r}_\varphi \times \mathbf{r}_\lambda) \\
&= \|\mathbf{r}_\varphi \times \mathbf{r}_\lambda\| > 0
\end{aligned} \tag{B.29}
$$

and that $E > 0$ and $G > 0$. For orthogonal curvilinear lines, we have $F = 0$ because $\mathbf{r}_\varphi \cdot \mathbf{r}_\lambda = 0$. Evaluating the fundamental coefficients for the ellipsoidal surface $[E]$ gives

$$E = M^2 \tag{B.30}$$

$$F = 0 \tag{B.31}$$

$$G = N^2 \cos^2 \varphi \tag{B.32}$$

$$ds^2 = M^2 \, d\varphi^2 + N^2 \cos^2 \varphi \, d\lambda^2 \tag{B.33}$$

The last term in (B.25),

$$\mathbf{p} = \frac{1}{2} \left\{ \mathbf{r}_{\varphi\varphi} \, d\varphi^2 + 2\mathbf{r}_{\varphi\lambda} \, d\varphi \, d\lambda + \mathbf{r}_{\lambda\lambda} \, d\lambda^2 \right\} \tag{B.34}$$

represents the deviation of a second-order surface approximation from the tangent plane. The vectors $\mathbf{r}_{\varphi\varphi}$ and $\mathbf{r}_{\lambda\lambda}$ contain the respective second partial derivative with respect to λ and φ, and $\mathbf{r}_{\varphi\lambda}$ contains the mixed derivatives. Introducing the surface normal **e** as

$$\mathbf{e} = \frac{\mathbf{r}_\varphi \times \mathbf{r}_\lambda}{\|\mathbf{r}_\varphi \times \mathbf{r}_\lambda\|} = \frac{\mathbf{r}_\varphi \times \mathbf{r}_\lambda}{\sqrt{EG - F^2}} \tag{B.35}$$

then the orthogonal distance of the second-order approximation to the tangent plane is

$$
\begin{aligned}
d &= -\mathbf{e} \cdot \mathbf{p} \\
&= \frac{1}{2} \left\{ -\mathbf{e} \cdot \mathbf{r}_{\varphi\varphi} \, d\varphi^2 - 2\mathbf{e} \cdot \mathbf{r}_{\varphi\lambda} \, d\varphi \, d\lambda - \mathbf{e} \cdot \mathbf{r}_{\lambda\lambda} \, d\lambda^2 \right\} \\
&= \frac{1}{2} \left\{ D \, d\varphi^2 + 2D' \, d\varphi \, d\lambda + D'' \, d\lambda^2 \right\}
\end{aligned} \tag{B.36}
$$

Expression (B.36) is the second fundamental form and the elements (D, D', D'') are called, since Gauss, the second fundamental coefficients. For the ellipsoid these coefficients have the simple form

$$D = N\cos^2\varphi \tag{B.37}$$

$$D' = 0 \tag{B.38}$$

$$D'' = M \tag{B.39}$$

The partial derivatives (B.18) to (B.21) are very helpful in verifying this simple form.

B.2.2 Gauss Curvature

At every point of a smooth surface there are two perpendicular directions along which the curvature attains a maximum and a minimum value. These are the principal directions. Denoting the respective principal radius of curvatures by R_1 and R_2, a deeper study of differential geometry reveals

$$K \equiv \frac{1}{R_1 R_2} = \frac{DD'' - D'^2}{EG - F^2} = \frac{1}{MN} \tag{B.40}$$

where K is called Gauss curvature. The latter part of (B.40) expresses the value of K for the ellipsoid. In general, if the curvilinear lines also happen to coincide with the principal directions, then $D' = 0$. It can be shown that the numerator $DD'' - D'^2$ can be expressed as a function of the first fundamental coefficients and their partial derivatives.

Since the denominator in (B.40) is always positive, the numerator determines the sign of K. A point is called *elliptic* if $K > 0$. In the neighborhood of an elliptic point, the surface lies on one side of the tangent plane. A point is called *hyperbolic* if $K < 0$. In the neighborhood of a hyperbolic point, the surface lies on both sides of the tangent plane. A point is *parabolic* if $K = 0$, in which case the surface may lie on both sides of the tangent plane.

For $K = 0$ one of the values R_1 and R_2 must be infinite as follows from (B.40). If this occurs at every point of the surface one family of the principal directions must be straight lines. Examples are cylinders or cones. Such surfaces are called developable surfaces and can be reshaped into a plane without stretching and tearing.

B.2.3 Elliptic Arc

If s denotes the length of the arc of the ellipse from the equator, or simply the ecliptic arc, then

$$s = \int_0^\varphi \sqrt{E}\, d\varphi = \int_0^\varphi M\, d\varphi \tag{B.41}$$

There is no closed expression for the integral in (B.41). The following series expansion (Snyder, 1979) is frequently used

$$s = a\left[\left(1 - \frac{e^2}{4} - \frac{3e^4}{64} - \frac{5e^6}{256}\right)\varphi - \left(\frac{3e^2}{8} + \frac{3e^4}{32} + \frac{45e^6}{1024}\right)\sin 2\varphi + \left(\frac{15e^4}{256} + \frac{45e^6}{1024}\right)\sin 4\varphi\right] \tag{B.42}$$

The inverse solution, i.e., given the elliptic arc with respect to the equator and computing the geodetic latitude, is available iteratively starting with the initial value

$$\varphi_{\text{initial}} = \frac{s}{a} \tag{B.43}$$

B.2.4 Angle

An angle on a surface is defined as the angle between two tangents. The angle, therefore, is a measure in the tangent plane. Figure B.5 shows two curves, f_1 and f_2, on the surface that could be implicitly defined as $f_1(\varphi, \lambda) = 0$ and $f_2(\varphi, \lambda) = 0$. The differentials $(d\varphi_1, d\lambda_1)$ and $(d\varphi_2, d\lambda_2)$, which follow from differentiating these two functions, determine the tangent vectors as

$$d\mathbf{r}_1 = \mathbf{r}_\varphi\, d\varphi_1 + \mathbf{r}_\lambda\, d\lambda_1 \tag{B.44}$$

$$d\mathbf{r}_2 = \mathbf{r}_\varphi\, d\varphi_2 + \mathbf{r}_\lambda\, d\lambda_2 \tag{B.45}$$

Thus the expression for the angle becomes

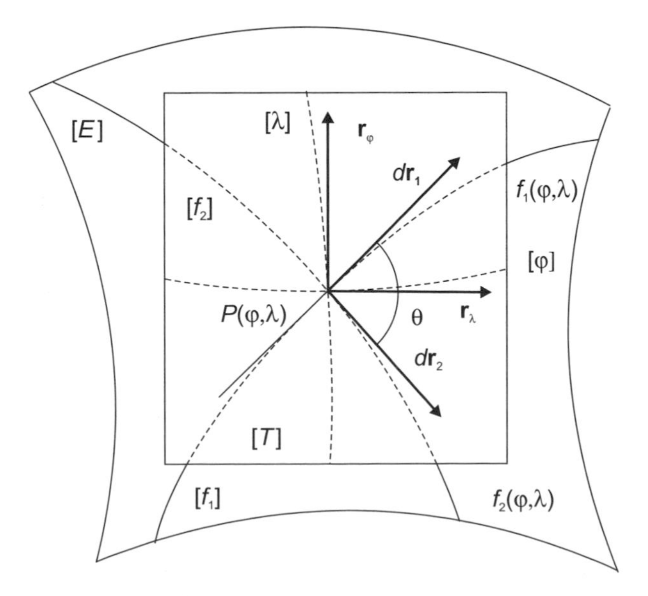

Figure B.5 Definition of surface angle.

$$\begin{aligned}\cos\theta &= \frac{d\mathbf{r}_1 \cdot d\mathbf{r}_2}{\|d\mathbf{r}_1\| \, \|d\mathbf{r}_2\|} \\ &= \frac{E\, d\varphi_1\, d\varphi_2 + F(d\varphi_1\, d\lambda_2 + d\varphi_2\, d\lambda_1) + G\, d\lambda_1\, d\lambda_2}{\sqrt{E\, d\varphi_1^2 + 2F\, d\varphi_1\, d\lambda_1 + G\, d\lambda_1^2}\sqrt{E\, d\varphi_2^2 + 2F\, d\varphi_2\, d\lambda_2 + G\, d\lambda_2^2}}\end{aligned} \tag{B.46}$$

Equation (B.46) is useful in verifying conformality in mapping.

B.2.5 Isometric Latitude

The first fundamental form (B.33) relates a differential change of curvilinear coordinates to the corresponding surface distance within first-order approximation. One can readily visualize that on the equator a respective change in φ or λ by one arc second traces about the same distance. This is not the case close to the pole because of the convergence of the meridians. Consider a new curvilinear parameter q, which is defined by the differential relation

$$dq \equiv \frac{M}{N\cos\varphi}\, d\varphi \tag{B.47}$$

Substituting (B.47) in first fundamental form (B.33) gives

$$ds^2 = N^2\cos^2\varphi\left(dq^2 + d\lambda^2\right) \tag{B.48}$$

Equation (B.48) clearly shows that the same changes in dq and $d\lambda$ cause the same change ds at a given point. Integrating (B.47) gives

$$q = \ln\left[\tan\left(45^\circ + \frac{\varphi}{2}\right)\left(\frac{1 - e\sin\varphi}{1 + e\sin\varphi}\right)^{e/2}\right] \tag{B.49}$$

The new parameter q is called the isometric latitude. It is a function of the geodetic latitude and reaches infinity at the pole. See Figure B.6. Because q is constant when φ is constant, the lines $q =$ constant are parallels on the ellipsoid. Equal incremented q parallels are spaced increasingly closer as one approaches the pole. The pair q and λ are called isometric curvilinear coordinates which trace respectively a lattice of isometric *curvilinear* lines $[q]$ and $[\lambda]$ on the ellipsoid.

The inverse solution, i.e., given the isometric latitude q and computing the geodetic latitude φ, is solved though iterations. Equation (B.49) can be written as

$$\tan\left(45^\circ + \frac{\varphi}{2}\right) = \varepsilon^q\left(\frac{1 + e\sin\varphi}{1 - e\sin\varphi}\right)^{e/2} \tag{B.50}$$

The symbol ε denotes the base of the natural system of logarithms ($\varepsilon = 2.71828\cdots$). It must not be confused with the eccentricity of the ellipsoid, which is assigned the

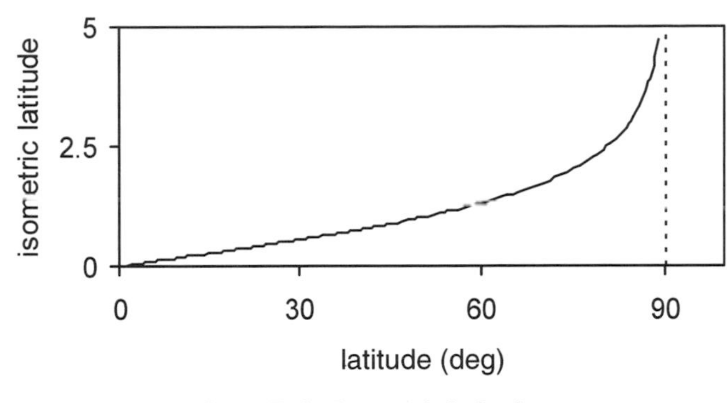

Figure B.6 Isometric latitude.

symbol e in this book. The iteration begins by taking $e = 0$ on the right side of (B.50), giving

$$\varphi_{\text{initial}} = 2\tan^{-1}\left(\varepsilon^{q}\right) - \frac{\pi}{2} \tag{B.51}$$

B.2.6 Differential Equation of the Geodesic

Probably the best-known property of the geodesic line (or simply the geodesic) is that it is the shortest surface line between two points on the surface. This property determines the differential equations of the geodesic. Differential geometry offers other equivalent definitions of the geodesic. Consider Figure B.7, which shows a general surface $[S]$, the tangent plane $[T]$, and surface normal $\mathbf{e}$ at a point $P(\varphi, \lambda)$. Let there be a curve $[g]$ on $[S]$ that passes through $P(\varphi, \lambda)$. The tangent on this space curve is denoted by $\mathbf{t}$. This tangent is located in the tangent plane spanned by $\mathbf{r}_\varphi$ and $\mathbf{r}_\lambda$. Next project curve $[g]$ orthogonally on the tangent plane in the differential neighborhood of $P(\varphi, \lambda)$. This generates a curve $[c]$ that is located in the tangent plane and has the tangent $\mathbf{t}$ in common with $[g]$. Like any plane curve, the curve $[c]$ has a curvature at the point $P(\varphi, \lambda)$, which is denoted here by κ_g. This is the geodesic curvature. It is related to the geodesic radius of curvature R_g by

$$\kappa_g = \frac{1}{R_g} \tag{B.52}$$

It can be shown that the geodesic curvature κ_g is a function of the first fundamental coefficients and their derivatives.

The situation described above and depicted in Figure B.7 for $P(\varphi, \lambda)$ can be conceptually repeated for every point of the curve $[g]$, i.e., for every point one can visualize the tangent plane and the orthogonal projection of $[g]$ in the differential neighborhood of the point of tangency. The curve $[g]$ is a geodesic if the geodesic curvature is zero at all these points, or, equivalently, the radius of the geodesic curvature is infinite.

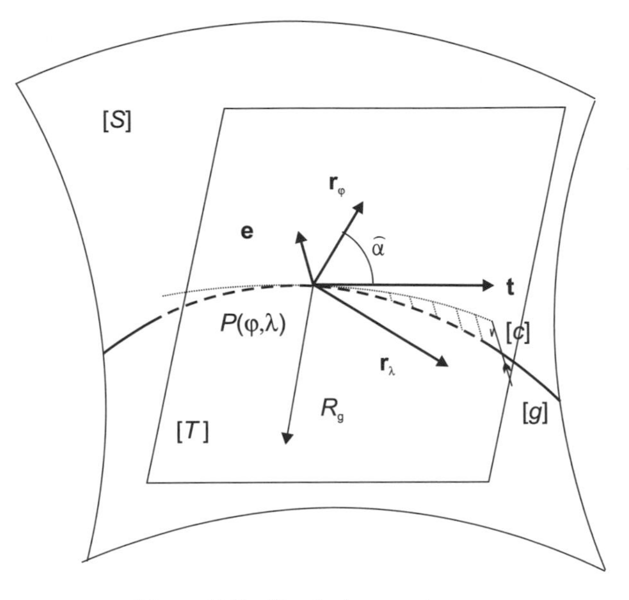

Figure B.7 Geodesic curvature.

Because the geodesic radius of curvature is infinite, the projection of the geodesic on the tangent plane is a straight line in the differential neighborhood of $P(\varphi, \lambda)$. This geometric definition of the geodesic is also sufficient to determine the differential equations of $[g]$.

Differential geometry offers yet another definition of the geodesic that is frequently stated. Assume that expressions for the three Cartesian coordinates of $[g]$ are given as a function of some free parameter s. Differentiating each component once with respect to s gives the tangent vector **t**; differentiating twice gives another vector called the principal normal of the curve $[g]$. It can be shown that the tangent vector and the principal normal of the curve are perpendicular. Next, the curves $[g]$ and $[c]$ can be viewed as curves on a general cylinder that is perpendicular to the tangent plane. Viewed like that, the curves $[c]$ and $[g]$ represent a normal section and a general section on the cylinder that have the tangent **t** in common. The respective radii of curvature are related by Meusnier's theorem. In this view the radius of curvature of the normal section $[c]$ is R_g. If R_g is to go to infinity, then Meusnier's theorem implies that the principal normal of $[g]$ and the surface normal **e** coincide.

The definition of the geodesic does not restrict the geodesics to plane curves. In fact, the geodesic will have, in general, curvature and torsion. However, the definition lends itself to some interpretation of "straightness." Consider a virtual surveyor who operates a virtual theodolite on the ellipsoidal surface. A first step in operating an actual theodolite is to level it, i.e., to align the vertical axis with the plumb line. In this example, the virtual surveyor will align the vertical axis with the surface normal. He is then tasked to stake out a straight line using differentially short sightings. He would begin setting up the instrument at the initial (first) point and stake out the second point using the azimuth $\widehat{\alpha}$. Next he would set up at the second point, backsight to the first point, and turn an angle of 180° to stake out the third point, and so on. In the mind of the virtual surveyor, he is staking out a straight line whereas he actually stakes out a geodesic using differentially short sightings.

Let $\widehat{\alpha}$ denote the azimuth of the geodesic, i.e., the angle between the tangent on the λ curvilinear line and the tangent on the geodesic as shown in Figure B.7, and let $\widehat{s}$ denote the length of the geodesic on [S]. Using the definition of the geodesic given above, the differential equations for the geodesic on a general surface can be developed as

$$\frac{d\varphi}{d\widehat{s}} = \frac{\sin\widehat{\alpha}}{\sqrt{E}} \tag{B.53}$$

$$\frac{d\lambda}{d\widehat{s}} = \frac{\cos\widehat{\alpha}}{\sqrt{G}} \tag{B.54}$$

$$\frac{d\widehat{\alpha}}{d\widehat{s}} = \frac{1}{\sqrt{EG}}\left(\frac{\partial\sqrt{G}}{\partial\varphi}\cos\widehat{\alpha} - \frac{\partial\sqrt{E}}{\partial\lambda}\sin\widehat{\alpha}\right) \tag{B.55}$$

In case of the ellipsoid [E] the respective equations are,

$$\frac{d\varphi}{d\widehat{s}} = \frac{\cos\widehat{\alpha}}{M} \tag{B.56}$$

$$\frac{d\lambda}{d\widehat{s}} = \frac{\sin\widehat{\alpha}}{N\cos\varphi} \tag{B.57}$$

$$\frac{d\widehat{\alpha}}{d\widehat{s}} = \frac{1}{N}\tan\varphi\sin\widehat{\alpha} \tag{B.58}$$

Figure B.8 shows a geodesic triangle whose corners consist of the pole $P(\varphi = 90°)$ and the points $P_1(\varphi_1, \lambda_1)$ and $P_2(\varphi_2, \lambda_2)$. The sides of this triangle are the meridians, which can be readily identified as geodesic lines, and the geodesic line

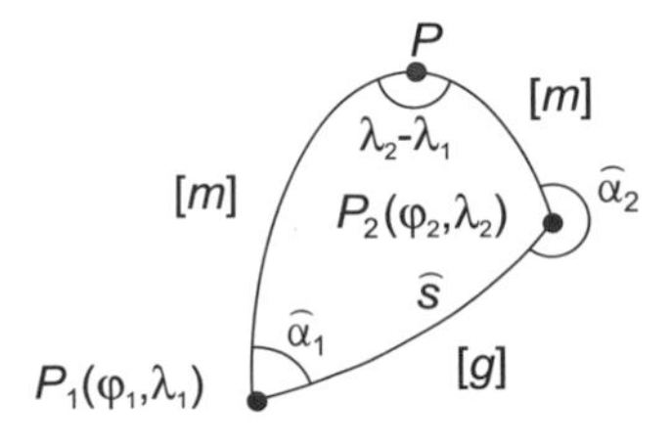

Figure B.8 Geodesic triangle.

from P_1 to P_2. At the heart of the ellipsoidal computations are the so-called direct and inverse problems. In the case of the direct problem, the geodetic latitude and longitude at one station, say, $P_1(\varphi_1, \lambda_1)$, and the geodesic azimuth and distance $(\widehat{\alpha}_1, \widehat{s})$ to another point, are known; the geodetic latitude φ_2, longitude λ_2, and back azimuth $\widehat{\alpha}_2$ are required. Formerly, the direct solution is written as

$$\begin{bmatrix} \varphi_2 \\ \lambda_2 \\ \widehat{\alpha}_2 \end{bmatrix} = \begin{bmatrix} d_1(\varphi_1, \lambda_1, \widehat{\alpha}_1, \widehat{s}) \\ d_2(\varphi_1, \lambda_1, \widehat{\alpha}_1, \widehat{s}) \\ d_3(\varphi_1, \lambda_1, \widehat{\alpha}_1, \widehat{s}) \end{bmatrix} \tag{B.59}$$

For the inverse problem, the geodetic latitudes and longitudes of $P_1(\varphi_1, \lambda_1)$ and $P_2(\varphi_2, \lambda_2)$ are given, and the forward and back azimuths and the length of the geodesic are required, i.e.,

$$\begin{bmatrix} \widehat{s} \\ \widehat{\alpha}_1 \\ \widehat{\alpha}_2 \end{bmatrix} = \begin{bmatrix} i_1(\varphi_1, \lambda_1, \varphi_2, \lambda_2) \\ i_2(\varphi_1, \lambda_1, \varphi_2, \lambda_2) \\ i_3(\varphi_1, \lambda_1, \varphi_2, \lambda_2) \end{bmatrix} \tag{B.60}$$

Most solutions of (B.56) to (B.58) rely on extensive series expansions with intermittent truncation of small terms. Various innovative approaches have been implemented to keep the number of significant terms small and yet achieve accurate solutions. Some solutions are valid only for short lines, while others apply to intermediary long lines, or even to lines that go all around the ellipsoid.

B.2.7 The Gauss Midlatitude Solution

Table B.2 summarizes the Gauss midlatitude solution (Grossman, 1976, pp. 101–106). The term *midlatitude* indicates that the point of expansion in the series developments is mean latitude and/or longitude between $P_1(\varphi_1, \lambda_1)$ and $P_2(\varphi_2, \lambda_2)$. The inverse solution begins by first evaluating the auxiliary expressions shown in the first row of the table, followed by the expressions in the second row. The first step for the direct solution requires the computation of approximate geodetic latitude and longitude for station $P_2(\varphi_2, \lambda_2)$ as indicated in row 3. These initial coordinates are used to evaluate the auxiliary quantities of the first row, which, in turn, are used to compute improved coordinates for station P_2 from the remaining expressions of the third row. The direct solution is iterated until convergence is achieved.

The linearized form of the inverse solution is important when computing (adjusting) networks on the ellipsoid. The truncated expressions of the partial derivatives in

$$d\widehat{s} = \frac{\partial i_1}{\partial \varphi_1} d\varphi_1 + \frac{\partial i_1}{\partial \lambda_1} d\lambda_1 + \frac{\partial i_1}{\partial \varphi_2} d\varphi_2 + \frac{\partial i_1}{\partial \lambda_2} d\lambda_2 \tag{B.61}$$

TABLE B.2 The Gauss Midlatitude Solution

Auxiliary Terms: $\Delta\varphi = \varphi_2 - \varphi_1$; $\Delta\lambda = \lambda_2 - \lambda_1$

$$\varphi = \frac{\varphi_1 + \varphi_2}{2}; \quad t = \tan\varphi; \quad \eta^2 = \frac{e^2}{1-e^2}\cos^2\varphi; \quad V^2 = 1 + \eta^2; \; f_1 = 1/M; \; f_2 = 1/N$$

$$f_3 = 1/24; \quad f_4 = \frac{1 + \eta^2 - 9\eta^2 t^2}{24V^4}, \quad f_5 - \frac{1 - 2\eta^2}{24}; \quad f_6 = \frac{\eta^2(1 - t^2)}{8V^4}; \quad f_7 = \frac{1 + \eta^2}{12};$$

$$f_8 = \frac{3 + 8\eta^2}{24V^4}$$

Inverse Solution: Given $(\varphi_1, \lambda_1, \varphi_2, \lambda_2)$, compute $(\widehat{s}, \widehat{\alpha}_1, \widehat{\alpha}_2)$

$$\widehat{s}\sin\widehat{\alpha} = \frac{1}{f_2}\Delta\lambda\cos\varphi\left[1 - f_3(\Delta\lambda\sin\varphi)^2 + f_4\,\Delta\varphi^2\right] \tag{a}$$

$$\widehat{s}\cos\widehat{\alpha} = \frac{1}{f_1}\Delta\varphi\cos\frac{\Delta\lambda}{2}\left[1 + f_5(\Delta\lambda\cos\varphi)^2 + f_6\,\Delta\varphi^2\right] \tag{b}$$

$$\Delta\widehat{\alpha} = \Delta\lambda\sin\varphi\left[1 + f_7(\Delta\lambda\cos\varphi)^2 + f_8\,\Delta\varphi^2\right] \tag{c}$$

$$\widehat{s} = \sqrt{(\widehat{s}\sin\widehat{\alpha})^2 + (\widehat{s}\cos\widehat{\alpha})^2} \tag{d}$$

$$\widehat{\alpha} = \tan^{-1}\left(\frac{\widehat{s}\sin\widehat{\alpha}}{\widehat{s}\cos\widehat{\alpha}}\right) \tag{e}$$

$$\widehat{\alpha}_1 = \widehat{\alpha} - \frac{\Delta\widehat{\alpha}}{2} \tag{f}$$

$$\widehat{\alpha}_2 = \widehat{\alpha} + \frac{\Delta\widehat{\alpha}}{2} \pm \pi \tag{g}$$

Direct Solution: Given $(\varphi_1, \lambda_1, \widehat{s}, \widehat{\alpha}_1)$, compute $(\varphi_2, \lambda_2, \widehat{\alpha}_2)$

$$\lambda_2 \approx \lambda_1 + \frac{\widehat{s}\sin\widehat{\alpha}_1}{N_1\cos\varphi_1} \tag{h}$$

$$\varphi_2 \approx \varphi_1 + \frac{\widehat{s}\cos\widehat{\alpha}_1}{M_1} \tag{i}$$

Iteration

$(\varphi_1, \lambda_1, \varphi_2, \lambda_2)$; reevaluate auxiliary terms

$$\Delta\widehat{\alpha} = \Delta\lambda\sin\varphi\left[1 + f_7(\Delta\lambda\cos\varphi)^2 + f_8\,\Delta\varphi^2\right] \tag{j}$$

$$\widehat{\alpha} = \widehat{\alpha}_1 + \frac{\Delta\widehat{\alpha}}{2} \tag{k}$$

$$\widehat{\alpha}_2 = \widehat{\alpha} + \frac{\Delta\widehat{\alpha}}{2} \pm \pi \tag{l}$$

$$\lambda_2 = \lambda_1 + f_2\frac{\widehat{s}\sin\widehat{\alpha}}{\cos\varphi}\left[1 + f_3(\Delta\lambda\sin\varphi)^2 - f_4\,\Delta\varphi^2\right] \tag{m}$$

$$\varphi_2 = \varphi_1 + f_1\frac{\widehat{s}\cos\widehat{\alpha}}{\cos(\Delta\lambda/2)}\left[1 - f_5(\Delta\lambda\cos\varphi)^2 - f_6\,\Delta\varphi^2\right] \tag{n}$$

TABLE B.3 Partial Derivatives of the Geodesic on the Ellipsoid

	$d\varphi_1$	$d\lambda_1$	$d\varphi_2$	$d\lambda_2$
$d\widehat{s}$	$-M_1 \cos\widehat{\alpha}_1$	$N_2 \cos\varphi_2 \sin\widehat{\alpha}_2$	$-M_2 \cos\widehat{\alpha}_2$	$-N_2 \cos\varphi_2 \sin\widehat{\alpha}_2$
$d\widehat{\alpha}_1$	$\dfrac{M_1 \sin\widehat{\alpha}_1}{\widehat{s}}$	$\dfrac{N_2 \cos\varphi_2 \cos\widehat{\alpha}_2}{\widehat{s}}$	$\dfrac{M_2 \sin\widehat{\alpha}_2}{\widehat{s}}$	$-\dfrac{N_2 \cos\varphi_2 \cos\widehat{\alpha}_2}{\widehat{s}}$

$$d\widehat{\alpha}_1 = \frac{\partial i_2}{\partial \varphi_1} d\varphi_1 + \frac{\partial i_2}{\partial \lambda_1} d\lambda_1 + \frac{\partial i_2}{\partial \varphi_2} d\varphi_2 + \frac{\partial i_2}{\partial \lambda_2} d\lambda_2 \tag{B.62}$$

are listed in Table B.3.

B.2.8 Angular Excess

The Gauss-Bonnet theorem of differential geometry provides an expression for the sum of interior angles $\widehat{\delta}_i$ of a general polygon (continuous curvature) on a surface

$$\sum_{i=1}^{\nu} \widehat{\delta}_i = (\nu - 2) \cdot \pi + \int_C \kappa_g \, dS + \iint_{\text{area}} K \, dA \tag{B.63}$$

For the sum of the interior angles of a geodesic triangle one readily obtains

$$\widehat{\delta}_1 + \widehat{\delta}_2 + \widehat{\delta}_3 = \pi + \varepsilon \tag{B.64}$$

with

$$\varepsilon = \iint_{\text{area}} K \, dA \tag{B.65}$$

because $\kappa_g = 0$. The sum of the angles of the geodesic triangle differs from π by the double integral of the Gauss curvature taken over the area of the triangle. The sum of the interior angles of a geodesic triangle is greater than, less than, or equal to π, depending on whether the Gauss curvature is positive, negative, or zero. There is angular excess for the geodesic triangle on the ellipsoid because $K > 0$. On the unit sphere, the excess in angular measurement is called the spherical excess. It equals the area of the triangle, i.e., $\varepsilon = A$, because $K = 1$ on the unit sphere.

B.2.9 Transformation in a Small Region

The following is an example of what might be called a "similarity transformation" on the ellipsoid. Consider a cluster of stations, $i = 1, \ldots, m$, each having two sets of coordinates $(\varphi_{o,i}, \lambda_{o,i})$ and $(\varphi_{n,i}, \lambda_{n,i})$ on the same ellipsoid. The subscripts o and n may be interpreted as "old" and "new." The goal is to establish a simple transformation between the coordinates.

The two-dimensional transformation is done with the tools developed in this appendix. First, we compute the center of figure (φ_c, λ_c) of the stations in the n set by simply averaging latitudes and longitudes, respectively. Next, consider the geodesics that connect the center of figure (φ_c, λ_c) with the positions $(\varphi_{n,i}, \lambda_{n,i})$. The discrepancies $(\varphi_{o,i} - \varphi_{n,i})$ and $(\lambda_{o,i} - \lambda_{n,i})$ take on the role of observation to be used to compute the transformation parameters by least-squares. We define four transformation parameters as follows: the translation of the center of figure $(d\varphi_c, d\lambda_c)$, the common azimuth rotation $d\widehat{\alpha}_c$ at the center of figure, and common scale factor $1 - \Delta$ for all geodesics going from the center of figure to the individual points. Thus,

$$\mathbf{x} = \begin{bmatrix} d\lambda_c & d\varphi_c & \Delta & d\widehat{\alpha}_c \end{bmatrix}^{\mathrm{T}} \tag{B.66}$$

Since the discrepancies $(\varphi_{o,i} - \varphi_{n,i})$ and $(\lambda_{o,i} - \lambda_{n,i})$ are small quantities, the coefficients listed in Table B.3 represent the linear mathematical model of the adjustment. The observation equations for the mixed adjustment model are

$$\begin{aligned} \Delta\, \widehat{s}_{ci} &= -M_i \cos\widehat{\alpha}_{ic} \left(\varphi_{o,i} - \varphi_{n,i}\right) - M_c \cos\widehat{\alpha}_{ci}\, d\varphi_c \\ &\quad - N_i \cos\varphi_i \sin\widehat{\alpha}_{ic} \left(\lambda_{o,i} - \lambda_{n,i}\right) + N_i \cos\varphi_i \sin\widehat{\alpha}_{ic}\, d\lambda_c \end{aligned} \tag{B.67}$$

$$\begin{aligned} d\widehat{\alpha}_c &= \frac{M_c}{\widehat{s}_{ci}} \sin\widehat{\alpha}_{ci}\, d\varphi_c + \frac{M_i}{\widehat{s}_{ci}} \sin\widehat{\alpha}_{ic} \left(\varphi_{o,i} - \varphi_{n,i}\right) \\ &\quad - \frac{N_i}{\widehat{s}_{ci}} \cos\varphi_i \cos\widehat{\alpha}_{ic} \left(\lambda_{o,i} - \lambda_{n,i}\right) + \frac{N_i}{\widehat{s}_{ci}} \cos\varphi_i \cos\widehat{\alpha}_{ic}\, d\lambda_c \end{aligned} \tag{B.68}$$

The respective submatrices of **B**, **A**, and **w** for station i are

$$\begin{matrix} & \varphi_{n,i} & \lambda_{n,i} & \varphi_{o,i} & \lambda_{o,i} \end{matrix}$$

$$\mathbf{B} = \begin{bmatrix} M_i \cos\widehat{\alpha}_{ic} & N_i \cos\varphi_i \sin\widehat{\alpha}_{ic} & -M_i \cos\widehat{\alpha}_{ic} & -N_i \cos\varphi_i \sin\widehat{\alpha}_{ic} \\ -\dfrac{M_i}{\widehat{s}_{ci}} \sin\widehat{\alpha}_{ic} & \dfrac{N_i}{\widehat{s}_{ic}} \cos\varphi_i \cos\widehat{\alpha}_{ci} & \dfrac{M_i}{\widehat{s}_{ci}} \sin\widehat{\alpha}_{ic} & -\dfrac{N_i}{\widehat{s}_{ci}} \cos\varphi_i \cos\widehat{\alpha}_{ic} \end{bmatrix} \tag{B.69}$$

$$\begin{matrix} & d\varphi_c & d\lambda_c & \Delta & d\widehat{\alpha}_c \end{matrix}$$

$$\mathbf{A} = \begin{bmatrix} -M_c \cos\widehat{\alpha}_{ci} & N_i \cos\varphi_i \sin\widehat{\alpha}_{ic} & -\widehat{s}_{ci} & 0 \\ \dfrac{M_c}{\widehat{s}_{ci}} \sin\widehat{\alpha}_{ci} & \dfrac{N_i}{\widehat{s}_{ci}} \cos\varphi_i \cos\widehat{\alpha}_{ic} & 0 & -1 \end{bmatrix} \tag{B.70}$$

$$\mathbf{w} = \begin{bmatrix} -M_i \cos \widehat{\alpha}_{ic} \left(\varphi_{o,i} - \varphi_{n,i} \right) - N_i \cos \varphi_i \sin \widehat{\alpha}_{ic} \left(\lambda_{o,i} - \lambda_{n,i} \right) \\ \dfrac{M_i}{\widehat{s}_{ci}} \sin \widehat{\alpha}_{ic} \left(\varphi_{o,i} - \varphi_{n,i} \right) - \dfrac{N_i}{\widehat{s}_{ci}} \cos \varphi_i \cos \widehat{\alpha}_{ic} \left(\lambda_{o,i} - \lambda_{n,i} \right) \end{bmatrix} \tag{B.71}$$

Once the adjusted transformation parameters are available, we can compute the position of the adjusted center of figure and the length and azimuth for any geodesics as follows:

$$\varphi_{o,c} = \varphi_{n,c} + d\varphi_c \tag{B.72}$$

$$\lambda_{o,c} = \lambda_{n,c} + d\lambda_c \tag{B.73}$$

$$\widehat{s}_{o,ci} = \widehat{s}_{n,ci} + \Delta\, \widehat{s}_{ci} \tag{B.74}$$

$$\widehat{\alpha}_{o,ci} = \widehat{\alpha}_{n,ci} + d\widehat{\alpha}_c \tag{B.75}$$

With (B.72) through (B.75) the positions of stations in the o system can be computed by using the direct solution given in Table B.2.

APPENDIX C

CONFORMAL MAPPING

The conformal property means that an angle between lines on the original equals the angle of their images. One must keep in mind that an angle is defined as the angle between tangents.

This section begins with conformal mapping of planes using complex functions. It serves two purposes. First, it demonstrates in a rather simple but clear manner the difference between conformality and similarity transformation. Second, it gives the technique for transforming the isometric plane into one of the desired standard conformal mappings, such as the ones by Mercator or Lambert. This is followed by the general formulation of conformality between general surfaces, making use of the first fundamental coefficients. We then deal with those conformal mappings that are generally used in surveying. The most important ones are the transverse Mercator mapping and the Lambert conformal mapping. For example, all but one of the U.S. state plane coordinate systems are based on these mappings. The lone exception is a system in Alaska that uses the oblique Mercator mapping. The latter is not discussed here.

Clearly, conformal mapping has a long history with many individuals having made significant contributions. The historically inclined reader may consult the specialized literature for a full exposition of this interesting aspect. It might not be easy to delineate individual contributions in all cases. This is in part true because concepts were sometimes formulated before the appropriate mathematical tools became available.

C.1 CONFORMAL MAPPING OF PLANES

The complex number z can be given in one of the following three well-known equivalent forms,

$$z = \lambda + iq = r(\cos\theta + i\sin\theta) = re^{i\theta} \tag{C.1}$$

The symbols λ and q denote the real and imaginary parts, respectively, and are typically graphed as Cartesian coordinates. The polar form, the middle part of (C.1), is specified by the magnitude r and the argument θ. The third part of (C.1) is the Euler form. The reader is referred to the mathematical literature to brush up on the algebra with complex numbers, if necessary. A function of complex numbers such as

$$w = f(z) \tag{C.2}$$

is called complex mapping. The variable $z = \lambda + iq$ represents points on the original which are to be mapped, and $w = x + iy$ represents the respective images or the map.

$$x + iy = f(\lambda + iq) \tag{C.3}$$

Separating the real and imaginary parts, we can write

$$x = x(\lambda, q) \tag{C.4}$$

$$y = y(\lambda, q) \tag{C.5}$$

The derivative of the complex function (C.2) plays a key role in assuring that the complex mapping is conformal. The image of the increment Δz is

$$\Delta w = f(z + \Delta z) - f(z) \tag{C.6}$$

Analogous to computing the derivative for real functions, the derivative of a complex function follows from the limit

$$\frac{dw}{dz} \equiv f'(z) = \lim_{\Delta z \to 0} \frac{f(z + \Delta z) - f(z)}{\Delta z} = \lim_{\Delta z \to 0} \frac{\Delta w}{\Delta z} \tag{C.7}$$

In contrast to the case of real functions, the increment Δz has a direction; one has virtually an infinite number of possibilities of letting Δz go to zero. If the limit exists and is independent of the manner in which Δz approaches zero, then the function $f(z)$ is called differentiable. It is proven in the mathematical literature that the Cauchy-Riemann equations

$$\frac{\partial x}{\partial \lambda} = \frac{\partial y}{\partial q} \tag{C.8}$$

$$\frac{\partial x}{\partial q} = -\frac{\partial y}{\partial \lambda} \tag{C.9}$$

represent necessary and sufficient conditions for the derivative to exist. In that case, the actual derivative is given by

$$f'(z) = \frac{\partial x}{\partial \lambda} + i\frac{\partial y}{\partial \lambda} = \frac{\partial y}{\partial q} - i\frac{\partial x}{\partial q} \tag{C.10}$$

In terms of interpreting (C.2) as conformal mapping, it is advantageous to rewrite (C.7) using Euler's form of complex numbers, i.e.,

$$\Delta z = |\Delta z|\, e^{i\theta} \tag{C.11}$$

$$\Delta w = |\Delta w|\, e^{i\varphi} \tag{C.12}$$

$$f'(z) = \lim_{\Delta z \to 0} \frac{|\Delta w|\, e^{i\varphi}}{|\Delta z|\, e^{i\theta}} = \lim_{\Delta z \to 0} \frac{|\Delta w|}{|\Delta z|} e^{i(\varphi-\theta)} = \left|f'(z)\right| e^{i\gamma} \tag{C.13}$$

The symbols θ and φ denote here the arguments of the respective differential numbers Δz and Δw. Since the derivative exists (we will consider only functions that fulfill the Cauchy-Riemann conditions), both the magnitude $\left|f'(z)\right|$ and the argument of the derivative

$$\gamma = \varphi - \theta \tag{C.14}$$

are independent of the manner in which Δz approaches zero. The mapping (C.2) in the differential neighborhood of z is

$$|\Delta w| = \left|f'(z)\right| |\Delta z| \tag{C.15}$$

According to (C.14), the argument of the image is

$$\arg \Delta w = \arg \Delta z + \arg f'(z) \tag{C.16}$$

Equations (C.15) and (C.16) allow the following interpretation: for complex mapping $w = f(z)$, assuming that the derivative exists, the length of an infinitesimal distance $|\Delta z|$ on the original is scaled by the factor $\left|f'(z)\right|$. This factor is solely a function of z and is independent of the direction of Δz. Similarly, the difference in the direction of the original Δz and its image Δw, $\arg \Delta w - \arg \Delta z$, is independent of the direction of the original Δz because the argument $\arg f'(z)$ is independent of Δz. Consequently, two infinitesimal segments at z will be mapped into two images that enclose the same angle, $\varphi_1 - \varphi_2 = \theta_1 - \theta_2$, or

$$\varphi_1 - \theta_1 = \gamma \tag{C.17}$$

$$\varphi_2 - \theta_2 = \gamma \tag{C.18}$$

Figure C.1 shows the mapping of two points in the differential neighborhood of z. The differential figures $(z_2 - z - z_1)$ and $(w_2 - w - w_1)$ differ by translation, rotation, and scale. The conformal mapping $f(z)$ does not change the angles between differentially located points; consequently, infinitesimally small figures are similar.

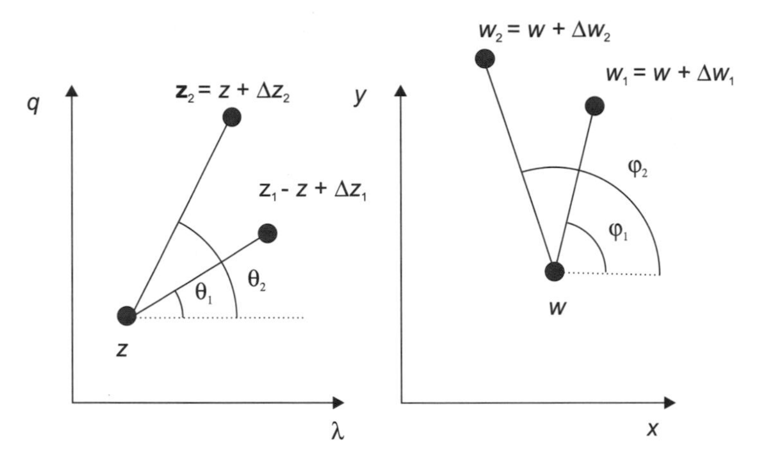

Figure C.1 Conformal mapping in differential neighborhood.

The scale factor of the mapping follows from (C.10)

$$k = \left|f'(z)\right| = \sqrt{\left(\frac{\partial x}{\partial \lambda}\right)^2 + \left(\frac{\partial y}{\partial \lambda}\right)^2} = \sqrt{\left(\frac{\partial x}{\partial q}\right)^2 + \left(\frac{\partial y}{\partial q}\right)^2} \tag{C.19}$$

The rotation angle γ, which will later be identified as the meridian convergence, follows from

$$\tan\gamma = \frac{\partial y/\partial \lambda}{\partial x/\partial \lambda} = -\frac{\partial x/\partial q}{\partial y/\partial q} \tag{C.20}$$

The following example should demonstrate the idea of conformal mapping. Using $z = \lambda + iq$ and $w = x + iy$ the simple mapping function

$$w = z^2 \tag{C.21}$$

gives

$$x = \lambda^2 - q^2 \tag{C.22}$$

$$y = 2\lambda q \tag{C.23}$$

Thus, the coordinates are $x = \lambda^2 - q^2$ and $y = 2\lambda q$. The partial derivatives

$$\frac{\partial x}{\partial \lambda} = \frac{\partial y}{\partial q} = 2\lambda \tag{C.24}$$

$$\frac{\partial x}{\partial q} = -\frac{\partial y}{\partial \lambda} = -2q \tag{C.25}$$

satisfy the Cauchy-Riemann equations and are continuous over the (λ, q) plane. The derivative is

$$f'(z) = \frac{\partial x}{\partial \lambda} + i\frac{\partial y}{\partial \lambda} = \frac{\partial y}{\partial q} - i\frac{\partial x}{\partial q} = 2\lambda + i2q \tag{C.26}$$

Images of the lines λ = constant = c_1 follow from the mapping equations (C.22) and (C.23) upon setting $\lambda = c_1$ and eliminating q,

$$y = \pm\sqrt{4c_1^4 - 4c_1^2 x} \tag{C.27}$$

Similarly we obtain for the images of the lines q = constant = c_2 as

$$y = \pm\sqrt{4c_2^4 + 4c_2^2 x} \tag{C.28}$$

The scale in the differential neighborhood of z follows from (C.19) and (C.26) as

$$k = |f'(z)| = \sqrt{4\lambda^2 + 4q^2} \tag{C.29}$$

The rotation in the same differential neighborhood is, according to (C.16) and (C.26),

$$\arg f'(z) = \tan^{-1}\frac{q}{\lambda} \tag{C.30}$$

Figure C.2 shows this mapping. Mathematically, any lines parallel to the q or λ axes map into parabolas. Using differential calculus it can be readily verified that the images of the parametric curves map into a family of orthogonal curves. The same tools can be used to verify that the angle between the general lines $f_1(\lambda, q) = 0$ and $f_2(\lambda, q) = 0$ will be the same on the map. Note that the scale and the rotation angle vary continuously with location. The figure (A, B, C, D) and its image (a, b, c, d) cannot be related by a similarity transformation.

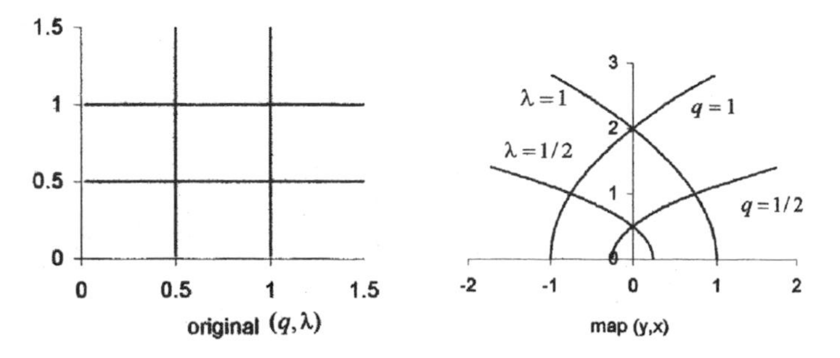

Figure C.2 Simple conformal mapping between planes.

C.2 CONFORMAL MAPPING OF GENERAL SURFACES

The approach is to find conditions for the first fundamental coefficients to assure that conformality is achieved. This general formulation is valid for conformal mapping of any surface, e.g., mapping the ellipsoid on the sphere, sphere on a plane, ellipsoid on a plane, etc.

Let the surfaces $[S]$ be expressed in terms of curvilinear coordinates (u, v),

$$\left.\begin{aligned} x &= x(u, v) \\ y &= y(u, v) \\ z &= z(u, v) \end{aligned}\right\} \tag{C.31}$$

This surface is to be mapped conformally on the surface $[\bar{S}]$,

$$\left.\begin{aligned} \bar{x} &= x'(\bar{u}, \bar{v}) \\ \bar{y} &= y'(\bar{u}, \bar{v}) \\ \bar{z} &= z'(\bar{u}, \bar{v}) \end{aligned}\right\} \tag{C.32}$$

whose curvilinear coordinates are denoted by $(\bar{u}, \bar{v})$. The mapping equations

$$\bar{u} = \bar{u}(u, v) \tag{C.33}$$

$$\bar{v} = \bar{v}(u, v) \tag{C.34}$$

relate both sets of curvilinear coordinates. These mapping equations of course are not arbitrary but must eventually be derived such that the mapping is conformal. Substituting these equations in the surface representation (C.32) gives

$$\left.\begin{aligned} \bar{x} &= \bar{x}(u, v) \\ \bar{y} &= \bar{y}(u, v) \\ \bar{z} &= \bar{z}(u, v) \end{aligned}\right\} \tag{C.35}$$

Equations (C.35) express the image surface $[\bar{S}]$ as a function of the curvilinear coordinates of the original surface $[S]$. The first fundamental forms (B.28) for both surfaces are

$$ds^2 = E\,du^2 + 2F\,du\,dv + G\,dv^2 \tag{C.36}$$

$$d\bar{s}^2 = \bar{E}\,du^2 + 2\bar{F}\,du\,dv + \bar{G}\,dv^2 \tag{C.37}$$

The conformal property is given in terms of the condition on the first fundamental conditions

$$k^2(u, v) \equiv \frac{\bar{E}}{E} = \frac{\bar{F}}{F} = \frac{\bar{G}}{G} \tag{C.38}$$

That conditions (C.38) indeed assure conformality as can be verified by computing the angle between the two curves $f_1(u, v) = 0$ and $f_2(u, v) = 0$ on $[S]$ and between the respective images on $[\bar{S}]$. Equation (B.46) gives the angle on the original as

$$\cos(ds_1, ds_2) = \frac{E\, du_1\, du_2 + F(du_1\, dv_2 + du_2\, dv_1) + G\, dv_1\, dv_2}{\sqrt{E\, du_1^2 + 2F\, du_1\, dv_1 + G\, dv_1^2}\sqrt{E\, du_2^2 + 2F\, du_2\, dv_2 + G\, dv_2^2}} \tag{C.39}$$

Since the image surface has been expressed in terms of the curvilinear coordinates (u, v) of the original, and since the functions $f_1(u, v) = 0$ and $f_2(u, v) = 0$ apply to the mapped lines as well, it follows that the angle on the image is given by

$$\cos(d\bar{s}_1, d\bar{s}_2) = \frac{\bar{E}\, du_1\, du_2 + \bar{F}(du_1\, dv_2 + du_2\, dv_1) + \bar{G}\, dv_1\, dv_2}{\sqrt{\bar{E}\, du_1^2 + 2\bar{F}\, du_1\, dv_1 + \bar{G}\, dv_1^2}\sqrt{\bar{E}\, du_2^2 + 2\bar{F}\, du_2\, dv_2 + \bar{G}\, dv_2^2}} \tag{C.40}$$

Replacing $\bar{E}$, $\bar{F}$, and $\bar{G}$ with k^2E, k^2G, and k^2G, respectively, following (C.38), one readily sees that

$$\cos(d\bar{s}_1, d\bar{s}_2) = \cos(ds_1, ds_2) \tag{C.41}$$

and therefore the angle enclosed by the tangents on f_1 and f_2 is preserved. The point scale factor for the mapping is

$$k(u, v) = \frac{d\bar{s}}{ds} \tag{C.42}$$

As an example, one might verify the general condition (C.38) for the simple conformal mapping (C.21) between two planes. Following the general notation, the equations for the original (C.31) have the simple form $y = q$ and $x = \lambda$. The respective first fundamental coefficients are $E = G = 1$ and $F = 0$. The expressions for the image surface (C.32) are $\bar{x} = x$ and $\bar{y} = y$. Substituting the mapping equations (C.22) and (C.23) into the image surface expressions gives $\bar{x} = \lambda^2 - q^2$ and $\bar{y} = 2\lambda q$. The first fundamental coefficients are $\bar{E} = \bar{G} = 4\lambda^2 + 4q^2$ and $\bar{F} = 0$. It follows that the condition (C.38) is indeed fulfilled for this simple mapping.

C.3 THE ISOMETRIC PLANE

An especially simple situation arises if the curvilinear coordinates (u, v) on the original are isometric and orthogonal. The curvilinear coordinates (q, λ), where q denotes the isometric latitude given in (B.49), form such an isometric net on the ellipsoid. The first fundamental form becomes, according to (B.48),

$$ds^2 = N^2 \cos^2 \varphi \left(dq^2 + d\lambda^2\right) \tag{C.43}$$

which implies that $E = G = N^2 \cos^2 \varphi$ and $F = 0$. The first step in utilizing the isometric curvilinear coordinates (q, λ) for conformal mapping is to consider the mapping equations

$$x = \lambda \tag{C.44}$$

$$y = q \tag{C.45}$$

and interpret (x, y) as Cartesian coordinates, i.e., the expressions for the image surface simply are

$$\bar{x} = \lambda \tag{C.46}$$

$$\bar{y} = q \tag{C.47}$$

and $\bar{E} = \bar{G} = 1$ and $\bar{F} = 0$. The first fundamental coefficients meet the condition (C.38). The point scale factor for this mapping is

$$k^2 = \frac{dq^2 + d\lambda^2}{E\left(dq^2 + d\lambda^2\right)} = \frac{1}{N^2 \cos^2 \varphi} \tag{C.48}$$

We may, therefore, conclude that one way of creating a conformal mapping of a general surface to a plane is to establish an isometric net on the original and then interpret the isometric coordinates as Cartesian coordinates and call the result the isometric mapping plane.

In a subsequent step, the isometric plane can be mapped conformally onto another mapping plane by the analytic function

$$x + iy = f(\lambda + iq) \tag{C.49}$$

The implied mapping equations are

$$x = x(q, \lambda) \tag{C.50}$$

$$y = y(q, \lambda) \tag{C.51}$$

where (x, y) denote the coordinates in the final map. The point scale factor of such a sequential conformal mapping equals the product of that of the individual mappings. According to (C.42) and (C.19), we have

$$\begin{aligned} k &= \frac{dS_{\text{IP}}}{dS} \frac{d\bar{S}}{dS_{\text{IP}}} = k_{\text{IP}} \cdot k_{\text{IP}\to\text{Map}} \\ &= \frac{\sqrt{(\partial x/\partial \lambda)^2 + (\partial y/\partial \lambda)^2}}{N \cos \varphi} = \frac{\sqrt{(\partial x/\partial q)^2 + (\partial y/\partial q)^2}}{N \cos \varphi} \end{aligned} \tag{C.52}$$

Additional specifications that the complex function must fulfill will assure that a conformal map with the desired properties will be obtained.

C.4 POPULAR CONFORMAL MAPPINGS

The transverse Mercator and Lambert conformal mappings are the most popular mappings used in geodetic computations. Not only do they serve as the basis for the U.S. state plane coordinate systems but they are also widely used by other countries as the national mapping system. Since the respective mapping equations can be easily programmed, these mappings are suitable for local mapping as well. The equatorial Mercator mapping is presented first because it follows in such a straightforward manner from the isometric plane. The transverse Mercator and the Lambert conformal mapping will then be discussed. Finally, the polar conformal mapping is specified. Most of the derivations related to this appendix are compiled in Leick (2002).

C.4.1 Equatorial Mercator

The equatorial Mercator mapping (EM) is a linear mapping of the isometric plane such that the equator of the ellipsoid and its image on the map are of the same length. See Figure C.3. This is accomplished by

$$x + iy = a(\lambda + iq) \tag{C.53}$$

where the symbol a denotes the semimajor axis of the ellipsoid. The mapping equations become

$$x = a\lambda \tag{C.54}$$

$$y = aq \tag{C.55}$$

This map is simply a magnification of the isometric plane. The meridians map into straight lines that are parallel to the y axis; y is zero at the equator. The mapped parallels are also straight lines and are parallel to the x axis, but spacing increases toward the poles for the same latitude increment. The equator is mapped equidistantly.

Figure C.3 Equatorial Mercator map.

The meridian convergence is zero because mapped meridians are parallel to the y axis. Zero meridian convergence can be readily verified by applying Expression (C.20) to the mapping equations (C.54) and (C.55). The point scale factor is, according to Equation (C.52),

$$k = \frac{a}{N \cos \varphi} \tag{C.56}$$

This point scale factor does not depend on the longitude; $k = 1$ on the equator and the value increases latitude. This makes this mapping attractive for use in regions close to the equator.

Any meridian can serve as central meridian or zero meridian, with which the y axis coincides. For example, the meridian, which passes through the middle of the mapping area, can be the zero meridian at which $x = 0$. Furthermore, the point scale factor must not be confused with the scale of a conventional map, which is the ratio of a plotted distance over the mapped distance. The point scale factor is a characteristic of the mapping and changes with location, whereas the scale of a map is dictated by the size of the plotting paper and the area to be plotted.

The loxodrome is a curve that intersects consecutive meridians at the same azimuth. It can be readily visualized that the loxodrome maps into a straight line for the equatorial Mercator mapping.

C.4.2 Transverse Mercator

The specifications for the transverse Mercator mapping (TM) are as follows:

1. Apply conformal mapping conditions.
2. Adopt a central meridian λ_0 that passes more or less through the middle of the area to be mapped. For reasons of convenience, relabel the longitudes starting with $\lambda = 0$ at the central meridian.
3. Let the mapped central meridian coincide with the y axis of the map. Assign $x = 0$ to the image of the central meridian.
4. The length of the mapped central meridian should be k_0 times the length of the corresponding elliptic arc, i.e., at the central meridian $y = k_0\, s_\varphi$.

The derivation of the transverse Mercator mapping begins with the isometric plane and a suitable complex function f for (C.49). Condition (4) specifies the image of the central meridian and implies

$$0 + ik_0 s_\varphi = f(0 + iq) \tag{C.57}$$

This function can be expanded in a Taylor series, which provides an opportunity to impose the Cauchy-Riemann conditions on the partial derivatives. The general picture of the transverse Mercator map is shown in Figure C.4. The image of the central meridian is a straight line; all other meridians are curved lines coming together at the

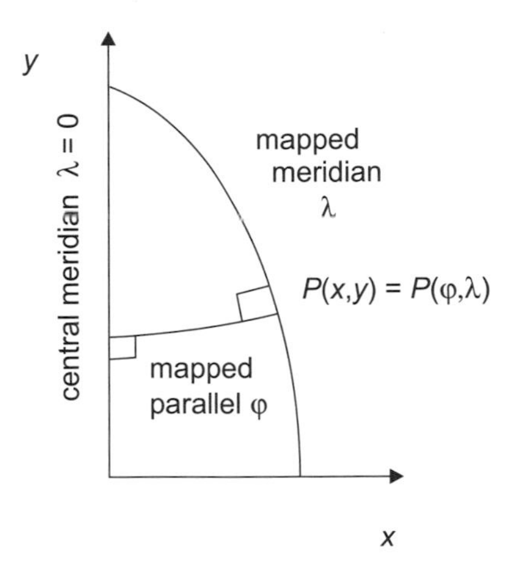

Figure C.4 Transverse Mercator map.

pole and being perpendicular to the image of the equator. The latter coincides with the x axis. The mapped parallels are, of course, perpendicular to the mapped meridians; however, they are not circles but mathematically complex curves.

The mapping equations for the direct mapping from $P(\varphi, \lambda)$ to $P(x, y)$ are listed in Table C.1. In these expressions, the longitude λ is counted positive to the east, starting at the central meridian. All quantities that depend on the latitude must be evaluated at φ. Equation (B.6) gives the expression for the radius of curvature N of the prime vertical section. The symbol s denotes the length of the elliptic arc from the equator to φ as given by (B.42). The symbols t and η are used for brevity and mean:

$$t = \tan\varphi \tag{C.58}$$

$$\eta^2 = \frac{e^2}{1-e^2}\cos^2\varphi \tag{C.59}$$

TABLE C.1 Transverse Mercator Direct Mapping

$$\frac{x}{k_0 N} = \lambda\cos\varphi + \frac{\lambda^3\cos^3\varphi}{6}\left(1-t^2+\eta^2\right) + \frac{\lambda^5\cos^5\varphi}{120}\left(5-18t^2+t^4+14\eta^2-58t^2\eta^2\right) \quad \text{(a)}$$

$$\begin{aligned}\frac{y}{k_0 N} &= \frac{s}{N} + \frac{\lambda^2}{2}\sin\varphi\cos\varphi + \frac{\lambda^4}{24}\sin\varphi\cos^3\varphi\left(5-t^2+9\eta^2+4\eta^4\right)\\ &+ \frac{\lambda^6}{720}\sin\varphi\cos^5\varphi\left(61-58t^2+t^4+270\eta^2-330t^2\eta^2\right)\end{aligned} \quad \text{(b)}$$

$$\gamma = \lambda\sin\varphi\left\{1+\frac{\lambda^2\cos^2\varphi}{3}\left(1+3\eta^2+2\eta^4\right)+\frac{\lambda^4\cos^4\varphi}{15}\left(2-t^2\right)\right\} \quad \text{(c)}$$

The inverse solution for mapping $P(x, y)$ to $P(\varphi, \lambda)$ is given in Table C.2. All latitude-dependent terms in this table must be evaluated for the so-called footpoint latitude φ_f. The footpoint is a point on the central meridian obtained by drawing a parallel to the x axis through the point $P(x, y)$. Given the y coordinate, the footpoint latitude can be computed iteratively from (B.42). Because of condition (4), the following relation holds,

$$s_f = \frac{y}{k_0} \tag{C.60}$$

where s_f is the length of the central meridian from the equator to the footpoint. Substitute (C.60) in (B.43) and solve φ_f iteratively.

The expression for the point scale factor is

$$\begin{aligned}\frac{k}{k_0} &= 1 + \frac{\lambda^2}{2}\cos^2\varphi\,(1+\eta^2)\\ &\quad + \frac{\lambda^4}{24}\cos^4\varphi\,(5 - 4t^2 + 14\eta^2 + 13\eta^4 - 28t^2\eta^2 + 4\eta^6 - 48t^2\eta^4 - 24t^2\eta^6)\\ &\quad + \frac{\lambda^6}{720}\cos^6\varphi\,(61 - 148t^2 + 16t^4)\end{aligned} \tag{C.61}$$

This equation shows that the scale factor k increases primarily with longitude. In fact, isoscale lines run more or less parallel to the image of the central meridian. Since the mapping distortions increase as k departs from 1, the factor k_0 is an important element of design. By selecting $k_0 < 1$, one allows some distortion at the central meridian for the benefit of having less distortion away from the central meridian. In this way, the longitudinal coverage of the area of a map can be extended given a level of acceptable distortion.

This appearance of the TM mapping expressions reveals the fact that they have been obtained from series expansions. Consequently, the expressions are accurate

TABLE C.2 Transverse Mercator Inverse Mapping

$$\begin{aligned}\varphi = \varphi_f &- \frac{t}{2}(1+\eta^2)\left(\frac{x}{k_0N}\right)^2 + \frac{t}{24}(5 + 3t^2 + 6\eta^2 - 6\eta^2t^2 - 3\eta^4 - 9t^2\eta^4)\left(\frac{x}{k_0N}\right)^4\\ &- \frac{t}{720}(61 + 90t^2 + 45\eta^4 + 107\eta^2 - 162\eta^2t^2 - 45t^4\eta^2)\left(\frac{x}{k_0N}\right)^6\end{aligned} \tag{a}$$

$$\begin{aligned}\lambda\cos\varphi_f &= \frac{x}{k_0N} - \frac{1}{6}\left(\frac{x}{k_0N}\right)^3(1 + 2t^2 + \eta^2)\\ &\quad + \frac{1}{120}\left(\frac{x}{k_0N}\right)^5(5 + 28t^2 + 24t^4 + 6\eta^2 + 8t^2\eta^2)\end{aligned} \tag{b}$$

only as long as the truncation errors are negligible. Note the symmetries with respect to the central meridian, $x(-\lambda) = -x(\lambda)$ and $y(-\lambda) = y(\lambda)$, and the equator, $y(-\varphi) = -y(\varphi)$ and $x(-\varphi) = x(\varphi)$. These TM expressions are also given in Thomas (1952, pp. 96–103), who lists some additional higher-order terms.

The transverse Mercator mapping of the ellipsoidal as given above is attributed to Gauss, who used his extensive developments in differential geometry to study conformal mapping of general surfaces. Other scientists further refined Gauss's basic developments in order to produce expressions suitable for calculation, which was a necessity before computers became available. Most notable are contributions by L. Krüger. Lee (1976) presents closed or exact formulas for the transverse Mercator mapping with respect to the ellipsoid; these elliptical expressions were programmed by Dozier (1980). Lee further discusses other variations of the transverse Mercator mapping, in addition to one with constant scale factor along the mapped central meridian presented here. Finally, it should be emphasized that Lambert (1772) already gave expressions for the transverse Mercator mapping with respect to the sphere.

C.4.3 Lambert Conformal

The specifications for the Lambert conformal (LC) mapping are:

1. Apply conformal mapping conditions.
2. Adopt a central meridian λ_0 that passes more or less through the middle of the area to be mapped. For reasons of convenience, relabel the longitudes, starting with $\lambda = 0$ at the central meridian.
3. Let the mapped central meridian coincide with the y axis of the map. Assign $x = 0$ for the image of the central meridian.
4. Map the meridians into straight lines passing through the image of the pole; map the parallels into concentric circles around the image of the pole. Select a standard parallel φ_0 that passes more or less through the middle of the area to be mapped. The length of the mapped standard parallel is k_0 times the length of the corresponding ellipsoidal parallel. The point scale factor along any mapped parallel is constant. Start counting $y = 0$ at the image of the standard parallel.

The general picture of the Lambert conformal map is shown in Figure C.5. The mapping is singular at the pole, which is the reason why the angle of the mapped meridian is λ', and not λ, at the pole. Denoting the distance from the mapped parallel to the pole by r, the pair (λ', r) are polar coordinates that form a set of orthogonal curvilinear lines on the map. The first fundamental form for this choice of coordinates is

$$d\bar{s}^2 = dr^2 + r^2\, d\lambda'^2 = r^2 \left(\frac{dr^2}{r^2} + d\lambda'^2 \right) \tag{C.62}$$

We observe that $\left(\lambda', r\right)$ is not an isometric net. The same increments in dr and $d\lambda'$ result in different changes of $d\bar{s}$. If we define the auxiliary coordinate

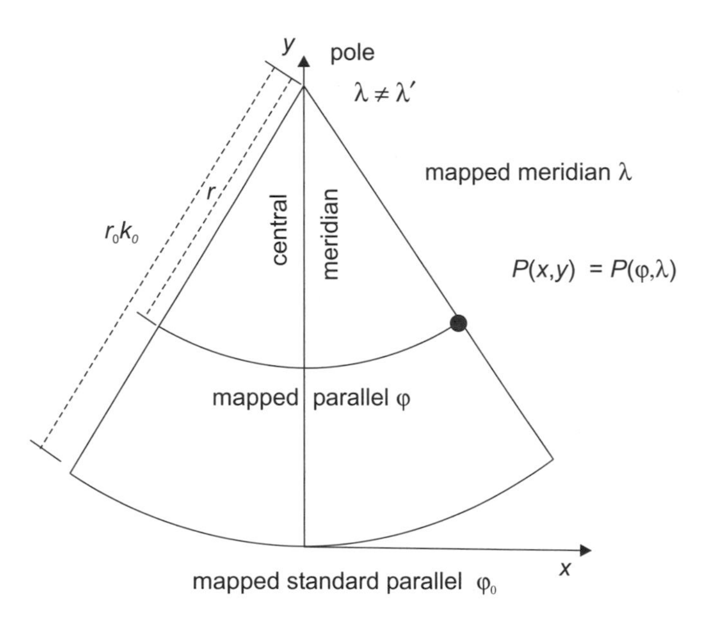

Figure C.5 Lambert conformal mapping.

$$dq' \equiv -\frac{dr}{r} \tag{C.63}$$

then (λ', q') indeed constitutes an isometric net in the mapping plane. The integration of (C.63) gives

$$q' = -\int_{k_0 r}^{r} \frac{dr}{r} = -(\ln\ r - \ \ln\ k_0 r_0) = \ -\ln\ \frac{r}{k_0 r_0} \tag{C.64}$$

At the standard parallel φ_0 we have $r = k_0 r_0$ and $q' = 0$. The negative sign in (C.63) takes care of the fact that q' increases toward the pole, whereas r decreases. Equal incremented quadrilaterals of (q', λ') decrease as the pole is approached. The Lambert conformal mapping is now specified by

$$\lambda' + iq' = \alpha\,\{\lambda + i(q - q_0)\} \tag{C.65}$$

where $\alpha = \sin\varphi_0$ and q_0 is the isometric latitude of the standard parallel. The value of the constant α is derived on the basis of condition (4).

The expressions for the direct and inverse mapping are listed in Tables C.3 and C.4. See Thomas (1952, p. 117) or Leick (2002) for a complete derivation. The symbol $\varepsilon = 2.71828\ldots$ denotes the base of the natural system of logarithm and should not be confused with the eccentricity e of the ellipsoid. The inverse solution gives the isometric latitude first, which can then be readily converted to the geodetic latitude.

TABLE C.3 Lambert Conformal Direct Mapping

$x = k_0 N_0 \cot \varphi_0 \, \varepsilon^{-\Delta q \sin \varphi_0} \sin(\lambda \sin \varphi_0)$	(a)
$y = k_0 N_0 \cot \varphi_0 \left[1 - \varepsilon^{-\Delta q \sin \varphi_0} \cos(\lambda \sin \varphi_0)\right]$	(b)
$\gamma \equiv \lambda' = \lambda \sin \varphi_0$	(c)

There is no series expansion involved. However, attention must be given to numerical accuracy when converting q to φ. The point scale factor is

$$k = \frac{k_0 N_0 \cot \varphi_0}{N \cos \varphi} \varepsilon^{-(q-q_0) \sin \varphi_0} \tag{C.66}$$

Note that (k_0, φ_0) or, equivalently, (k_0, q_0), specifies the expressions for the Lambert conformal mapping. The area of smallest distortion is along the image of the standard parallel in the east-west direction; as one departs from the standard parallel, the distortions increase in the north-south direction. By selecting $k_0 < 1$ it is possible to reduce the distortions at the northern and southern extremities of the mapping area by allowing some distortions in the vicinity of the standard parallel. Whenever $k_0 < 1$ there are two parallels, one south and one north of the standard parallel, along which the point scale factor k equals 1; i.e., these two parallels are mapped without distortion in length.

The designer of the map has the choice of either specifying k_0 and φ_0 or the two parallels for which $k = 1$. In the latter case, one speaks of a two-standard-parallel Lambert conformal mapping. If the Lambert conformal mapping is specified by two standard parallels φ_1 and φ_2 with $k_1 = k_2 = 1$, then k_0 and φ_0 follow from the expression of Table C.5.

TABLE C.4 Lambert Conformal Inverse Mapping

$\tan \lambda' = \dfrac{x}{k_0 N_0 \cot \varphi_0 - y}$	(a)
$r = \dfrac{k_0 N_0 \cot \varphi_0 - y}{\cos \lambda'}$	(b)
$\lambda = \dfrac{\lambda'}{\sin \varphi_0}$	(c)
$\Delta q = -\dfrac{1}{\sin \varphi_0} \ln\left(\dfrac{r}{k_0 N_0 \cot \varphi_0}\right)$	(d)
$q = q_0 + \Delta q$	(e)

TABLE C.5 Conversion from Two Standard Parallels to One Standard Parallel

$\varphi_0 = \sin^{-1}\left[\dfrac{\ln(N_1 \cos\varphi_1) - \ln(N_2 \cos\varphi_2)}{q_2 - q_2}\right]$	(a)
$k_0 = \dfrac{N_1 \cos\varphi_1}{N_0 \cos\varphi_0}\,\varepsilon^{(q_1 - q_0)\sin\varphi_0} = \dfrac{N_2 \cos\varphi_2}{N_0 \cos\varphi_0}\,\varepsilon^{(q_2 - q_0)\sin\varphi_0}$	(b)

In the special case of $\varphi_0 = 90°$, the Lambert conformal mapping becomes the polar conformal mapping. The expressions are obtained by noting the following mathematical limit

$$F \equiv \lim_{\varphi_0 \to 90°} N_0(\cos\varphi_0)\varepsilon^{q_0} = \frac{2a^2}{b}\left(\frac{1-e}{1+e}\right)^{e/2} \tag{C.67}$$

where the symbols a and b denote the semiaxis of the ellipsoid and the general relation $b/a = \sqrt{1-e^2}$ has been used. Using (C.67), the equations for the polar conformal mapping become

$$x = k_0 F \varepsilon^{-q} \sin\lambda \tag{C.68}$$

$$y = k_0 F \varepsilon^{-q} \cos\lambda \tag{C.69}$$

$$\gamma = \lambda \tag{C.70}$$

$$k = \frac{k_0 F \varepsilon^{-q}}{N \cos\varphi} \tag{C.71}$$

As is the case with the Lambert conformal mapping, the meridians are straight lines radiating from the image of the pole and the parallels are concentric circles around the pole. The y axis coincides with the 180° meridian. See Figure C.6 for details. There is no particular advantage in selecting the central meridian of the area to be mapped as the zero meridian. One, therefore, will usually select the Greenwich meridian. The point scale factor is k_0 at the pole.

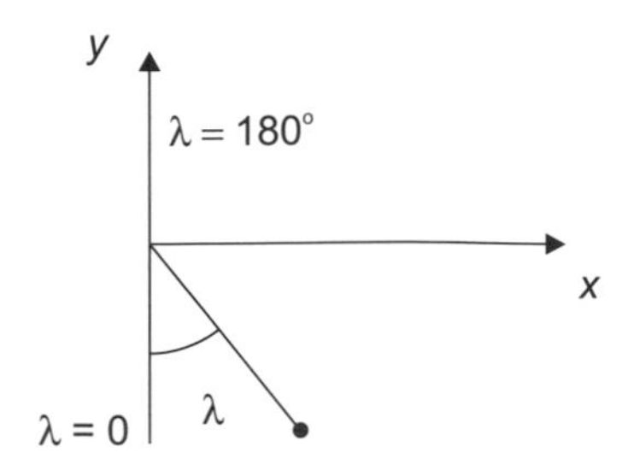

Figure C.6 Polar conformal mapping.

The polar conformal mapping is not a stereographic projection of the ellipsoid. Only in the special case of $e = 0$ do the expressions given above transition to those of the stereographic projection of the sphere (polar aspect) with the point of perspective being at the South Pole. The mapping is not only stereographic in the case of the sphere (because it is projected from a single point of perspective), but it is also azimuthal because the sphere is projected on the tangent plane. In the case of the oblique aspect of the stereographic projection, the sphere is projected on the tangent plane other than at the pole, with the center of projection being located on the sphere diametrically opposite the point of tangency. The ellipsoidal form of the oblique aspect is not perspective (no one single center of projection) in order to meet the conformal property. Readers desiring information on oblique aspect mappings are referred to the specialized literature.

In order to emphasize that the mappings discussed in this section are derived from the conformality condition, the term *mapping* has been used consistently. For example, we prefer to speak of transverse Mercator or Lambert conformal mapping instead of the transverse Mercator or Lambert conformal projection. There is not one single point of perspective for these mappings of the ellipsoid.

C.4.4 SPC and UTM

Each state and U.S. possession has a state plane coordinate system (SPC) defined for use in surveying and mapping (Stem, 1989). Many of the state plane coordinate systems use the transverse Mercator mapping. States with large east-west extent use the Lambert conformal mapping. Many states divide their region into zones and use both the transverse Mercator and the Lambert conformal mapping for individual zones. The exception to this scheme is the state plane coordinate system for the panhandle of Alaska, for which the oblique Mercator mapping is used.

The defining constants for the U.S. state plane coordinate system of 1983 are given in Tables C.6 and C.7. Table C.7 also contains the adopted values of false north and false east and a four-digit code to identify the projection. The "origin" as given in Table C.7 is not identical with the origin of the coordinate system in Figures C.4 and C.5. The state plane coordinates refer to the NAD83 ellipsoid. The specifications of the UTM mapping are given in Table C.8. These specifications must be taken into account when using the mapping equations given in this appendix. The National Geodetic Survey and other mapping agencies make software available on the Internet for computing coordinates for the officially adopted mappings.

It is the surveyor's choice to use a state plane coordinate system or to generate a local mapping by merely specifying k_0 (usually 1) and the central meridian and/or the standard parallel. In the latter case the mapping reductions are small. If, in addition, a local ellipsoid is specified, then most of the reductions can be neglected for small surveys. While these specifications might lead to a reduction in the computational load, which, in view of modern computer power is not critical any longer, it increases the probability that reductions are inadvertently neglected when they should not be.

TABLE C.6 Legend for U.S. State Plane Coordinate System Defining Constants

Mapping	T	Transverse Mercator
	L	Lambert conformal
	O	Oblique Mercator
	UTM	Universal Transverse Mercator
	1:M	Scale reduction at central meridian

	Conversion factors	
Meters	U.S. survey feet	International feet
152,400.3048	500,000.0	
213,360.0		700,000.0
304,800.6096	1,000,000.0	
609,600.0		2,000,000.0
609,601.2192	2,000.000.0	
914,401.8289	3,000,000.0	
1.	3.28083333333	
1.		3.28083989501
0.3048		1.
1200/3937	1.	
0.30480060960	1.	

TABLE C.7 U.S. State Plane Coordinate Systems. West longitudes are listed.

State Zone	SPCS Zone	Proj Type T/L/O	Latitude of Origin (DD-MM)	Longitude of Origin (DDD-MM)	False North (M)	False East (M)	$\frac{1}{1-k_0}$	Standard South (DD-MM)	Parallels North (DD-MM)
Alabama									
East	0101	T	30 30	85 50	0.0	200000.0	25000		
West	0102	T	30 00	87 30	0.0	600000.0	15000		
Alaska									
Zone 1	5001	0	57 00	133 40	−5000000.0	5000000.0	10000	Axis Az	$\text{Tan}^{-1}(-3/4)$
Zone 2	5002	T	54 00	142 00	0.0	500000.0	10000		
Zone 3	5003	T	54 00	146 00	0.0	500000.0	10000		
Zone 4	5004	T	54 00	150 00	0.0	500000.0	10000		
Zone 5	5005	T	54 00	154 00	0.0	500000.0	10000		
Zone 6	5006	T	54 00	158 00	0.0	500000.0	10000		
Zone 7	5007	T	54 00	162 00	0.0	500000.0	10000		
Zone 8	5008	T	54 00	166 00	0.0	500000.0	10000		
Zone 9	5009	T	54 00	170 00	0.0	500000.0	10000		
Zone 10	5010	L	51 00	176 00	0.0	1000000.0		51 50	53 50
Arizona									
East	0201	T	31 00	110 10	0.0	213360.0	10000		
Central	0202	T	31 00	111 55	0.0	213360.0	10000		
West	0203	T	31 00	113 45	0.0	213360.0	15000		
Arkansas									
North	0301	L	34 20	92 00	0.0	400000.0		34 56	36 14
South	0302	L	32 40	92 00	400000.0	400000.0		33 18	34 46

(continued)

TABLE C.7 U.S. State Plane Coordinate Systems (*Continued*)

State Zone	SPCS Zone	Proj Type T/L/O	Latitude of Origin (DD-MM)	Longitude of Origin (DDD-MM)	False North (M)	False East (M)	$\frac{1}{1-k_0}$	Standard South (DD-MM)	Parallels North (DD-MM)
California									
Zone 1	0401	L	39 20	122 00	500000.0	2000000.0		40 00	41 40
Zone 2	0402	L	37 40	122 00	500000.0	2000000.0		38 20	39 50
Zone 3	0403	L	36 30	120 30	500000.0	2000000.0		37 04	38 26
Zone 4	0404	L	35 20	119 00	500000.0	2000000.0		36 00	37 15
Zone 5	0405	L	33 30	118 00	500000.0	2000000.0		34 02	35 28
Zone 6	0406	L	32 10	116 15	500000.0	2000000.0		32 47	33 53
Colorado									
North	0501	L	39 20	105 30	304800.6096	914401.8289		39 43	40 47
Central	0502	L	37 50	105 30	304800.6096	914401.8289		38 27	39 45
South	0503	L	36 40	105 30	304800.6096	914401.8289		37 14	38 26
Connecticut	0600	L	40 50	72 45	152400.3048	304800.6096		41 12	41 52
Delaware	0700	T	38 00	75 25	0.0	200000.0	200000		
Florida									
East	0901	T	24 20	81 00	0.0	200000.0	17000		
West	0902	T	24 20	82 00	0.0	200000.0	17000		
North	0903	L	29 00	84 30	0.0	600000.0		29 35	30 45
Georgia									
East	1001	T	30 00	82 10	0.0	200000.0	10000		
West	1002	T	30 00	84 10	0.0	700000.0	10000		
Hawaii									
Zone 1	5101	T	18 50	155 30	0.0	500000.0	30000		

Zone 2	5102	T	20 20	156 40	0.0	500000.0	30000		
Zone 3	5103	T	21 10	158 00	0.0	500000.0	100000		
Zone 4	5104	T	21 50	159 30	0.0	500000.0	100000		
Zone 5	5105	T	21 40	160 10	0.0	500000.0	∞		
Idaho									
East	1101	T	41 40	112 10	0.0	200000.0	19000		
Central	1102	T	41 40	114 00	0.0	500000.0	19000		
West	1103	T	41 40	115 45	0.0	800000.0	15000		
Illinois									
East	1201	T	36 40	88 20	0.0	300000.0	40000		
West	1202	T	36 40	90 10	0.0	700000.0	17000		
Indiana									
East	1301	T	37 30	85 40	250000.0	100000.0	30000		
West	1302	T	37 30	87 05	250000.0	900000.0	30000		
Iowa									
North	1401	L	41 30	93 30	1000000.0	1500000.0		42 04	43 16
South	1402	L	40 00	93 30	0.0	500000.0		40 37	41 47
Kansas									
North	1501	L	38 20	98 00	0.0	400000.0		38 43	39 47
South	1502	L	36 40	98 30	400000.0	400000.0		37 16	38 34
Kentucky									
North	1601	L	37 30	84 15	0.0	500000.0		37 58	38 58
South	1602	L	36 20	85 45	500000.0	500000.0		36 44	37 56
Louisiana									
North	1701	L	30 30	92 30	0.0	1000000.0		31 10	32 40
South	1702	L	28 30	91 20	0.0	1000000.0		29 18	30 42
Offshore	1703	L	25 30	91 20	0.0	1000000.0		26 10	27 50

(continued)

TABLE C.7 U.S. State Plane Coordinate Systems (*Continued*)

State Zone	SPCS Zone	Proj Type T/L/O	Latitude of Origin (DD-MM)	Longitude of Origin (DDD-MM)	False North (M)	False East (M)	$\frac{1}{1-k_0}$	Standard South (DD-MM)	Parallels North (DD-MM)
Maine									
East	1801	T	43 40	68 30	0.0	300000.0	10000		
West	1802	T	42 50	70 10	0.0	900000.0	30000		
Maryland	1900	L	37 40	77 00	0.0	400000.0		38 18	39 27
Massachusetts									
Mainland	2001	L	41 00	71 30	750000.0	200000.0		41 43	42 41
Island	2002	L	41 00	70 30	0.0	500000.0		41 17	41 29
Michigan									
North	2111	L	44 47	87 00	0.0	8000000.0		45 29	47 05
Central	2112	L	43 19	84 22	0.0	6000000.0		44 11	45 42
South	2113	L	41 30	84 22	0.0	4000000.0		42 06	43 40
Minnesota									
North	2201	L	46 30	93 06	100000.0	800000.0		47 02	48 38
Central	2202	L	45 00	94 15	100000.0	800000.0		45 37	47 03
South	2203	L	43 00	94 00	100000.0	800000.0		43 47	45 13
Mississippi									
East	2301	T	29 30	88 50	0.0	300000.0	20000		
West	2302	T	29 30	90 20	0.0	700000.0	20000		
Missouri									
East	2401	T	35 50	90 30	0.0	250000.0	15000		
Central	2402	T	35 50	92 30	0.0	500000.0	15000		
West	2403	T	36 10	94 30	0.0	850000.0	17000		

Montana	2500	L	44 15	109 30	0.0	600000.0		45 00	49 00
Nebraska	2600	L	39 50	100 00	0.0	500000.0		40 00	43 00
Nevada									
East	2701	T	34 45	115 35	8000000.0	200000.0	10000		
Central	2702	T	34 45	116 40	6000000.0	500000.0	10000		
West	2703	T	34 45	118 35	4000000.0	800000.0	10000		
New Hampshire	2800	T	42 30	71 40	0.0	300000.0	30000		
New Jersey	2900	T	38 50	74 30	0.0	150000.0	10000		
New Mexico									
East	3001	T	31 00	104 20	0.0	165000.0	11000		
Central	3002	T	31 00	106 15	0.0	500000.0	10000		
West	3003	T	31 00	107 50	0.0	830000.0	12000		
New York									
East	3101	T	38 50	74 30	0.0	150000.0	10000		
Central	3102	T	40 00	76 35	0.0	250000.0	16000		
West	3103	T	40 00	78 35	0.0	350000.0	16000		
Long Island	3104	L	40 10	74 00	0.0	300000.0		40 40	41 02
North Carolina	3200	L	33 45	79 00	0.0	609601.22		34 20	36 10
North Dakota									
North	3301	L	47 00	100 30	0.0	600000.0		47 26	48 44
South	3302	L	45 40	100 30	0.0	600000.0		46 11	47 29
Ohio									
North	3401	L	39 40	82 30	0.0	600000.0		40 26	41 42
South	3402	L	38 00	82 30	0.0	600000.0		38 44	40 02

(continued)

TABLE C.7 U.S. State Plane Coordinate Systems (*Continued*)

State Zone	SPCS Zone	Proj Type T/L/O	Latitude of Origin (DD-MM)	Longitude of Origin (DDD-MM)	False North (M)	False East (M)	$\frac{1}{1-k_0}$	Standard South (DD-MM)	Parallels North (DD-MM)
Oklahoma									
North	3501	L	35 00	98 00	0.0	600000.0		35 34	36 46
South	3502	L	33 20	98 00	0.0	600000.0		33 56	35 14
Oregon									
North	3601	L	43 40	120 30	0.0	2500000.0		44 20	46 00
South	3602	L	41 40	120 30	0.0	1500000.0		42 20	44 00
Pennsylvania									
North	3701	L	40 10	77 45	0.0	600000.0		40 53	41 57
South	3702	L	39 20	77 45	0.0	600000.0		39 56	40 58
Rhode Island	3800	T	41 05	71 30	0.0	100000.0	160000		
South Carolina	3900	L	31 50	81 00	0.0	609600.0		32 30	34 50
South Dakota									
North	4001	L	43 50	100 00	0.0	600000.0		44 25	45 41
South	4002	L	42 20	100 20	0.0	600000.0		42 50	44 24
Tennessee	4100	L	34 20	86 00	0.0	600000.0		35 15	36 25
Texas									
North	4201	L	34 00	101 30	1000000.0	200000.0		34 39	36 11
North Central	4202	L	31 40	98 30	2000000.0	600000.0		32 08	33 58
Central	4203	L	29 40	100 20	3000000.0	700000.0		30 07	31 53
South Central	4204	L	27 50	99 00	4000000.0	600000.0		28 23	30 17
South	4205	L	25 40	98 30	5000000.0	300000.0		26 10	27 50

Utah									
North	4301	L	40 20	111 30	10000000.0	500000.0		40 43	41 47
Central	4302	L	38 20	111 30	20000000.0	500000.0		39 01	40 39
South	4303	L	36 40	111 30	30000000.0	500000.0		37 13	38 21
Vermont	4400	T	42 30	72 30	0.0	500000.0	28000		
Virginia									
North	4501	L	37 40	78 30	20000000.0	3500000.0		38 02	39 12
South	4502	L	36 20	78 30	10000000.0	3500000.0		36 46	37 58
Washington									
North	4601	L	47 00	120 50	0.0	500000.0		47 30	48 44
South	4602	L	45 20	120 30	0.0	500000.0		45 50	47 20
West Virginia									
North	4701	L	38 30	79 30	0.0	600000.0		39 00	40 15
South	4702	L	37 00	81 00	0.0	600000.0		37 29	38 53
Wisconsin									
North	4801	L	45 10	90 00	0.0	600000.0		45 34	46 46
Central	4802	L	43 50	90 00	0.0	600000.0		44 15	45 30
South	4803	L	42 00	90 00	0.0	600000.0		42 44	44 04
Wyoming									
East	4901	T	40 30	105 10	0.0	200000.0	16000		
East Central	4902	T	40 30	107 20	100000.0	400000.0	16000		
West Central	4903	T	40 30	108 45	0.0	600000.0	16000		
West	4904	T	40 30	110 05	100000.0	800000.0	16000		
Puerto Rico	5200	L	17 50	66 26	200000.0	200000.0		18 02	18 26
Virgin Islands	5200	L	17 50	66 26	200000.0	200000.0		18 02	18 26

1983 defining constants.

TABLE C.8 UTM Mapping System Specifications

UTM zones	6° in longitude (exceptions exist)
Limits in latitude	$-80° <$ latitude $< 80°$
Longitude of origin	Central meridian of each zone
Latitude of origin	0° (equator)
Units	Meter
False northing	0 m at equator, increases northward for northern hemisphere 10,000,000 m at equator, decreases southward for southern hemisphere
False easting	500,000 m at the central meridian, decreasing westward
Central meridian scale	0.9996 (exceptions exist)
Zone numbers	Starting with 1 centered at 177° W and increasing eastward to zone 60 entered at 177° E (exceptions exist)
Limits of zones and overlap	Zones are bounded by meridians that are multiples of 6° W and 6° E of Greenwich
Reference ellipsoid	Depending on region and datum, e.g., GRS80 in United States for NAD83

REFERENCES

Allen, L. W. (1999). United States' Nationwide Differential Positioning System. *GPS Solutions*, **2**(4):1–6.

Argus, D. F., and R. G. Gordon. (1991). No-net-rotation Model of Current Plate Velocities Incorporating Plate Rotation Model NUVEL-1. *Geophysical Research Letter,* **18**(11):2039–2042.

Ashby, N. (1993). Relativity and GPS. *GPS World,* **4**(11):42–47.

Ashkenazi, V., C. J. Hill, and J. Nagel. (1992). Wide Area Differential GPS: A Performance Study. *Proceedings of the International Technical Meeting of the Satellite Division of the Institute of Navigation, ION-GPS-92,* 589–598.

Askne, J., and H. Nordius. (1987). Estimation of Tropospheric Delay for Microwaves from Surface Weather Data. *Radio Science*, **22**(3):379–386.

Awange, J., and E. Grafarend. (2002a). Algebraic Solutions of GPS Pseudo-ranging Equations. *GPS Solutions*, **5**(4):20–32.

Awange, J., and E. Grafarend. (2002b). Nonlinear Adjustment of GPS Observations of Type Pseudorange. *GPS Solutions*, **5**(4):80–96.

Baarda, W. (1967). *Statistical Concepts in Geodesy.* Publication on Geodesy, New Series, **2**(4), Netherlands Geodetic Commission.

Baarda, W. (1968). *A Testing Procedure for Use in Geodetic Networks.* Publication on Geodesy, New Series, **2**(5), Netherlands Geodetic Commission.

Badekas, J. (1969). *Investigations Related to the Establishment of a World Geodetic System.* Department of Geodetic Science, Ohio State University, OSUR 124.

Bancroft, S. (1985). An Algebraic Solution of the GPS Equations. *IEEE Transactions on Aerospace and Electronic Systems*, **21**(7):56–59.

Bar-Sever, Y. (1996). A New Model for GPS Yaw Attitude. *Journal of Geodesy,* **70**(11):714–723.

Bar-Sever, Y., P. M. Kroger, and J. A. Borjesson. (1998). Estimating Horizontal Gradients of Tropospheric Delay with a Single GPS Receiver. *Journal of Geophysical Research,* **103**(B3):5019–5035.

Bazlov Y. A., V. F. Galazin, B. L. Kaplan, V. G. Maksimov, and V. P. Rogozin. (1999a). GLONASS to GPS—A New Coordinate Transformation. *GPS World*, **10**(1):54–58.

Bazlov, Y. A., V. F. Galazin, B. L. Kaplan, V. G. Maksimov, and V. P. Rogozin. (1999b). Propagating PZ90 and WGS84 Transformation Parameters. *GPS Solutions*, **3**(1):13–16.

Bevis, M., S. Businger, T. A. Herring, C. Rocken, R. A. Anthes, and R. H. Ware. (1992). GPS Meteorology: Remote Sensing of Water Vapor Using the Global Positioning System. *Journal of Geophysical Research,* **97**(D14):15,787–15,801.

Bevis, M., S. Businger, S. Chriswell, T. Herring, R. Anthes, C. Rocken, and R. Ware. (1994). GPS Meteorology: Mapping Zenith Wet Delay onto Perceivable Water. *Journal of Applied Meteorology*, **33**(3):379–386.

Bergquist J. C., S. R. Jefferts, and D. J. Wineland. (2001, March). Time Measurement at the Millennium. *Physics Today*, 37–42.

Betz, J. W. (2002). Binary Offset Carrier Modulations for Radionavigation. *Navigation*, **48**(4): 227–246.

Bishop, G. L., J. A. Klobuchar, and P. H. Doherty. (1985). Multipath Effects on the Determination of Absolute Ionospheric Time Delay from GPS Signals. *Radio Science,* **20**(3):388–396.

Blewitt, G. (1990). An Automatic Editing Algorithm for GPS Data. *Geophysical Research Letters,* **17**(3):199–202.

Blomenhofer, H., G. Hein, and D. Walsh. (1993). On-the-fly Phase Ambiguity Resolution for Precise Aircraft Landing. *Proceedings of the International Technical Meeting of the Satellite Division of the Institute of Navigation, ION-GPS-93,* **2**:821–830.

Bock, Y., R. I. Abbot, C. C. Counselman, S. A. Gourevitch, and R. W. King. (1985). Establishment of Three-dimensional Geodetic Control by Interferometry with the Global Positioning System. *Journal of Geophysical Research,* **90**(B9):7689–7703.

Brown, A. (1989). Extended Differential GPS. *Navigation*, **36**(3):265–285.

Byun, S., G. A. Hajj, and L. E. Young. (2002). Development and Application of GPS Signal Multipath Simulator. *Radio Science*, **37**(6):1–23.

Chaffee, J., and J. Abel. (1994). On the Exact Solutions of Pseudorange Equations. *IEEE Transactions on Aerospace and Electronic Systems*, **30**(4):1021–1030.

Chen, D., and G. Lachapelle. (1995). A Comparison of the FASF and Least Squares Search Algorithms for Ambiguity Resolution On The Fly. *Navigation*, **42**(2):371–390.

Collins, J., and A. Leick. (1985). Analysis of Macrometer Networks with Emphasis on the Montgomery (PA) County Survey. *Proc. Positioning with GPS-1985*, 667–693. NGS.

Cook, B. (2000). The United States Nationwide Differential Global Positioning System. *Proceedings of the National Technical Meeting of the Satellite Division of the Institute of Navigation, ION-NTM-2000,* 297–305.

Coster A. J., E. M. Gaposchkin, and L. E. Thornton. (1992). Real-time Ionospheric Monitoring System Using the GPS. *Navigation*, **39**(2):191–204.

Couselman, C. C. (1999). Multipath-rejection GPS Antenna. *Proceedings of the IEEE*, **87**(1): 86–91.

Counselman, C. C., and S. A. Gourevitch. (1981). Miniature Interferometer Terminals for Earth

Surveying: Ambiguity and Multipath with Global Positioning System. *IEEE Transactions on Geoscience and Remote Sensing,* **GE-19**(4):244–252.

Coyne, G. V., M. A. Hoskin, and O. Pedersen (eds.). (1983). Gregorian Reform of the Calendar. *Proceedings of the Vatican Conference to Commemorate Its 400th Anniversary 1582–1982.* Vatican City: Pontifica Academia Scientiarum.

Craymer, M. R., and N. Beck. (1992). Session versus Single-baseline GPS Processing. *Proceedings of the 6th International Geodetic Symposium on Satellite Positioning*, 995–1004, DMA.

Davies, K. (1990). *Ionospheric Radio.* IEE Electromagnetic Waves Series 31. Published by Peter Peregrinus Ltd, London, UK.

Davis, J. L., T. A. Herring, I. I. Shapiro, A. E. E. Roger, and G. Elgered. (1985). Geodesy by Radio Interferometry: Effects of Atmospheric Modeling Errors on Estimates of Baseline Length. *Radio Science*, **20**(6):1593–1607.

DeMets, C., R. G. Gordon, D.F. Argus, and S. Stein. (1994). Effect of Recent Revisions to the Geomagnetic Reversal Time Scale on Estimates of Current Plate Motions. *Geophysical Research Letter,* **21**(20):2191–2194.

van Dierendonck, A. J., P. C. Fenton, and T. J. Ford. (1992). Theory and Performance of Narrow Correlator Spacing in a GPS Receiver. *Navigation*, **39**(3):265–283.

Doherty, P. H., J. A. Klobuchar, and J. M. Kunches. (2000). The Correlation between Solar 10.7 cm Radio Flux and Ionospheric Range Delay. *GPS Solutions*, **3**(4):75–79.

Dozier, J. (1980). Improved Algorithm for Calculation of UTM and Geodetic Coordinates. NOAA TR NESSS 81, NGS.

Engelis, T., R. Rapp, and Y. Bock. (1985). Measuring Orthometric Height Differences with GPS and Gravity Data. *Manuscripta Geodaetica*, **10**(3):187–194.

Eren, K. (1987). *Geodetic Network Adjustment Using GPS Triple Difference Observations and a priori Stochastic Information.* TR 1, Institute of Geodesy, University of Stuttgart.

Escobal, P. R. (1965). *Methods of Orbit Determination.* Wiley, New York.

Euler, H. J., and C. C. Goad. (1991). On Optimal Filtering of GPS Dual-frequency Observations without Orbit Information. *Bulletin Géodésique*, **65**(2):130–143.

Euler, H. J., and H. Landau. (1992). Fast Ambiguity Resolution On-the-fly for Real-time Applications. *Proceedings of the 6th International Geodetic Symposium on Satellite Positioning*, 650–658, DMA.

Euler, H. J., and B. Schaffrin. (1990). On a Measure for the Discernibility between Different Ambiguity Solutions in the Static-kinematic GPS Mode. *Proceedings of the Kinematic Systems in Geodesy, Surveying, and Remote Sensing*, 285–295. Springer Verlag. Alberta, Canada.

Fenton, P. C., W. H. Falkenberg, T. J. Ford, K. K. Ng, and A. J. van Dierendonck. (1991). NovAtel's GPS Receiver: The High Performance OEM Sensor of the Future. *Proceedings of the International Technical Meeting of the Satellite Division of the Institute of Navigation, ION-GPS-91,* 49–58.

Ferland, R. (2002). IGS Reference Frame Coordination and Working Group Activities. 2001 IGS Annual Report, JPL, Pasadena, 24–27.

van Flandern, T. C., and K. F. Pulkkinen. (1979). Low Precision Formulae for Planetary Positions. *The Astronomical Journal Supplement Series*, **41**:391–411.

Fliegel, H. F., and T. E. Gallini. (1989). Radiation Pressure Models for Block II GPS Satellites. *Proceedings of the 5th International Geodetic Symposium on Satellite Positioning*, 789–798, DMA.

Fliegel, H. F., W. A. Fees, W. C. Layton, and N. W. Rhodus. (1985). The GPS Radiation Force Model. *Proc. Positioning with GPS-1985*, 113–119, NGS.

Fliegel, H. F., T. E. Gallini, and E. R. Swift. (1992). Global Positioning System Radiation Force Model for Geodetic Applications. *Journal of Geophysical Research*, **97**(B1):559–568.

Fontana, R. D., W. Cheung, P. M. Novak, and T. A. Stansell. (2001a). The New L2 Civil Signal. *Proceedings of the International Technical Meeting of the Satellite Division of the Institute of Navigation, ION-GPS-01*, 617–631.

Fontana, R. D., W. Cheung, and T. A. Stansell. (2001b). The Modernized L2 Civil Signal. *GPS World*, **12**(9):28–34.

Fortes, L.P.S. (2002). Optimizing the Use of GPS Multi-reference Stations for Kinematic Positioning. Doctoral dissertation, Department of Geomatics Engineering, University of Calgary, Canada.

Frei, E., and G. Beutler. (1990). Rapid Static Positioning Based on the Fast Ambiguity Resolution Approach "FARA." Theory and First Results, *Manuscripta Geodaetica*, **15**(6):325–356.

Fu, Z., A. Hornbostel, J. Hammesfahr, and A. Konovaltsev. (2003). Suppression of Multipath and Jamming Signals by Digital Beamforming for GNSS/Galileo Applications. *GPS Solutions* **6**(4):257–264.

Galileo. (2002). Galileo Home Page at www.galileo-pgm.org.

Gelb, A. (1974). *Applied Optimal Estimation*, Cambridge, MA: The MIT Press.

George, A., and J.W-H. Liu. (1981). *Computer Solutions for Large Sparse Positive Definite Matrices*. Englewood Cliffs, NJ: Prentice-Hall.

Georgiadou, Y., and A. Kleusberg. (1988). On Carrier Signal Multipath Effects in Relative GPS Positioning. *Manuscripta Geodaetica*, **13**(3):172–179.

GLONASS. (2002). GLONASS Home Page at the Ministry of Defense of the Russian Federation Coordination Scientific Information Center, www.rssi.ru/SFCSIC/SFCSIC_main.html.

GLONASS. (1998). GLONASS Interface Control Document, fourth revision. Coordination Scientific Information Center, Russia 117279, Moscow, P.O. Box 14. Web address: www.rssi.ru/SFCSIC/SFCSIC_main.html.

Goad, C. C. (1990). Optimal Filtering of Pseudoranges and Phases from Single-frequency GPS Receivers. *Navigation*, **37**(3):191–204.

Goad, C. C. (Ed.). (1985). *Proc. Positioning with GPS-1985*, NGS.

Goad, C. C. (1998). Single-Site GPS Models. In *GPS for Geodesy* (P. J. G. Teunissen and A. Kleusberg, Eds.), Chapter 10, 446–449, Springer Verlag, Wien.

Goad, C. C., and A. Mueller. (1988). An Automated Procedure for Generating an Optimum Set of Independent Double Difference Observables Using Global Positioning System Carrier Phase Measurements, *Manuscripta Geodaetica*, **13**(6):365–369.

Gourevitch, S. A., S. Sila-Novitsky, and F. van Diggelen. (1996). The GG24 Combined GPS+GLONASS Receiver. *Proceedings of the International Technical Meeting of the Satellite Division of the Institute of Navigation, ION-GPS-96*, 141–145.

GPS. (2002). GPS Home Page at U.S. Coast Guard. www.navcen.uscg.gov/default.htm.

Grafarend, E. (1992). The Modeling of Free Satellite Networks in Spacetime. *Proc. Global*

Positioning System in Geosciences, 1–20, Department of Mineral Resources, Technical University of Crete, Greece.

Grafarend, E., and J. Shan. (2002). GPS Solutions: Closed Forms, Critical and Special Configurations of P4P. *GPS Solutions*, **5**(3):29–41.

Greenwalt, C. R., and M. E. Shultz. (1962). *Principles of Error Theory and Cartographic Applications,* ACIC Technical Report No. 96, Aeronautical Chart and Information Center, St. Louis, Missouri.

Grossman, W. (1976). *Geodätische Rechnungen und Abbildungen in der Landesvermessung.* Witter Verlag, Stuttgart, Germany.

GSFC. (2002). EGM96 Website at Goddard Space Flight Center cddisa.gsfc.nasa.gov/926/egm96/egm96.html.

Hajj, G. A., and L. J. Romans. (1998). Ionospheric Electron Density Profiles Obtained with the Global Positioning System: Results from the GPS/MET Experiment. *Radio Science*, **33**(1):175–190.

Hargreaves, J. K. (1992). *The Solar-terrestrial Environment.* Cambridge University Press, Cambridge, UK.

Hatch, R. R. (1990). Instantaneous Ambiguity Resolution. *Proc. Kinematic Systems in Geodesy, Surveying and Remote Sensing*, 299–308, IAG symposium 107, Springer Verlag.

Hatch, R. R. (1992). *Escape from Einstein.* Kneat Kompany, Wilmington, CA.

Hatch, R. R., R. Keegan, and T. A. Stansell. (1992). Kinematic Receiver Technology from Magnavox. *Proceedings of the 6th International Geodetic Symposium on Satelite Positioning*, 174–181, DMA.

Hatch, R. R., J. Jung, P. Enge, and B. Pervan. (2000). Civilian GPS: The Benefits of Three Frequencies. *GPS Solutions*, **3**(4):1–9.

Hatch, R. R., T. Sharpe, and P. Galyean. (2002). StarFire: A Global High Accuracy Differential System. *Proc. GPS Symposium 2002*, Japan Institute of Navigation, Tokyo University of Mercantile Marine, November, 57–68.

Heck, B. (1987). *Rechenverfahren und Auswertemodelle der Landesvermessung.* Herbert Wichmann Verlag, Karlsruhe.

Heflin, M., D. Argus, D. Jefferson, D. Webb, and J. Zumberge. (2002). Comparison of a GPS Defined Global Reference Frame with ITRF2000. *GPS Solutions*, **6**(2):72–75.

Hein, G. W., J. Godet, J. L. Issler, J. C. Martin, P. Erhard, R. Lucas-Rodriguez, and T. Pratt. (2002). Status of Galileo Frequency and Signal Design. *Proceedings of the International Technical Meeting of the Satellite Division of the Institute of Navigation, ION-GPS-2002,* 266–277.

Heiskanen, W. A., and H. Moritz. (1967). *Physical Geodesy.* San Francisco: Freeman.

Hilla, S., and M. Jackson. (2000). GPS Toolbox: Spanning Trees. *GPS Solutions*, **3**(3):65–68.

Hofmann-Wellenhof, B., H. Lichtenegger, and J. Collins (2001). *Global Positioning System, Theory and Practice.* Springer Verlag, Wien.

Hopfield, H. S. (1969). Two-quartic Tropospheric Refractivity Profile for Correcting Satellite Data. *Journal of Geophysical Research,* **74**(18):4487–4499.

Hristow, W. K. (1955). *Die Gausschen und Geographischen Koordinaten auf dem Ellipsoid von Krassowsky.* VEB Verlag Technik Berlin.

ICD-GPS-200C. (2000). Interface Control Document ICD-GPS-200, Revision C, IRN-200C-004, April 12. ARINC, Incorporated, El Segundo, CA. Available at GPS (2002) for downloading.

ICD-GPS-705. (2002). Interface Control Document ICD-GPS-705 (Initial Release, March 29). ARINC, Incorporated, El Segundo, CA. Available at GPS (2002) for downloading.

IERS. (2002). International Earth Rotation Service website www.iers.org.

IGDG. (2002). Internet based Global Differential GPS. Website gipsy.jpl.nasa.gov/igdg.

IGS. (2002). International GPS Service website igscb.jpl.nasa.gov.

Janssen, M. A. (Ed.). (1993). *Atmospheric Remote Sensing by Microwave Radiometry.* Wiley, New York.

de Jonge, P. J., and C.C.J.M. Tiberius. (1996). The LAMBDA Method for Integer Ambiguity Estimation: Implementation Aspects. Delft Geodetic Computing Center LGR series, No. 12; available at www.geo.tudelft.nl/mgp.

Jorgensen, P. S. (1986). Relativity Correction in GPS User Equipment. *Proceedings of the Position Location and Navigation System 1986 (PLANS),* 177–183, IEEE.

Kaplan E. D. (Ed.). (1996). *Understanding GPS Principles and Applications.* Artech House, Norwood, MA.

Kaula, W. K. (1962). Development of the Lunar and Solar Disturbing Functions for a Close Satellite. *The Astronomical Journal*, **67**(5):300–303.

Kaula, W. M. (1966). *Theory of Satellite Geodesy.* Blaisdell Publishing Company. Waltham, MA.

Kee, C., B. W. Parkinson, and P. Axelrad. (1991). Wide Area Differential GPS. *Navigation*, **38**(2):123–143.

Klobuchar, J. A. (1987). Ionospheric Time-Delay Algorithm for Single-frequency GPS Users. *IEEE Transactions on Aerospace and Electronic Systems*, AES-**23**(3):325–331.

Kneissl, M. (1959). *Mathematische Geodäsie. Handbuch der Vermessungskunde (Jordan—Eggert—Kneissl).* Metzler, Stuttgart, Germany.

Koch, K. R. (1988). *Parameter Estimation and Hypothesis Testing in Linear Models.* Springer Verlag, New York.

Kok, J. (1984). *On Data Snooping and Multiple Outlier Testing.* NOAA Technical Report NOS NGS 30, National Geodetic Information Center, NOAA.

Kouba, J. (2001). ITRF Transformations. *GPS Solutions*, **5**(3):88–90.

Kozlov, D., and M. Tkachenko. (1998). Instant RTK cm with Low Cost GPS and GLONASS C/A Receivers. *Navigation*, **45**(2):137–147.

Krabill, W. B., and C. F. Martin. (1987). Aircraft Positioning Using Global Positioning System Carrier Phase Data. *Navigation*, **34**(1):1–21.

Kunches J. M., and J. A. Klobuchar. (2000). Some Aspects of the Variability of Geomagnetic Storms. *GPS Solutions*, **4**(1):77–78.

Kursinski, E. R. (1994). Monitoring the Earth's Atmosphere with GPS. *GPS World* **5**(3):50–54.

Kursinski, E. R., G. A. Hajj, W. I. Bertiger, S. S. Leroy, T. K. Meehan, L. J. Romans, J. T. Schofield, D. J. McCleese, W. G. Melbourne, C. L. Thornton, T. P. Yunck, J. R. Eyre, and R. N. Nagatani. (1996). Initial Results of Radio Occultation Observations of Earth's Atmosphere Using the Global Positioning System. *Science*, **271**:1107–1110.

Kursinski, E. R., G.A. Hajj, J. T. Schofield, R. P. Linfield, and K. R. Hardy. (1997). Observing the Earth's Atmosphere with Radio Occultation Measurements Using the Global Positioning System. *Journal of Geophysical Research,* **102**(D19):23429–23465.

Lachapelle, G., M. E. Cannon, and C. Erickson. (1992). High Precision C/A-code Technology for Rapid Static DGPS Surveys. *Proceedings of the 6th International Geodetic Symposium on Satellite Positioning*, 165–173, DMA.

Lachapelle, G., H. Sun, M. E. Cannon, and G. Lu. (1994). Precise Aircraft-to-aircraft Positioning Using a Multiple Receiver Configuration. *Proceedings of the ION Technical Meeting*, 793–799.

Lachapelle, G., M. E. Cannon, W. Qui, and C. Varner. (1996). Precise Aircraft Single-point Positioning Using GPS Post-mission Orbits and Satellite Clock Corrections. *Journal of Geodesy*, **70**(6):562–571.

Lachapelle, G., P. Alves, L. P. S. Fortes, M. E. Cannon, and B. Townsend. (2000). DGPS RTK Positioning using a Reference Network. *Proceedings of the International Technical Meeting of the Satellite Division of the Institute of Navigation,* 1165–1171.

Ladd, L. W., C. C. Counselman, and S. A. Gourevitch. (1985). The Macrometer II Dual-band Interferometric Surveyor. *Proc. Positioning with GPS-1985*, 175–180, NGS.

Lambert, J. H. (1772). *Notes and Comments on the Composition of Terrestrial and Celestial Maps.* Michigan Geographical Publication No. 8. Translated by Waldo R. Tobler, 1972, University of Michigan, Ann Arbor.

Lee, L. P. (1976). *Conformal Projections Based on Elliptic Function.* Monograph No. 16, supplement No. 1 to Canadian Cartographer, Vol. 13. Department of Geography, York University, Toronto, Canada. University of Toronto Press.

Leick, A. (2002). Lectures in GPS, Geodesy, and Adjustments—A Supplement. Unpublished notes, Department of Spatial Information Science and Engineering, University of Maine.

Leick, A., and M. Emmons. (1994). Quality Control with Reliability for Large GPS Networks. *Surveying Engineering*, **120**(1):26–41.

Leick, A. and B. H. W. van Gelder (1975). On Similarity Transformations and Geodetic Network Distortions Based on Doppler Satellite Observations. Dept. of Geodetic Science, Ohio State University, OSUR 235.

Leick, A., J. Li, J. Beser, and G. Mader. (1995). Processing GLONASS Carrier Phase Observations—Theory and First Experience. *Proceedings of the International Technical Meeting of the Satellite Division of the Institute of Navigation, ION-GPS-95,* 1041–1047.

Leick, A., J. Beser, P. Rosenboom, and B. Wiley. (1998). Assessing GLONASS Observation. *Proceedings of the International Technical Meeting of the Satellite Division of the Institute of Navigation, ION-GPS-98,* 1605–1612.

Lemoine, F. G., S. C. Kenyon, J. K. Factor, R. G. Trimmer, N. K. Pavlis, D. S. Chinn, C. M. Cox, S. M. Klosko, S. B. Lutchcke, M. H. Torrence, Y. M. Wang, R. G. Williamson, E. C. Pavlis, R. H. Rapp, and T. R. Olson. (1998). *The Development of the Joint NASA GSFC and the National Imagery and Mapping Agency (NIMA) Geopotential Model EGM96. NASA/TP-1998–206861.* NASA Goddard Space Flight Center, Greenbelt, Maryland.

Lichten, S. M., and J. S. Border. (1987). Strategies for High Precision GPS Orbit Determination. *Journal of Geophysical Research,* **92**(B12):12751–12762.

Liu, Y. (1999). Remote Sensing of the Atmospheric Water Vapor Using GPS Data in the Hong Kong Region. Ph.D. thesis, Department of Land Surveying and Geo-Informatics. The Hong Kong Polytechnic University.

Loh R., V. Wullschleger, B. Elrod, M. Lage, and F. Haas. (1995). The U.S. Wide-Area Augmentation System (WAAS). *Navigation*, **42**(3):435–465.

Mader, G. L. (1986). Dynamic Positioning Using GPS Carrier Phase Measurements. *Manuscripta Geodaetica*, **11**(4):272–277.

Mader, G. L. (1999). GPS Antenna Calibration at the National Geodetic Survey. *GPS Solutions*, **3**(1):50–58.

Mader, G. L., and F. Czopek. (2001). Calibrating the L1 and L2 Phase Centers of a Block IIA Antenna. *Proceedings of the International Technical Meeting of the Satellite Division of the Institute of Navigation, ION-GPS-2001,* 1979–1984.

Mannucci, A. J., B. D. Wilson, D. N. Yuan, C. H. Ho, U. J. Lindqwister, and T. F. Runge. (1998). A Global Mapping Technique for GPS-derived Ionospheric Total Electron Content Measurements. *Radio Science*, **33**(3):565–582.

McCarthy, D. D. (Ed.). (1996). IERS Conventions 1996. IERS Technical Note 21, Paris Observatory.

McCarthy, D. D., and W. J. Klepczynski. (1999). GPS and Leap Seconds. *GPS World*, **10**(11): 50–57.

McKinnon, J. A. (1987). *Sunspot Numbers: 1610–1985 Based on the Sunspot Activity in the Years 1620–1960,* National Academy of Sciences, Report UAG-95.

Meehan, T. K., and L. E. Young. (1992). On-receiver Signal Processing for GPS Multipath Reduction. *Proceedings of the 6th International Geodetic Symposium on Satellite Positioning*, 200–208, DMA.

Melbourne, W. G. (1985). The Case for Ranging in GPS-based Geodetic Systems. *Proc. Positioning with GPS-1985*, 373–386, NGS.

Mendes, V. B. (1999). Modeling the Neutral-atmosphere Propagation Delay in Radiometric Space Techniques. Ph.D. dissertation, Department of Geodesy and Geomatics Engineering Technical Report No. 199, University of New Brunswick, Fredericton, NB, Canada.

Mendes, V. B., and R. B. Langley. (1999). Tropospheric Zenith Delay Prediction Accuracy for High-precision GPS Positioning and Navigation. *Navigation*, **46**(1):25–34.

Milbert, D. G. (1984). Heard of Gold: Computer Routines for Large Space, Least Squares Computations. NOAA TM NOS NGS-39.

Misra, P., and P. Enge. (2001). *Global Positioning System Signals, Measurements, and Performance.* Ganga-Jamuna Press, Lincoln, MA.

Mohr, P. J., and B. N. Taylor. (2001, March). Adjusting the Value of the Fundamental Constants. *Physics Today*, 29–34.

Molodenskii, M. S., V. F. Eremeev, and M. I. Yurkina. (1962). *Methods for Study of the External Gravitational Field and Figure of the Earth.* Translation from Russian. National Technical Information Services, Springfield, VA.

Moritz, H. (1984). Geodetic Reference System 1980. *Bulletin Géodésique*, **58**(3):388–398.

Mueller, I. I. (1964). *Introduction to Satellite Geodesy.* F. Ungar, New York.

Mueller, I. I. (1969). *Spherical and Practical Astronomy as Applied to Geodesy.* F. Ungar, New York.

Muellerschoen, R. J., W. I. Bertiger, M. F. Lough, D. Stowers, and D. Dong. (2000). An Internet-based Global Differential GPS System, Initial Results. *Proceedings of the ION Technical Meeting*, 220–225.

Muellerschoen, R. J., A. Reichert, D. Kuang, M. B. Heflin, W. I. Bertiger, and Y. E. Bar-Sever. (2001). Orbit Determination with NASA's High Accuracy Real-time Global Differential GPS System. *Proceedings of the International Technical Meeting of the Satellite Division of the Institute of Navigation, ION-GPS-2001,* 2294–2303.

Niell, A. E. (1996). Global Mapping Functions for the Atmospheric Delay at Radio Wavelengths. *Journal of Geophysical Research,* **101**(B2):3227–3246.

Niell A. E. (2000). Improved Atmospheric Mapping Functions for VLBI and GPS. *Earth Planets Space,* **52**(10):699–702.

NGS. (2002). National Geodetic Survey website www.ngs.noaa.gov.

NIMA. (2002). National Imagery and Mapping Agency website www.nima.mil.

Odijk, D. (2002). Fast Precise GPS Positioning in the Presence of Ionospheric Delays. Doctoral dissertation, Department of Mathematical Geodesy and Positioning, Delft University of Technology, The Netherlands.

Øvstedal, O. (2002). Absolute Positioning with Single-frequency GPS Receivers. *GPS Solutions,* **5**(4):33–44.

Parkinson, B. W., and J. J. Spilker (Eds.), P. Axelrad and P. Enge (Assoc. Eds.). (1996). Global Positioning System: Theory and Applications. *Progress in Aeronautics and Astronautics,* Vols. 163 and 164.

Pope, A. J. (1971). Transformation of Covariance Matrices Due to Changes in Minimal Control. Paper presented at the American Geophysical Union Fall Meeting, San Francisco.

Pope, A. J. (1976). *The Statistics of Residuals and the Detection of Outliers.* NOAA TR NOS 65 NGS 1, NOAA, Silver Spring, MD.

Povalyaev, A. (1997). Using Single Differences for Relative Positioning in GLONASS. *Proceedings of the International Technical Meeting of the Satellite Division of the Institute of Navigation, ION-GPS-97,* 929–934.

PPIRN-200C-007. (2001). Pre-Proposed Interface Revision Notice (PPIRN) to ICD-GPS-200C for L2 Civil (L2C) Signal, May 31, labeled Draft. ARINC, Incorporated, El Segundo, CA. Available at GPS (2002) for downloading.

Pratt, M., B. Burke, and P. Misra. (1997). Single Epoch Integer Ambiguity Resolution with GPS-GLONASS L1 Data. *Proceedings of 53rd Annual Meeting of the ION,* Albuquerque, NM, 691–699.

Qin, X., S. Gourevitch, and M. Kuhl. (1992). Very Precise GPS—Development Status and Results. *Proceedings of the International Technical Meeting of the Satellite Division of the Institute of Navigation, ION-GPS-92,* 615–624.

Raby, P., and P. Daly. (1993). Using the GLONASS System for Geodetic Surveys. *Proceedings of the International Technical Meeting of the Satellite Division of the Institute of Navigation, ION-GPS-93,* 1129–1138.

Raquet, J. F. (1998). Development of a Method for Kinematic GPS Carrier Phase Ambiguity Resolution. Ph.D. thesis. Department of Geomatics Engineering, The University of Calgary.

Rapoport, L. (1997). General-purpose Kinematic/static GPS/GLONASS Post-processing Engine. *Proceedings of the International Technical Meeting of the Satellite Division of the Institute of Navigation, ION-GPS-97,* 1757–1772.

Remondi, B. W. (1984). Using the Global Positioning System (GPS) Phase Observable for Relative Geodesy: Modeling, Processing, and Results. NOAA, reprint of doctoral dissertation. Center for Space Research, University of Texas at Austin.

Remondi, B. W. (1985). Performing Centimeter-level Surveys in Seconds with GPS Carrier Phase: Initial Results. *Navigation,* **32**(4):386–400.

Rocken, C., S. Sokolovsky, J. Johnson, and D. Hunt. (2001). Improved Mapping of Tropospheric Delays. *Atmospheric and Oceanic Technology,* **18**(7):1205–1213.

Rosenkranz, P. W. (1998). Water Vapor Microwave Continuum Absorption: A Comparison of Measurement and Models. *Radio Science,* **33**(4):919–928.

Roßbach, U. (2001). *Positioning and Navigation Using the Russian Satellite System GLONASS.* University FAF Munich, Section Geodesy and Geo Information, Publication 70.

RTCM. (2002). Website of the Radio Technical Commission for Maritime Services www.rtcm.org/index2.html.

Ruland, R., and A. Leick. (1985). Application of GPS to a High Precision Engineering Survey Network. *Proceedings of Positioning with GPS-1985*, 483–493, NGS.

Saastamoinen, J. (1972). *Atmospheric Correction for the Troposphere and Stratosphere in Radio Ranging of Satellites.* Geophysical Monograph 15, Use of Artificial Satellites for Geodesy, 247–251. American Geophysical Union.

Sardón, E., and N. Zarraoa. (1997). Estimation of Total Electron Content Using GPS Data: How Stable are the Differential Satellite and Receiver Instrumental Biases? *Radio Science,* **32**(5):1899–1910.

Sardón, E., A. Rius, and N. Zarraoa. (1994). Estimation of the Transmitter and Receiver Differential Biases and the Ionospheric Total Electron Content from Global Positioning System Observations. *Radio Science*, **29**(3):577–586.

Schmitz M., G. Wübbena, and G. Boettcher. (2002). Test of Phase Center Variations of Various GPS Antennas and Some Results. *GPS Solutions*, **6**(1,2):18–27.

Schomaker, M. C., and R. M. Berry. (1981). *Geodetic Leveling.* NOAA Manual 3, NGS.

Schupler, B. R., and T. A. Clark. (2001). Characterizing the Behavior of Geodetic GPS Antennas. *GPS World*, **12**(2): 48–52.

Schüler, T. (2001). *On Ground-based GPS Tropospheric Delay Estimation.* Schriftenreihe, Vol. 73, Studiengang Geodäsie und Geoinformation, Univeristät der Bundeswehr München.

Schwarze, V. S., T. Hartmann, M. Leins, and M. H. Soffel. (1993). Relativistic Effects in Satellite Positioning. *Manuscripta Geodaetica*, **18**(5):306–316.

Seeber, G. (2003). *Satellite Geodesy.* De Gruyter, New York.

Seeber, G., and G. Wübbena. (1989). Kinematic Positioning with Carrier Phases and "On-the-way" Ambiguity Resolution. *Proceedings of the 5th International Geodetic Symposium on Satellite Positioning*, 600–609, DMA.

Sigl, R. (1977). *Ebene und Sphärische Trigonometrie mit Anwendungen auf Kartographie, Geodäsie und Astronomie.* Herbert Wichmann Verlag, Karlsruhe, Germany.

Snyder, J. P. (1979). Calculating Map Projections for the Ellipsoid. *The American Cartographer*, **6**(1):67–76.

Snyder, J. P. (1982). *Map Projections Used by the U.S. Geological Survey.* Geological Survey Bulletin 1532. United States Printing Office, Washington.

Soler, T., and B.H.W. van Gelder. (1987). On Differential Scale Changes and the Satellite Doppler System Z-Shift. *Geophys. J. R. Astr. Soc.*, **91**(3):639–656.

Soler, T., and J. Marshall. (2002). Rigorous Transformation of Variance-Covariance Matrices of GPS Derived Coordinates and Velocities. *GPS Solutions*, **6**(2):76–90.

Solheim, F. S. (1993). Use of Pointed Water Vapor Radiometer Observations to Improve Vertical GPS Surveying Accuracy. Doctoral thesis, University of Colorado.

Spilker, J. J. (1996). Tropospheric Effects on GPS. GPS Theory and Applications **1**, 517–546, Spilker and Parkinson, eds. *Progress in Astronautics and Aeronautics,* Vol. 163.

Springer, T. A., G. Beutler, and M. Rothacher. (1999). A New Solar Radiation Pressure Model for GPS Satellites. *GPS Solutions*, **2**(3):50–62.

SPS. (2001). Global Positioning System Standard Positioning Service Performance Standard. OASD/C3I)/ODASD(C3ISR&S)/Space Systems. Attn: Assistant for GPS, Positioning and Navigation, Pentagon. Available at GPS (2002) for downloading.

Stem, J. E. (1989). *State Plane Coordinate Systems of 1983*. NOAA Manual NOS NGS 5.

Strang, G., and K. Borre. (1997). *Linear Algebra, Geodesy, and GPS.* Wellesley–Cambridge Press, Wellesley, MA.

Talbot, N. C. (1993). Centimeters in the Field: A User's Perspective of Real-time Kinematic Positioning in a Production Environment. *Proceedings of the International Technical Meeting of the Satellite Division of the Institute of Navigation, ION-GPS-93,* 589–598.

Tetewsky, A. K., and F. E. Mullen. (1997). Carrier Phase Wrap-up Induced by Rotating GPS Antennas. *GPS World,* **8**(2):51–57.

Teunissen, P.J.G. (1993). Least Squares Estimation of Integer GPS Ambiguities. Invited Lecture, Section IV Theory and Methodology. IAG General Meeting, Beijing, China, August 1993.

Teunissen, P.J.G. (1994). A New Method for Fast Carrier Phase Ambiguity Resolution. *Proc. IEEE Position, Location and Navigation Symposium PLANS'94*, 662–673.

Teunissen, P.J.G. (1997). On the Widelane and Its Decorrelating Property. *Journal of Geodesy*, **71**(9):577–587.

Teunissen P.J.G. (1998). Success Probability of Integer GPS Ambiguity Rounding and Bootstrapping. *Journal of Geodesy*, **72**(10):606–612.

Teunissen, P.J.G. (1999). An Optimality Property of the Integer Least-squares Estimator. *Journal of Geodesy*, **73**(5):587–593.

Thayer, G. D. (1974). An Improved Equation for Radio Refractive Index of Air. *Radio Science*, **9**(10):803–807.

Thomas, P. D. (1952). *Conformal Projections in Geodesy and Cartography.* Special Publication No. 251, U.S. Department of Commerce.

USNO. (2002). U.S. Naval Observatory website maia.usno.navy.mil.

Vaniček, P., and D. E. Wells. (1974). Positioning of Horizontal Geodetic Datums. *The Canadian Surveyor*, **28**(5):531–538.

Veis, G. (1960). Geodetic Use of Artificial Satellites. *Smithsonian Contributions to Astrophysics,* **3**(9):95–161.

Veitsel V. A., A. V. Zhdanov, and M. I. Zhodzishsky. (1998). The Mitigation of Multipath Errors by Strobe Correlators in GPS/GLONASS Receivers. *GPS Solutions*, **2**(2):38–45.

Vincenty, T. (1979). The HAVAGO Three-dimensional Adjustment Program. NOAA TM NOS NGS 17.

Vollath U., A. Buecherl, H. Landau, C. Pagels, and B. Wagner. (2000). Multi-base RTK Positioning Using Virtual Reference Stations. *Proceedings of the International Technical Meeting of the Satellite Division of the Institute of Navigation, ION-GPS-2000,* 123–131.

Wang, J., M. P. Stewart, and M. Tsakiri. (1998). A Discrimination Test Procedure for Ambiguity Resolution On-the-fly. *Journal of Geodesy,* **72**(11):644–653.

Wang, J., C. Rizos, M. P. Stewart, and A. Leick. (2001). GPS and GLONASS Integration: Modeling and Ambiguity Resolution Issues. *GPS Solutions,* **5**(1):55–64.

Webb, D. F., and R. A. Howard. (1994). The Solar Cycle Variation of Coronal Mass Ejections and the Solar Wind Mass Flux. *Journal of Geophysical Research,* **99**(A4):4201–4220.

Westwater, E. R. (1978). The Accuracy of Water Vapor and Cloud Liquid Determination by Dual-frequency Ground-based Microwave Radiometry. *Radio Science*, **13**(4):677–685.

Whitehead, M. L., G. Penno, W. J. Feller, I. C. Messinger, W. L. Bertiger, R. J. Muellerschoen, B. A. Iijima, and G. Piesinger. (1998). A Close Look at Satloc's Real-time WADGPS System. *GPS Solutions*, **2**(2):45–63.

Wilson, B. D., C. H. Yinger, W. A. Feess, and C. Shank. (1999). New and Improved: The Broadcast Interfrequency Biases. *GPS World*, **10**(9):56–66.

Witchayangkoon, B. (2000). Elements of GPS Precise Point Positioning. Dissertation, Department of Spatial Information Science and Engineering, University of Maine.

Witchayangkoon, B., and P.C.L. Segantine. (1999). Testing JPL's PPP Service. *GPS Solutions*, **3**(1):73–76.

WMO. (1961). *Guide to Meteorological Instrument and Observing Practices.* Second Edition, World Meteorological Organization, No. 8.

Wolf, H. (1963). Die Grundgleichungen der dreidimensionalen Geodäsie in elementarer Darstellung. *Zeitschrift für Vermessungswesen,* **88**(6):225–233.

Wu, J. T., S. C. Wu, G. A. Hajj, W. I. Bertiger, and S. M. Lichen. (1993). Effects of Antenna Orientation on GPS Carrier Phase. *Manuscripta Geodaetica*, **18**(2):91–98.

Wübbena, G. (1990) Zur Modellierung von GPS-Beobachtungen für die hochgenaue Positionsbestimmung. Dissertation, Universität Hannover.

Wübbena, G., A. Bagge, G. Seeber, V. Böder, and P. Hankemeier. (1996). Reducing Distance Dependent Errors in Real-time Precise DGPS Applications by Establishing Reference Station Networks. *Proceedings of the International Technical Meeting of the Satellite Division of the Institute of Navigation, ION-GPS-96,* 1845–1852.

Wübbena, G., M. Schmitz, F. Menge, V. Böder, and G. Seeber. (2000). Automated Absolute Field Calibration of GPS Antennas in Real Time. *Proceedings of the International Technical Meeting of the Satellite Division of the Institute of Navigation, ION-GPS-2000,* 2512–2522.

Zebhauser, B. E., H. J. Euler, and C. R. Keenan. (2002). A Novel Approach for the Use of Information from Reference Station Networks Conforming to RTCM V2.3 and Future V3.0. *Proceedings of the International Technical Meeting of the Satellite Division of the Institute of Navigation, ION-GPS-2002,* 863–876.

Zhdanov, A. V., M. I. Zhodzishsky, V. A. Veitsel, and J. Ashjaee. (2001). Evolution of Multipath Error Reduction with GPS Signal Processing. *GPS Solutions*, **5**(1):19–28.

Ziebart, M., P. Cross, and S. Adhya. (2002). Modeling Photon Pressure: The Key to High-precision GPS Satellite Orbits. *GPS World,* **13**(1):43–50.

Zumberge, J. F. (1998). Automated GPS Data Analysis Service. *GPS Solutions*, **2**(3):76–78.

Zumberge, J. F., M. B. Heflin, D. C. Jefferson, M. M. Watkins, and F. H. Webb (1998a). Precise Point Processing for the Efficient and Robust Analysis of GPS Data from Large Networks. *Journal of Geophysical Research,* **102**(B3):5005–5017.

Zumberge, J. F., M. M. Watkins, and F. H. Webb. (1998b). Characteristics and Application of Precise GPS Clock Solutions Every 30 Seconds. *Navigation*, **44**(4):449–456.

AUTHOR INDEX

SUBJECT INDEX